Reliability Engineering

WILEY SERIES IN SYSTEMS ENGINEERING AND MANAGEMENT

Andrew P. Sage, Editor

A complete list of the titles in this series appears at the end of this volume.

Reliability Engineering

Kailash C. Kapur
Michael Pecht

Copyright © 2014 by John Wiley & Sons, Inc. All rights reserved

Published by John Wiley & Sons, Inc., Hoboken, New Jersey
Published simultaneously in Canada

No part of this publication may be reproduced, stored in a retrieval system, or transmitted in any form or by any means, electronic, mechanical, photocopying, recording, scanning, or otherwise, except as permitted under Section 107 or 108 of the 1976 United States Copyright Act, without either the prior written permission of the Publisher, or authorization through payment of the appropriate per-copy fee to the Copyright Clearance Center, Inc., 222 Rosewood Drive, Danvers, MA 01923, (978) 750-8400, fax (978) 750-4470, or on the web at www.copyright.com. Requests to the Publisher for permission should be addressed to the Permissions Department, John Wiley & Sons, Inc., 111 River Street, Hoboken, NJ 07030, (201) 748-6011, fax (201) 748-6008, or online at http://www.wiley.com/go/permissions.

Limit of Liability/Disclaimer of Warranty: While the publisher and author have used their best efforts in preparing this book, they make no representations or warranties with respect to the accuracy or completeness of the contents of this book and specifically disclaim any implied warranties of merchantability or fitness for a particular purpose. No warranty may be created or extended by sales representatives or written sales materials. The advice and strategies contained herein may not be suitable for your situation. You should consult with a professional where appropriate. Neither the publisher nor author shall be liable for any loss of profit or any other commercial damages, including but not limited to special, incidental, consequential, or other damages.

For general information on our other products and services or for technical support, please contact our Customer Care Department within the United States at (800) 762-2974, outside the United States at (317) 572-3993 or fax (317) 572-4002.

Wiley also publishes its books in a variety of electronic formats. Some content that appears in print may not be available in electronic formats. For more information about Wiley products, visit our web site at www.wiley.com.

Library of Congress Cataloging-in-Publication Data:
Kapur, Kailash C., 1941–
  Reliability engineering / Kailash C. Kapur, Michael Pecht.
    pages cm
  Includes index.
  ISBN 978-1-118-14067-3 (cloth)
  1. Reliability (Engineering) I. Pecht, Michael. II. Title.
  TA169.K37 2014
  620'.00452–dc23
                                          2013035518

Printed in the United States of America

10 9 8 7 6 5 4 3 2 1

Contents

Preface .... **xv**

1 Reliability Engineering in the Twenty-First Century **1**
1.1 What Is Quality? 1
1.2 What Is Reliability? 2
    1.2.1 The Ability to Perform as Intended 4
    1.2.2 For a Specified Time 4
    1.2.3 Life-Cycle Conditions 5
    1.2.4 Reliability as a Relative Measure 5
1.3 Quality, Customer Satisfaction, and System Effectiveness 6
1.4 Performance, Quality, and Reliability 7
1.5 Reliability and the System Life Cycle 8
1.6 Consequences of Failure 12
    1.6.1 Financial Loss 12
    1.6.2 Breach of Public Trust 13
    1.6.3 Legal Liability 15
    1.6.4 Intangible Losses 15
1.7 Suppliers and Customers 16
1.8 Summary 16
    Problems 17

2 Reliability Concepts **19**
2.1 Basic Reliability Concepts 19
    2.1.1 Concept of Probability Density Function 23
2.2 Hazard Rate 26
    2.2.1 Motivation and Development of Hazard Rate 27
    2.2.2 Some Properties of the Hazard Function 28
    2.2.3 Conditional Reliability 31
2.3 Percentiles Product Life 33
2.4 Moments of Time to Failure 35
    2.4.1 Moments about Origin and about the Mean 35
    2.4.2 Expected Life or Mean Time to Failure 36
    2.4.3 Variance or the Second Moment about the Mean 36
    2.4.4 Coefficient of Skewness 37
    2.4.5 Coefficient of Kurtosis 37

| 2.5 | Summary | 39 |
| | Problems | 40 |

3 Probability and Life Distributions for Reliability Analysis 45
- 3.1 Discrete Distributions. 45
  - 3.1.1 Binomial Distribution. 46
  - 3.1.2 Poisson Distribution. 50
  - 3.1.3 Other Discrete Distributions 50
- 3.2 Continuous Distributions. 51
  - 3.2.1 Weibull Distribution 55
  - 3.2.2 Exponential Distribution 61
  - 3.2.3 Estimation of Reliability for Exponential Distribution ... 64
  - 3.2.4 The Normal (Gaussian) Distribution 67
  - 3.2.5 The Lognormal Distribution 73
  - 3.2.6 Gamma Distribution. 75
- 3.3 Probability Plots 77
- 3.4 Summary 83
- Problems 84

4 Design for Six Sigma 89
- 4.1 What Is Six Sigma? 89
- 4.2 Why Six Sigma? 90
- 4.3 How Is Six Sigma Implemented? 91
  - 4.3.1 Steps in the Six Sigma Process 92
  - 4.3.2 Summary of the Six Sigma Steps 97
- 4.4 Optimization Problems in the Six Sigma Process 98
  - 4.4.1 System Transfer Function. 99
  - 4.4.2 Variance Transmission Equation 100
  - 4.4.3 Economic Optimization and Quality Improvement 101
  - 4.4.4 Tolerance Design Problem 102
- 4.5 Design for Six Sigma 103
  - 4.5.1 Identify (I). 105
  - 4.5.2 Characterize (C) 106
  - 4.5.3 Optimize (O) 106
  - 4.5.4 Verify (V). 106
- 4.6 Summary 108
- Problems 108

5 Product Development 111
- 5.1 Product Requirements and Constraints. 112
- 5.2 Product Life Cycle Conditions. 113
- 5.3 Reliability Capability 114
- 5.4 Parts and Materials Selection 114
- 5.5 Human Factors and Reliability 115
- 5.6 Deductive versus Inductive Methods. 117
- 5.7 Failure Modes, Effects, and Criticality Analysis. 117
- 5.8 Fault Tree Analysis. 119
  - 5.8.1 Role of FTA in Decision-Making 121
  - 5.8.2 Steps of Fault Tree Analysis. 122

	5.8.3	Basic Paradigms for the Construction of Fault Trees 122
	5.8.4	Definition of the Top Event 122
	5.8.5	Faults versus Failures 122
	5.8.6	Minimal Cut Sets 127
5.9	Physics of Failure 128	
	5.9.1	Stress Margins 128
	5.9.2	Model Analysis of Failure Mechanisms 129
	5.9.3	Derating 129
	5.9.4	Protective Architectures 130
	5.9.5	Redundancy 131
	5.9.6	Prognostics 131
5.10	Design Review 131	
5.11	Qualification 132	
5.12	Manufacture and Assembly 134	
	5.12.1	Manufacturability 134
	5.12.2	Process Verification Testing 136
5.13	Analysis, Product Failure, and Root Causes 137	
5.14	Summary 138	
	Problems 138	

6	**Product Requirements and Constraints** 141	
6.1	Defining Requirements 141	
6.2	Responsibilities of the Supply Chain 142	
	6.2.1	Multiple-Customer Products 142
	6.2.2	Single-Customer Products 143
	6.2.3	Custom Products 144
6.3	The Requirements Document 144	
6.4	Specifications 144	
6.5	Requirements Tracking 146	
6.6	Summary 147	
	Problems 147	

7	**Life-Cycle Conditions** 149	
7.1	Defining the Life-Cycle Profile 149	
7.2	Life-Cycle Events 150	
	7.2.1	Manufacturing and Assembly 151
	7.2.2	Testing and Screening 151
	7.2.3	Storage 151
	7.2.4	Transportation 151
	7.2.5	Installation 151
	7.2.6	Operation 152
	7.2.7	Maintenance 152
7.3	Loads and Their Effects 152	
	7.3.1	Temperature 152
	7.3.2	Humidity 155
	7.3.3	Vibration and Shock 156
	7.3.4	Solar Radiation 156
	7.3.5	Electromagnetic Radiation 157

		7.3.6	Pressure	157
		7.3.7	Chemicals	158
		7.3.8	Sand and Dust	159
		7.3.9	Voltage	159
		7.3.10	Current	159
		7.3.11	Human Factors	160
	7.4	Considerations and Recommendations for LCP Development		160
		7.4.1	Extreme Specifications-Based Design (Global and Local Environments)	160
		7.4.2	Standards-Based Profiles	161
		7.4.3	Combined Load Conditions	161
		7.4.4	Change in Magnitude and Rate of Change of Magnitude	165
	7.5	Methods for Estimating Life-Cycle Loads		165
		7.5.1	Market Studies and Standards Based Profiles as Sources of Data	165
		7.5.2	In Situ Monitoring of Load Conditions	166
		7.5.3	Field Trial Records, Service Records, and Failure Records	166
		7.5.4	Data on Load Histories of Similar Parts, Assemblies, or Products	166
	7.6	Summary		166
		Problems		167
8	**Reliability Capability**			**169**
	8.1	Capability Maturity Models		169
	8.2	Key Reliability Practices		170
		8.2.1	Reliability Requirements and Planning	170
		8.2.2	Training and Development	171
		8.2.3	Reliability Analysis	172
		8.2.4	Reliability Testing	172
		8.2.5	Supply-Chain Management	173
		8.2.6	Failure Data Tracking and Analysis	173
		8.2.7	Verification and Validation	174
		8.2.8	Reliability Improvement	174
	8.3	Summary		175
		Problems		175
9	**Parts Selection and Management**			**177**
	9.1	Part Assessment Process		177
		9.1.1	Performance Assessment	178
		9.1.2	Quality Assessment	179
		9.1.3	Process Capability Index	179
		9.1.4	Average Outgoing Quality	182
		9.1.5	Reliability Assessment	182
		9.1.6	Assembly Assessment	185
	9.2	Parts Management		185
		9.2.1	Supply Chain Management	185
		9.2.2	Part Change Management	186
		9.2.3	Industry Change Control Policies	187

9.3	Risk Management	188
9.4	Summary	190
	Problems	191

10 Failure Modes, Mechanisms, and Effects Analysis ... 193
10.1	Development of FMMEA	193
10.2	Failure Modes, Mechanisms, and Effects Analysis	195
	10.2.1 System Definition, Elements, and Functions	195
	10.2.2 Potential Failure Modes	196
	10.2.3 Potential Failure Causes	197
	10.2.4 Potential Failure Mechanisms	197
	10.2.5 Failure Models	197
	10.2.6 Life-Cycle Profile	198
	10.2.7 Failure Mechanism Prioritization	198
	10.2.8 Documentation	200
10.3	Case Study	201
10.4	Summary	205
	Problems	206

11 Probabilistic Design for Reliability and the Factor of Safety ... 207
11.1	Design for Reliability	207
11.2	Design of a Tension Element	208
11.3	Reliability Models for Probabilistic Design	209
11.4	Example of Probabilistic Design and Design for a Reliability Target	211
11.5	Relationship between Reliability, Factor of Safety, and Variability	212
11.6	Functions of Random Variables	215
11.7	Steps for Probabilistic Design	219
11.8	Summary	219
	Problems	220

12 Derating and Uprating ... 223
12.1	Part Ratings	223
	12.1.1 Absolute Maximum Ratings	224
	12.1.2 Recommended Operating Conditions	224
	12.1.3 Factors Used to Determine Ratings	225
12.2	Derating	225
	12.2.1 How Is Derating Practiced?	225
	12.2.2 Limitations of the Derating Methodology	231
	12.2.3 How to Determine These Limits	238
12.3	Uprating	239
	12.3.1 Parts Selection and Management Process	241
	12.3.2 Assessment for Uprateability	241
	12.3.3 Methods of Uprating	242
	12.3.4 Continued Assurance	245
12.4	Summary	245
	Problems	246

Contents

13 Reliability Estimation Techniques **247**
13.1 Tests during the Product Life Cycle 247
    13.1.1 Concept Design and Prototype 247
    13.1.2 Performance Validation to Design Specification 248
    13.1.3 Design Maturity Validation 248
    13.1.4 Design and Manufacturing Process Validation 248
    13.1.5 Preproduction Low Volume Manufacturing 248
    13.1.6 High Volume Production 249
    13.1.7 Feedback from Field Data 249
13.2 Reliability Estimation 249
13.3 Product Qualification and Testing 250
    13.3.1 Input to PoF Qualification Methodology 250
    13.3.2 Accelerated Stress Test Planning and Development 255
    13.3.3 Specimen Characterization 257
    13.3.4 Accelerated Life Tests 259
    13.3.5 Virtual Testing 260
    13.3.6 Virtual Qualification 261
    13.3.7 Output 262
13.4 Case Study: System-in-Package Drop Test Qualification 263
    13.4.1 Step 1: Accelerated Test Planning and Development 263
    13.4.2 Step 2: Specimen Characterization 265
    13.4.3 Step 3: Accelerated Life Testing 266
    13.4.4 Step 4: Virtual Testing 270
    13.4.5 Global FEA 271
    13.4.6 Strain Distributions Due to Modal Contributions 272
    13.4.7 Acceleration Curves 273
    13.4.8 Local FEA 273
    13.4.9 Step 5: Virtual Qualification 274
    13.4.10 PoF Acceleration Curves 275
    13.4.11 Summary of the Methodology for Qualification 276
13.5 Basic Statistical Concepts 276
    13.5.1 Confidence Interval 277
    13.5.2 Interpretation of the Confidence Level 277
    13.5.3 Relationship between Confidence Interval and Sample Size 279
13.6 Confidence Interval for Normal Distribution 279
    13.6.1 Unknown Mean with a Known Variance for
           Normal Distribution 279
    13.6.2 Unknown Mean with an Unknown Variance for
           Normal Distribution 280
    13.6.3 Differences in Two Population Means with
           Variances Known 281
13.7 Confidence Intervals for Proportions 282
13.8 Reliability Estimation and Confidence Limits for
    Success–Failure Testing 283
    13.8.1 Success Testing 286
13.9 Reliability Estimation and Confidence Limits for
    Exponential Distribution 287
13.10 Summary 292
    Problems 292

14 Process Control and Process Capability 295
14.1 Process Control System 295
    14.1.1 Control Charts: Recognizing Sources of Variation 297
    14.1.2 Sources of Variation 297
    14.1.3 Use of Control Charts for Problem Identification 297
14.2 Control Charts 299
    14.2.1 Control Charts for Variables 306
    14.2.2 X-Bar and R Charts 306
    14.2.3 Moving Range Chart Example 308
    14.2.4 X-Bar and S Charts 311
    14.2.5 Control Charts for Attributes 312
    14.2.6 p Chart and np Chart 312
    14.2.7 np Chart Example 313
    14.2.8 c Chart and u Chart 314
    14.2.9 c Chart Example 315
14.3 Benefits of Control Charts 316
14.4 Average Outgoing Quality 317
    14.4.1 Process Capability Studies 318
14.5 Advanced Control Charts 323
    14.5.1 Cumulative Sum Control Charts 323
    14.5.2 Exponentially Weighted Moving Average Control Charts 324
    14.5.3 Other Advanced Control Charts 325
14.6 Summary 325
Problems 326

15 Product Screening and Burn-In Strategies 331
15.1 Burn-In Data Observations 332
15.2 Discussion of Burn-In Data 333
15.3 Higher Field Reliability without Screening 334
15.4 Best Practices 335
15.5 Summary 336
Problems 337

16 Analyzing Product Failures and Root Causes 339
16.1 Root-Cause Analysis Processes 341
    16.1.1 Preplanning 341
    16.1.2 Collecting Data for Analysis and Assessing Immediate Causes 343
    16.1.3 Root-Cause Hypothesization 344
    16.1.4 Analysis and Interpretation of Evidence 348
    16.1.5 Root-Cause Identification and Corrective Actions 348
    16.1.6 Assessment of Corrective Actions 350
16.2 No-Fault-Found 351
    16.2.1 An Approach to Assess NFF 353
    16.2.2 Common Mode Failure 355
    16.2.3 Concept of Common Mode Failure 356
    16.2.4 Modeling and Analysis for Dependencies for Reliability Analysis 360

	16.2.5 Common Mode Failure Root Causes 362
	16.2.6 Common Mode Failure Analysis 364
	16.2.7 Common Mode Failure Occurrence and Impact Reduction 366
16.3	Summary 373
	Problems 374

17 System Reliability Modeling 375
17.1 Reliability Block Diagram 375
17.2 Series System 376
17.3 Products with Redundancy 381
    17.3.1 Active Redundancy 381
    17.3.2 Standby Systems 385
    17.3.3 Standby Systems with Imperfect Switching 387
    17.3.4 Shared Load Parallel Models 390
    17.3.5 (k, n) Systems 391
    17.3.6 Limits of Redundancy 393
17.4 Complex System Reliability 393
    17.4.1 Complete Enumeration Method 393
    17.4.2 Conditional Probability Method 395
    17.4.3 Concept of Coherent Structures 396
17.5 Summary 401
    Problems 402

18 Health Monitoring and Prognostics 409
18.1 Conceptual Model for Prognostics 410
18.2 Reliability and Prognostics 412
18.3 PHM for Electronics 414
18.4 PHM Concepts and Methods 417
    18.4.1 Fuses and Canaries 418
18.5 Monitoring and Reasoning of Failure Precursors 420
    18.5.1 Monitoring Environmental and Usage Profiles for Damage Modeling 424
18.6 Implementation of PHM in a System of Systems 429
18.7 Summary 431
    Problems 431

19 Warranty Analysis 433
19.1 Product Warranties 434
19.2 Warranty Return Information 435
19.3 Warranty Policies 436
19.4 Warranty and Reliability 437
19.5 Warranty Cost Analysis 439
    19.5.1 Elements of Warranty Cost Models 440
    19.5.2 Failure Distributions 440
    19.5.3 Cost Modeling Calculation 440
    19.5.4 Modeling Assumptions and Notation 441
    19.5.5 Cost Models Examples 442
    19.5.6 Information Needs 444
    19.5.7 Other Cost Models 446

19.6	Warranty and Reliability Management	448
19.7	Summary	449
	Problems	449

Appendix A: Some Useful Integrals 451

Appendix B: Table for Gamma Function 453

Appendix C: Table for Cumulative Standard Normal Distribution 455

**Appendix D: Values for the Percentage Points $t_{\alpha,\nu}$ of the
t-Distribution** 457

**Appendix E: Percentage Points $\chi^2_{\alpha,\nu}$ of the Chi-Square
Distribution** 461

Appendix F: Percentage Points for the F-Distribution 467

Bibliography 473

Index 487

Preface

Humans have come to depend on engineered systems to perform their daily tasks. From homes and offices to cars and cell phones, the context in which we live our lives has been largely constructed by engineers who have designed systems and brought their ideas to the marketplace.

While engineered systems have many benefits, they also present risks. How do we know that a building is safe and reliable? How do we know that a sensor in a train will work? How do we know that airbags and brakes will function in an emergency? No matter how many experts were involved in designing systems, the chance for failure always lingers. Thus, all engineering disciplines need reliability.

Today, reliability engineering is a sophisticated and demanding interdisciplinary field. All engineers must ensure the reliability of their designs and products. Moreover, they must be able to analyze a product and assess which parts of the system might be prone to failure. This requires a wide-ranging body of knowledge in the basic sciences, including physics, chemistry, and biology, and an understanding of broader issues within system integration and engineering, while at the same time considering costs and schedules.

The purpose of this book is to present an integrated approach for the design, engineering, and management of reliability activities throughout the life cycle of a product. This book is for those who are interested in gaining fundamental knowledge of the practical aspects of reliability to design, manufacture, and implement tests to ensure product reliability. It is equally helpful for those interested in pursuing a career in reliability, as well as for maintainability, safety, and supportability teams. We have thus written this book to provide students and practitioners with a comprehensive understanding of reliability engineering.

The book is organized into 19 chapters. Each chapter consists of a number of numerical examples and homework problems. References on the topics covered are presented to help the reader delve into more detail.

Chapter 1 provides an overview and discussion of the relevance of reliability engineering for the twenty-first century. This chapter presents a definition of reliability and describes the relationship between reliability, quality, and performance. The consequences of having an unreliable product, that is, a product that fails, are presented with examples. The chapter concludes with a discussion of supplier–customer reliability objectives and responsibilities. It also discusses various stakeholders in product reliability. Principles for designing and managing a reliability program for the twenty-first century are presented.

Preface

Chapter 2 presents the fundamental mathematical theory for reliability. Useful reliability measures for communicating reliability are presented. The focus is on reliability and unreliability functions, the probability density function, the hazard rate, the conditional reliability function, and key time-to-failure metrics, such as mean time to failure, median time to failure, percentiles of life, various moments of a random variable, and their usefulness in quantifying and assessing reliability. The bathtub curve and its characteristics and applications in reliability are discussed.

Chapter 3 covers basic concepts in probability related to reliability, including statistical distributions and their applications in reliability analysis. Two discrete distributions (binomial and Poisson) and five continuous distributions (exponential, normal, lognormal, gamma, and Weibull) that are commonly used in reliability modeling and hazard rate assessments are presented. The concepts of probability plotting and the graphical method for reliability estimation are also presented with examples.

Chapter 4 gives a comprehensive review of the Six Sigma methodology, including Design for Six Sigma. Six Sigma provides a set of tools to use when a focused technical breakthrough approach is required to resolve complicated technical issues, including reliability in design and manufacturing. In this chapter, an introduction to Six Sigma is provided, and the effect of process shift on long-term and short-term capabilities and process yield is explained. A historical overview of Six Sigma is provided, including a thorough discussion of the phases of quality improvement and the process of Six Sigma implementation. Optimization problems in Six Sigma quality improvement, transfer function, variance transmission, and tolerance design are presented. The chapter concludes with a discussion of the implementation of Design for Six Sigma.

Chapter 5 discusses the role of reliability engineering in product development. Product development is a process in which the perceived need for a product leads to the definition of requirements, which are then translated into a design. The chapter introduces a wide range of essential topics, including product life-cycle concepts; organizational reliability capability assessment; parts and materials selection; product qualification methods; and design improvement through root cause analysis methods such as failure modes effects and criticality analysis, fault tree analysis, and the physics-of-failure approach.

Chapter 6 covers methods for preparing and documenting the product requirements for meeting reliability targets and the associated constraints. The definition of requirements is directly derived from the needs of the market and the possible constraints in producing the product. This chapter discusses requirements, specifications, and risk tracking. The discussion also includes methods of developing qualified component suppliers and effective supply chains, product requirement specifications, and requirements tracking to achieve the reliability targets.

Chapter 7 discusses the characteristics of the life-cycle environment, definition of the life-cycle environmental profile (LCEP), steps in developing an LCEP, life-cycle phases, environmental loads and their effects, considerations and recommendations for LCEP development, and methods for developing product life-cycle profiles, based on the possible events, environmental conditions, and various types of loads on the product during its life cycle. Methods for estimating life-cycle loads and their effects on product performance are also presented.

Chapter 8 provides a discussion on the reliability capability of organizations. Capability maturity models and the eight key reliability practices, namely reliability requirements and planning, training and development, reliability analysis, reliability testing, supply chain management, failure data tracking and analysis, verification and validation, and reliability improvement, are presented.

Chapter 9 discusses parts selection and management. The key elements to a practical selection process, such as performance analysis of parts for functional adequacy, quality analysis of the production process through process capability, and average outgoing quality assessment, are presented. Then, the practices necessary to ensure continued acceptability over the product life cycle, such as the supply chain, parts change, industry change, control policies, and the concepts of risk management, are discussed.

Chapter 10 presents a new methodology called failure modes, mechanisms, and effects analysis (FMMEA) which is used to identify the potential failure mechanisms and models for all potential failures modes and prioritize the failure mechanisms. Knowledge of failure mechanisms that cause product failure is essential for the implementation of appropriate design practices for the design and development of reliable products. FMMEA enhances the value of failure mode and effects analysis (FMEA) and failure mode, effects, and criticality analysis (FMECA) by identifying the "high priority failure mechanisms" to help create an action plan to mitigate their effects. Knowledge of the causes and consequences of mechanisms found through FMMEA helps to make product development efficient and cost effective. A case study describing the FMMEA process for a simple electronic circuit board assembly is presented. Methods for the identification of failure mechanisms, their prioritization for improvement and risk analysis, and a procedure for documentation are discussed. The FMMEA procedure is illustrated by a case study.

Chapter 11 covers basic models and principles to quantify and evaluate reliability during the design stage. Based on the physics of failure, the designer can understand the underlying stress and strength variables, which are random variables. This leads us to consider the increasingly popular probabilistic approach to design. Thus, we can develop the relationships between reliability and different types of safety factors. This chapter provides a review of statistical tolerances, and develops the relationship between tolerances and the characteristics of the parts and reliability.

Chapter 12 discusses the concepts of derating and uprating. This chapter demonstrates that the way in which a part is used (i.e., the part's stress level) has a direct impact on the performance and reliability of parts. This chapter introduces how users can modify the usage environment of parts based on ratings from the manufacturer, derating, and uprating. The discussion includes factors considered for determining part rating, and the methods and limitations of derating. Stress balancing is also presented.

Chapter 13 covers reliability estimation techniques. The purpose of reliability demonstration and testing is to determine the reliability levels of a product. We have to design tests in such a manner that the maximum amount of information can be obtained from the minimum amount of testing. For this, various statistical techniques are used. A major problem for the design of adequate tests is simulating the real-world environment. The product is subjected to many environmental factors during its lifetime, such as temperature, vibrations and shock, and rough handling. These stresses may be encountered individually, simultaneously, or sequentially, and there are other random factors. Methods to determine the sample size required for testing and its relationship to confidence levels are presented. Reliability estimation and the confidence intervals for success-failure tests and when the time to failure is an exponential distribution are also discussed with numerical examples. A case study is also presented for reliability test qualification.

Chapter 14 describes statistical process control and process capability. Quality in manufacturing is a measure of a product's ability to meet the design specifications and workmanship criteria of the manufacturer. Process control systems, sources of variation, and attributes that define control charts used in industry for process control are introduced. Several numerical examples are provided.

Chapter 15 discusses methods for product screening and burn-in strategies. If the manufacturing or assembly processes cannot be improved, screening and burn-in strategies are used to eliminate the weak items in the population. The chapter demonstrates the analysis of burn-in data and discusses the pros and cons of implementing burn-in tests. A case study demonstrates that having a better manufacturing process and quality control system is preferable to 100% burn-in of products.

Chapter 16 discusses root cause analysis and product failure mechanisms, presents a methodology for root cause analysis, and provides guidance for decision-making. A root cause is the most basic causal factor (or factors) that, if corrected or removed, will prevent the recurrence of a problem. It is generally understood that problem identification and correction requires the identification of the root cause. This chapter presents what exactly a root cause analysis is, what it entails, and at what point in the investigation one should stop. This chapter also reviews the possible causes and effects for no-fault-found observations and intermittent failures, and summarizes them into cause-and-effect diagrams. The relationships between several techniques for root-cause identification, such as Ishikawa diagrams, fault tree analysis, and failure mode, mechanisms, and effects analysis, are covered.

Chapter 17 describes how to combine reliability information from the system architecture to compute system-level reliability. Reliability block diagrams are preferred as a means to represent the logical system architecture and develop system reliability models. Both static and dynamic models for system reliability and their applications are presented in this chapter. Reliability block diagrams, series, parallel, stand-by, k-out-of-n, and complex system reliability models are discussed. Methods of enumeration, conditional probability, and the concepts of coherent structures are also presented.

Chapter 18 highlights the significance of health monitoring and prognostics. For many products and systems, especially those with long life-cycle reliability requirements, high in-service reliability can be a means to ensure customer satisfaction and remain competitive. Achieving higher field reliability and operational availability requires knowledge of in-service use and life-cycle operational and environmental conditions. In particular, many data collection and reliability prediction schemes are designed before in-service operational and environmental aspects of the system are entirely understood. This chapter discusses conceptual models for prognostics, the relationship between reliability and prognostics, the framework for prognostics and health management (PHM) for electronics, monitoring and reasoning of failure precursors, the application of fuses and canaries, monitoring usage profiles for damage modeling, estimation of remaining useful life, uncertainties associated with PHM, and the implementation of these concepts in complex systems.

Chapter 19 discusses warranty analysis and its relationship to reliability. A warranty is a guarantee from a manufacturer defining a responsibility with respect to the product or service provided. A warranty is a commitment to repair or replace a product or re-perform that service in a commercially acceptable manner if it fails to meet certain standards in the marketplace. Customers value a good warranty as economic protection, but a product is generally not considered good if it fails during the product's useful life (as perceived by the customer), regardless of the warranty. The chapter covers warranty return information, types of warranty policies and cost analyses, the effect of burn-in on warranty, simplified system characterization, and managerial issues with regard to warranty.

The authors are grateful to several people for their help with this book. Dr. Diganta Das, a research scientist at CALCE at the University of Maryland, provided insights in improving all aspects of the text. Dr. Vallayil N. A. Naikan, professor of reliability engineering at IIT Kharagpur, and Dr. P. V. Varde of the Bhabha Atomic Research Centre in India were also instrumental in the development of the book in terms of critically reviewing several chapters. We also thank Professor Sanborn and Yan Ning for their insights into warranties; and Professor Abhijit Dasgupta, Dr. Carlos Morillo, and Elviz George for their perspectives on accelerated testing, screening, and burn-in.

Kailash C. Kapur
Michael Pecht

1 Reliability Engineering in the Twenty-First Century

Institutional and individual customers have increasingly better and broader awareness of products (and services) and are increasingly making smarter choices in their purchases. In fact, because society as a whole continues to become more knowledgeable of product performance, quality, reliability, and cost, these attributes are considered to be market differentiators.

People are responsible for designing, manufacturing, testing, maintaining, and disposing of the products that we use in daily life. Perhaps you may agree with Neville Lewis, who wrote, "Systems do not fail, parts and materials do not fail—people fail!" (Lewis 2003) It is the responsibility of people to have the knowledge and skills to develop products that function in an acceptably reliable manner. These concepts highlight the purpose of this book: to provide the understanding and methodologies to efficiently and cost effectively develop reliable products and to assess and manage the operational availability of complex products, processes, and systems.

This chapter presents the basic definitions of reliability and discusses the relationship between quality, reliability, and performance. Consequences of having an unreliable product are then presented. The chapter concludes with a discussion of supplier–customer reliability objectives and responsibilities.

1.1 What Is Quality?

The word *quality* comes from the Latin *qualis*, meaning "how constituted." Dictionaries define *quality* as the essential character or nature of something, and as an inherent characteristic or attribute. Thus, a product has certain qualities or characteristics, and a product's overall performance, or its effectiveness, is a function of these qualities.

Juran and Gryna (1980) looked at multiple elements of fitness for use and evaluated various quality characteristics (or "qualities"), such as technological characteristics (strength, weight, and voltage), psychological characteristics (sensory characteristics, aesthetic appeal, and preference), and time-oriented characteristics (reliability and

---

Reliability Engineering, First Edition. Kailash C. Kapur and Michael Pecht.
© 2014 John Wiley & Sons, Inc. Published 2014 by John Wiley & Sons, Inc.

Figure 1.1 The relationship of quality, customer satisfaction, and target values.

maintainability). Deming (1982) also investigated several facets of quality, focusing on quality from the viewpoint of the customer.

The American Society for Quality (ASQC Glossary and Tables for Statistical Quality Control 1983) defines *quality* as the "totality of features and characteristics of a product or service that bear on its ability to satisfy a user's given needs." Shewhart (1931) stated it this way:

> The first step of the engineer in trying to satisfy these wants is, therefore, that of translating as nearly as possible these wants into the physical characteristics of the thing manufactured to satisfy these wants. In taking this step, intuition and judgment play an important role, as well as a broad knowledge of the human element involved in the wants of individuals. The second step of the engineer is to set up ways and means of obtaining a product which will differ from the arbitrary set standards for these quality characteristics by no more than may be left to chance.

One of the objectives of quality function deployment (QFD) is to achieve the first step proposed by Shewhart. QFD is a means of translating the "voice of the customer" into substitute quality characteristics, design configurations, design parameters, and technological characteristics that can be deployed (horizontally) through the whole organization: marketing, product planning, design, engineering, purchasing, manufacturing, assembly, sales, and service.

Products have several characteristics, and the "ideal" state or value of these characteristics is called the target value (Figure 1.1). QFD (Figure 1.2) is a methodology to develop target values for substitute quality characteristics that satisfy the requirements of the customer. Mizuno and Akao (Shewhart 1931) have developed the necessary philosophy, system, and methodology to achieve this step.

1.2 What Is Reliability?

Although there is a consensus that reliability is an important attribute of a product, there is no universally accepted definition of *reliability*. Dictionaries define *reliability* (noun) as the state of being reliable, and *reliable* (adjective) as something that can be relied upon or is dependable.

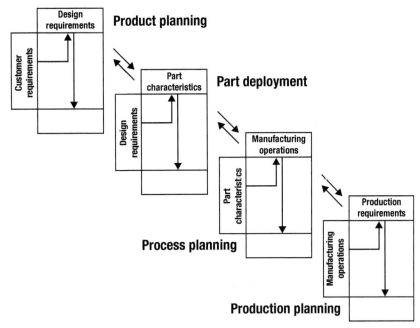

Figure 1.2 Illustration of the steps in QFD.

When we talk about reliability, we are talking about the future performance or behavior of the product. Will the product be dependable in the future? Thus, reliability has been considered a time-oriented quality (Kapur 1986; O'Conner 2000). Some other definitions for reliability that have been used in the past include:

- Reduction of things gone wrong (Johnson and Nilsson 2003).
- An attribute of a product that describes whether the product does what the user wants it to do, when the user wants it to do so (Condra 2001).
- The capability of a product to meet customer expectations of product performance over time (Stracener 1997).
- The probability that a device, product, or system will not fail for a given period of time under specified operating conditions (Shishko 1995).

As evident from the listing, various interpretations of the term *reliability* exist and usually depend on the context of the discussion. However, in any profession, we need an operational definition for reliability, because for improvement and management purposes, reliability must be precisely defined, measured, evaluated, computed, tested, verified, controlled, and sustained in the field.

Since there is always uncertainty about the future performance of a product, the future performance of a product is a random variable, and the mathematical theory of probability can be used to qualify the uncertainty about the future performance of a product. Probability can be estimated using statistics, and thus reliability needs both probability and statistics. Phrases such as "perform satisfactorily" and "function normally" suggest that a product must function within certain performance limits in order to be reliable. Phrases such as "under specified operating conditions" and "when used according to specified conditions" imply that reliability is dependent upon the environmental and application conditions in which a product is used. Finally, the terms "given period of time" and "expected lifetime" suggest that a product must properly function for a certain period of time.

In this book, reliability is defined as follows:

> Reliability is the ability of a product or system to perform as intended (i.e., without failure and within specified performance limits) for a specified time, in its life cycle conditions.

This definition encompasses the key concepts necessary for designing, assessing, and managing product reliability. This definition will now be analyzed and discussed further.

1.2.1 The Ability to Perform as Intended

When a product is purchased, there is an expectation that it will perform as intended. The intention is usually stated by the manufacturer of the product in the form of product specifications, datasheets, and operations documents. For example, the product specifications for a cellular phone inform the user that the cell phone will be able to place a call so long as the user follows the instructions and uses the product within the stated specifications.[1] If, for some reason, the cell phone cannot place a call when turned on, it is regarded as not having the ability to perform as intended, or as having "failed" to perform as intended.

In some cases, a product might "work," but do so poorly enough to be considered unreliable. For example, the cell phone may be able to place a call, but if the cell phone speaker distorts the conversation and inhibits understandable communication, then the phone will be considered unreliable. Or consider the signal problems reported for Apple's iPhone 4 in 2010. The metal bands on the sides of the iPhone 4 also acted as antennas for the device. Some users reported diminished signal quality when gripping the phone in their hands and covering the black strip on the lower left side of the phone. The controversy caused Apple to issue free protective cases for the iPhone 4 for a limited time to quell consumer complaints (Daniel Ionescu 2010).

1.2.2 For a Specified Time

When a product is purchased, it is expected that it will operate for a certain period of time.[2] Generally, a manufacturer offers a warranty, which states the amount of time during which the product should not fail, and if it does fail, the customer is guaranteed a replacement. For a cell phone, the warranty period might be 6 months, but customer expectations might be 2 years or more. A manufacturer that only designs

---

[1]The specifications for a product may also state conditions that must be satisfied to guarantee that the product will operate in a reliable manner. These conditions can include mechanical, electrical, and chemical limits. For example, a product might have voltage or temperature limits that should not be exceeded to guarantee the reliable operation of the product. The specifications usually depend on the design, materials, and processes used to make the product and the expected conditions of use.

[2]Time may be expressed as the total age of a product, the number of hours of operation, the number of miles, or some other metric of use or age.

for the warranty can have many unhappy customers if the expectations are not met. For example, most customers expect their car to be able to operate at least 10 years with proper maintenance.

1.2.3 Life-Cycle Conditions

The reliability of a product depends on the conditions (environmental and usage loads) that are imposed on the product. These conditions arise throughout the life cycle of the product, including in manufacture, transport, storage, and operational use.[3] If the conditions are severe enough, they can cause an immediate failure. For example, if we drop or sit on a cell phone, we may break the display. In some cases, the conditions may only cause a weakening of the product, such as a loosening of a screw, the initiation of a crack, or an increase in electrical resistance. However, with subsequent conditions (loads), this may result in the product not functioning as intended. For example, the product falls apart due to a missing screw, causing a connection to separate; cracking results in the separation of joined parts; and a change in electrical resistance causes a switch to operate intermittently or a button to fail to send a signal.

1.2.4 Reliability as a Relative Measure

Reliability is a relative measure of the performance of a product. In particular, it is relative to the following:

- Definition of function from the viewpoint of the customer
- Definition of unsatisfactory performance or failure from the viewpoint of the customer
- Definition of intended or specified life
- Customer's operating and environmental conditions during the product life cycle.

Furthermore, the reliability of a product will be dependent, as a probability, on the following:

- Intended definition of function (which may be different for different applications)
- Usage and environmental conditions
- Definition of satisfactory performance
- Time.

Many organizations have a document called "Failure Definitions and Scoring Criteria." Such a document delineates how each incident or call for attention in a product will be handled with regard to reliability, maintainability, or safety.

---

[3] A good analogy to products is people. A person's physical reliability will depend on the conditions (loads and stresses) "imposed" on him/her, starting from birth. These conditions can include, but are not limited to, diseases, lifestyle, and accidents. Such conditions can cause the body to wear out or fail in a catastrophic manner.

1 Reliability Engineering in the Twenty-First Century

1.3 Quality, Customer Satisfaction, and System Effectiveness

For consumer products, quality has been traditionally associated with customer satisfaction or happiness. This interpretation of quality focuses on the total value or the utility that the customer derives from the product. This concept has also been used by the U.S. Department of Defense, focusing on system effectiveness as the overall ability of a product to accomplish its mission under specified operating conditions.

There are various characteristics (e.g., engineering, technological, psychological, cost, and delivery) that impact customer satisfaction. Thus, quality (Q) may be modeled as:

$$Q = \text{Customer Satisfaction} = \phi(x_1, x_2, \ldots, x_i, \ldots, x_n, \ldots), \qquad (1.1)$$

where x_i is the ith characteristic ($i = 1, 2, \ldots, n, \ldots$).

These qualities will impact the overall value perceived by the customer, as shown in Figure 1.3. In the beginning, we have ideal or target values of the characteristics $x_1, x_2, \ldots, x_i, \ldots, x_n, \ldots$ These values result in some measure of customer satisfaction. With time, changes in these qualities will impact customer satisfaction. Reliability as a "time-oriented" quality impacts customer satisfaction.

The undesirable and uncontrollable factors that cause a functional characteristic to deviate from its target value are called *noise factors*. Some examples of noise factors are:

- Outer noise: environmental conditions, such as temperature, humidity, dust, and different customer usage conditions.
- Inner noise: changes in the inherent properties of the product, such as deterioration, wear, fatigue, and corrosion—all of which may be a result of the outer noise condition.
- Product noise: piece-to-piece variation due to manufacturing variation and imperfections.

A reliable product must be robust over time, as demonstrated in Figure 1.4.

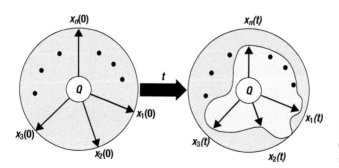

Figure 1.3 Time-oriented qualities and customer satisfaction.

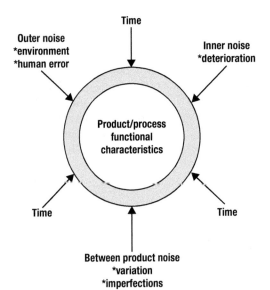

Figure 1.4 A reliable product/process is robust over time.

1.4 Performance, Quality, and Reliability

Performance is usually associated with the functionality of a product—what the product can do and how well it can do it. For example, the functionality of a camera involves taking pictures. How well it can take pictures and the quality of the pictures involves performance parameters such as pixel density, color clarity, contrast, and shutter speed.

Performance is related to the question, "How well does a product work?" For example, for a race car, speed and handling are key performance requirements. The car will not win a race if its speed is not fast enough. Of course, the car must finish the race, and needs sufficiently high reliability to finish the race. After the race, the car can be maintained and even replaced, but winning is everything.[4]

For commercial aircraft, the safe transportation of humans is the primary concern. To achieve the necessary safety, the airplane must be reliable, even if its speed is not the fastest. In fact, other than cost, reliability is the driving force for most commercial aircraft design and maintenance decisions, and is generally more important than performance parameters, which may be sacrificed to achieve the required reliability.

Improving the performance of products usually requires adding technology and complexity. This can make the required reliability more difficult to achieve.

Quality is associated with the workmanship of the product. For example, the quality metrics of a camera might include defects in its appearance or operation, and the camera's ability to meet the specified performance parameters when the customer first receives the product. Quality defects can result in premature failures of the product.

---

[4]If the racing car were only used in normal commuter conditions, its miles to failure (reliability) might be higher since the subsystems (e.g., motor and tires) would be less "stressed."

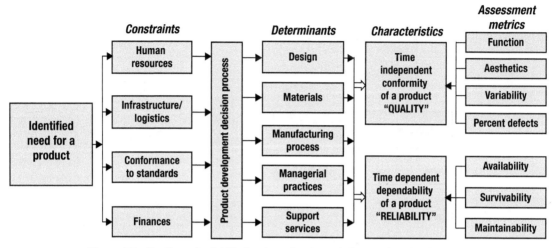

Figure 1.5 Quality and reliability inputs and outputs during product development.

Reliability is associated with the ability of a product to perform as intended (i.e., without failure and within specified performance limits) for a specified time in its life cycle. In the case of the camera, the customer expects the camera to operate properly for some specified period of time beyond its purchase, which usually depends on the purpose and cost of the camera. A low-cost, throwaway camera may be used just to take one set of pictures. A professional camera may be expected to last (be reliable) for decades, if properly maintained.

"To measure quality, we make a judgment about a product today. To measure reliability, we make judgments about what the product will be like in the future" (Condra 2001). Quality in this way of thinking is associated primarily with manufacturing, and reliability is associated mostly with design and product operation. Figure 1.5 shows the role of quality and reliability in product development.

Product quality can impact product reliability. For example, if the material strength of a product is decreased due to defects, the product reliability may also be decreased, because lower than expected life-cycle conditions could cause failures. On the other hand, a high-quality product may not be reliable, even though it conforms to workmanship specifications. For example, a product may be unable to withstand environmental or operational conditions over time due to the poor selection of materials, even though the materials meet workmanship specifications. It is also possible that the workmanship specifications were not properly selected for the usage requirements.

1.5 Reliability and the System Life Cycle

Reliability activities should span the entire life cycle of the system. Figure 1.6 shows the major points of reliability practices and activities for the life cycle of a typical system. The activities presented in Figure 1.6 are briefly explained in the following sections.

1.5 Reliability and the System Life Cycle

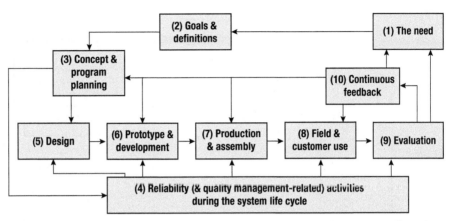

Figure 1.6 Reliability (and quality management related activities) during system life cycle.

Step 1: Need. The need for reliability must be anticipated from the beginning. A reliability program can then be justified based on specific system requirements in terms of life-cycle costs and other operational requirements, including market competitiveness, customer needs, societal requirements in terms of safety and public health, liability, and statutory needs.

Step 2: Goals and Definitions. Requirements must be specified in terms of well-defined goals. Chapter 2 covers some of the useful ways to quantitatively measure reliability. Additional material given in Chapters 3 and 4 can be used for this. Chapter 3 covers useful life distributions to model time to failure, and Chapter 17 covers topics related to modeling and analysis of system reliability.

Step 3: Concept and Program Planning. Based on reliability and other operational requirements, reliability plans must be developed. Concept and program planning is a very important phase in the life cycle of the system. Figure 1.7 illustrates that 60–70% of the life cycle may be determined by the decisions made at the concept stage. Thus, the nature of the reliability programs will also determine the overall effectiveness of the total program.

Step 4: Reliability and Quality Management Activities. The plans developed in step 3 are implemented, and the total program is continuously monitored in the organization for the life-cycle phases. An organizational chart for the implementation of these plans must exist with well-defined responsibilities. Some guiding principles that can be used for any reliability program and its processes and management include:

- *Customer Focus.* Quality, and reliability as one of its qualities, is defined and evaluated by the customer, and the organization has a constancy of purpose to meet and/or exceed the needs and requirements of the customer.[5]

---

[5] We use the word *customer* in a very broad sense. Anything the system affects is the customer. Thus, in addition to human beings and society, the environmental and future impacts of the product are considered in the program.

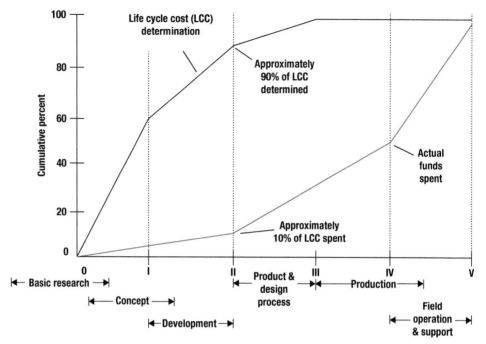

Figure 1.7 Conceptual relationship of life-cycle cost and different phases of life cycle.

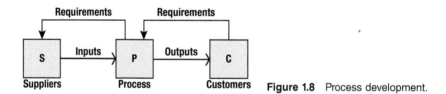

Figure 1.8 Process development.

- *System Focus.* Emphasis is on system integration, synergy, and the interdependence and interactions of all the parts of the system (hardware, software, human, and other elements). All the tools and methodologies of systems engineering and some of the developments in Design for Six Sigma (DFSS) (Chapter 4 in this book) are an integral part of this focus.
- *Process Focus.* Design and management of reliability processes should be well developed and managed using cross-functional teams using the methodology of concurrent design and engineering (Figure 1.8).
- *Structure.* The reliability program must understand the relationships and interdependence of all the components, assemblies, and subsystems. High reliability is not an end in itself but is a means to achieve higher levels of customer satisfaction, market share, and profitability. Thus, we should be able to translate reliability metrics to financial metrics that management and customers can understand and use for decision-making processes.
- *Continuous Improvement and Future Focus.* Continuous, evolutionary, and breakthrough improvement is an integral part of any reliability process.

1.5 Reliability and the System Life Cycle

The organization should have a philosophy of never-ending improvement and reliance on long-term thinking.

- *Preventive and Proactive Strategies.* The real purpose of reliability assurance processes is to prevent problems from happening. Throughout the book, we will present many design philosophies and methodologies to achieve this objective.
- *Scientific Approach.* Reliability assurance sciences are based on mathematical and statistical approaches in addition to using all the other sciences (such as the physics, chemistry, and biology of failure). We must understand the causation (cause–effect and means–end relationships), and we should not depend on anecdotal approaches. Data-driven and empirical methods are used for the management of reliability programs.
- *Integration.* Systems thinking includes broader issues related to the culture of the organization. Thus, the reliability program must consider the integration of cultural issues, values, beliefs, and habits in any organization for a quality and productivity improvement framework.

Step 5: Design. Reliability is a design parameter, and it must be incorporated into product development at the design stage. Figure 1.9 illustrates the importance of design in terms of cost to address or fix problems in the future of the life cycle of the product.

Step 6: Prototype and Development. Prototypes are developed based on the design specifications and life-cycle requirements. The reliability of the design is verified through development testing. Concepts, such as the design and development of reliability test plans, including accelerated testing, are used in this step. If the design has deficiencies, they are corrected by understanding the root failure causes and their effect on the design. After the product has achieved the required levels of reliability, the design is released for production.

Step 7: Production and Assembly. The product is manufactured and assembled based on the design specifications. Quality control methodologies, such as statistical process control (SPC), are used. The parts, materials, and processes are controlled based on the quality assurance methodologies covered in Chapter

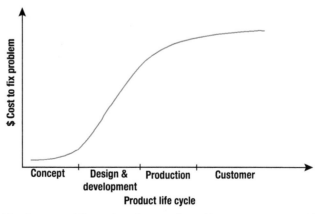

Figure 1.9 Conceptual illustration of cost to fix problems versus product life cycle.

14 of this book. Product screening and burn-in strategies are also covered in Chapter 15. One of the objectives of quality assurance programs during this phase of the system is to make sure that the product reliability is not degraded and can be sustained in the field.

Step 8: Field and Customer Use. Before the product is actually shipped and used in the field by customers, it is important to develop handling, service, and, if needed, maintenance instructions. If high operational availability is needed, then a combination of reliability and maintainability will be necessary.

Step 9: Continuous System Evaluation. The product in the field is continuously evaluated to determine whether the required reliability goals are actually being sustained. For this purpose, a reliability monitoring program and field data collection program are established. Topics related to warranty analysis and prognostics and system health management are covered in Chapters 18 and 19.

Step 10: Continuous Feedback. There must be continuous feedback among all the steps in the life cycle of the product. A comprehensive data gathering and information system is developed. A proper communication system is also developed and managed for all the groups responsible for the various steps. This way, all field deficiencies can be reported to the appropriate groups. This will result in continuous improvement of the product. Some useful material for this step is also covered in Chapters 13, 18, and 19.

1.6 Consequences of Failure

There is always a risk of a product failing in the field. For some products, the consequences of failure can be minor, while for others, it can be catastrophic. Possible consequences include financial loss, personal injury, and various intangible costs. Under U.S. law, consequences of product failure may also include civil financial penalties levied by the courts and penalties under statutes, such as the Consumer Product Safety Act, building codes, and state laws. These penalties can include personal sanctions such as removal of professional licenses, fines, and jail sentences.

1.6.1 Financial Loss

When a product fails, there is often a loss of service, a cost of repair or replacement, and a loss of goodwill with the customer, all of which either directly or indirectly involve some form of financial loss. Costs can come in the form of losses in market share due to damaged consumer confidence, increases in insurance rates, warranty claims, or claims for damages resulting from personal injury. If negative press follows a failure, a company's stock price or credit rating can also be affected.

Often, costs are not simple to predict. For example, a warranty claim may include not only the cost of replacement parts, but also the service infrastructure that must be maintained in order to handle failures (Dummer et al. 1997). Repair staff must be trained to respond to failures. Spare parts may be required, which increases inventory levels. Service stations must be maintained in order to handle product repairs.

As an example of a financial loss, in July 2000, a month after the release of its new 1.13 GHz Pentium III microprocessors, Intel was forced to make a recall (Jayant 2000). The chips had a hardware glitch that caused computers to freeze or crash under certain conditions. Although fewer than 10,000 units were affected, the recall was an embarrassment and Intel's reputation was called into question at a time when competition in the microprocessor market was fierce.

In January 2011, Intel discovered a design flaw in its 6 Series Cougar Point support chips. Intel found that some of the connection ports in those chipsets could degrade over time and interrupt the flow of data from disk drives and DVD drives. By the time it discovered this problem, Intel had already shipped over 8 million defective chips to customers. As a result, Intel expected its revenue for the first quarter of 2011 to be cut by $300 million, and expected to spend $700 million for repair and replacement of the affected chips. This problem was the costliest in Intel's history and affected products from top manufacturers, including Dell, Hewlett-Packard, and Samsung (Tibken 2011).

Another example was problematic graphics processing units that were made by Nvidia. Customers began observing and reporting intermittent failures in their computers to companies such as Hewlett-Packard, Toshiba, and Dell. However, the absence of an effective reliability process caused a delay in understanding the problems, the failure mechanisms, the root causes, and the available corrective actions. These delays resulted in the continued production and sale of defective units, ineffective solutions, consumer and securities lawsuits, and costs to Nvidia of at least $397 million.[6]

In December 2011, Honda announced a recall of over 300,000 vehicles due to a defect in the driver's airbag. This was the latest in a series of recalls that had taken place in November 2008, June 2009, and April 2011, and involved nearly 1 million vehicles. The defective airbags were recalled because they could deploy with too much pressure, possibly endangering the driver (Udy 2011).

Between 2009 and 2011, Toyota had a string of recalls totaling 14 million vehicles. The problems included steering problems and the highly publicized sudden acceleration problem. In 2010 alone, Toyota paid three fines totaling $48.8 million. As a result of these safety concerns and damage to its reputation, Toyota had the lowest growth of the major automakers in the United States during 2010, growing 0.2 percent in a year when the U.S. auto market grew by 11.2 percent. Between July and September 2011, Toyota's profits declined 18.5 percent to around $1 billion (Foster 2011; Roland 2010a). In November 2011, Toyota recalled 550,000 vehicles worldwide due to possible steering problems caused by misaligned rings in the vehicles' engines.

The cost of failure also often includes financial losses for the customer incurred as a result of failed equipment not being in operation. For some products, this cost may greatly exceed the actual cost of replacing or repairing the equipment. Some examples are provided in Table 1.1 (Washington Post 1999).

1.6.2 Breach of Public Trust

The National Society of Professional Engineers notes that "Engineers, in the fulfillment of their professional duties, shall hold paramount the safety, health, and welfare

---
[6]U.S. Securities and Exchange Commission, May 2, 2010.

Table 1.1 Cost of lost service due to a product failure

Type of business	Average hourly impact
Retail brokerage	$6,450,000
Credit card sales authorization	$2,600,000
Home shopping channels	$113,750
Catalog sales center	$90,000
Airline reservation centers	$89,500
Cellular service activation	$41,000
Package shipping service	$28,250
Online network connect fees	$22,250
ATM service fees	$14,500

of the public" (National Society of Professional Engineers 1964). In many cases, public health, safety, and welfare are directly related to reliability.

On July 17, 1981, the second- and fourth-floor suspended walkways within the atrium of the Kansas City Hyatt Regency Hotel collapsed. This was the single largest structural disaster in terms of loss of life in U.S. history at that time. The hotel had only been open for a year. The structural connections supporting the ceiling rods that supported the walkways across the atrium failed and both walkways collapsed onto the crowded first-floor atrium below. One hundred fourteen people were killed, and over 200 were injured. Millions of dollars in damages resulted from the collapse (University of Utah, Mechanical Engineering Department 1981). The accident occurred due to improper design of the walkway supports: the connections between the hanger rods and the main-carrying box beams of the walkways failed. Two errors contributed to the deficiency: a serious error in the original design of the connections, and a change in the hanger rod arrangement during construction, which doubled the load on the connection.

Another significant failure occurred on April 28, 1988, when a major portion of the upper crown skin of the fuselage of a 19-year-old Aloha Airlines 737 blew open at 24,000 ft. The structure separated in flight, causing an explosive decompression of the cabin that killed a flight attendant and injured eight other people. The airplane was determined to be damaged beyond repair. The National Transportation Security Board (NTSB), which investigated the Aloha accident, concluded the jet's roof and walls tore off in flight because there were multiple fatigue cracks in the jet's skin that had not been observed in maintenance. The cracks developed because the lap joints, which connect two overlapping metal sheets of the fuselage and were supposed to hold the fuselage together, corroded and failed (Stoller 2001).

In September 2011, the Federal Aviation Administration (FAA) fined Aviation Technical Services Inc. (ATS), a maintenance provider for Southwest Airlines, $1.1 million for making improper repairs to 44 Southwest Boeing 737-300 jetliners. The FAA had provided directives for finding and repairing fatigue cracks in the fuselage skins of the planes. The FAA alleged that ATS failed to properly install fasteners in all the rivet holes of the fuselage skins. In April 2011, a 5-ft hole was torn in the fuselage of a Southwest 737-300 in midflight at 34,000 ft. The pilot was able to make an emergency landing in Arizona, and none of the 122 people on board were seriously injured. While this plane was not among the ones repaired by ATS, this near-disaster highlighted the need for correct maintenance practices. After the incident, Southwest inspected 79 other Boeing 737s and found that five of them had fuselage cracks requiring repairs (Carey 2011).

On July 23, 2011, a high-speed train collided with a stalled train near the city of Wenzhou in southeastern China. It was reported that 40 people were killed and nearly 200 wounded. When he visited the scene of the accident, Chinese Premier Wen Jiabao said, "The high-speed railway development should integrate speed, quality, efficiency and safety. And safety should be in the first place. Without safety, high-speed trains will lose their credibility" (Dean et al. 2011).

1.6.3 Legal Liability

There are a number of legal risks associated with product reliability and failure. A company can be sued for damages resulting from failures. A company can also be sued if they did not warn users of defects or reliability problems. In extreme cases of negligence, criminal charges can be brought in addition to civil damages.

Most states in the United States operate on the theory of strict liability. Under this law, a company is liable for damages resulting from a defect for no reason other than that one exists, and a plaintiff does not need to prove any form of negligence to win their case. Companies have a duty to exercise "ordinary and reasonable care" to make their products safe and reliable. If a plaintiff can prove that a defect or risk existed with a product, that this defect or risk caused an injury, that this defect or risk was foreseeable, and that the company broke their duty of care, damages can be assessed. A defect, for legal purposes, can include manufacturing flaws, design oversights, or inadequacies in the documentation accompanying a product. Thus, almost every job performed by a designer or an engineer can be subjected to legal scrutiny.

An example of failure resulting in legal liability occurred with 22 million Ford vehicles built between 1983 and 1995 that had defective thick film ignition (TFI) modules. The TFI module was the electronic control in the ignition system that controlled the spark in the internal combustion process. Defects in the TFI could cause vehicles to stall and die on the highway at any time. Failure at highway speeds could cause the driver to lose control or result in a stalled vehicle being hit by another vehicle. In October 2001, Ford agreed to the largest automotive class-action settlement in history, promising to reimburse drivers for the faulty ignition modules. The settlement was estimated to have cost Ford as much as $2.7 billion (Castelli et al. 2003).

In 1999, Toshiba was sued for selling defective laptop computers (Pasztor and Landers 1999). More than five million laptops were built with a defective floppy disk drive controller chip that would randomly corrupt data without warning. Toshiba agreed to a $2.1 billion settlement to prevent the case from going to trial, as Toshiba felt that a verdict as high as $9 billion might have been imposed.

Another example of liability occurred with Toyota's vehicles. Toyota had a host of recalls in 2010, and it was required to pay over $32 million in fines because of the late timing of the recalls (Roland 2010b).

1.6.4 Intangible Losses

Depending on the expectations that customers have for a product, relations with customers can be greatly damaged when they experience a product failure. Failures can also damage the general reputation of a company. A reputation for poor reliability can discourage repeat and potential future customers from buying a product, even if the causes of past failures have been corrected.

In some cases, the effects of a lack of reliability can hurt the national psyche, for example, failures in space, military, and transportation applications. The higher the profile of a failure event, the greater the effect is on society. Failures that affect public health and the environment can also create discontent with government and regulatory bodies.

1.7 Suppliers and Customers

The rapid pace of technological developments and the globalization of supply chains have made customers dependent upon worldwide suppliers who provide parts (materials), subassemblies, and final products. When customers have to wait until they receive their parts, subassemblies, or products to assess if they are reliable, this can be an expensive iterative process. An upfront evaluation of suppliers is a beneficial alternative. Measuring the reliability capability of a supplier yields important information about the likelihood that a reliable product can be produced (Tiku et al. 2007). Reliability capability can be defined as follows:

> Reliability capability is a measure of the practices within an organization that contribute to the reliability of the final product, and the effectiveness of these practices in meeting the reliability requirements of customers.

To obtain optimal reliability and mutually beneficial results, suppliers and customers in the supply chain should cooperate. The IEEE Reliability Program Standard 1332 (IEEE Standards Project Editors 1998) identifies three reliability objectives between suppliers and customers:

- The supplier, working with the customer, should determine and understand the customer's requirements and product needs so that a comprehensive design specification can be generated.
- The supplier should structure and follow a series of engineering activities so that the resulting product satisfies the customer's requirements and product needs with regard to product reliability.
- The supplier should include activities that assure the customer that reliability requirements and product needs have been satisfied.

1.8 Summary

Reliability pertains to the ability of a product to perform without failure and within specified performance limits for a specified time in its life-cycle application conditions. Performance and quality are related to reliability. Performance parameters typically describe the functional capabilities of a product. Quality parameters are commonly used to assess the manufacturing goodness and the ability of a product to work when first received by the customer.

Reliability engineering deals with preventing, assessing, and managing failures. The tools of reliability engineers include statistics, probability theory, and many fields of engineering and the sciences related to the problem domain.

Problems

1.1 Pick an example product and explain the differences between performance, quality, and reliability. Select a datasheet for a product and check what is listed in terms of performance, quality, and reliability. Document your observations.

1.2 Identify the reliability metrics provided in the specification sheets of a part or product. Discuss the relevance of these metrics.

1.3 Find an example of an actual product failure. Why did it occur? What was the root cause of the failure? What were the consequences? Can you put a value (e.g., time and money) on the consequences?

1.4 In some situations, the definition of failure may depend on both the performance specifications and expectations. Can you think of a past experience where you considered a product to have failed but it may not have been considered a failure according to the product specifications? Describe the situation. If you cannot think of a situation, report a hypothetical case.

1.5 Prepare a one-page brief on the "engineer's responsibility" laws and statutes of your country. If your country is the United States, choose another country's laws to report on.

1.6 Once a product leaves the manufacturer, it will be used in many different applications; some may not be for what the product was designed. From whom should product reliability/failure information be gathered? How should the many sources of reliability information be weighted?

1.7 In Section 1.6, four consequences of failure were introduced. Which of these do you think is most important? Why?

2 Reliability Concepts

In Chapter 1, the reliability of a product was defined as "the ability of a product to perform as intended (i.e., without failure and within specified performance limits) for a specified time, in its life cycle conditions." This chapter presents the fundamental definitions and measures needed for quantifying and communicating the reliability of a product. The focus in this chapter is on reliability and unreliability functions, the probability density function, hazard rate, conditional reliability function, percentiles of life, and time-to-failure metrics.

The purpose of, and the need for, a particular product determines the kind of reliability measures that are most meaningful and most useful. In general, a product may be required to perform various functions, each having a different reliability. In addition, at any given time (or number of cycles, or any other measure of the use of a product), the product may have a different probability of successfully performing the required function under the stated conditions.

2.1 Basic Reliability Concepts

For a constant sample size, n_0, of identical products that are tested or being monitored, if n_f products have failed and the remaining number of products, n_S, are still operating satisfactorily at any time, t, then

$$n_S(t) + n_f(t) = n_0. \tag{2.1}$$

The factor t in Equation 2.1 can pertain to age, total time elapsed, operating time, number of cycles, distance traveled, or be replaced by a measured quantity that could range from $-\infty$ to ∞ for any general random variable. This quantity is called a variate in statistics. Variates may be discrete (for the life of a product, the range is from 0 to ∞; e.g., number of cycles) or continuous when they can take on any real value within a certain range of real numbers.

Reliability Engineering, First Edition. Kailash C. Kapur and Michael Pecht.
© 2014 John Wiley & Sons, Inc. Published 2014 by John Wiley & Sons, Inc.

2 Reliability Concepts

The ratio of failed products per sample size is an estimate of the unreliability, $\hat{Q}(t)$, of the product at any time t:

$$\hat{Q}(t) = \frac{n_f(t)}{n_0}, \tag{2.2}$$

where the caret above the variable indicates that it is an estimate. Similarly, the estimate of reliability, $\hat{R}(t)$, of a product at time t is given by the ratio of operating (not failed) products per sample size or the underlying frame of reference:

$$\hat{R}(t) = \frac{n_S(t)}{n_0} = 1 - \hat{Q}(t). \tag{2.3}$$

As fractional numbers, $\hat{R}(t)$ and $\hat{Q}(t)$ range in value from zero to unity; multiplied by 100, they give the estimate of the probability as a percentage.

Example 2.1

A semiconductor fabrication plant has an average output of 10 million devices per week. It has been found that over the past year 100,000 devices were rejected in the final test.

(a) What is the unreliability of the semiconductor devices according to the conducted test?

(b) If the tests reject 99% of all defective devices, what is the chance that any device a customer receives will be defective?

Solution:
The total number of devices produced in a year is:

(a) $n_0 = 52 \times 10 \times 10^6 = 520 \times 10^6$
The number of rejects (failures), n_f, over the same period is:

$$n_f = 1 \times 10^5.$$

Therefore, from Equation 2.2, an estimate for device unreliability is:

$$\hat{Q}(t) = \frac{n_f(t)}{n_0} = \frac{1 \times 10^5}{520 \times 10^6} \approx 1.92 \times 10^{-4},$$

or 1 chance in 5200.

(b) If the rejected devices represent 99% of all the defective devices produced, then the number of defectives that passed testing is:

$$x_d = \left[\frac{1 \times 10^5}{0.99} - (1 \times 10^5) \right] \approx 1010.$$

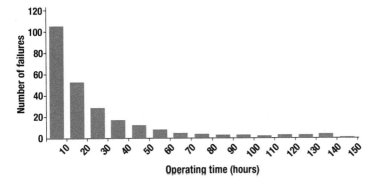

Figure 2.1 Frequency histogram or life characteristic curve for data from Table 2.2.

Therefore, the probability of a customer getting a defective device, or the unreliability of the supplied devices on first use, is:

$$\hat{Q}(t) = \frac{1010}{(520 \times 10^6) - (1 \times 10^5)} \approx 1.94 \times 10^{-6},$$

or 1 chance in 515,000.

Reliability estimates obtained by testing or monitoring samples in the field generally exhibit variability. For example, light bulbs designed to last for 10,000 hours of operation that are all installed at the same time in the same room are unlikely to fail at exactly the same time, let alone at exactly 10,000 hours. Variability in both the measured product response as well as the time of operation is expected. In fact, product reliability assessment is often associated with the measurement and estimation of this variability.

The accuracy of a reliability estimate at a given time is improved by increasing the sample size, n_0. The requirement of a large sample is analogous to the conditions required in experimental measurements of probability associated with coin tossing and dice rolling. This implies that the estimates given by Equation 2.2 and Equation 2.3 approach actual values for $R(t)$ and $Q(t)$ as the sample size becomes infinitely large. Thus, the practical meanings of reliability and unreliability are that in a large number of repetitions, the proportional frequency of occurrence of success or failure will be approximately equal to the $\hat{R}(t)$ and $\hat{Q}(t)$ estimates, respectively.

The response values for a series of measurements on a certain product parameter of interest can be plotted as a histogram in order to assess the variability. For example, Table 2.1 lists a series of time to failure results for 251 samples that were tested in 11 different groups. These data are summarized as a frequency table in the first two columns of Table 2.2, and a histogram was created from those two columns (Figure 2.1). In the histogram, each rectangular bar represents the number of failures in the interval. This histogram represents the life distribution curve for the product.

The ratios of the number of surviving products to the total number of products (i.e., the reliability at the end of each interval) are calculated in the fourth column of Table 2.2 and are plotted as a histogram in Figure 2.2. As the sample size increases, the intervals of the histogram can be reduced, and often the plot will approach a smooth curve.

2 Reliability Concepts

Table 2.1 Measured time to failure data (hours) for 251 samples

Group number										
1	2	3	4	5	6	7	8	9	10	11
Data										
1	1	1	1	1	1	1	1	1	1	1
1	1	2	2	2	2	2	2	2	2	2
2	3	2	2	3	3	3	3	3	3	3
3	3	3	3	3	3	3	4	4	4	4
4	4	4	4	4	4	4	4	4	4	4
5	5	5	5	5	5	5	5	5	5	5
6	6	6	6	6	6	6	6	6	6	6
6	6	7	7	7	7	7	7	7	7	7
8	8	8	8	8	8	8	8	9	9	9
9	9	9	9	10	10	11	11	11	11	11
11	12	12	12	12	12	12	13	13	13	13
13	14	14	14	14	15	15	15	15	15	15
16	16	16	16	17	17	17	17	17	18	18
18	18	18	18	18	18	19	19	19	19	20
20	20	20	21	21	22	22	23	23	24	24
25	25	26	26	27	27	27	28	28	28	28
28	28	29	29	29	29	29	29	30	31	31
32	32	33	33	34	34	35	35	36	36	36
36	37	38	39	41	41	42	42	43	44	45
46	47	48	49	49	51	52	53	54	55	56
58	59	62	64	65	66	67	69	72	76	78
79	83	85	89	93	97	99	105	107	111	115
117	120	125	126	131	131	137	140	142	–	–

Table 2.2 Grouped and analyzed data from Table 2.1

Operating time interval (hours)	Number of failures in the interval	Number of surviving products at the end of the interval	Relative frequency	Estimate of reliability at the end of the interval	Estimate of hazard rate in each interval (failures/hour)
0–10	105	146	0.418	0.582	0.042
11–20	52	94	0.207	0.375	0.036
21–30	28	66	0.112	0.263	0.030
31–40	17	49	0.068	0.195	0.026
41–50	12	37	0.048	0.147	0.024
51–60	8	29	0.032	0.116	0.022
61–70	6	23	0.024	0.092	0.021
71–80	4	19	0.016	0.076	0.017
81–90	3	16	0.012	0.064	0.016
91–100	3	13	0.012	0.052	0.019
101–110	2	11	0.008	0.044	0.015
111–120	3	8	0.012	0.032	0.027
121–130	3	5	0.012	0.020	0.038
131–140	4	1	0.016	0.004	0.080
Over 140	1	0	0.004	0.000	–

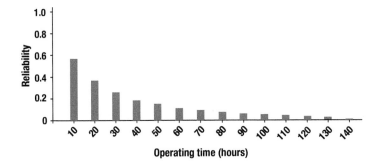

Figure 2.2 Reliability histogram of data from Table 2.1.

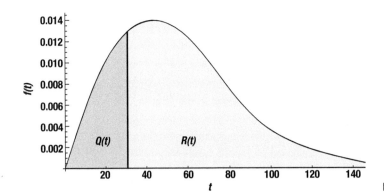

Figure 2.3 Probability density function.

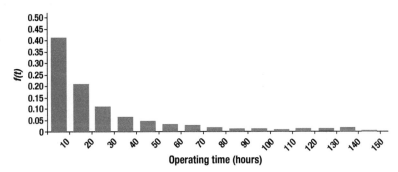

Figure 2.4 Probability density function for the data in Table 2.1.

2.1.1 Concept of Probability Density Function

One reliability concern is the life of a product from a success and failure point of view. The random variable used to measure reliability is the time to failure (T) random variable. If we assume time t as continuous, the time to failure random variable has a probability density function $f(t)$. Figure 2.3 shows an example of a probability density function (pdf).

The ratio of the number of product failures in an interval to the total number of products gives an estimate of the probability density function corresponding to the interval. For the data in Table 2.1, the estimate of the probability density function for each interval is evaluated in the fourth column of Table 2.2. Figure 2.4 shows the

2 Reliability Concepts

estimate of the probability density function for the data in Table 2.1. The sum of all values in the pdf is equal to unity (e.g., the sum of all values in column four of Table 2.2 is equal to 1).

The probability density function is given by:

$$f(t) = \frac{1}{n_0}\frac{d[n_f(t)]}{dt} = \frac{d[Q(t)]}{dt}. \qquad (2.4)$$

Integrating both sides of this equation gives the relation for unreliability in terms of $f(t)$,

$$Q(t) = \frac{n_f(t)}{n_0} = \int_0^t f(\tau)d\tau, \qquad (2.5)$$

where the integral is the probability that a product will fail in the time interval $0 \leq \tau \leq t$. The integral in Equation 2.5 is the area under the probability density function curve to the left of the time line at some time t (see Figure 2.3). The reliability at any point in time, called the reliability function, is

$$\begin{aligned} R(t) &= \text{Probability [Product life} > t] = P[T > t] \\ &= 1 - P[T \leq t]. \end{aligned} \qquad (2.6)$$

$P[T \leq t]$ is the cumulative probability of failure, denoted by $F(t)$, and is called the cumulative distribution function (cdf), as explained above.

Similarly, the percentage of products that have not failed up to time t is represented by the area under the curve to the right of t by

$$R(t) = \int_t^\infty f(\tau)d\tau. \qquad (2.7)$$

Since the total probability of failures must equal 1 at the end of life for a population, we have

$$\int_0^\infty f(t)dt = 1. \qquad (2.8)$$

Figure 2.5 gives an example of the cdf and the reliability function and their relationships. The cdf is a monotonically nondecreasing function, and thus $R(t)$ is a monotonically nonincreasing function.

Example 2.2

From the histogram in Figure 2.4:

(a) Calculate the unreliability of the product at a time of 30 hours.
(b) Also calculate the reliability.

2.1 Basic Reliability Concepts

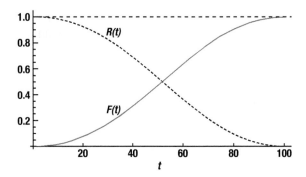

Figure 2.5 Example of F(t) and R(t).

Solution:

(a) For the discrete data represented in this histogram, the unreliability is the sum of the failure probability density function values from $t = 0$ to $t = 30$. This sum, as a percentage, is 73.7%.

(b) The reliability is equal to 26.3% and can be read from column 5 of Table 2.2. The sum of reliability and unreliability must always be equal to 100%.

Example 2.3

A product has a maximum life of 100 hours, and its pdf is given by a triangular distribution, as shown in the figure below. Develop the pdf, cdf, and the reliability function for this product.

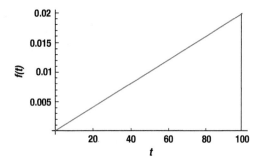

Solution:
Its pdf, cdf, and reliability function, respectively, are given below:

$$f(t) = \begin{cases} \dfrac{t}{5{,}000}, & \text{for } 0 \leq t \leq 100 \\ 0, & \text{otherwise} \end{cases}$$

$$F(t) = \int_0^t f(\tau)d\tau = \int_0^t \dfrac{\tau}{5{,}000} d\tau = \begin{cases} 0, & \text{for } t < 0 \\ \dfrac{t^2}{10{,}000}, & \text{for } 0 \leq t \leq 100 \\ 1, & \text{for } t > 100 \end{cases}$$

2 Reliability Concepts

$$R(t) = 1 - F(t) = \begin{cases} 1, & \text{for } t < 0 \\ 1 - \dfrac{t^2}{10,000}, & \text{for } 0 \leq t \leq 100 \\ 0, & \text{for } t > 100. \end{cases}$$

2.2 Hazard Rate

The failure of a population of fielded products can arise from inherent design weaknesses, manufacturing- and quality control-related problems, variability due to customer usage, the maintenance policies of the customer, and improper use or abuse of the product. The hazard rate, $h(t)$, is the number of failures per unit time per number of nonfailed products remaining at time t. An idealized (though rarely occurring) shape of the hazard rate of a product is the bathtub curve (Figure 2.6). A brief description of each of the three regions is given in the following:

1. *Infant Mortality Period.* The product population exhibits a hazard rate that decreases during this first period (sometimes called "burn-in," "infant mortality," or the "debugging period"). This hazard rate stabilizes at some value at time t_1 when the weak products in the population have failed. Some manufacturers provide a burn-in period for their products, as a means to eliminate a high proportion of initial or early failures.

2. *Useful Life Period.* The product population reaches its lowest hazard rate level and is characterized by an approximately constant hazard rate, which is often referred to as the "constant failure rate." This period is usually considered in the design phase.

3. *Wear-Out Period.* Time t_2 indicates the end of useful life and the start of the wear-out phase. After this point, the hazard rate increases. When the hazard rate becomes too high, replacement or repair of the population of products should be conducted. Replacement schedules are based on the recognition of this hazard rate.

Optimizing reliability must involve the consideration of the actual life-cycle periods. The actual hazard rate curve will be more complex in shape and may not even exhibit all of the three periods.

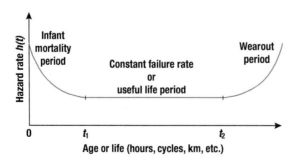

Figure 2.6 Idealized bathtub hazard rate curve.

2.2.1 Motivation and Development of Hazard Rate

Suppose N items are put on test at time $t = 0$. Let $N_S(t)$ be the random variable denoting the number of products functioning at time t. $N_S(t)$ follows the binomial distribution (see Chapter 3) with parameters N and $R(t)$, where $R(t)$ is the reliability of a product at time t. Denoting the expected value of $N_S(t)$ by $\overline{N_S}(t)$, we have

$$E[N_S(t)] = \overline{N_S}(t) = NR(t) \tag{2.9}$$

or

$$R(t) = \frac{\overline{N_S}(t)}{N}. \tag{2.10}$$

Also, we have

$$F(t) = 1 - R(t) = \frac{N - \overline{N_S}(t)}{N}. \tag{2.11}$$

And by differentiating, we have

$$\begin{aligned} f(t) &= \frac{dF(t)}{dt} = -\frac{1}{N}\frac{d\overline{N_S}(t)}{dt} \\ &= \lim_{\Delta t \to 0} \frac{\overline{N_S}(t) - \overline{N_S}(t+\Delta t)}{N\Delta t}. \end{aligned} \tag{2.12}$$

Equation 2.12 illustrates that the failure pdf is normalized in terms of the size of the original population, N. However, it is often more meaningful to normalize the rate with respect to the average number of units successfully functioning at time t, since this indicates the hazard rate for those surviving units. If we replace N with $\overline{N_S}(t)$, we have the hazard rate or "instantaneous" failure rate, which is given by Equation 2.13:

$$\begin{aligned} h(t) &= \lim_{\Delta t \to 0} \frac{\overline{N_s}(t) - \overline{N_s}(t+\Delta t)}{\overline{N_s}(t)\Delta t} \\ &= \frac{N}{\overline{N_s}(t)} f(t) = \frac{f(t)}{R(t)}. \end{aligned} \tag{2.13}$$

Thus, the hazard rate is the rate at which failures occur in a certain time interval for those items that are working at the start of the interval. If N_1 units are working at the beginning of time t, and after the time increment Δt, N_2 units are working, that is, if $(N_1 - N_2)$ units fail during Δt, then the failure rate $\hat{h}(t)$ at time t is given by:

$$\hat{h}(t) \approx \frac{N_1 - N_2}{N_1 \Delta t}. \tag{2.14}$$

Or, in words,

2 Reliability Concepts

$$\text{Hazard rate} = \frac{\text{of failures in the given time interval}}{\text{of survivors at the start of interval} \times \text{interval length}}.$$

Hazard rate is thus a relative rate of failure, in that it does not depend on the original sample size. From Equation 2.13, a relation for the hazard rate in terms of the reliability is:

$$h(t) = \frac{-1}{R(t)} \frac{dR(t)}{dt} \qquad (2.15)$$

because

$$f(t) = -\frac{dR(t)}{dt}. \qquad (2.16)$$

Integrating Equation 2.15 over an operating time from 0 to t and noting that $R(t=0) = 1$ gives:

$$\int_0^t h(\tau) d\tau = -\int_0^t \frac{1}{R(\tau)} dR(\tau) = -\ln R(t) \qquad (2.17)$$

$$R(t) = e^{-\int_0^t h(\tau) d\tau}. \qquad (2.18)$$

2.2.2 Some Properties of the Hazard Function

Some properties of the hazard rate are valuable for understanding reliability. We can prove that

$$\int_0^t h(\tau) d\tau \xrightarrow[t \to \infty]{} \infty. \qquad (2.19)$$

In order to prove it, first note that

$$h(t) = \frac{f(t)}{R(t)} = \frac{1}{R(t)} \left[-\frac{d}{dt} R(t) \right]. \qquad (2.20)$$

Hence,

$$\begin{aligned} \int_0^t h(\tau) d\tau &= -\int_0^t \frac{1}{R(\tau)} \left[\frac{d}{d\tau} R(\tau) \right] d\tau \\ &= -\ln[R(\tau)]\big|_0^t \\ &= -\ln[R(\tau)] + \ln[R(0)]. \end{aligned} \qquad (2.21)$$

Now, $R(t) \to 0$ as $t \to \infty$, hence $-\ln[R(t)] \to \infty$ as $t \to \infty$, and $\ln[R(0)] = \ln[1] = 0$. Thus,

$$\int_0^\infty h(t) dt \to \infty. \qquad (2.22)$$

We also note:

$$\int_0^{t\to\infty} h(\tau)d\tau = \int_0^{t\to\infty} \frac{f(\tau)}{R(\tau)}d\tau = \int_0^{t\to\infty} \frac{f(\tau)}{1-F(\tau)}d\tau. \quad (2.23)$$

We can let $u = 1 - F(\tau)$, and then we have

$$du = -f(\tau)d\tau. \quad (2.24)$$

So,

$$-\int_1^0 \frac{du}{u} = -\ln u \Big|_1^0 \to \infty. \quad (2.25)$$

The rate at which failures occur in a certain time interval $[t_1, t_2]$ is called the hazard (or failure) rate during that interval. This time-dependent function is a conditional probability defined as the probability that a failure per unit time occurs in the interval $[t_1, t_2]$ given that a failure has not occurred prior to t_1. Thus, the hazard rate is

$$\frac{R(t_1) - R(t_2)}{(t_2 - t_1)R(t_1)}. \quad (2.26)$$

If we redefine the interval as $[t, t + \Delta t]$, the above expression becomes:

$$\frac{R(t) - R(t + \Delta t)}{\Delta t \cdot R(t)}. \quad (2.27)$$

The "rate" in the above definitions is expressed as failures per unit "time," where "time" is generic in the sense that it denotes units of product usage, which might be expressed in hours, cycles, or kilometers of usage.

The hazard function, $h(t)$, is defined as the limit of the failure rate as Δt approaches zero:

$$h(t) = \lim_{\Delta t \to 0} \frac{R(t) - R(t + \Delta t)}{\Delta t \cdot R(t)} = \frac{1}{R(t)}\left(-\frac{d}{dt}R(t)\right) = \frac{f(t)}{R(t)}. \quad (2.28)$$

Thus, $h(t)$ can be interpreted as the rate of change of the conditional probability of failure given that the system has survived up to time t.

The importance of the hazard function is that it indicates the change in failure rate over the life of a population of devices. For example, two designs may provide the same reliability at a specific point in time; however, the hazard rates can differ over time. Accordingly, it is often useful to evaluate the cumulative hazard function, $H(t)$. $H(t)$ is given by:

$$H(t) = \int_{\tau=0}^{t} h(\tau)d\tau. \quad (2.29)$$

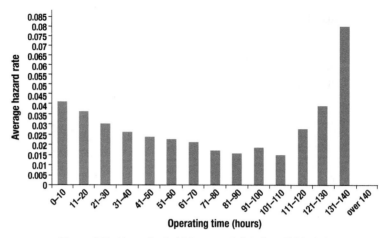

Figure 2.7 Hazard rate histogram of data from Table 2.1.

Both $R(t)$ and $F(t)$ are related to $h(t)$ and $H(t)$, and we can develop the following relationships:

$$h(t) = \frac{f(t)}{R(t)} = \frac{1}{R(t)}\left(-\frac{d}{dt}R(t)\right) = -\frac{d\ln[R(t)]}{dt} \tag{2.30}$$

or

$$-d\ln(R(t)) = h(t)dt. \tag{2.31}$$

Integrating both sides leads to the following relationship:

$$-\ln[R(t)] = \int_{\tau=0}^{t} h(\tau)d\tau = H(t) \tag{2.32}$$

or

$$R(t) = \exp\left(-\int_{\tau=0}^{t} h(\tau)d\tau\right) = \exp(-H(t)). \tag{2.33}$$

Using the data from Table 2.1 and Equation 2.14, an estimate (over Δt) of the hazard rate is calculated in the last column of Table 2.2. Figure 2.7 is the histogram of hazard rate versus time.

Example 2.4

The failure or hazard rate of a component is given by (life is in hours):

$$h(t) = \begin{cases} 0.015, & t \leq 200 \\ 0.025, & t > 200. \end{cases}$$

Thus, the hazard rate is piecewise constant.
Find an expression for the reliability function of the component.

Solution:
Using Equation 2.18 or Equation 2.33, we have

$$R(t) = \exp\left[-\int_0^t h(\tau)d\tau\right].$$

For

$$0 \leq t \leq 200 : R(t) = \exp\left[-\int_0^t 0.015 d\tau\right] = \exp[-0.015t].$$

For

$$t > 200 : R(t) = \exp\left[-\left(\int_0^{200} 0.015 d\tau + \int_{200}^t 0.025 d\tau\right)\right]$$
$$= [-(0.015(200) + 0.025t - 0.025(200))]$$
$$= \exp[-(0.025t - 2)] = \exp[2 - 0.025t].$$

The four functions $f(t)$, $F(t)$, $R(t)$, and $h(t)$ are all related. If we know any one of these four functions, we can develop the other three using the following equations:

$$h(t) = \frac{f(t)}{R(t)} \tag{2.34}$$

$$R(t) = \exp\left[-\int_0^t h(u)du\right] \tag{2.35}$$

$$f(t) = h(t)\exp\left[-\int_0^t h(u)du\right] \tag{2.36}$$

$$Q(t) = F(t) = 1 - R(t). \tag{2.37}$$

2.2.3 Conditional Reliability

The conditional reliability function $R(t, t_1)$ is defined as the probability of operating for a time interval of duration, t, given that the nonrepairable system has operated for a time t_1 prior to the beginning of the interval. The conditional reliability can be expressed as the ratio of the reliability at time $(t + t_1)$ to the reliability at t_1, where t_1 is the "age" of the system at the beginning of a new test or mission. That is,

$$R(t, t_1) = P[(t+t_1) > T \mid T > t_1] = \frac{P[(t+t_1) > T]}{P[T > t_1]} \qquad (2.38)$$

or

$$R(t, t_1) = \frac{R(t+t_1)}{R(t_1)}. \qquad (2.39)$$

For a product with a decreasing hazard rate, the conditional reliability will increase as the age, t_1, increases. The conditional reliability will decrease for a product with an increasing hazard rate. The conditional reliability of a product with a constant rate of failure is independent of age. This suggests that a product with a constant failure rate can be treated "as good as new" at any time.

Example 2.5

The reliability function for a system is assumed to be an exponential distribution (see Chapter 3) and is given by

$$R(t) = e^{-\lambda_0 t},$$

where λ_0 is a constant (i.e., a constant hazard rate).

Calculate the reliability of the system for mission time, t, given that the system has already been used for 10 years.

Solution:
Using Equation 2.39

$$R(t, 10) = \frac{R(t+10)}{R(10)} = \frac{e^{-\lambda_0(t+10)}}{e^{-\lambda_0 10}} = e^{-\lambda_0 t} = R(t).$$

That is, the system reliability is "as good as new," regardless of the age of the system.

Example 2.6

If T is a random variable representing the hours to failure for a device with the following pdf:

$$f(t) = t \exp\left(\frac{-t^2}{2}\right), \quad t \geq 0.$$

(a) Find the reliability function.

Solution:
To develop the reliability function, $R(t)$, we have

$$R(t) = \int_t^\infty f(\tau) d\tau = \int_t^\infty \tau \exp(-\tau^2/2) d\tau.$$

Let $u = \tau^2/2$, $du = \tau d\tau$; then we have

$$R(t) = \int_{\frac{t^2}{2}}^{\infty} \exp(-u)\,du = \exp\left(\frac{-t^2}{2}\right), \quad t \geq 0.$$

(b) Find the hazard function.

Solution:
To develop the hazard function $h(t)$, we have

$$h(t) = f(t)/R(t) = t, \quad t \geq 0.$$

Thus the hazard rate is linearly increasing with a slope of 1.

(c) If 50 devices are placed in operation and 27 are still in operation 1 hour later, find approximately the expected number of failures in the time interval from 1 to 1.1 hours using the hazard function.

Solution:
To answer this question, we can use the information in Section 2.2.1, and we have

$$N_S(0) = 50, \; N_S(1 \text{ hour}) = 27, \; \Delta N = N_S(1) - N_S(1.1) = ?$$

For small Δt, the expected number failing can be calculated using Equation 2.14:

$$\Delta N = N_S(t) - N_S(t + \Delta t) \approx h(t) \Delta t \, N_S(t)$$
$$= 1.0 \times 0.1 \times 27 = 2.7.$$

Note that by the using the concept of conditional reliability, we also get,

$$P[T > 1.1 | T > 1.0] = \frac{R(1.1)}{R(1.0)} = \frac{0.54607}{0.60653} = 0.8904$$

or

$$\Delta N = 27 \times (1 - 0.9) = 2.7.$$

2.3 Percentiles Product Life

The reliability of a product can be experienced in terms of percentiles of life. Because this approach was originally used to specify the life of bearings, the literature often uses the symbol B_α, where the B_α life is the time by which α percent of the products fail, or:

$$F(B_\alpha) = \frac{\alpha}{100} \qquad (2.40)$$

or

$$R(B_\alpha) = 1 - \frac{\alpha}{100}. \qquad (2.41)$$

For example, B_{10} life is the 10th percentile of life of the product. Thus,

$$F(B_{10}) = \frac{10}{100} = 0.10. \qquad (2.42)$$

Similarly, B_{95} is the 95th percentile of life of the product and is given by

$$F(B_{95}) = \frac{95}{100} = 0.95 \qquad (2.43)$$

or

$$R(B_{95}) = 1 - \frac{95}{100} = 0.05. \qquad (2.44)$$

Median life is the 50th percentile of life and is denoted by B_{50}. Thus, the median life, M, of a probability distribution is the time at which the area under the distribution is divided in half (i.e., the time to reach 50% reliability). That is,

$$\int_0^M f(t)\,dt = 0.50. \qquad (2.45)$$

Example 2.7

The failure rate or hazard rate of a component is:

$$h(t) = 0.02 t^{1.7}, \quad t \geq 0.$$

The failure rate is in failures per year.

(a) What is the reliability function of this component and what is the value of the reliability for a period of 2 years?

Solution:

$$R(t) = \exp\left[-\int_0^t h(\tau)\,d\tau\right] = \exp\left[-\int_0^t 0.02\tau^{1.7}\,d\tau\right] = e^{-0.007407 t^{2.7}}$$

$$R(2) = e^{-0.007407 \times 2^{2.7}} = e^{-0.048131} = 0.953009.$$

(b) What is the median life or B_{50} life of this component?

Solution:

$$R(B_{50}) = 0.50 = e^{-0.007407(B_{50})^{2.7}}$$

$$\ln 0.50 = -0.007407(B_{50})^{2.7}$$

$$B_{50} = \left(\frac{-\ln 0.50}{0.007407}\right)^{1/2.7} = 5.37 \text{ years.}$$

2.4 Moments of Time to Failure

The mean or expected value of T, a measure of the central tendency of the random variable, also known as the first moment, is denoted as $E[T]$ or μ, and given by

$$E[T] = \mu = \int_{-\infty}^{\infty} t f(t) dt. \tag{2.46}$$

Higher order moments are discussed in the following section.

2.4.1 Moments about Origin and about the Mean

The kth moment about the origin of the random variable T is

$$\mu'_k = E[T^k] = \int_{-\infty}^{\infty} t^k f(t) dt, \quad k = 1, 2, 3, \ldots. \tag{2.47}$$

Notice that the first moment about the origin is just the mean. That is,

$$E[T] = \mu'_1 = \mu. \tag{2.48}$$

The kth moment about the mean of the random variable T is

$$\mu_k = E[(T-\mu)^k] = \int_{-\infty}^{\infty} (t-\mu)^k f(t) dt \tag{2.49}$$

$$k = 2, 3, 4, \ldots.$$

For large k, the above integration can be tedious. The equation to derive the kth moment about the mean is:

$$\mu_k = \sum_{j=0}^{k} (-1)^j \binom{k}{j} \mu^j \mu'_{k-j} \tag{2.50}$$

where

$$C_j^k = \binom{k}{j} = \frac{k!}{j!(k-j)!}. \tag{2.51}$$

2.4.2 Expected Life or Mean Time to Failure

For a given underlying probability density function, the mean time to failure (MTTF) is the expected value for the time to failure. It is defined as

$$E[T] = \text{MTTF} = \int_0^\infty tf(t)dt. \tag{2.52}$$

It can also be shown that MTTF is equivalent to

$$\text{MTTF} = \int_0^\infty R(t)dt. \tag{2.53}$$

Thus, $E[T]$ is the first moment or the center of gravity of the probability density function (like the fulcrum of a seesaw). $E[T]$ is also called the mean time between failures (MTBF), when the product exhibits a constant hazard rate; that is, the failure probability density function is an exponential.

The MTTF should be used only when the failure distribution function is specified, because the value of the reliability function at a given MTTF depends on the probability distribution function used to model the failure data. Furthermore, different failure distributions can have the same MTTF while having very different reliability functions.

The first few failures that occur in a product or system often have the biggest impact on safety, warranty, and supportability, and consequently on the profitability of the product. Thus, the beginning of the failure distribution is a much more important concern for reliability than the mean.

2.4.3 Variance or the Second Moment about the Mean

Information on the dispersion of the values with respect to the mean is expressed in terms of variance, standard deviation, or coefficient of variation. The variance of the random variable T, a measure of variability or spread in the data about the mean, is also known as the second central moment and is denoted as $V[T]$. It can be calculated as

$$\mu_2 = V[T] = E\left[(T - E[T])^2\right] = \int_{-\infty}^\infty \left(t - E[T]\right)^2 f(t)dt. \tag{2.54}$$

Using Equation 2.50, we have

$$\mu_2 = \mu_2' - 2\mu\mu_1' + \mu^2\mu_0' = \mu_2' - \mu^2$$
$$\text{because} \quad \mu_0' = \int_0^\infty t^0 f(t)dt = 1 \quad \text{and} \quad \mu = \mu_1'. \tag{2.55}$$

Since the second moment about the origin is $E[T^2] = \mu_2'$, we can write the variance of a random variable in terms of moments about the origin as follows:

$$V[T] = E[T^2] - \{E[T]\}^2 = \mu_2' - \mu^2. \tag{2.56}$$

The positive square root of the variance is called the standard deviation, denoted by σ, and is written as

$$\sigma = \sqrt[+]{V[T]}. \tag{2.57}$$

Although the standard deviation value is expressed in the same units as the mean value, its value does not directly indicate the degree of dispersion or variability in the random variable, except in reference to the mean value. Since the mean and the standard deviation values are expressed in the same units, a nondimensional term can be introduced by taking the ratio of the standard deviation and the mean. This is called the coefficient of variation and is denoted as $CV[T]$:

$$\alpha_2 = CV[T] = \frac{\mu_2^{1/2}}{\mu} = \frac{\sigma}{\mu}. \tag{2.58}$$

2.4.4 Coefficient of Skewness

The degree of symmetry in the probability density function can be measured using the concept of skewness, which is related to the third moment, μ_3. Since it can be positive or negative, a nondimensional measure of skewness, known as the coefficient of skewness, can be developed to avoid dimensional problems as given below:

$$\alpha_3 = \frac{\mu_3}{\mu_2^{3/2}}. \tag{2.59}$$

If α_3 is zero, the distribution is symmetrical about the mean; if α_3 is positive, the dispersion is more above the mean than below the mean; and if it is negative, the dispersion is more below the mean. If a distribution is symmetrical, then the mean and the median are the same. If the distribution is negatively skewed, then the median is greater than the mean. And if the distribution is positively skewed, then the mean is greater than the median.

For reliability, we want products to last longer and hence we should design products so that the life distribution is negatively skewed. For maintainability, we want to restore the function of the system in a small amount of time, and hence the time to repair or restoration should follow a positively skewed distribution.

2.4.5 Coefficient of Kurtosis

Skewness describes the amount of asymmetry, while kurtosis measures the concentration (or peakedness) of data around the mean and is measured by the fourth central moment. To find the coefficient of kurtosis, divide the fourth central moment by the square of the variance to get a nondimensional measure. The coefficient of kurtosis represents the peakedness or flatness of a distribution and is defined as:

$$\alpha_4 = \frac{\mu_4}{\mu_2^2}. \tag{2.60}$$

2 Reliability Concepts

The normal distribution (see Chapter 3) has $\alpha_4 = 3$, and hence sometimes we define a coefficient of kurtosis as

$$\alpha_4 - 3 = \frac{\mu_4}{\mu_2^2} - 3, \qquad (2.61)$$

to compare the peakness or flatness of the distribution with a normal distribution.

Example 2.8

For the triangular life distribution given in Example 2.3, calculate the $E[T]$, $V[T]$, and standard deviation.

Solution:
We have

$$f(t) = \begin{cases} \dfrac{t}{5,000}, & \text{for } 0 \leq t \leq 100 \\ 0, & \text{otherwise.} \end{cases}$$

Now

$$E[T] = \int_0^\infty t f(t) \, dt$$
$$= \int_0^{100} t \frac{t}{5,000} \, dt = \frac{1}{5,000} \frac{t^3}{3} \bigg|_0^{100} = \frac{1}{5,000} \frac{100^3}{3}$$
$$= \frac{2}{3} \cdot 100 = 66.67 \text{ hours}$$

and

$$E[T^2] = \int_0^\infty t^2 f(t) \, dt$$
$$= \int_0^{100} t^2 \frac{t}{5,000} \, dt = \frac{1}{5,000} \frac{t^4}{4} \bigg|_0^{100} = \frac{100^4}{20,000} = 5,000$$

so,

$$V[T] = E[T^2] - (E[T])^2$$
$$= 5,000 - \left(\frac{200}{3}\right)^2 = \frac{5,000}{9} = 555.55.$$

The standard deviation, σ, is 23.57, and the coefficient of variation is $23.57/66.67 = 0.354$.

Example 2.9

The failure rate per year of a component is given by:

$$h(t) = 0.003t^2, \quad t \geq 0.$$

(a) Find an expression for the reliability function and the probability density function for the time to failure of the component.

Solution:

$$R(t) = \exp\left(-\int_0^t h(\tau)d\tau\right) = \exp\left(-\int_0^t 0.003\tau^2 d\tau\right)$$
$$= \exp(-0.001t^3)$$

and for the probability density function, we have

$$f(t) = h(t)R(t) = 0.003t^2 \exp(-0.001t^3).$$

(b) Find the B_{20} (the 20th percentile) for the life of the component.

Solution:
We have

$$0.80 = \exp(-0.001 B_{20}^3)$$

$$B_{20} = \left(\frac{\ln 0.80}{-0.001}\right)^{1/3} = 6.065 \text{ years}.$$

(c) Find the expected life (MTTF) for the component.

Solution:

$$E[T] = \int_0^\infty R(t)dt = \int_0^\infty t \cdot f(t)dt = \int_0^\infty 0.003t^3 \exp(-0.001t^3)dt.$$

Let $u = 0.001t^3$, $du = 0.003t^2 dt$

$$E[T] = \frac{1}{0.001^{1/3}} \int_0^\infty u^{(1/3+1)-1} e^{-u} du = \frac{1}{0.001^{1/3}} \Gamma(1.333) = 10 \times 0.89302 = 8.9302 \text{ years}$$

where the value of the gamma function is found from the table in Appendix B.

2.5 Summary

The fundamental reliability concepts presented in this chapter include reliability and unreliability functions, the probability density function, hazard rate, conditional reliability function, percentiles of life, and time-to-failure metrics. The proper reliability

2 Reliability Concepts

measure for a product is determined by the specific purpose, and need, for a product. A single product may perform separate functions that each have a different level of reliability. In addition, a single product can have different reliability values at different times during its lifetime, depending on various operational and environmental conditions. The concepts presented in this chapter represent the basis for successful implementation of a reliability program in an engineering system.

Problems

2.1 Following the format of Table 2.1, record and calculate the different reliability metrics after bending 30 paper clips 90° back and forth to failure. Thus, the number of bending cycles is the underlying random variable. Plot the life characteristics curve, the estimate of the probability density function, the reliability and unreliability, and the hazard rate. Do you think your results depend on the amount of bend in the paper clip? Explain.

2.2 A warranty reporting system reports field failures. For the rear brake drums on a particular pickup truck, the following (coded) data were obtained. For the data provided, plot the hazard rate, the failure probability density function, and the reliability function. Assume that the population size is 2680 and that the data represent all of the failures.

Kilometer interval	Number of failures
$M < 2000$	707
$2000 \leq M < 4000$	532
$4000 \leq M < 6000$	368
$6000 \leq M < 8000$	233
$8000 \leq M < 10{,}000$	231
$10{,}000 \leq M < 12{,}000$	136
$12{,}000 \leq M < 14{,}000$	141
$14{,}000 \leq M < 16{,}000$	78
$16{,}000 \leq M < 18{,}000$	101
$18{,}000 \leq M < 20{,}000$	46
$20{,}000 \leq M < 22{,}000$	51
$22{,}000 \leq M < 24{,}000$	56

2.3 Consider the piecewise linear bathtub hazard function defined over the three regions of interest given below. The constants in the expressions are determined so that they satisfy the normal requirements for $h(t)$ to be a hazard function.

$$h(t) = \begin{cases} h_1(t) = b_1 - c_1 t, & 0 \leq t \leq t_1 \\ h_2(t) = b_1 - c_1 t_1 - c_2(t - t_1), & t_1 \leq t \leq t_2 \\ h_3(t) = b_1 - c_1 t_1 - c_2(t_2 - t_1) + c_3(t - t_2), & t_2 \leq t \leq \infty. \end{cases}$$

Develop the equations for the reliability function and the probability density function for the time to failure random variable based on the above hazard function.

2.4 Consider the following functions:

(a) e^{-at}
(b) e^{at}
(c) ct^5
(d) dt^{-3}

where a, c, and d are positive constants.
   Which of the above functions can serve as hazard function models? Also, develop mathematical expressions for the probability density function and the reliability function for the valid hazard functions.

2.5 Prove that

$$\text{MTTF} = \int_0^\infty tf(t)\,dt = \int_0^\infty R(t)\,dt.$$

2.6 The time to failure random variable, t, for a product follows the following probability density function, where time is in years:

$$f(t) = \begin{cases} \dfrac{t}{200}, & 0 \le t \le 20 \\ 0, & \text{otherwise} \end{cases}.$$

(a) Find the standard deviation for the time to failure random variable.
(b) Find the B_{10} and B_{50} life of the product based on the above probability density function.
(c) Draw the failure rate (or hazard rate) curve for the above product by evaluating it at $t = 0, 1, 2, 5, 10, 15, 20$.
(d) Find the coefficient of skewness, α_3, for this life distribution of the product.

2.7 The hazard rate or failure rate of a product is given by

$$h(t) = 0.002t, \quad t \ge 0.$$

The failure rate is in failures per year.

(a) Find an expression for the reliability function and the probability density function for the time to failure of the product.
(b) Find the B_{10} (the 10th percentile) life of the product.
(c) Find the expected value for the life of the product.

2.8 The failure rate of a component is given by:

$$h(t) = 0.006t^2, \quad t \ge 0.$$

The failure rate is in failures per year.

(a) Find an expression for the reliability function and the probability density function for the time to failure for the component.

(b) Find the B_{20} (the 20th percentile) for the life of the component.

2.9 Calculate the MTTF for a failure probability density function given by:

$$f(t) = \begin{cases} 0, & \text{for } (t < t_1) \\ \dfrac{1}{t_2 - t_1}, & \text{for } (t_1 \leq t \leq t_2) \\ 0, & \text{for } (t > t_2) \end{cases}.$$

2.10 The failure or hazard rate of a component is given by (life is in hours):

$$h(t) = \begin{cases} 0.015, & t \leq 200 \\ 0.025, & t > 200. \end{cases}$$

Find the expected life or MTTF for the component.

2.11 The failure density function for a group of components is:

$$f(t) = 0.25 - \left(\dfrac{0.25}{8}\right)t, \quad \text{for } 0 \leq t \leq 8$$

($f(t)$ is 0 otherwise). Time is in years.

(a) Show how this is a valid pdf.
(b) Find $F(t)$, $h(t)$, and $R(t)$.
(c) Find MTTF.
(d) Find B_{10} and B_{90} for the life of the components.
(e) Find the probability that this component fails within the first year of operation.

2.12 Assume that the system in Example 2.5 is a car. Do the results in the example 2.5 make sense? Why? Provide some examples of systems where the results may be more appropriate.

2.13 What does the conditional reliability reduce to if the hazard rate is a constant?

2.14 The time to failure random variable T of a product follows the following probability density function:

$$f(t) = \begin{cases} \dfrac{t}{80,000}, & 0 \leq t \leq 400 \\ 0, & \text{otherwise.} \end{cases}$$

(a) Find the standard deviation for the time to failure random variable.
(b) Find the coefficient of skewness for the distribution for the time to failure random variable.
(c) Find the B_5 and B_{50} life of the product based on the above probability density function.

Draw the failure rate (or hazard rate) curve for the above product by evaluating it at:

$$t = 0, 50, 100, 300, 400.$$

2.15 The failure rate or hazard rate of a component is:

$$h(t) = 0.02t^{1.7}, \quad t \geq 0.$$

The failure rate is in failures per year.

(a) What is the reliability function of this component for a period of 2 years?
(b) What is the median life or B_{50} life of this component?
(c) What is the expected life for this component?

2.16 Calculate the coefficient of kurtosis for the probability density function given in Problem 2.9, where $t_1 = 3$ and $t_2 = 10$.

2.17 If the unreliability for a part is given as:

$$F(t) = \begin{cases} 0, & t < 0 \\ 0.5t^2 + 0.5t, & 0 \leq t \leq 1 \\ 1, & 1 < t \end{cases}.$$

What is the hazard rate as a function of time?

3 Probability and Life Distributions for Reliability Analysis

In reliability engineering, data are often collected from analysis of incoming parts and materials, tests during and after manufacturing, fielded products, and warranty returns. If the collected data can be modeled, then properties of the model can be used to make decisions for product design, manufacture, reliability assessment, and logistics support (e.g., maintainability and operational availability).

In this chapter, discrete and continuous probability models (distributions) are introduced, along with their key properties. Two discrete distributions (binomial and Poisson) and five continuous distributions (exponential, normal, lognormal, Weibull, and gamma) that are commonly used in reliability modeling and hazard rate assessments are presented.

3.1 Discrete Distributions

A discrete random variable is a random variable with a finite (or countably infinite) set of values. If a discrete random variable (X) has a set of discrete possible values $(x_1, x_2, \ldots x_n)$, a probability mass function (pmf), $f(x_i)$, is a function such that

$$f(x_i) \geq 0, \quad \text{for all } i$$
$$\sum_{i=1}^{n} f(x_i) = 1 \qquad (3.1)$$
$$f(x_i) = P\{X = x_i\}.$$

The cumulative distribution function (cdf) is written as:

$$F(x_i) = P\{X \leq x_i\}. \qquad (3.2)$$

Reliability Engineering, First Edition. Kailash C. Kapur and Michael Pecht.
© 2014 John Wiley & Sons, Inc. Published 2014 by John Wiley & Sons, Inc.

The mean, μ, and variance, σ^2, of a discrete random variable are defined using the pmf as (see also Chapter 2):

$$\mu = E[X] = \sum_i x_i f(x_i) \tag{3.3}$$

$$\sigma^2 = V[X] = \sum_i (x_i - \mu)^2 f(x_i) = \sum_i x_i^2 f(x_i) - \mu^2. \tag{3.4}$$

3.1.1 Binomial Distribution

The binomial distribution is a discrete probability distribution applicable in situations where there are only two mutually exclusive outcomes for each trial or test. For example, for a roll of a die, the probability is one to six that a specified number will occur (success) and five to six that it will not occur (failure). This example, known as a "Bernoulli trial," is a random experiment with only two possible outcomes, denoted as "success" or "failure." Of course, success or failure is defined by the experiment. In some experiments, the probability of the result not being a certain number may be defined as a success.

The pmf, $f(x)$, for the binomial distribution gives the probability of exactly k successes in m attempts:

$$f(k) = \binom{m}{k} p^k q^{m-k}, \quad 0 \le p \le 1, \quad q = 1 - p, \quad k = 0, 1, 2, \ldots, m, \tag{3.5}$$

where p is the probability of the defined success, q (or $1 - p$) is the probability of failure, m is the number of independent trials, k is the number of successes in m trials, and the combinational formula is defined by

$$\binom{m}{k} \equiv C_k^m = \frac{m!}{k!(m-k)!}, \tag{3.6}$$

where ! is the symbol for factorial. Since $(p + q)$ equals 1, raising both sides to a power j gives

$$(p+q)^j = 1. \tag{3.7}$$

The general equation is

$$\sum_{k=0}^{m} f(k) = F(m) = (p+q)^m = 1. \tag{3.8}$$

The binomial expansion of the term on the left in Equation 3.7 gives the probabilities of j or less number of successes in j trials, as represented by the binomial distribution. For example, for three components or trials each with equal probability of success (p) or failure (q), Equation 3.7 becomes:

$$(p+q)^3 = p^3 + 3p^2q + 3pq^2 + q^3 = 1. \tag{3.9}$$

The four terms in the expansion of $(p+q)^3$ give the values of the probabilities for getting 3, 2, 1, and no successes, respectively. That is, for $m = 3$ and the *probability of success* $= p$, $f(3) = p^3$, $f(2) = 3p^2q$, $f(1) = 3pq^2$, and $f(0) = q^3$.

The binomial expansion is also useful when there are products with different success and failure probabilities. The formula for the binomial expansion in this case is

$$\prod_{i=1}^{m}(p_i + q_i) = 1, \tag{3.10}$$

where i pertains to the ith component in a system consisting of m components. For example, for a system of three different components, the expansion takes the form

$$(p_1 + q_1)(p_2 + q_2)(p_3 + q_3) = p_1p_2p_3 + (p_1p_2q_3 + p_1q_2p_3 + q_1p_2p_3)$$
$$+ (p_1q_2q_3 + q_1p_2q_3 + q_1q_2p_3) + q_1q_2q_3 = 1, \tag{3.11}$$

where the first term on the right side of the equation gives the probability of success of all three components, the second term (in parentheses) gives the probability of success of any two components, the third term (in parentheses) gives the probability of success of any one component, and the last term gives the probability of failure for all components.

The cdf for the binomial distribution, $F(k)$, gives the probability of k or fewer successes in m trials. It is defined by using the pmf for the binomial distribution,

$$F(k) = \sum_{i=0}^{k} \binom{m}{i} p^i q^{(m-i)}. \tag{3.12}$$

For a binomial distribution, the mean, μ is given by

$$\mu = mp \tag{3.13}$$

and the variance is given by

$$\sigma^2 = mp(1-p). \tag{3.14}$$

Example 3.1

An engineer wants to select four capacitors from a large lot of capacitors in which 10 percent are defective. What is the probability of selecting four capacitors with:

(a) Zero defective capacitors?
(b) Exactly one defective capacitor?
(c) Exactly two defective capacitors?
(d) Two or fewer defective capacitors?

Solution:
Let success be defined as "getting a good capacitor." Therefore, $p = 0.9$, $q = 0.1$, and $m = 4$. Using Equation 3.5 and Equation 3.6, f (4) is the probability of all four being

3 Probability and Life Distributions for Reliability Analysis

good (no defectives)—that is, based on four components (trials), the values of p and q are equal for all the capacitors.

$$f(4) = \binom{4}{4}(0.9)^4(0.1)^0 = 0.6561.$$

Another way to solve this problem is by defining success as "getting a certain number of defective capacitors" with $p = 0.1$ and thus $q = 0.9$. In this case, $f(0)$ gives the probability that there will be no defectives in the four selected samples. That is,

(a) $f(0) = \binom{4}{0}(0.1)^0(0.9)^4 = 0.6561$

Continuing with the latter approach, the solution to problems (b), (c), and (d), respectively, are:

(b) $f(1) = \binom{4}{1}(0.1)^1(0.9)^3 = 0.2916$

(c) $f(2) = \binom{4}{2}(0.1)^2(0.9)^2 = 0.0486$

(d) $F(2) = f(0) + f(1) + f(2) = 0.9963.$

Example 3.2

Consider a product with a probability of failure in a given test of 0.1. Assume 10 of these products are tested.

(a) What is the expected number of failures that will occur in the test?
(b) What is the variance in the number of failures?
(c) What is the probability that no product will fail?
(d) What is the probability that two or more products will fail?

Solution:
Here $m = 10$, and $p = 0.1$.

(a) The expected number of failures is the mean,

$$\mu = mp = (10 \times 0.1) = 1.$$

(b) The variance is:

$$\sigma^2 = mp(1-p) = [10 \times 0.1 \times (1-0.1)] = 0.9.$$

(c) The probability of having no failures is the pmf with $k = 0$. That is,

$$f(0) = \binom{10}{0} \times 0.1^0 \times (1-0.1)^{10} = 0.349.$$

(d) The probability of having two or more failures is the same as 1 minus the probability of having zero or one failures. It is given by:

$$\Pr(\text{two or more failures}) = [1 - \{f(0) + f(1)\}]$$
$$= \left[1 - 0.349 - \left\{10 \times 0.1 \times (1-0.1)^9\right\}\right]$$
$$= 0.264.$$

Example 3.3

An electronic automotive control module consists of three identical microprocessors in parallel. The microprocessors are independent of each other and fail independently. For successful operation of the module, at least two microprocessors must operate normally. The probability of success of each microprocessor for the duration of the warranty is 0.95. Determine the failure probability of the control module during warranty.

Solution:
The module fails when two or more microprocessors fail. In other words, the module fails when only one or none of the microprocessors is working. So the probability of failure of the module during warranty will be given by:

$$\Pr(\text{module fails during warranty}) = [f(0) + f(1)],$$

where $m = 3$ components, $k = 0$ or 1 is the total number of working components, $p = 0.95$, and $q = 0.05$. Therefore:

$$\Pr(\text{module fails during warranty}) = \left[(0.05)^3 + \left\{3 \times 0.95 \times (0.05)^2\right\}\right]$$
$$= 0.00725.$$

Example 3.4

The probability of a Black Hawk helicopter surviving a mission is 0.91. If seven helicopters are sent on a mission and five must succeed for mission success, what is the probability of mission success?

Solution:
This is also called a 5-out-of-7 system in reliability (see Chapter 17). If the number of successes is five or more, the mission will be a success. Hence, the probability of mission success or mission reliability is

$$R_S = \sum_{i=k}^{m} \binom{m}{i} R^i (1-R)^{m-i} = \sum_{i=5}^{7} \binom{7}{i} R^i (1-R)^{7-i}$$
$$= \binom{7}{5} 0.91^5 (0.09)^2 + \binom{7}{6} 0.91^6 (0.09)^1 + \binom{7}{7} 0.91^7 (0.09)^0$$
$$= 0.1061 + 0.3577 + 0.5168 = 0.9806.$$

3 Probability and Life Distributions for Reliability Analysis

3.1.2 Poisson Distribution

In situations where the probability of success (p) is very low and the number (m) of samples tested (i.e., the number of Bernoulli trials conducted) is large, it is cumbersome to evaluate the binomial coefficients. A Poisson distribution is useful in such cases.

The pmf of the Poisson distribution is given as:

$$f(k) = \frac{\mu^k}{k!} e^{-\mu}; k = 0, 1, 2, \ldots, \quad (3.15)$$

where μ is the mean and also the variance of the Poisson random variable.

For a Poisson distribution for m Bernoulli trials, with the probability of success in each trial equal to p, the mean and the variance are given by:

$$\mu = mp, \sigma^2 = mp. \quad (3.16)$$

The Poisson distribution is widely used in industrial and quality engineering applications. It is also the foundation of some of the attribute control charts. For example, it is used in applications such as determination of particles of contamination in a manufacturing environment, number of power outages, and flaws in rolls of polymers.

Example 3.5

Solve Example 3.2 using the Poisson distribution approximation.

Solution:
The expected number of failures is the same as the mean,

$$\mu = (10)(0.1) = 1.$$

The variance is also equal to 1.
The probability of obtaining no failures is the same as the pmf with $k = 0$,

$$f(0) = e^{-\mu} = e^{-1} = 0.3678.$$

The probability of getting two or more failures is the same as 1 minus the probability of obtaining zero or one failures. It is given by:

$$\Pr(\text{two or more failures}) = [1 - \{f(0) + f(1)\}]$$
$$= [1 - \{0.3678 + e^{-1}\}] = 0.2642.$$

Note the differences from Example 3.2, because m is not very large.

3.1.3 Other Discrete Distributions

Other discrete distributions that are used in reliability analysis include the geometric distribution, the negative binomial distribution, and the hypergeometric distribution.

With the geometric distribution, the Bernoulli trials are conducted until the first success is obtained. The geometric distribution has the "lack of memory" property, implying that the count of the number of trials can be started at any trial without affecting the underlying distribution. In this regard, this distribution is similar to the continuous exponential distribution, which will be described later.

With the negative binomial distribution (a generalization of the geometric distribution), the Bernoulli trials are conducted until a certain number of successes are obtained. Negative binomial distribution is, however, conceptually different from the binomial distribution, since the number of successes is predetermined, and the number of trials is a random variable.

With the hypergeometric distribution, testing or sampling is conducted without replacement from a population that has a certain number of defective products. The hypergeometric distribution differs from the binomial distribution in that the population is finite and the sampling from the population is made without replacement.

3.2 Continuous Distributions

If the range of a random variable, X, extends over an interval (either finite or infinite) of real numbers, then X is a continuous random variable. The cdf is given by:

$$F(x) = P\{X \leq x\}. \tag{3.17}$$

The probability density function (pdf) is analogous to pmf for discrete variables, and is denoted by $f(x)$, where $f(x)$ is given by (if $F(x)$ is differentiable):

$$f(x) = \frac{d}{dx} F(x), \tag{3.18}$$

which yields

$$F(x) = \int_{-\infty}^{x} f(u)\,du. \tag{3.19}$$

The mean, μ, and variance, σ^2, of a continuous random variable are defined over the interval from $-\infty$ to $+\infty$ in terms of the probability density function as (see Chapter 2):

$$\mu = \int_{-\infty}^{+\infty} x f(x)\,dx \tag{3.20}$$

$$\sigma^2 = \int_{-\infty}^{+\infty} (x-\mu)^2 f(x)\,dx = \int_{-\infty}^{+\infty} x^2 f(x)\,dx - \mu^2. \tag{3.21}$$

Reliability is concerned with the time to failure random variable T and thus X is replaced by T. Thus, Equation 3.19 corresponds to Equation 2.5 and Equation 3.20 corresponds to Equation 2.46.

3 Probability and Life Distributions for Reliability Analysis

Example 3.6

The pdf for the time to failure of an appliance is given by:

$$f(t) = \frac{1}{16} t \cdot e^{-t/4},$$

where t is in years, and $t > 0$.

(a) What is the probability of failure in the first year?
(b) What is the probability of the appliance lasting at least 5 years?
(c) If no more than 5% of the appliances will require warranty service, what is the maximum number of months for which the appliance should be warranted?

Solution:

(a) For the given pdf, the cdf is

$$F(t) = \frac{1}{16} \int_0^t \tau \cdot e^{-\tau/4} \cdot d\tau = 1 - \left(\frac{t}{4} + 1\right) e^{-t/4}.$$

The probability of failure during the first year = $F(1) = 0.0265$.

(b) The probability of lasting more than 5 years is = $[1 - F(5)] = [1 - 0.3554] = 0.6446$.

(c) For this case, $F(t_0)$ has to be less than or equal to 0.05, where t_0 is the warranty period. From the above results, we find that the time has to be more than 1 year. Also, $F(2)$ is equal to 0.09, hence the warranty period should be between 1 and 2 years. We can find that for no more than 5% warranty service, $t_0 = 1.42$ years. Therefore, the warranty should be set at no greater than 17 months.

Example 3.7

The time-to-failure random variable, T, of a product follows the following probability density function:

$$F(t) = \frac{t}{80,000}, \quad 0 \le t \le 400$$
$$= 0, \quad \text{otherwise.}$$

We give solutions to the following four parts.

(a) Find the standard deviation for the time-to-failure random variable.

Solution:

$$E[T] = \int_0^{400} t \cdot \frac{t}{80,000} dt = \frac{t^3}{240,000}\Big|_0^{400} = 266.67$$

$$E[T^2] = \int_0^{400} t^2 \cdot \frac{t}{80,000} dt = \frac{t^4}{320,000}\Big|_0^{400} = 80,000,$$

then variance V[T] and the standard deviation are given by (see Eq. 2.35 in Chapter 2)

$$V[T] = E[T^2] - (E[T])^2 = 8888.89$$
$$\text{Standard deviation} = \sqrt{V[T]} = 94.28.$$

(b) Find the coefficient of skewness of the distribution for the time-to-failure random variable.

Solution:

$$\mu_1 = \mu = \mu_1' = E[T]$$
$$\mu_2' = E[T^2] \quad \mu_2 = V[T]$$
$$\mu_3' = \int_0^{400} t^3 \frac{t}{80,000} dt = \frac{t^5}{400,000}\Big|_0^{400} = 25.6 \times 10^6.$$

Using Equation 2.52 (Chapter 2), we have

$$\mu_3 = \mu_3' - 3\mu\mu_2' + 3\mu^2\mu_1' - \mu^3$$
$$= \mu_3' - 3\mu\mu_2' + 2\mu^3 = -474,074.$$

Hence,

$$\alpha_3 = \frac{\mu_3}{\mu_2^{3/2}} = \frac{-474,074}{(8888.89)^{1.5}} = -0.5657.$$

The above triangular distribution is negatively skewed, which is good in terms of reliability because the time to failure is a "larger the better" characteristic for the product.

(c) Find the B_5 and B_{50} life of the product based on the above probability density function.

Solution:
Using Equation 2.5 and Equation 2.40, give

3 Probability and Life Distributions for Reliability Analysis

$$F(t) = \int_0^t \frac{u}{80,000} du = \frac{t^2}{160,000}$$

$$F(B_5) = 0.05 = \frac{B_5^2}{160,000}$$

$$B_5 = 89.44$$

$$F(B_{50}) = 0.5 = \frac{B_{50}^2}{160,000}$$

$$B_{50} = 282.843.$$

(d) Draw the failure rate (or hazard rate) curve for the above product by evaluating it at $t = 0, 50, 100, 300, 400$.

Solution:
Using Equation 2.5, the following table can be developed and the hazard rate function $h(t)$ is drawn as shown in Figure 3.1.

	Value of h(t) vs. t		
t	f(t)	R(t)	h(t)
0	0	1	0
50	0.000625	0.984375	0.000635
100	0.00125	0.9375	0.001333
300	0.00375	0.4375	0.008571
400	0.005	0	Infinity

The values of the hazard function are given in Figure 3.1. It is clear that for the above triangular distribution, the failure rate is increasing; such distributions have the property of an increasing failure rate (IFR). Many products that wear or deteriorate with time will exhibit IFR behavior.

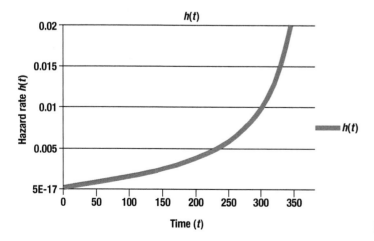

Figure 3.1 Hazard rate function, $h(t)$.

3.2.1 Weibull Distribution

The Weibull distribution is a continuous distribution developed in 1939 by Waloddi Weibull (1939), and who presented it in detail in 1951 (Weibull 1951). The Weibull distribution is widely used for reliability analyses because a wide diversity of hazard rate curves can be modeled with it. The distribution can also be approximated to other distributions under special or limiting conditions. The Weibull distribution has been applied to life distributions for many engineered products, and has also been used for reliability testing, material strength, and warranty analysis.

The probability density function for a three-parameter Weibull probability distribution function is

$$f(t) = \beta \eta^{-\beta} (t-\gamma)^{\beta-1} e^{-\left(\frac{t-\gamma}{\eta}\right)^{\beta}}, \qquad (3.22)$$

where $\beta > 0$ is the shape parameter, $\eta > 0$ is the scale parameter, which is also denoted by θ in many references and books, and γ is the location or time delay parameter. The reliability function is given by

$$R(t) = \int_{t}^{\infty} f(\tau) d\tau = e^{-\left(\frac{t-\gamma}{\eta}\right)^{\beta}}. \qquad (3.23)$$

It can be shown that Equation 3.23 gives, for a duration $t = \gamma + \eta$, starting at time $t = 0$, a reliability value of $R(t) = 36.8\%$, regardless of the value of β. Thus, for any Weibull failure probability density function, 36.8% of the products survive for $t = \gamma + \eta$.

The time to failure of a product with a specified reliability, R, is given by

$$t = \gamma + \eta [-\ln R(t)]^{1/\beta}. \qquad (3.24)$$

The hazard rate function for the Weibull distribution is given by

$$h(t) = \frac{f(t)}{R(t)} = \frac{\beta}{\eta} \left[\frac{t-\gamma}{\eta}\right]^{\beta-1}. \qquad (3.25)$$

The conditional reliability function is (see Eq. 2.39):

$$R(t, t_1) = \frac{R(t+t_1)}{R(t_1)}$$
$$= \exp\left\{-\left[\frac{t+t_1-\gamma}{\eta}\right]^{\beta} + \left[\frac{t_1-\gamma}{\eta}\right]^{\beta}\right\}. \qquad (3.26)$$

Equation 3.26 gives the reliability for a new mission of duration t for which t_1 hours of operation were previously accumulated up to the beginning of this new mission. It is seen that the Weibull distribution is generally dependent on both the age at the beginning of the mission and the mission duration (unless $\beta = 1$). In fact, this is true for most distributions, except for the exponential distribution (discussed later).

3 Probability and Life Distributions for Reliability Analysis

Table 3.1 Weibull distribution parameters

Location	γ
Shape parameter	β
Scale parameter	η
Mean (arithmetic average)	$\gamma + \eta + \Gamma(1/\beta + 1)$
Median (B_{50}, or time at 50% failure)	$\gamma + \eta(\ln 2)^{1/\beta}$
Mode (highest value of $f(t)$)	for $\beta > 1$ $\gamma + \eta(1 - 1/\beta)^{1/\beta}$ for $\beta = 1$ γ
Standard deviation	$\eta\sqrt{\Gamma\left(\dfrac{2}{\beta}+1\right)-\Gamma^{2}\left(\dfrac{1}{\beta}+1\right)}$

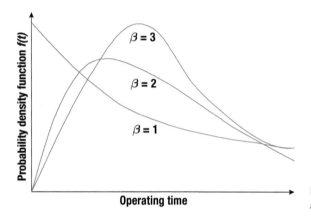

Figure 3.2 Effects of shape parameter β on probability density function, where $\eta = 1$ and $\gamma = 0$.

Table 3.1 lists the key parameters for a Weibull distribution and values for mean, median, mode, and standard deviation. The function Γ is the gamma function, for which the values are available from statistical tables and also are provided in Appendix B.

The shape parameter of a Weibull distribution determines the shape of the hazard rate function. With $0 < \beta < 1$, the hazard rate decreases as a function of time, and can represent early life failures (i.e., infant mortality). A $\beta = 1$ indicates that the hazard rate is constant and is representative of the "useful life" period in the "idealized" bathtub curve (see Figure 2.6). A $\beta > 1$ indicates that the hazard rate is increasing and can represent wearout. Figure 3.2 shows the effects of β on the probability density function curve with $\eta = 1$ and $\gamma = 0$. Figure 3.3 shows the effect of β on the hazard rate curve with $\eta = 1$ and $\gamma = 0$.

The scale parameter η has the effect of scaling the time axis. Thus, for a fixed γ and β, an increase in η will stretch the distribution to the right while maintaining its starting location and shape (although there will be a decrease in the amplitude, since the total area under the probability density function curve must be equal to unity). Figure 3.4 shows the effect of η on the probability density function for $\beta = 2$ and $\gamma = 0$.

The location parameter locates the distribution along the time axis and thus estimates the earliest time to failure. For $\gamma = 0$, the distribution starts at $t = 0$. With $\gamma > 0$, this implies that the product has a failure-free operating period equal to γ. Figure 3.5 shows the effects of γ on the probability density function curve for $\beta = 2$ and $\eta = 1$. Note that if γ is positive, the distribution starts to the right of the $t = 0$

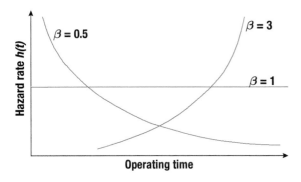

Figure 3.3 Dependence of hazard rate on shape parameter, where $\eta = 1$ and $\gamma = 0$.

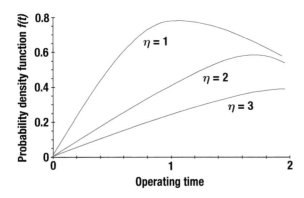

Figure 3.4 Effects of scale parameter η on the pdf of a Weibull distribution, where $\beta = 2$ and $\gamma = 0$.

line, or the origin. If γ is negative, the distribution starts to the left of the origin, and could imply that failures had occurred prior to the time $t = 0$, such as during transportation or storage. Thus, there is a probability mass $F(0)$ at $t = 0$, and the rest of the distribution for $t > 0$ is given in Figure 3.5. The Weibull distribution can also be formulated as a two-parameter distribution with $\gamma = 0$.

The reliability function for the two-parameter Weibull distribution is

$$R(t) = \int_t^\infty f(\tau)d\tau = e^{-\left(\frac{t}{\eta}\right)^\beta} \qquad (3.27)$$

and the hazard rate function is

$$h(t) = \frac{f(t)}{R(t)} = \frac{\beta}{\eta}\left[\frac{t}{\eta}\right]^{\beta-1}. \qquad (3.28)$$

The *two-parameter Weibull distribution* can be used to model skewed data. When $\beta < 1$, the failure rate for the Weibull distribution is decreasing and hence can be used to model infant mortality or a debugging period, situations when the reliability in terms of failure rate is improving, or reliability growth. When $\beta = 1$, the Weibull distribution is the same as the exponential distribution. When $\beta > 1$, the failure rate is increasing, and hence can model wearout and the end of useful life. Some examples of this are corrosion life, fatigue life, or the life of antifriction bearings, transmission gears, and electronic tubes.

3 Probability and Life Distributions for Reliability Analysis

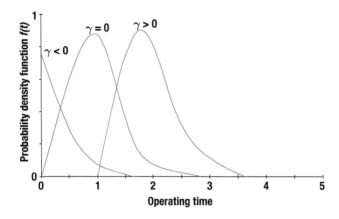

Figure 3.5 Effects of location parameter γ, where $\beta = 2$ and $\eta = 1$.

The *three-parameter Weibull distribution* is a model when there is a minimum life or when the odds of the component failing before the minimum life are close to zero. Many strength characteristics of systems do have a minimum value significantly greater than zero. Some examples are electrical resistance, capacitance, and fatigue strength.

Example 3.8

Assume that the time to failure of a product can be described by the Weibull distribution, with estimated parameter values of $\eta = 1000$ hours, $\gamma = 0$, and $\beta = 2$. Estimate the reliability of the product after 100 hours of operation. Also determine the MTTF.

Solution:
From Equation 3.27, we have:

$$R(100) = e^{-(100/1000)^2} = 0.990.$$

And from Table 3.1 we have

$$\text{MTTF} = 1000\Gamma(1/2+1) = 1000\Gamma(1.50) = 886 \text{ hours}.$$

where the value of $\Gamma(1.50)$ can be found from the table in Appendix B.

Example 3.9

Suppose that the life distribution for miles to failure for a give failure mode for the transmission of a GM Cadillac model follows the two-parameter Weibull distribution with $\eta = 150{,}000$ mi, $\beta = 4.5$.

(a) Find the mean miles between failures or the expected life in miles of these transmissions.

58

Solution:
Using the table in Appendix B, we have

$$E[T] = \eta \Gamma\left(1+\frac{1}{\beta}\right) = 150,000 \Gamma\left(1+\frac{1}{4.5}\right)$$
$$= 150,000 \times \Gamma(1.2222) = 150,000 \times 0.912573$$
$$= 136,886 \text{ mi.}$$

(b) Find the standard deviation for the miles to failure random variable.

Solution:
Again, using the table in Appendix B,

$$V[T] = \eta^2 \left[\Gamma\left(1+\frac{2}{\beta}\right) - \Gamma^2\left(1+\frac{1}{\beta}\right)\right]$$
$$= 150,000^2 \left[\Gamma(1.4444) - \Gamma(1.2222)^2\right]$$
$$= 150,000^2 \times (0.8858 - 0.912573^2) = 1.19 \times 10^9$$

$$\text{Standard deviation} = \sqrt{V(T)} = \sqrt{1.19 \times 10^9} = 34,513 \text{ mi.}$$

(c) If GM gives a warranty for 70,000 mi on these transmissions, what percent of these transmissions will fail during the warranty period?

Solution:

$$1 - R(70,000) = 1 - e^{-\left(\frac{70,000}{150,000}\right)^{4.5}} = 0.03188.$$

Thus, 3.188% of the transmissions will fail during the warranty period for the given failure mode.

Example 3.10

Suppose that the life distribution (life in years of continuous use) of hard disk drives for a computer system follows a two-parameter Weibull distribution with the following parameters: $\beta = 3.10$ and $\eta = 5$ years.

(a) The manufacturer gives a warranty for 1 year. What is the probability that a disk drive will fail during the warranty period?

Solution:

$$F(1) = 1 - R(1) = 1 - \exp\left[-\left(\frac{1}{5}\right)^{3.10}\right]$$
$$= 1 - 0.993212 = 0.006788.$$

(b) Find the mean life and the median life (B_{50}) for the disk drive.

Solution:

$$\text{mean} = 5\Gamma\left(1+\frac{1}{3.10}\right) = 5\Gamma(1.32258)$$
$$= 5 \times 0.89431 = 4.47155 \text{ years.}$$

And to find the median life, we have

$$R(B_{50}) = \exp\left[-\left(\frac{B_{50}}{5}\right)^{3.10}\right] = 0.5$$

$$\left(\frac{B_{50}}{5}\right)^{3.10} = -\ln(0.5) = 0.693147$$

$$B_{50} = 0.888492 \times 5 = 4.44246 \text{ years.}$$

(c) By what time will 95% of the disk drives fail? (Find the B_{95} life).

Solution:

$$R(B_{95}) = \exp\left[-\left(\frac{B_{95}}{5}\right)^{3.10}\right] = 0.05$$

$$\left(\frac{B_{95}}{5}\right)^{3.10} = -\ln(0.05) = 2.99573$$

$$B_{95} = 1.42466 \times 5 = 7.12329 \text{ years.}$$

Example 3.11

The failure rate of a component, in failures per year, is given by:

$$h(t) = 0.003t^2, \quad t \geq 0.$$

(a) Find an expression for the reliability function and the probability density function for the time to failure of the component.

Solution:
Using Equation. 2.35 and Equation 2.36 in Chapter 2, we have

$$h(t) = 0.003t^2$$

$$H(t) = \int_0^t 0.003x^2 dx = 0.001t^3$$

$$R(t) = \exp\left[-0.001t^3\right]$$

$$f(t) = 0.003t^2 \exp\left[-0.001t^3\right].$$

This is easily recognizable as a Weibull distribution with the following values of the parameters:

$$\text{Weibull: } \beta = 3, \eta = 10.$$

(b) Find the expected life (MTTF) for the component.

Solution:

$$E[T] = \theta \Gamma\left(1 + \frac{1}{\beta}\right)$$
$$= 10\Gamma\left(1 + \frac{1}{3}\right) = 10 \times 0.89298 = 8.9298 \text{ years}.$$

(c) Find the B_{10} (the 10th percentile) for the life of the component.

Solution:
We need to find the value of t, such that the item has a 10% chance of failing. This is equivalent to finding the point at which $R(t) = 0.9$. Solving for t:

$$0.9 = e^{-0.001(B_{10})^3}$$
$$\ln 0.9 = -0.001(B_{10})^3$$
$$B_{10} = \sqrt[3]{\frac{-\ln(0.9)}{0.001}} = 4.723 \text{ years}.$$

3.2.2 Exponential Distribution

The exponential distribution is a single-parameter distribution that can be viewed as a special case of a Weibull distribution, where $\beta = 1$. The probability density function has the form

$$f(t) = \lambda_0 e^{-\lambda_0 t}, \quad t \geq 0, \tag{3.29}$$

where λ_0 is a positive real number, often called the constant failure rate. The parameter λ_0 is typically an unknown that must be calculated or estimated based on statistical methods discussed later in this section. Figure 3.6 gives a graph for an exponential distribution, with $\lambda_0 = 0.10$. Table 3.2 summarizes the key parameters for the exponential distribution.

Once λ_0 is known, the reliability can be determined from the probability density function as

$$R(t) = \int_t^\infty f(\tau) d\tau = \int_t^\infty \lambda_0 e^{-\lambda_0 \tau} d\tau = e^{-\lambda_0 t}. \tag{3.30}$$

The cdf or unreliability is given by

$$F(t) = Q(t) = 1 - \exp[-\lambda_0 t]. \tag{3.31}$$

3 Probability and Life Distributions for Reliability Analysis

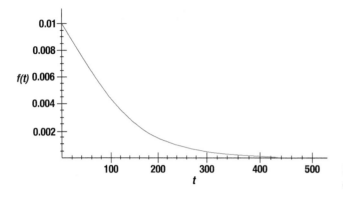

Figure 3.6 An example of an exponential distribution.

Table 3.2 Exponential distribution parameter

Scale parameter	$1/\lambda_0$
Median (B_{50})	$0.693/\lambda_0$
Mode (highest value of f(t))	0
Standard deviation	$1/\lambda_0$
Mean	$1/\lambda_0$

As mentioned, the hazard rate for the exponential distribution is constant:

$$h(t) = \frac{f(t)}{R(t)} = \frac{1}{e^{-\lambda_0 t}}\left(\lambda_0 e^{-\lambda_0 t}\right) = \lambda_0. \tag{3.32}$$

The conditional reliability is

$$R(t, t_1) = \frac{R(t+t_1)}{R(t_1)} = e^{-\lambda_0(t+t_1)} / e^{-\lambda_0 t_1} = e^{-\lambda_0 t}. \tag{3.33}$$

Equation 3.33 shows that previous usage (e.g., tests or missions) do not affect future reliability. This "as good as new" result stems from the fact that the hazard rate is a constant and the probability of a product failing is independent of the past history or use of the product.

The mean time to failure (MTTF) for an exponential distribution, also denoted by θ, is determined from the general equation for the mean of a continuous distribution:

$$\text{MTTF} = \int_0^\infty R(t)dt = \int_0^\infty e^{-\lambda_0 t}dt = \frac{1}{\lambda_0}. \tag{3.34}$$

Thus, the MTTF or the MTBF is inversely proportional to the constant failure rate, and thus the reliability can be expressed as

$$R(t) = e^{-t/\text{MTBF}}. \tag{3.35}$$

The MTBF is sometimes misunderstood to be the life of the product or the time by which 50% of products will fail. For a mission time of $t = \text{MTBF}$, the reliability calculated from Equation 3.30 gives $R(\text{MTBF}) = 0.368$. Thus, only 36.8% of the products survive a mission time equal to the MTBF.

Example 3.12

Show that the exponential distribution is a special case of the Weibull distribution.

Solution:
From Equation 3.22, set $\beta = 1$ and $\gamma = 0$

$$f(t) = \frac{1}{\eta} e^{-\frac{t}{\eta}}.$$

Thus, in this case, the Weibull distribution reduces to the single-parameter exponential distribution with $\lambda_0 = 1/\eta$. The reliability and the hazard rate functions simplify to:

$$R(t) = e^{-\frac{t}{\eta}}$$

$$h(t) = \frac{1}{\eta},$$

where

$$\eta = \frac{1}{\lambda_0}.$$

If $\beta = 1$ and $\gamma > 0$, then the Weibull distribution is the same as the exponential distribution with minimum life, γ, or is also a two-parameter exponential distribution.

Example 3.13

Consider an electronic product that exhibits a constant hazard rate. If the MTBF is 5 years, at what time will 10% of the products fail?

Solution:
Using Equation 3.35 with $R = 0.90$ and $\text{MTBF} \approx 43{,}800$ hours (5 years), we solve for t, where t is in hours. Thus, $t = -[(\text{MTBF}) \times \ln(R)] \approx 4600$ hours, or nearly half a year.

Example 3.14

Here we consider a mixture of exponential distributions. The pdf for the life of a device is given by the following probability density function, which is a mixture of two exponential distributions.

$$f(t) = \frac{1}{4} e^{-t} + \frac{3}{2} e^{-2t}, \quad t \geq 0.$$

(a) Prove that the above function is a valid pdf.

3 Probability and Life Distributions for Reliability Analysis

Solution:

$$\int_0^\infty \left(\frac{1}{4}e^{-t} + \frac{3}{2}e^{-2t}\right)dt = -\frac{1}{4}e^{-t} - \frac{3}{2} \times \frac{1}{2}e^{-2t}\bigg|_0^\infty$$

$$= -\frac{1}{4}(0-1) - \frac{3}{4}(0-1) = 1.$$

Therefore, the above function is a valid pdf.

(b) Find the probability that a device will last at least 3 hours.

Solution:

$$f(t) = \frac{1}{4}e^{-t} + \frac{3}{2}e^{-2t}, t \geq 0.$$

$$R(t) = \int_t^\infty f(\tau)d\tau = \int_t^\infty \left(\frac{1}{4}e^{-\tau} + \frac{3}{2}e^{-2\tau}\right)d\tau$$

$$= \left(-\frac{1}{4}e^{-\tau} - \frac{3}{4}e^{-2\tau}\right)\bigg|_t^\infty = \frac{1}{4}e^{-t} + \frac{3}{4}e^{-2t}$$

$$R(3) = \frac{1}{4}e^{-3} + \frac{3}{4}e^{-6} = 0.01431.$$

Alternatively: $R(t) = 1 - F(t)$.

$$F(t) = \int_0^t \left(\frac{1}{4}e^{-\tau} + \frac{3}{2}e^{-2\tau}\right)d\tau = \left(-\frac{1}{4}e^{-\tau} - \frac{3}{4}e^{-2\tau}\right)\bigg|_0^t$$

$$= -\frac{1}{4}e^{-t} - \frac{3}{4}e^{-2t} + 1$$

$$F(3) = -\frac{1}{4}e^{-3} - \frac{3}{4}e^{-6} + 1 = 0.98569$$

$$R(3) = 1 - F(3) = 1 = 1 - 0.98569 = 0.01431.$$

(c) Find the expected life or the MTBF of the device.

Solution:

$$\text{MTBF} = \int_0^\infty R(t)dt = \int_0^\infty \left(\frac{1}{4}e^{-t} + \frac{3}{4}e^{-2t}\right)dt$$

$$= \left(-\frac{1}{4}e^{-t} - \frac{3}{8}e^{-2t}\right)\bigg|_0^\infty = \frac{1}{4} + \frac{3}{8} = \frac{5}{8} \text{ hours.}$$

3.2.3 Estimation of Reliability for Exponential Distribution

For reliability tests in which the hazard rate is assumed to be constant, and the time to failure can be assumed to follow an exponential distribution, the constant failure

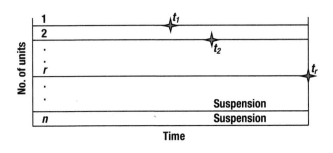

Figure 3.7 Failure-truncated test. Failures are denoted by ✢.

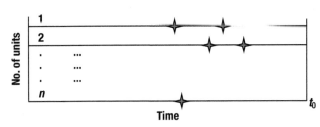

Figure 3.8 Time-truncated test. ✢, Failures.

rate can be estimated by life testing. There are various ways to test the items. Figure 3.7 gives an example of a failure-truncated test, in which n items on individual test stands are monitored to failure. The test ends as soon as there are r failures (without replacement $r \leq n$), as shown in Figure 3.7.

The total time on test, T_T, considering both failed and unfailed (or suspended) units, is calculated by the following equation:

$$T_T = \sum_{i=1}^{r} t_i + (n-r)t_r. \tag{3.36}$$

Another test situation is called time-truncated testing. In Figure 3.8, there are n test stands (or n items on test in a test chamber). The units are monitored and replaced as soon as they fail. Testing for these units continues until some predetermined time, t_0. In this case, the total time on test is

$$T_T = nt_0. \tag{3.37}$$

Then the point estimator (minimum variance unbiased estimator) for θ, the MTBF, is

$$\hat{\theta} = \frac{T_T}{r}. \tag{3.38}$$

Further details are given in Chapter 13. Also, the point estimator for λ is

$$\hat{\lambda} = \frac{1}{\hat{\theta}}. \tag{3.39}$$

Chapter 13 will present the methodology for the point estimation and confidence interval for several test situations and underlying life distributions.

Example 3.15

Seven prototypes are monitored for some failure during development testing or as fielded products. The failures on the products are fixed (it is assumed that we can renew the product) and testing continues. Then testing is stopped at the times given below for each product:

Product no.	Hours when failures are recorded	Hours when testing is stopped
01	2467; 3128; 3283; 7988	8012
02	None	6147
03	1870; 6121; 6175	9002
04	3721; 4393; 5848; 6425; 6353	11,000
05	498	4651
06	184; 216; 561; 2804	5012
07	2342; 4213	12,718

Estimate the MTBF for this product.

Solution:
In this case, T_T, the total time on test, is obtained by adding all the hours when testing was stopped:

$$T_T = 8012 + \cdots + 12{,}718 = 56{,}542 \text{ hours.}$$

During this total test period, there were 19 failures. Thus, the point estimator for the MTBF, under the assumption that the time between failures follows the exponential distribution, is

$$\hat{\theta} = 56{,}542/19 = 2975 \text{ hours.}$$

Example 3.16

Estimate the MTBF (point estimator) or ($\hat{\theta}$) for the following reliability test situations:

(a) Failure terminated, with no replacement. Twelve items were tested until the fourth failure occurred, with failures at 200, 500, 625, and 800 hours.
(b) Time terminated, with no replacement. Twelve items were tested up to 1000 hours, with failures at 200, 500, 625, and 800 hours.
(c) Failure terminated, with replacement. Eight items were tested until the third failure occurred, with failures at 150, 400, and 650 hours.
(d) Time terminated, with replacement. Eight items were tested up to 1000 hours, with failures at 150, 400, and 650 hours.
(e) Mixed replacement/nonreplacement. Six items were tested through 1000 hours on six different test stands. The first failure on the test stand occurred at 300 hours, and its replacement failed after an additional 400 hours. On the second test stand, failure occurred at 350 hours, and its replacement failed after an additional 500 hours. On the third test stand, failure occurred at 600 hours, and

its replacement did not fail up to the completion of the test. The items on the other three test stands did not fail for the duration of the test.

Solution:

(a) $\text{MTBF}(e) = \hat{\theta} = (200 + 500 + 625 + 800 + 8(800))/4 = 2{,}131$ hours
(b) $\text{MTBF}(e) = \hat{\theta} = (200 + 500 + 625 + 800 + 8(1000))/4 = 2{,}531$ hours
(c) $\text{MTBF}(e) = \hat{\theta} = (8)(650)/3 = 1{,}733$ hours
(d) $\text{MTBF}(e) = \hat{\theta} = (8)(1000)/3 = 2{,}667$ hours
(e) $\text{MTBF}(e) = \hat{\theta} = (700 + 850 + 1000 + (3)(1000))/5 = 1{,}110$ hours.

Example 3.17

Forty modules were placed on life test for 20 days (24 hours per day). Failed boards were replaced on the test stands with new ones. The test produced two failures. Estimate the MTBF or the failure rate for the modules.

Solution:
In this case, the total time on test is

$$T_T = 20 \times 24 \times 40 = 19{,}200 \text{ hours}$$

$$\hat{\theta} = T_T/r = 19{,}200/2 = 9600 \text{ hours}$$

$$\hat{\lambda} = r/T_T = 2/19{,}200 = 1.04 \times 10 - 4 \text{ failures per hour.}$$

3.2.4 The Normal (Gaussian) Distribution

The normal distribution occurs whenever a random variable is affected by a sum of random effects, such that no single factor dominates. This motivation is based on central limit theorem, which states that under mild conditions, the sum of a large number of random variables is approximately normally distributed. It has been used to represent dimensional variability in manufactured goods, material properties, and measurement errors. It has also been used to assess product reliability.

The normal distribution has been used to model various physical, mechanical, electrical, or chemical properties of systems. Some examples are gas molecule velocity, wear, noise, the chamber pressure from firing ammunition, the tensile strength of aluminum alloy steel, the capacity variation of electrical condensers, electrical power consumption in a given area, generator output voltage, and electrical resistance.

The probability density function for the normal distribution is based on the following Gaussian function:

$$f(t) = \frac{1}{\sigma\sqrt{2\pi}} \exp\left[-\frac{1}{2}\left(\frac{t-\mu}{\sigma}\right)^2\right], \quad -\infty \leq t \leq +\infty, \tag{3.40}$$

where the parameter μ is the mean or the MTTF, and σ is the standard deviation of the distribution. The parameters for a normal distribution are listed in Table 3.3.

3 Probability and Life Distributions for Reliability Analysis

Table 3.3 Normal distribution parameters

Mean (arithmetic average)	μ
Median (B_{50} or 50th percentile)	μ
Mode (highest value of $f(t)$)	μ
Location parameter	μ
Shape parameter/standard deviation	σ
s (an estimate of σ)	$B_{50} - B_{16}$

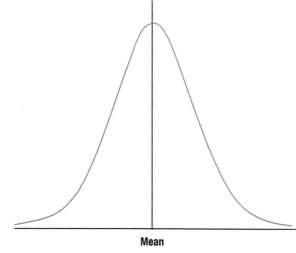

Figure 3.9 Probability density function for normal distribution.

Figure 3.9 shows the shape of the probability density function for the normal distribution.

The cdf, or unreliability, for the normal distribution is:

$$F(t) = \frac{1}{\sigma\sqrt{2\pi}} \int_{-\infty}^{t} \exp\left[-\frac{1}{2}\left(\frac{x-\mu}{\sigma}\right)^2\right] dx. \tag{3.41}$$

A normal random variable with mean equal to zero and variance of 1 is called a standard normal variable (Z), and its pdf is given by

$$\phi(z) = \frac{1}{\sqrt{2\pi}} e^{-z^2/2}, \tag{3.42}$$

where $z \equiv (t - \mu)/\sigma$.

The properties of the standard normal variable, in particular the cumulative probability distribution function, are tabulated in statistical tables (provided in Appendix C). Table 3.4 provides the percentage values of the areas under the normal curve at different distances from the mean in terms of multiples of σ. For example,

$$P[X \leq \mu - 3\sigma] = 0.00135 \tag{3.43}$$

and

Table 3.4 Areas under the normal curve

$\mu - 1\sigma = 15.87\%$	$\mu + 1\sigma = 84.130\%$
$\mu - 2\sigma = 2.28\%$	$\mu + 2\sigma = 97.720\%$
$\mu - 3\sigma = 0.135\%$	$\mu + 3\sigma = 99.865\%$
$\mu - 4\sigma = 0.003\%$	$\mu + 4\sigma = 99.997\%$

$$P[X \leq \mu + 3\sigma] = 0.99865. \qquad (3.44)$$

There is no closed-form solution to the integral of Equation 3.41, and, therefore, the values for the area under the normal distribution curve are obtained from the standard normal tables by converting the random variable, t, to a random variable, z, using the transformation:

$$z = \frac{t-\mu}{\sigma}, \qquad (3.45)$$

given by Equation 3.42. We have

$$F(t) = \Phi(z) = \Phi\left(\frac{t-\mu}{\sigma}\right) \qquad (3.46)$$

or

$$R(t) = 1 - \Phi\left(\frac{t-\mu}{\sigma}\right) \qquad (3.47)$$

and

$$h(t) = \frac{\phi[(t-\mu)/\sigma]}{\sigma R(t)}, \qquad (3.48)$$

where $\phi(.)$ is the pdf for the standard normal distribution and $\Phi(z)$ is the cdf for the standard normal random variable Z.

From Equation 3.48, we can prove that the normal distribution has an increasing hazard rate (IHR). The normal distribution has been used to describe the failure distribution for products that show wearout and that degrade with time. The life of tire tread and the cutting edges of machine tools fit this description. In these situations, life is given by a mean value of μ, and the variability about the mean value is defined through standard deviation. When the normal distribution is used, the probabilities of a failure occurring before or after this mean time are equal because the mean is the same as the median.

Example 3.18

A machinist estimates that there is a 90% probability that the washer in an air compressor will fail between 25,000 and 35,000 cycles of use. Assuming a normal distribution for washer degradation, find the mean life and standard deviation of the life of the washers.

3 Probability and Life Distributions for Reliability Analysis

Solution:
Assuming that 5% of the failures are at fewer than 25,000 cycles and 5% are at more than 35,000 cycles, the mean of the distribution will be centered at 30,000 cycles of use, that is, $\mu = 30{,}000$.
In this condition:

$$\Phi(z_1) = 0.05,\ z_1 = \frac{25{,}000 - \mu}{\sigma}$$

$$\Phi(z_2) = 0.95,\ z_2 = \frac{35{,}000 - \mu}{\sigma}.$$

From the normal distribution table, $z_1 = -1.65$, and $z_2 = 1.65$. Hence, $-1.65\sigma = 25{,}000 - \mu$ and $1.65\sigma = 35{,}000 - \mu$.

Solving the above two equations with the mean value of 30,000 cycles results in a σ of 3030 cycles.

Example 3.19

The time for failure due to fungi growth is normally distributed with mean $\mu = 2.8$ hours and standard deviation $\sigma = 0.6$ hours.

(a) What is the probability that the failure due to fungi growth will occur in 1.5 hrs?

(b) If we accept a probability of failure due to fungi growth of only 10%, after what time from the start should the fungi be analyzed?

Solution:

(a) The probability that the fungi will grow in less than 1.5 hours is given by:

$$P\{T < 1.5\} = Q(1.5) = \Phi(z),$$

$$z = (t - \mu)/\sigma = (1.5 - 2.8)/0.6 = -2.1667.$$

From the standard normal table, $\Phi(-2.1667) = 0.0151$.

(b) For this condition, $F(t) = 0.1 = \Phi(z)$, then from the standard normal table, z is approximately -1.28. Therefore, $-t + \mu = 1.28\sigma$, hence $t = 2.03$ hours.

Example 3.20

A component has the normal distribution for time to failure, with $\mu = 20{,}000$ hours and $\sigma = 3000$ hours.

(a) Find the probability that the component will fail between 14,000 hours and 15,000 hours.

3.2 Continuous Distributions

Solution:

$$P[14{,}000 \leq T \leq 15{,}000]$$
$$= P\left[\frac{14{,}000-20{,}000}{3000} \leq \frac{T-\mu}{\sigma} \leq \frac{15{,}000-20{,}000}{3000}\right]$$
$$= P[-2 \leq Z \leq -1.667]$$
$$= \Phi(-1.667) - \Phi(-2)$$
$$= (1-0.95225) - (1-0.97725)$$
$$= 0.04775 - 0.02275 = 0.025.$$

(b) Find the failure rate of a component that has been working for 14,000 hours.

Solution:

$$f(14{,}000) = \phi\left(\frac{14{,}000-20{,}000}{3000}\right)\bigg/3000$$
$$= \frac{1.7997 \times 10^{-5}}{3000} = 5.999 \times 10^{-9}$$
$$\text{pdf } \phi(14{,}000) = \phi\left(\frac{14{,}000-20{,}000}{3000}\right) = 1.7997 \times 10^{-5}$$

based on MS Excel evaluation

$$R(t) = P\left[Z > \frac{14{,}000-20{,}000}{3000}\right] = P[Z > -2] = \Phi(2) = 0.977249$$
$$h(t) = \frac{f(t)}{R(t)} = 6.13865 \times 10^{-9} \text{ failures/cycle.}$$

Example 3.21

The time to failure random variable for a light bulb made by Company X follows a normal distribution, with $\mu = 1600$ hours and $\sigma = 250$ hours.

(a) Find the B_{10} life of these light bulbs.

Solution:
Setting the value of reliability at B_{10} equal to 0.90, we have

$$R(B_{10}) = 0.9$$
$$z = -1.28 = \frac{B_{10} - 1600}{250}$$
$$B_{10} = (-1.28)(250) + 1600 = 1280 \text{ hours.}$$

(b) Find the reliability of the light bulbs for 1100 hours.

Solution:

$$R(1100) = \Pr\left(Z \geq \frac{1100-1600}{250}\right) = \Pr(Z \geq -2.0) = 0.9773.$$

3 Probability and Life Distributions for Reliability Analysis

(c) What is the failure rate or hazard rate of a light bulb that has not failed for 1100 hours?

Solution:

$$h(1100) = \frac{f(1100)}{R(1100)} = \frac{\phi(-2.0)/250}{\Phi(-2.0)} = \frac{0.05399/250}{0.9773}$$
$$= 0.0002210 \text{ failures/hour}.$$

where:

$$\phi\left(\frac{1100-1600}{250}\right) = \phi(-2) = 0.05399.$$

based on MS Excel evaluation, where:

$$\phi(-2.0) = \frac{1}{\sqrt{2\pi}} e^{-\left[\frac{(-2)^2}{2}\right]} = (0.39894)(0.13533) = 0.05399.$$

Alternatively,

$$f(1100) = \frac{1}{\sqrt{2\pi}\sigma} e^{-\left[\frac{(-2)^2}{2}\right]}$$
$$= \frac{0.05399}{250} = 0.0002160$$
$$h(1100) = \frac{f(1100)}{R(1100)}$$
$$= \frac{0.0002160}{0.9773} = 0.0002210.$$

(d) Company X has 5 million light bulbs in the field that have been in use for 1100 hours and have not failed so far. How many light bulbs will fail in the next day (or 24 hours)? Assume that light bulbs are used on average for 10 hours per day.

Solution:
Using the concepts covered in Section 2.2.1 of Chapter 2, we have $N_s(1100) = 5 \times 10^6$ and $\Delta t = 10$ hours.

$$h(1100) = 0.0002210 =$$
$$\frac{N_S(1100) - N_S(1100+10)}{N_S(1100) \times \Delta t} = \frac{N_S(1100) - N_S(1100+10)}{5 \times 10^6 \times 10}.$$

Then the number of failures between 1100 and 1110 hours is given by:

$$(0.0002210)(5 \times 10^6)(10) = 11,050 \text{ light bulbs}.$$

3.2.5 The Lognormal Distribution

For a continuous random variable, there may be a situation in which the random variable is a product of a series of random variables. The lognormal distribution is a positively skewed distribution and has been used to model situations where large occurrences are concentrated at the tail (left) end of the range. Some examples are the amount of electricity used by different customers, the downtime of systems, the time to repair, the light intensities of light bulbs, the concentration of chemical process residues, and automotive mileage accumulation by different customers. For example, the wear on a system may be proportional to the product of the magnitudes of the loads acting on it. Thus, a random variable may be modeled as a lognormal random variable if it can be thought of as the multiplicative product of many independent random variables each of which is positive. If a random variable has lognormal distribution, then the logarithm of the random variable is normally distributed. If X is a random variable with a normal distribution, then $Y = e^X$ has a lognormal distribution; or if Y has lognormal distribution, then $X = \log Y$ has normal distribution.

Suppose Y is the product of n independent random variables given by

$$Y = Y_1 Y_2 Y_3 \ldots \ldots Y_n. \qquad (3.49)$$

Taking the natural logarithm of Equation 3.49 gives

$$\ln Y = \ln Y_1 + \ln Y_2 + \ln Y_3 + \cdots + \ln Y_n. \qquad (3.50)$$

Then $\ln Y$ may have approximately normal distribution based on the central limit theorem.

The lognormal distribution has been shown to apply to many engineering situations, such as the strengths of metals and the dimensions of structural elements, and to biological parameters, such as loads on bone joints. Lognormal distributions have been applied in reliability engineering to describe failures caused by fatigue and to model time to repair for maintainability analysis. The probability density function for the lognormal distribution is:

$$f(t) = \frac{1}{\sigma t \sqrt{2\pi}} \exp\left[-\left(\frac{1}{2}\right)\left(\frac{\ln t - \mu}{\sigma}\right)^2\right], \qquad (3.51)$$

where σ is the standard deviation of the logarithms of all times to failure, and μ is the mean of the logarithms of all times to failure. If random variable T follows a lognormal distribution with parameters μ and σ, then $\ln T$ follows a normal distribution so that

$$E[\ln T] = \mu \text{ and } V[\ln T] = \sigma^2. \qquad (3.52)$$

The cdf (unreliability) for the lognormal distribution is:

$$Q(t) = \frac{1}{\sigma\sqrt{2\pi}} \int_0^t \frac{1}{x} \exp\left[-\left(\frac{1}{2}\right)\left(\frac{\ln x - \mu}{\sigma}\right)^2\right] dx$$
$$= \Phi\left(\frac{\ln t - \mu}{\sigma}\right). \qquad (3.53)$$

3 Probability and Life Distributions for Reliability Analysis

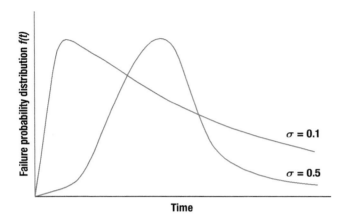

Figure 3.10 Lognormal probability density function where $\sigma = 0.1$ and $\sigma = 0.5$.

Table 3.5 Lognormal distribution parameters

Mean	$\exp[\mu + 0.5\sigma^2]$
Variance	$(e^{\sigma^2} - 1)e^{2\mu+\sigma^2}$
Median (B_{50} or time at 50% failures)	$B_{50} = e^\mu$
Mode (highest value of $f(t)$)	$t = \exp[\mu - \sigma^2]$
Location parameter	e^μ
Shape parameter	σ
s (estimate of σ)	$\ln(B_{50}/B_{16})$

The probability density function for two values of σ are as shown in Figure 3.10. The key parameters for the lognormal distribution are provided in Table 3.5.

The MTTF for a population for which the time to failure follows a lognormal distribution is given by

$$\text{MTTF} = \exp\left[\mu + \frac{\sigma^2}{2}\right] \qquad (3.54)$$

and the failure rate is given by

$$h(t) = \phi\left(\frac{\ln t - \mu}{\sigma}\right) \Big/ t\sigma R(t). \qquad (3.55)$$

The hazard rate for the lognormal distribution is neither always increasing nor always decreasing. It takes different shapes, depending on the parameters μ and σ. We can prove that the hazard rate of a lognormal distribution is increasing on average (called IHRA).

From the basic properties of the logarithm operator, it can be shown that if variables X and Y are distributed lognormally, then the product random variable $Z = XY$ is also lognormally distributed.

3.2 Continuous Distributions

Example 3.22

A population of industrial circuit breakers was found to have a lognormal failure distribution with parameters $\mu = 3$ and $\sigma = 1.8$. What is the MTTF of the population? What is the estimate of reliability of these circuit breakers for continuous operation over 30 years?

Solution:
From Equation 3.54 for the MTTF,

$$\text{MTTF} = \exp(3 + 0.5 \times (1.8)2) = 101.5 \text{ years}.$$

For a 30-year operation (from Eq. 3.53),

$$z = \frac{\ln(30) - 3}{1.8} = \frac{3.41 - 3}{1.8} = 0.223.$$

Hence, from the table of standard normal distribution, the estimate of reliability for a 30-year operation is given by:

$$R(30) = [1 - \Phi(z)] = [1 - \Phi(0.223)] = [1 - 0.588] = 0.412.$$

Example 3.23

The time to repair a copy machine follows the lognormal distribution with $\mu = 2.50$ and $\sigma = 0.40$. Time is in minutes.

(a) Find the probability that the copy machine will be repaired in 20 minutes.

Solution:

$$P[T \leq 20] = P[\ln T \leq \ln 20] = P\left[Z \leq \frac{\ln 20 - 2.5}{0.40}\right]$$
$$= P[Z \leq 1.23933] = 0.89239.$$

(b) Find the median value, or B50 life, for the time to repair a random variable.

$$P[T \leq B_{50}] = 0.5 = P[Z \leq 0]$$
$$0 = \frac{\ln T - 2.5}{0.40}$$
$$T = 12.185.$$

3.2.6 Gamma Distribution

The probability density function for the gamma distribution is given by

$$f(t) = \frac{\lambda^\eta}{\Gamma(\eta)} t^{\eta-1} e^{-\lambda t}, \quad t \geq 0, \tag{3.56}$$

where $\Gamma(\eta)$ is the gamma function (values for this function are given in Appendix B). The gamma distribution has two parameters, η and λ, where η is called the shape parameter and λ is called the scale parameter. The gamma distribution reduces to the exponential distribution if $\eta = 1$. Adding η exponential distributions, $\eta \geq 1$, with the same parameter λ, provides the gamma distribution. Thus, the gamma distribution can be used to model time to the ηth failure of a system if the underlying system/component failure distribution is exponential with parameter λ. We can also state that if T_i is exponentially distributed with parameter λ, $i = 1, 2, \ldots, \eta$, then $T = T_1 + T_2 + \cdots + T_\eta$ has a gamma distribution with parameters λ and η. This distribution could be used if we wanted to determine the system reliability for redundancy with identical components all having a constant failure rate.

From Equation 3.56, the cumulative distribution or the unreliability function is

$$F(t) = \int_0^t \frac{\lambda^\eta}{\Gamma(\eta)} \tau^{\eta-1} e^{-\lambda \tau} d\tau, \quad t \geq 0. \tag{3.57}$$

If η is an integer, then the gamma distribution is also called the Erlang distribution, and it can be shown by successive integration by parts that

$$F(t) = \sum_{k=\eta}^{\infty} \frac{(\lambda t)^k \exp[-\lambda t]}{k!}. \tag{3.58}$$

Then,

$$R(t) = \sum_{k=0}^{\eta-1} \frac{(\lambda t)^k \exp[-\lambda t]}{k!} \tag{3.59}$$

and

$$h(t) = \frac{f(t)}{R(t)}. \tag{3.60}$$

Also,

$$E(T) = \frac{\eta}{\lambda} \tag{3.61}$$

and

$$V(T) = \frac{\eta}{\lambda^2}. \tag{3.62}$$

The failure rate for the gamma distribution is decreasing when $\eta < 1$, is constant when $\eta = 1$ (because it is an exponential distribution), and is increasing when $\eta > 1$.

Example 3.24

The time to a major failure in hours for a copy machine follows a gamma distribution with parameters $\eta = 3$ and $\lambda = 0.002$.

(a) What is the expected life, or mean time between failures (MTBF), for the copy machine?

Solution:
Using Equation 3.61, we have

$$\text{MTBF} = \frac{\eta}{\lambda} = \frac{3}{0.002} = 1500 \text{ hours.}$$

(b) What is the reliability of the copy machine for 500 hours of continuous operation?

Solution:
Using Equation 3.59,

$$R(t) = \sum_{k=0}^{\eta-1} \frac{(\lambda t)^k e^{-\lambda t}}{k!}$$

$$R(500) = \sum_{k=0}^{3-1} \frac{(0.002 \times 500)^k e^{-0.002 \times 500}}{k!}$$

$$= 0.919698.$$

(c) What is the failure rate of a copy machine that has been working for 500 hours?

Solution:
Using Equation 3.56 and Equation 3.60, we have

$$f(t) = \frac{\lambda^\eta}{\Gamma(\eta)} t^{\eta-1} e^{-\lambda t}$$

$$f(500) = \frac{0.002^3}{\Gamma(3)} 500^{3-1} e^{-(0.002*500)} = 0.000368$$

$$h(500) = f(500)/R(500)$$
$$= 0.0004001 \text{ failures per hour.}$$

Thus, the failure rate is 0.0004 failures per hour, or 4 failures per 10,000 hours of total use.

3.3 Probability Plots

Probability plotting is a method for determining whether data (observations) conform to a hypothesized distribution. Typically, computer software is used to assess the

Table 3.6 Examples of cdf estimates for $N = 20$

Rank order (i)	Estimate of cumulative distribution function or unreliability			
	Midpoint plotting position	Expected plotting position	Median plotting position	Median rank
1	2.5	4.8	3.4	3.406
2	7.5	9.5	8.3	8.251
3	12.5	14.3	13.2	13.147
4	17.5	19.0	18.1	18.055
5	22.5	23.8	23.0	22.967
6	27.5	28.6	27.9	27.880
7	32.5	33.3	32.8	32.795
8	37.5	38.1	37.7	37.710
9	42.5	42.8	42.6	42.626
10	47.5	47.6	47.5	47.542
11	52.5	52.4	52.5	52.458
12	57.5	57.1	57.4	57.374
13	62.5	61.9	62.3	62.289
14	67.5	66.7	67.2	67.205
15	72.5	71.4	72.1	72.119
16	77.5	76.4	77.0	77.033
17	82.5	80.1	81.9	81.945
18	87.5	85.7	86.8	86.853
19	92.5	90.5	91.7	91.749
20	97.5	95.2	96.6	96.594

hypothesized distribution and determine the parameters of the underlying distribution. The method used by the software tools is analogous to using constructed probability plotting paper to plot data. The time-to-failure data is ordered from the smallest to the largest in value in an appropriate metric (e.g., time to failure and cycles to failure). An estimate of the percent of unreliability is selected. The data are plotted against a theoretical distribution in such a way that the points should form a straight line if the data come from the hypothesized distribution. The data are plotted on probability plotting papers (these are distribution specific), with ordered times to failure in the x-axis and the estimate of percent unreliability as the y-axis. A best-fit straight line is drawn through the plotted data points.

The time to failure data used for the x-axis is obtained from the field or testing. The estimate of unreliability against which to plot this time-to-failure data is not that obvious. Several different techniques, such as "midpoint plotting position," "expected plotting position," "median plotting position," "median rank," and Kaplan–Meier ranks (in software) are used for this estimate. Table 3.6 provides estimates for unreliability based on different estimation schemes for a sample size of 20.

The median rank value for the ith failure, Q_i, is given by the solution to the following equation:

$$\sum_{k=i}^{N} \frac{N!}{k!(N-k)!} (1-Q_i)^{N-k} Q_i^k = 0.5, \qquad (3.63)$$

where N is the sample size, i is the failure number, and Q_i is the median rank (or estimate of unreliability at the failure time of the ith failure). Equation 3.64, which estimates the median plotting positions, can be used in place of the median rank as an approximation:

$$Q_i = \frac{100 \times (i - 0.3)}{N + 0.4}. \tag{3.64}$$

The axes used for the plots are not linear. The axes are different for each probability distribution and are created by linearizing the cmf or unreliability function, typically by taking the logarithm of both sides repeatedly. For example, mathematical manipulation based on Equation 3.27 for a two-parameter Weibull distribution will result in an ordinate (y-axis) as log log reciprocal of $R(t) = 1 - Q(t)$ scale and the abscissa as a log scale of time to failure, and is derived below:

$$\begin{aligned} Q(t) &= 1 - e^{-\left(\frac{t}{\eta}\right)^\beta} \\ \ln(1 - Q(t)) &= \ln\left(e^{-\left(\frac{t}{\eta}\right)^\beta}\right) \\ \ln(-\ln(1 - Q(t))) &= \beta \ln\left(\frac{t}{\eta}\right) \\ y &= \beta x - \beta \ln(\eta), \end{aligned} \tag{3.65}$$

where $x = \ln(t)$ and $y = \ln(-\ln(1 - Q(t)))$.

Once the probability plots are prepared for different distributions, the goodness of fit of the plots is one factor in determining which distribution is the right fit for the data. Probability distributions for data analysis should be selected based on their ability to fit the data and for physics-based reasons. There should be a physics-based argument for selection of a distribution that draws from the failure model for the mechanism(s) that caused the failures. These decisions are not always clear-cut. For example, the lognormal and the Weibull distribution both model fatigue failure data well, and hence it is often possible for both to fit the failure data; thus, experience-based engineering judgments need to be made.

There is no reason to assume that all the time-to-failure data taken together need to fit only one failure distribution. Since the failures in a product can be caused by more than one mechanism, it is possible that some of the failures are caused by one mechanism and the others by a different mechanism. In that case, no single probability distribution will fit the data well. Even if it appears that one distribution fits all the data, that distribution may not have good predictive ability. That is why it may be necessary to separate the failures by mechanisms into sets and then fit separate distributions for each set.

Table 3.7 shows times to failure separated into two groups by failure mechanism. Figure 3.11 shows the Weibull probability plots for the competing failure mechanism data. Note that the shape and scale factors for the two sets are distinct, with one set having a decreasing hazard rate ($\beta = 0.67$) and the other set having an increasing hazard rate ($\beta = 4.33$). If the data are plotted together, the result shows an almost constant hazard rate. However, spare part and support decisions made based on results from a combined data analysis can be misleading and counterproductive.

3 Probability and Life Distributions for Reliability Analysis

Table 3.7 Time to failure data separated by failure mechanism

Ordered Data	State F or S	Time to F or S	Failure Mechanism Group
1	F	2	V
2	F	10	V
3	F	13	V
4	F	23	V
5	F	23	V
6	F	28	V
7	F	30	V
8	F	65	V
9	F	80	V
10	F	88	V
11	F	106	V
12	F	143	V
13	F	147	W
14	F	173	V
15	F	181	W
16	F	212	W
17	F	245	W
18	F	247	V
19	F	261	V
20	F	266	W
21	F	275	W
22	F	293	W
23	S	300	
24	S	300	
25	S	300	
26	S	300	
27	S	300	
28	S	300	
29	S	300	
30	S	300	

F, failure; S, suspension; V, failure mechanism 1; W, failure mechanism 2.

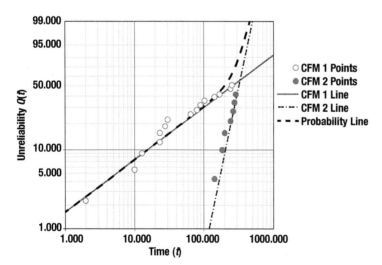

Figure 3.11 Weibull probability plot for competing failure mechanism data shown in Table 3.7. $\beta_1 = 0.67$, $\eta_1 = 450$; $\beta_2 = 4.33$, $\eta_2 = 340$.

3.3 Probability Plots

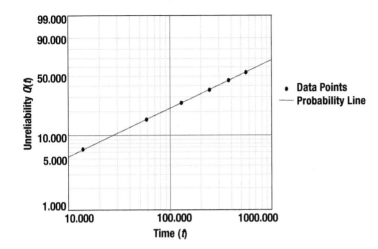

Figure 3.12 Two-parameter Weibull probability plot for time-to-failure data shown in Table 3.8. $\beta_1 = 0.65$, $\eta_1 = 825$.

Table 3.8 Test Data for Example 3.25

Sample number	Time to failure (hours)	Sample number	Time to failure (hours)
1	14	6	563
2	58	7	–
3	130	8	–
4	245	9	–
5	382	10	–

Example 3.25

Figure 3.12 shows reliability test data for 10 identical products out of which six products failed within the test duration of 600 hors. The time to failure is plotted on two-parameter Weibull probability plotting paper. Using the plot, estimate the following:

(a) The unreliability and reliability at the end of 50 hours.
(b) The reliability for a new period of 50 hours, starting after the end of the previous 50-hour period.
(c) The longest duration that will provide a reliability of 95% assuming the operation starts at 50 hours.

Solution:

(a) For this example, we find that $\beta = 0.65$, and η is estimated to be 825 hours. It is now possible to write the equation for the reliability and use it for analysis. The plotted straight line can also be used to determine the reliability values directly.

From Figure 3.12, the unreliability estimate for a mission time of 50 hours can be read directly from the straight line. The value is $Q(50) = 15\%$. Thus, the reliability for this duration is $R(50) = 1 - Q(50) = 85\%$.

81

(b) The reliability for a new 50-hour period starting with an age of 50 hours is given by the conditional reliability equation as

$$R(50, 50) = \frac{R(50+50)}{R(50)} = \frac{R(100)}{R(50)} = \frac{0.78}{0.85} = 91.7\%,$$

where $R(100) = 1 - Q(100)$ can be taken directly from the curve.

(c) For a mission time, t, that starts after a 50-hour period and must have a reliability of 95%,

$$R(t, 50) = \frac{R(t+50)}{R(50)} = \frac{R(t+50)}{0.85} = 0.95$$

or

$$R(t+50) = 0.95 \times 0.85 = 0.808.$$

To obtain this reliability, the unreliability is 0.192 or 19.2%. From the curve, the time to obtain this unreliability is about 75 hours. Thus, $50 + t = 75$ gives a maximum new mission time of 25 hours in order to have a reliability of 95%.

When the life data contains two or more life segments—such as infant mortality, useful life, and wearout—a mixed Weibull distribution can be used to fit parts of the data with different distribution parameters. A curved or S-shaped Weibull probability plot (in either two or three parameters) is an indication that a mixed Weibull distribution may be present.

Statistical analysis provides no magical way of projecting into the future. The results from an analysis are only as good as the assumed model and assumptions, including how failure is defined, the validity of the data, how the model is used, and taking into consideration the tail of the distribution and the limits of extrapolations and interpolations. The following example demonstrates the absurdity of extrapolating times to failure beyond their reasonable limits.

Example 3.26

A Weibull probability plot was made for a population collected over the first 10 years of its life containing failures (see Figure 3.13).

(a) Estimate the percentage of this population expected to fail by 300 years.
(b) Does the answer make sense if the time-to-failure data is for human mortality? Explain.

Solution:

(a) The results show that the probability of failure at 300 years is approximately 2%.
(b) The mortality data for over a billion people for a 10-year period from the time of birth fits a Weibull distribution very well. This looks impressive, but is nevertheless all wrong. It is clear that this data should not be used for making any

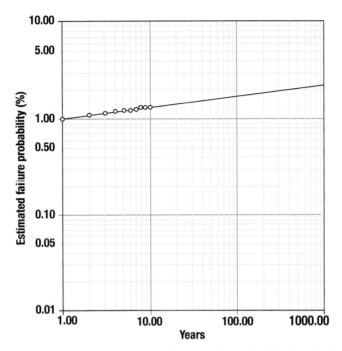

Figure 3.13 Weibull probability plot of time-to-failure data for Example 3.26.

judgment on human longevity, even though all the calculations are correct. The mortality pattern of humans in the first 10 years of life cannot be extrapolated, because the mortality pattern changes with age. This is also often true for engineered goods. Failures that occur in postmanufacturing tests are often caused by defects introduced in manufacturing. The first 10 years of time-to-failure data will result in a shape factor (β) of less than one. However, during early childhood through a large part of adulthood, the shape factor will be close to one, where most deaths can be considered random (e.g., caused by many causes such as accidents). Then the population will enter a wearout stage during which people die from old age. Complete human mortality data should be modeled using a mixed Weibull distribution.

3.4 Summary

The reliability function is used to describe the probability of successful system operation during a system's life. A natural question is then, "What is the shape of a reliability function for a particular system?" There are basically three ways in which this can be determined:

1. Test many systems to failure using a mission profile identical to use conditions. This would provide an empirical curve based on the histogram that can give some idea about the nature of the underlying life distribution.

2. Test many subsystems and components to failure under use conditions recreated in the test environment. This empirically provides the component reliability functions. Then derive analytically or numerically or through simulation the system reliability function. (Chapter 17 covers topics related to system reliability.)

3. Based on past experience with similar systems, hypothesize the underlying failure distribution. Fewer systems can be tested to determine the parameters needed to adapt the failure distribution to a particular situation. However, this will not account for new failure mechanisms or new use conditions.

In some cases, the failure physics involved in a particular situation may lead to the hypothesis of a particular distribution. For example, fatigue of certain metals tends to follow either a lognormal or Weibull distribution. Once a distribution is selected, the parameters for a particular application can be ascertained using statistical or graphical procedures.

In this chapter, various distributions were presented. However, the most appropriate distribution(s) for a particular failure mechanism or product that exhibits certain failure mechanisms must be determined by the actual data, and not guessed. The distribution(s) that best fit the data and that also make sense in terms of the failure processes should be used.

Problems

3.1 Prove that for a binomial distribution in which the number of trials is m and the probability of success in each trial is p, the mean and the variance are equal to mp and $mp(1-p)$, respectively.

3.2 Prove that for a Poisson distribution, the mean and the variance are equal to the Poisson parameter μ.

3.3 Compare the results of Examples 3.2 and 3.5. What is the reason for the differences?

3.4 Consider a system that has seven components; the system will work if any five of the seven components work. Each component has a reliability of 0.930 for a given period. Find the reliability of the system.

3.5 For an exponential distribution, show that the time to 50% failure is given by $0.693/\lambda_0$.

3.6 For an exponential distribution, show that the standard deviation is equal to $1/\lambda_0$.

3.7 Show that for a two-parameter Weibull distribution, for $t = \eta$, the reliability $R(t) = 0.368$, irrespective of β.

3.8 The front wheel roller bearing life for a car is modeled by a two-parameter Weibull distribution with the following two parameters: $\beta = 3.7$, $\theta = 145{,}000$ mi. What is the 100,000-mi reliability for a bearing?

3.9 The life distribution (life in years of continuous use) of hard disk drives for a computer system follows the Weibull distribution with the following parameters:

$$\beta = 2.7 \quad \text{and} \quad \theta = 5.5 \text{ years.}$$

(a) The manufacturer gives a warranty for 1 year. What is the probability that a disk drive will fail during the warranty period?
(b) Find the mean life and the median life (B_{50}) for the disk drive.
(c) By what time will 99% of the disk drives fail? (That is, find the B_{99} life.)

3.10 The life distribution for miles to failure for the engine of a Lexus car follows the Weibull distribution with

$$\beta = 3.8 \quad \text{and} \quad \theta = 185{,}000 \text{ mi.}$$

(a) Find the mean miles between failures, or the expected life for the engine.
(b) Find the standard deviation for miles to failure.
(c) What percent of these engines will fail by 100,000 mi?
(d) What is the failure rate of an engine that has a life of 100,000 mi?

If a certain model has 200,000 engines in the field with a life of 100,000 mi, how many engines on average will fail in the next 100 mi of use out of the 200,000 engines?

3.11 A component has the normal distribution for time to failure, with $\mu = 26{,}000$ hours and $\sigma = 3500$ hours.

(a) Find the probability that the component will fail between 22,000 hours and 23,000 hours.
(b) Find the failure rate of a component that has been working for 22,000 hours.

3.12 The time to failure random variable for a battery follows a normal distribution, with $\mu = 800$ hours and $\sigma = 65$ hours.

(a) Find the B_{10} life of these batteries.
(b) Find the probability that a battery will fail between 700 and 710 hours, given that it has not failed by 700 hours.
(c) What is the failure rate or hazard rate of a battery that has a life of
  (i) 700 hours
  (ii) 710 hours.

3.13 The time to failure for the hard disk drives for a computer system follows a normal distribution with

$$\mu = \text{mean life} = 14{,}000 \text{ hours}$$

$$\sigma = \text{standard deviation} = 1500 \text{ hours.}$$

(a) A manufacturer gives a warranty for 1 year of continuous use, or 365 × 24 hours of use. What percentage of hard disk drives will fail during this warranty period?

(b) What is the failure or hazard rate of a drive that has been working successfully for 1 year of continuous use?

(c) An IT manager of a large company, based on field surveys and inventory management, finds that the company has 250,000 of these drives on which the warranty has just expired—that is, they are working today after one year of continuous use. What is the expected number of these drives that will fail in the next 24 hours?

3.14 The time to repair a communication network system follows a lognormal distribution with $\mu = 3.50$ and $\sigma = 0.75$. The time is in minutes.

(a) What is the probability that the communication network will be repaired by 60 minutes?

(b) Find the B_{20} value (the 20th percentile) for the time-to-repair random variable.

(c) Find the mean time to repair (MTTR) for the communication network.

3.15 The time to repair a copy machine follows the lognormal distribution with $\mu = 2.70$ and $\sigma = 0.65$. Time is in minutes.

(a) Find the probability that the copy machine will be repaired in 30 minutes.

(b) Find the median value or B_{50} life for the time-to-repair random variable.

3.16 The time to failure for a copy machine follows a gamma distribution with parameters $\eta = 2$ and $\lambda = 0.004$.

(a) What is the expected or mean time between failures (MTBF) for the copy machine?

(b) What is the reliability of the copy machine for 200 hours of continuous operation?

(c) What is the failure rate of a copy machine that has been working for 200 hours?

3.17 Describe two examples of systems that require a failure-free operating period, without any maintenance. What are the timeframes involved?

3.18 Describe two examples of systems that require a failure-free operating period, but may allow a maintenance period. Discuss the timeframes.

3.19 Show that the mode of the three parameter Weibull distribution is for

$$t = \gamma + \eta(1 - 1/\beta)^{1/\beta}$$

for $\beta > 1$.

3.20 A company knows that approximately 3 out of every 1000 processors that it manufactures are defective. What is the probability that out of the next 20 processors selected (at random):

(a) All 20 are working processors?
(b) Exactly 2 defective processors?
(c) At most 2 defective processors?
(d) At least 18 are defective?

4 Design for Six Sigma

The concept of Six Sigma originated at Motorola in the 1980s by Dr. Mikel Harry, an engineer. He realized that systems can be improved by measuring and reducing variation. General Electric in the 1990s started implementing these concepts in all their divisions. Impressive quality improvements were experienced. Estimates are that cost savings due to the application of Six Sigma exceeded $300 million within the first 2 years and more than $1 billion by 1999.

Sigma (σ) is a Greek letter used to denote standard deviation, which is used to compare expected outcomes versus failures in a population. Six Sigma is the definition of outcomes as close as possible to perfection. With Six Sigma, the goal is to arrive at 3.4 defects per million opportunities, or 99.9997% perfection. As an example, with Six Sigma, an airline would lose only three to four pieces of luggage for every 1 million pieces that it handles. This is what the Six Sigma process strives to achieve. Over the last 25 years, Six Sigma has been successfully implemented in many industries, from large manufacturers to small businesses, from financial services and the insurance industry to healthcare systems (Barry et al. 2002; Harry and Schroeder 2000; Hoerl 1998; Pande et al. 2000; Pyzdek and Keller 2009).[1]

4.1 What Is Six Sigma?

In many organizations, Six Sigma is a business management process that provides tangible business results to the bottom line by continuous process improvement and variation reduction. As a data-driven, statistically based approach, Six Sigma aims to deliver near-zero defects for every product, process, and transaction within an organization.

The concept of Six Sigma was developed based on the assumption that the process characteristic follows a normal distribution. The objective of Six Sigma is to achieve a target of at most 3.4 defectives per million items, even if the process mean shifts by 1.5 times its standard deviation over a period of time. This is possible only if the

---

[1] http://www.ge.com/sixsigma/.

Reliability Engineering, First Edition. Kailash C. Kapur and Michael Pecht.
© 2014 John Wiley & Sons, Inc. Published 2014 by John Wiley & Sons, Inc.

4 Design for Six Sigma

Table 4.1 Defects per million for normal distribution

Quality level	Defects per million without any process shift	Defects per million with 1.5 Sigma shift
2 Sigma	45,500	308,771
3 Sigma	2,700	66,803
4 Sigma	63	6,200
5 Sigma	0.57	233
6 Sigma	0.002	3.4
7 Sigma	0.0000026	0.019

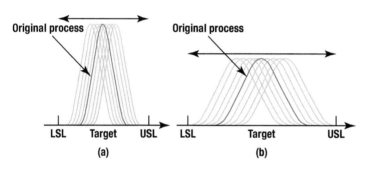

Figure 4.1 Effect of shifting process mean for (a) a Six Sigma process and (b) an ordinary process.

process variation is considerably reduced such that six times the standard deviation on either side of the process mean is within the specifications. Six Sigma actually aims at producing less than 0.002 defectives per million. However, considering the possible shifting of the process mean by 1.5 Sigma over time, it aims at less than 3.4 defectives per million. Table 4.1 and Figure 4.1 clarify these concepts. If the mean of a process which is not Six Sigma qualified shifts, it will produce a large number of defectives compared to a Six Sigma process.

A Six Sigma process can also be interpreted in terms of process capability. The typical definition for the process capability index, C_{pk}, is

$$C_{pk} = \min\left\{\frac{\text{USL} - \hat{\mu}}{3\hat{\sigma}}, \frac{\hat{\mu} - \text{LSL}}{3\hat{\sigma}}\right\}, \qquad (4.1)$$

where USL is the upper specification limit, LSL is the lower specification limit, $\hat{\mu}$, is the point estimator of the mean, and $\hat{\sigma}$ is the point estimator of the standard deviation for the underlying quality characteristic. If the process is centered at the middle of the specifications, which is also interpreted as the target value (for this model), then the Six Sigma process means that $C_{pk} = 2$. If the process shifts by 1.5 Sigma, the C_{pk} will be 1.5, leading to less than or equal to 3.4 defectives per million. Six Sigma is a continuous process and is a strategy to improve the present process capability (say $C_{pk} = 1$) to the Six Sigma capability ($C_{pk} = 2$).

4.2 Why Six Sigma?

Six Sigma is a methodology for structured, process-oriented, and systematic quality improvement. The primary reason for the success of Six Sigma is that it provides a

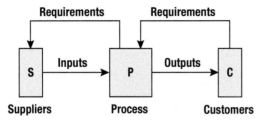

Figure 4.2 Process mapping.

systematic approach for quality and process improvement, rather than being just a collection of tools. The Six Sigma strategy is a good way to integrate such methods as design of experiments (DoE or DoX), statistical process control (SPC), failure mode, effects and criticality analysis (FMECA), fault tree analysis (FTA), and quality function deployment (QFD). Implemented project by project, Six Sigma provides an overall process that clearly shows how to link and sequence individual tools.

Many companies, such as Motorola,[2] GE,[3] and Honeywell,[4] began continuous process improvement with Six Sigma methodology. The method has a customer focus and is data-driven and analytically sound. Six Sigma is a rigorous, data-driven, decision-making approach to analyzing the root causes of problems and improving process capability.

4.3 How Is Six Sigma Implemented?

Improving processes is very important for businesses to stay competitive in today's marketplace, where customers are demanding better and better products and services.

Understanding the meaning of a process before trying to improve it is very important. A typical process is shown in Figure 4.2. There are many ways to visualize and model a process; one good model uses the Suppliers–Inputs–Process–Outputs–Customers (SIPOC) diagram. Other good models are Taguchi's P-diagram and the Ishikawa/fishbone diagram. In use, the model requires identification of the supplier(s), process inputs, process, associated outputs, and customer(s). The model also shows the feedback loop based on the requirements of the customer and the process.

A process consists of many input variables and one or more output variables. The input variables include both controllable factors and uncontrollable, or noise, factors. For instance, for an electric circuit designed to obtain a target output voltage, the designer can specify the nominal values of resistors and capacitors, but cannot control the influence of temperature or moisture, degradation over time, and measurement error. A typical process with one output variable is given in Figure 4.3, where X_1, $X_2, \ldots, X_n$ are the controllable variables and y is the realization of the random output variable Y.

---

[2]http://www.motorola.com/Business/US-EN/Motorola+University.
[3]http://www.ge.com/en/company/companyinfo/quality/whatis.htm.
[4]http://www51.honeywell.com/honeywell/our-culture-n3n4/continually-improving.html?c=11.

4 Design for Six Sigma

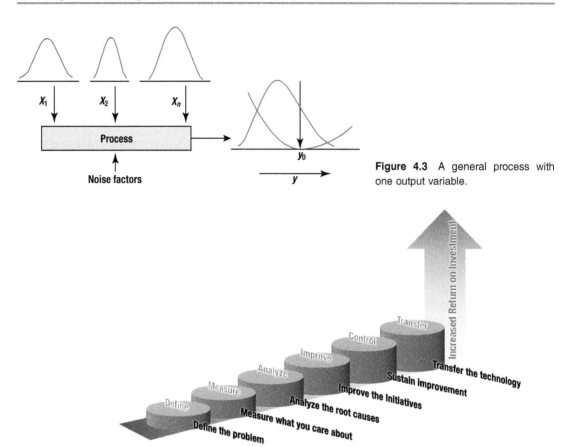

Figure 4.3 A general process with one output variable.

Figure 4.4 Six phases for the Six Sigma process.

4.3.1 Steps in the Six Sigma Process

At the strategic level, the goal of Six Sigma is to align an organization to its marketplace and deliver real improvements to the bottom line. At the operational level, Six Sigma strives to move product or process characteristics into the specifications required by customers, shrink process variation to a Six Sigma level, and reduce the causes of defects that negatively affect quality (Bertels 2003). A typical procedure for Six Sigma quality improvement has six well-known and highly utilized phases: define, measure, analyze, improve, control, and technology transfer (DMAICT), as shown in Figure 4.4. The process of DMAICT stays on track by establishing deliverables for each phase, by creating engineering models over time to reduce the process variation, and by continuously upgrading the predictability of system performance. Each of the six phases in the DMAICT process is critical to achieving success.

4.3.1.1 Step 1: Define—What Problem Needs to Be Solved? In this phase, there are three critical factors: the scope of the project, the customer, and issues that are critical to quality (CTQ) are identified and the core processes are defined. It is important to define the scope, expectations, resources, and timelines for the selected project.

Once an organization decides to launch a Six Sigma project, it needs to first define the improvement activities involved. Usually, the following two factors are considered in the define phase.

92

Identifying and Prioritizing Customer Requirements Methods such as benchmarking surveys, spider charts, and customer needs mapping are used to ensure that the customer's requirements are properly identified. The critical-to-quality (CTQ) characteristics (also called external CTQs) are specified. The external CTQs need to be translated into internal CTQs that are key process requirements. This translation is the key step in the measure phase.

Selecting the Project Six Sigma process improvement is a reactive approach that is initiated when a process does not deliver a satisfactory result (according to the customer affected by the process). Based on customer requirements, a target project is selected by analyzing the gap between the current process performance and the requirements of customers.

For a selected project, a charter must be developed that specifies project scope, expectations, resources, milestones, and core processes. The charter identifies and documents necessary information before the measurement (M) step is applied. Charter development companies follow or use the following steps.

STEP 1.1: DRAFT A PROJECT CHARTER Drafting a project charter is the first step in the Six Sigma methodology. The project charter should include the following: business case; goals and objectives of the project; milestones; project scope, constraints, and assumptions; team membership; roles and responsibilities; and a preliminary project plan.

STEP 1.2: IDENTIFY AND DOCUMENT THE PROCESS The Six Sigma approach focuses on one process at a time, either a core process (product development and customer service) or a support process (human resources and information system). One process is chosen for the project. After the process for improvement is identified, a process model (P-diagram (Phadke 1989) or SIPOC) is selected and used to model and analyze the process. Once the project is understood and the baseline performance is documented, it is time to do an analysis of the process. In this phase, Six Sigma applies statistical tools to validate the root causes of problems. The objective is to understand the process in sufficient detail so that options for improvement can be formulated.

STEP 1.3: IDENTIFY, ANALYZE, AND PRIORITIZE CUSTOMER REQUIREMENTS In Six Sigma, the customer is defined not so much as the traditional buyer, but as the environment, producer, seller, or buyer that is affected by the process. Because quality is measured from the customer's perspective, there has to be a link between the output that a process delivers and the quality that the customer expects. There are two types of customer requirements: product output requirements, which must be translated from voice of the customer (VOC) into design parameters; and service-level requirements, which involve establishing the service needs of the customer, often with some level of abstraction and subjectivity.

Elements of this step include selecting critical-to-quality (CTQ) characteristics using tools such as quality function development (QFD) (Akao 1989) and failure modes, effects, and criticality analysis (FMECA) to translate the external CTQs into internal requirements denoted by Ys. Some of the objectives for this step include defining performance standards by: defining, constructing, and interpreting the QFDs; participating in a customer needs mapping exercise; applying (FMECA) to the process of selecting CTQ characteristics; identifying CTQs and internal Ys; and analyzing and determining the priority of the customer requirements. Since the customer's expectations often include multiple requirements, these need to be ranked by importance.

STEP 1.4: DEVELOP APPROPRIATE MEASUREMENT SYSTEMS Once the SIPOC elements and the customer's functional requirements are identified, measurement tools are used to evaluate the current performance. (Note that Six Sigma strategy focuses only on an existing process that needs rework or improvement.) After the product requirements, Ys, and measurement standards for the Ys are defined, QFD, FMECA, and process mapping can be used to establish internal measurement standards.

4.3.1.2 Step 2: Measure—What Is the Current Capability of the Process?
Design for Six Sigma is a data-driven approach that requires quantifying and benchmarking the process using actual data. In this phase, the performance or process capability for the identified CTQ characteristics is evaluated.

Measurement is a very important element in the Six Sigma strategy. This step involves data collection and data processing before proceeding to the analysis step. Notice that if this step is wrongly executed, a statistical error could result in a measurement error, leading to an incorrect analysis and wrongly executed procedures. The first step in the measurement stage is to select which of the process elements needs to be measured. Generally, the relevant measures include both input and output measures.

Input measures may involve data stratification. One input variable may be the output of another input. Cause-and-effect relationships can lead to the lowest independent input variables that may influence the output values.

Output measures include CTQ data, such as the lower and upper specification limits and defect counts. It is necessary to develop a data collection strategy that defines sampling frequency, the method of measurement, the format of data collection forms, and the measurement instruments. The team also must consider the possibility of Type II statistical error (and also measurement error) and use a well-planned strategy to tackle it.

In addition, the team should consider the type of data (discrete vs. continuous) and the sampling method. Thus, the steps for the measurement phase may be summarized as follows:

Step 2.1: Establish Product Capability The current product capability, associated confidence levels, and sample size are established by statistical analysis. The typical definition for the process capability index, C_{pk}, is

$$C_{pk} = \min\left\{\frac{\text{USL} - \hat{\mu}}{3\hat{\sigma}}, \frac{\hat{\mu} - \text{LSL}}{3\hat{\sigma}}\right\}, \tag{4.2}$$

where USL is the upper specification limit, LSL is the lower specification limit, $\hat{\mu}$ is the point estimator of the mean, and $\hat{\sigma}$ is the point estimator of the standard deviation. If the process is centered at the middle of the specifications, which is also interpreted as the target value—that is,

$$\hat{\mu} = \frac{\text{USL} + \text{LSL}}{2} = y_0, \tag{4.3}$$

then the Six Sigma process means that $C_{pk} = 2$. If the process mean shifts by 1.50σ, which is typically assumed in the literature for Six Sigma methodology, then $C_{pk} = 1.50$. It is this 1.50σ shift that results in 3.4 defects per million opportunities (DPMO).

Step 2.2: Define Performance Objectives The performance objectives are defined to establish a balance between improving process capability (and thus, customer satisfaction) and the available technology capability.

Step 2.3: Identify Sources of Variation This step begins to identify the causal variables that affect the product requirements or the responses of the process. Some of these causal variables might be used to control the response Ys.

4.3.1.3 Step 3: Analyze—What Are the Root Causes of Process Variability? Once the project is understood and the baseline performance is documented, it is time to do an analysis of the process. In this phase, the Six Sigma approach applies statistical tools to validate the root causes of problems. The objective is to understand the process at a level sufficient to be able to formulate options for improvement. We should be able to compare the various options with each other to determine the most promising alternatives. In general, during the process of analysis, the collected data are analyzed and process maps are used to determine the root causes of defects and prioritize opportunities for improvement.

The collected data can be used to find patterns, trends, outliers, and other differences that could support or reject theories (hypothesis testing) about cause and effect. The methods frequently used include design of experiments (Hicks and Turner 1999; Montgomery 2001), the Shanin method,[5] root cause analysis, cause–effect diagrams, failure modes and effects analysis (FMEA), Pareto charts, and validation of root cause.

Process analysis uses tools such as value stream mapping, process management, and the process mapping technique to analyze nonvalue-adding steps that result in nonconformity. Thus, the steps for the analysis phase may be summarized in the following sections.

Step 3.1: Discover Variable Relationships In the previous stage, causal variables, Xs, are identified with a possible prioritization as to their importance in controlling Ys. In this step, the impact of each vital X on the response Ys is explored. A system transfer function (STF) is developed as an empirical model relating Ys and the vital Xs.

Step 3.2: Establish Operating Tolerances After understanding the functional relationship between the vital Xs and the response Ys, we need to establish the operating tolerances of Xs that optimize the performance of Ys. Mathematically, we develop a variance transmission equation (VTE) that transfers the variances of the vital Xs to variances of Ys.

Step 3.3: Optimize Variable Settings The STF and VTE are used to determine the key operating parameters and tolerances to achieve the desired performance of the Ys. Optimization models are developed to determine the optimum values for both the means and variances for these vital Xs.

---

[5]http://www.shainin.com/.

4.3.1.4 Step 4: Improve—How Can the Process Capability Be Improved?
During the improvement phase, ideas and solutions are established to initialize the needed changes. Based on the root causes discovered and validated for the existing opportunity, the target process is improved by designing creative solutions to fix and prevent problems. Some experiments and trials may be necessary in order to find the best solution. If a mathematical model is developed, then optimization methods are utilized to determine the best solution.

After completing the analysis step, the team should be able to identify the root causes of nonconformity. If the root cause is identified by data analysis tools, finding the solutions to fix the process could either be easy or hard, because analysis tools point directly to the nonconformity culprit. Sometimes, the solutions applied can fix the problem indicated by the analysis tools, but may also result in another problem caused by other variables. This is due to the interdependence of the variables. The team can use brainstorming or the theory of inventive problem-solving called TRIZ to tackle the problem. If the root cause is identified by the process analysis, the Six Sigma team can use process management techniques, such as process simplification, parallel processing, and bottleneck elimination.

4.3.1.5 Step 5: Control—What Controls Can Be Put in Place to Sustain the Improvement?
The key to the overall success of Six Sigma methodology is its sustainability, which seeks to make the process incrementally better on a continuous basis. The sum of all these incremental improvements can be substantial. Without continuous sustenance, over time, the process will worsen until finally it is time for another effort toward improvement. As part of the Six Sigma approach, performance tracking mechanisms and measurements are put in place to assure that the gains made in the project are not lost over time and that the process remains on the new course. The steps for the control phase may be summarized in the following sections.

Step 5.1: Validate the Measurement System The measurement system tools first applied in Step 1.4 are now used for the Xs.

Step 5.2: Implement Process Controls Statistical process control is a critical element in maintaining a Six Sigma level. Control charting is the major tool used to control the few vital Xs. Special causes of process variations are identified through the use of control charts, and corrective actions are implemented to reduce variations.

Step 5.3: Document the Improvement The project is not complete until the changes are documented in the appropriate quality management system, such as QS9000/ISO9000. A translation package and plan should be developed for possible technology transfer.

Once improvements have been made, proper documentation and standards should be established to monitor the process. If the process is improved by process management, new process standards should be established. If the process is improved by eliminating the root causes of bad performance, the new performance should be measured consistently by controlling the critical variable related to the chart.

4.3.1.6 Step 6: Technology Transfer
Ideas and knowledge developed in one part of the organization can be transferred to other parts of the organization. In addition, the methods and solutions developed for one product or process can be applied to

other similar products or processes. With technology transfer, the Six Sigma approach can create exponentially increasing returns.

4.3.2 Summary of the Six Sigma Steps

The DMAICT process stays on track by reducing process variations and establishing deliverables at each phase. In each phase, several quality improvement methods, tools, and techniques can be used. Each organization has different ways of summarizing these steps, based on the nature of its products, processes, and customers. Table 4.2 is one such summary. Another is given in Table 4.3 for comparison.

4.3.2.1 Future Trends of Six Sigma Although Six Sigma originated in the manufacturing industry, it has been successfully adopted throughout the public and private sectors in applications from financial services and health care to information technology and knowledge management. Its successful implementation over 20 years supports the hypothesis that the basic theory and methods of Six Sigma have lasting value, regardless of what names they are marketed under. These ideas can be integrated into other productivity improvement methods—for example, the recent

Table 4.2 Summary of key steps in Six Sigma process improvement and Six Sigma tools at each step

Key steps in Six Sigma process improvement	Six Sigma steps	Six Sigma tools
Define (D)	■ Draft project charter ■ Identify and document process ■ Identify VOC	■ SIPOC modeling ■ P-diagram ■ Ishikawa diagram ■ Kano analysis ■ Quality function deployment (QFD)
Measure (M)	■ Select measurement variables ■ Develop data collection plan ■ Calculate process Sigma level	■ Statistical process capability ■ DOE
Analyze (A)	■ Data analysis ■ Process analysis	■ Root cause analysis ■ Cause–effect diagram ■ FMEA ■ Pareto chart ■ DOE ■ Shainin method
Improve (I)	■ Statistical improvement: eliminate the root cause of inconsistency ■ Process improvement: increase the value-adding processes and decrease nonvalue-adding processes	■ Brainstorming ■ TRIZ
Control (C)	■ Statistical control: develop a data collection strategy to ensure consistency of performance ■ Process control: establish new standards	■ SPC ■ EPC ■ Documentation

Table 4.3 DMAICT framework

	Phase	Tools
D	Define the scope and objective of the project, the critical-to-quality (CTQ) issues, and the potential opportunities.	■ Project charter ■ Benchmarking surveys ■ Spider charts ■ Flowchart
M	Measure the process performance, especially the CTQ issues, to analyze the operations of the current system.	■ Quality function deployment (QFD) ■ Failure modes, effects, and criticality analysis (FMECA) ■ Gage R&R
A	Analyze data collected and use process maps to determine root causes of defects and prioritize opportunities for improvement. Apply statistical tools to guide the analysis.	■ Histogram/Pareto chart/run chart ■ Scatter plot/cause and effect diagram ■ Product capability analysis
I	Improve the process by designing creative solutions to fix and prevent problems. Some experiments may be performed in order to find the best solution. Optimization methods are utilized to determine the optimum solution.	■ Quality function deployment (QFD) ■ FMECA ■ Statistical experimental design and analysis ■ Simulation
C	Control the process on the new course. Performance tracking mechanisms and measurements are put in place to ensure that the gains are not lost over time. The key to overall success is sustainability.	■ Gage R&R ■ Statistical process control/control charts ■ QS9000/ISO9000
T	Transfer ideas and knowledge developed in one project to other sections of the organization. Transfer the methods and solutions developed for one product or process to other similar products or processes.	■ Project management ■ Collaborative team effort and cross-functional teams

emergence of Lean Six Sigma, based on Toyota's production system (Liker 2003; Ohno 1988; Shingo 1989).

4.4 Optimization Problems in the Six Sigma Process

Many optimization problems occur in the six phases of this methodology. In this section, optimization models to improve the quality of the system to the Six Sigma level are reviewed. Various methods and tools of probabilistic design, robust design, design of experiments, multivariable optimization, and simulation techniques can be used for this purpose. The methodology can be improved and extended for the analysis and improvement phases of the Six Sigma process (Kapur and Feng 2005). In the analysis phase, the system transfer function and variance transmission equations need to be developed to enable formulating options for improvement by understanding the system. Based on the system transfer function or variance transmission equation,

4.4 Optimization Problems in the Six Sigma Process

optimization models are formulated and solved to obtain the best decisions. These topics are briefly discussed in the following section.

4.4.1 System Transfer Function

A typical system consists of many input variables and one or more output variables. The input variables include both controllable factors and uncontrollable, or noise, factors. For a system with one output variable, as given in Figure 4.3, X_1, $X_2, \ldots, X_n$ are the controllable variables, and y is the realization of random output variable Y.

As we discussed, in the measurement phase of the DMAICT process, the critical-to-quality (CTQ) characteristics are developed. In order to understand the system, we need to analyze the functional relationship between the output variable and the input variables, which can be described as a system transfer function (STF):

$$y = g(x_1, x_2, \ldots, x_n) + \varepsilon, \qquad (4.4)$$

where ε is the system error caused by the noise factors. Let y, x_1, x_2, ..., x_n be the realization of random variables Y, X_1, X_2, ..., X_n, respectively.

The CTQ characteristics in the system are linked together through the system transfer functions. The CTQ flow-down tree (Kapur and Feng 2006) in Figure 4.5 illustrates how the system transfer functions establish the relationships among the CTQs at different levels.

The process can be improved during the design phase by reducing the bias or variance of the system output—that is, by changing the mean and variance of the quality characteristics of the output. Statistical methods for process optimization, such as experimental design, response surface methods (Myers and Montgomery 2002), and Chebyshev's orthogonal polynomials, can be used.

Integrated optimization models are developed to minimize the total cost to both producers and customers by determining the distribution of the controllable factors. For many complex systems, the analytical forms of the STF are explicitly known; even

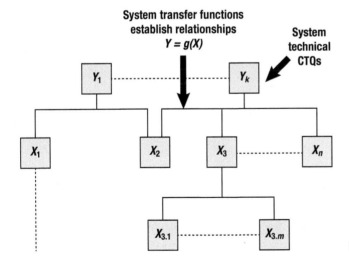

Figure 4.5 The CTQ flow-down tree diagram.

so, it is usually very complicated to work with them. Given a set of values for the input variables of the system, the corresponding values of the response variables can be obtained through computer simulations or actual experiments. Based on the simulated or experimental data, an empirical model of the system transfer function can be developed using the regression method. The mean and variance models can be obtained by applying conditional expectation and variance operators to the regression model. Myers and Montgomery discuss this approach to obtain the mean and variance response models.

4.4.2 Variance Transmission Equation

Six Sigma methodology strives to improve quality by reducing the variation of a process. Given a particular requirement of the system, one of the problems in the Six Sigma process is to determine the optimal variances of the input variables. Instead of finding the system transfer function, what must be found is the relationship of the variances between the input and output variables. Letting σ_Y^2 denote the variance of the output variable Y and $\sigma_1^2, \sigma_2^2, \ldots, \sigma_n^2$ denote the variances of the input variables X_1, $X_2, \ldots, X_n$, Six Sigma methodology strives to improve quality by reducing the variation of a process. Given a particular requirement of the system, one of the problems in the Six Sigma process is to determine the optimal variances of the input variables. Instead of finding the system transfer function, what must be found is the relationship of the variances between the input and output variables. Letting denote the variance of the output variable Y and denote the variances of the input variables X_1, X_2, $\ldots, X_n$, the functional relationship of the variances can be expressed as a variance transmission equation (VTE) as given below:

$$\sigma_Y^2 = h(\sigma_1^2, \ldots, \sigma_n^2) + \varepsilon, \quad (4.5)$$

where ε is the error. The VTE transfers the variances of the input variables to the variance of the response variable.

Different approaches can be used to develop the variance transmission equation based on the information we have. If the STF is known and differentiable, the VTE can be approximated using Taylor's expansion:

$$\sigma_Y^2 \approx \sum_{i=1}^{n} \left[\frac{\partial g(\mu_1, \mu_2, \ldots, \mu_n)}{\partial x_i}\right]^2 \sigma_i^2, \quad (4.6)$$

where μ_i is the expected value of x_i, $i = 1, 2, \ldots, n$, and we assume that $X_1, X_2, \ldots, X_n$ are independent variables. Kapur and Lamberson (1977) give an example commonly used in reliability design to analyze the error in this approximation method.

If the STF is not known in an analytical form, the VTE can be developed using statistical tools such as linear regression, design of experiments, and response surface methodology. Computer simulations or actual experiments are used to obtain data for the analysis of variance. The emphasis on fractional factorial design may limit the number of real experiments.

4.4.2.1 Taguchi's VTE for Fixed Effect Model Taguchi (1987) constructed a variance transmission equation with the assumption of no interaction between the components. The equation is intuitively appealing but has no solid theoretical basis, and the interactions between the components are overlooked. The VTE developed by Taguchi's methods has the advantage that it can be developed by experimentation or simulation even if the analytical form for STF is not known. Using Taguchi's "three-level factorial experiments," the total evaluations of the function are significantly fewer than that required by a Monte Carlo simulation.

In practice, it may be necessary to choose the levels of the design factors at random. For the random effects models, the VTE should be based on the expected mean square (EMS) values. An analysis of variance for a random effect model is used to develop the variance of treatment effects.

4.4.3 Economic Optimization and Quality Improvement

The ultimate objective of the Six Sigma strategy is to minimize the total cost to both the producer and the consumer, or the cost of the whole system. The cost to the consumer is related to the expected quality loss of the output variable, and it is caused by the deviation from the target value. The cost to the producer is associated with the changing probability distributions of input variables. If the system transfer function and the variance transmission equation are available, and if the cost functions for different grades of input factors are given, then the general optimization models can be developed and are briefly discussed.

We usually consider the first two moments of the probability distributions of input variables, and then the optimization models focus on the mean and variance values. Therefore, the expected quality loss to the consumer consists of two parts: the bias of the process and the variance of the process. The strategy to reduce bias is to find adjustment factors that do not affect variance and thus are used to bring the mean closer to the target value. Design of experiments can be used to find these adjustment factors, although it will incur some cost to the producer. In order to reduce the variance of Y, the designer should reduce the variances of the input variables, which will also increase cost. The problem is to balance the reduced expected quality loss with the increased cost for the reduction of the bias and variances of the input variables. Typically, the variance control cost for the ith input variable, X_i, is denoted by $\sum_{i=1}^{n} C_i(\sigma_i^2)$, and the mean control cost for the ith input variable, X_i, is denoted by $D_i(\mu_i)$. Focusing on the first two moments of the probability distributions of $X_1, X_2, \ldots, X_n$, the general optimization model is formulated as follows:

Minimize

$$TC = \sum_{i=1}^{n} C_i(\sigma_i^2) + \sum_{i=1}^{n} D_i(\mu_i) + k\left[\sigma_Y^2 + (\mu_Y - y_0)^2\right], \quad (4.7)$$

subject to

$$\mu_Y \approx m(\mu_1, \mu_2, \ldots, \mu_n) \quad (4.8)$$

$$\sigma_Y^2 \approx h(\sigma_1^2, \sigma_2^2, \ldots, \sigma_n^2). \quad (4.9)$$

In this objective function, the first two terms,

$$\sum_{i=1}^{n} C_i(\sigma_i^2) \text{ and } \sum_{i=1}^{n} D_i(\mu_i),$$

are the control costs on the variances and means of input variables, or the cost to the producer; the term $k\left[\sigma_Y^2 + (\mu_Y - y_0)^2\right]$ is the expected quality loss to the customer, where k is a constant in the quality loss function and y_0 is the target value of y. The first constraint, $\mu_Y \approx m(\mu_1, \mu_2, \ldots, \mu_n)$, is the mean model of the system, which can be obtained through the system transfer function. The second constraint, $\sigma_Y^2 \approx h(\sigma_1^2, \sigma_2^2, \ldots, \sigma_n^2)$, is the variance transmission equation.

4.4.4 Tolerance Design Problem

Assuming that the bias reduction has been accomplished, the general optimization problem given by Equation 4.7 can be simplified as a tolerance design problem, which is given below:
Minimize

$$TC = \sum_{i=1}^{n} C_i(\sigma_i^2) + k\sigma_Y^2, \tag{4.10}$$

subject to

$$\sigma_Y^2 \approx h(\sigma_1^2, \sigma_2^2, \ldots, \sigma_n^2). \tag{4.11}$$

The objective of the tolerance design is to determine the tolerances (which are related to variances) of the input variables to minimize the total cost, which consists of the expected quality loss due to variation, $k\sigma_Y^2$, and the control cost on the tolerances of the input variables, $\sum_{i=1}^{n} C_i(\sigma_i^2)$. Typically, $C_i(\sigma_i^2)$ is a nonincreasing function of each σ_i^2.

4.4.4.1 The Dual Problem of Tolerance Design
In addition, given the constraint on the cost of the control of tolerances, the dual problem of the tolerance design problem can be developed to minimize the variance of response as given below:
Minimize

$$\sigma_Y^2 \approx h(\sigma_1^2, \sigma_2^2, \ldots, \sigma_n^2), \tag{4.12}$$

subject to

$$\sum_{i=1}^{n} C_i(\sigma_i^2) \leq C^*, \tag{4.13}$$

where C^* is the maximum allowable cost to the producer, or the control cost on the tolerances of input variables.

4.5 Design for Six Sigma

While the Six Sigma process improvement approach leaves the fundamental structure of a process unchanged, Design for Six Sigma (DFSS) involves changing or redesigning the process at the early stages of the product and/or process life cycle. DFSS becomes necessary when the current process has to be replaced, rather than repaired or just improved; the required quality level cannot be achieved by just improving an existing process; when an opportunity is identified to offer a new process; and/or when there is a breakthrough and new disruptive technologies are available.

DFSS (Yang and El-Haik 2003) is the other strategy used to achieve Six Sigma process capability. However, the main difference between the Six Sigma process improvement strategy and DFSS is the approach taken to reach Six Sigma process capability. Six Sigma process improvement focuses on the improvement of the process after it has been developed and is in operation. Therefore, data are already available for measurement and analysis. DFSS focuses on the design steps that ensure the problem will not happen in the first place. DFSS is usually applied before the production routine operation is started or the product is in the field. Because of this, Six Sigma process improvement is easier to analyze in terms of cost–benefit analysis, because it works from the existing operation, from which data can be collected before and after the improvement strategy implementation and compared in terms of Six Sigma process levels. Consequently, DFSS is considered more of philosophical tool than a practical tool, because it can only be used at a preoperations level and its focus is on the research, development, and design phases. Table 4.4 shows an example of a product/process life cycle and Six Sigma tasks and tools.

Table 4.4 Product/process life cycle and Six Sigma tasks and tools

Product life cycle stages	Six Sigma tasks	Six Sigma strategy	Six Sigma tools
1. Impetus/ideation	■ Identify project scope, customers, suppliers, customer needs	DFSS	■ Customer research, process analysis, Kano analysis, QFD
2. Concept development	■ Develop new process concept to come up with right functional requirements ■ Ensure that new concept can lead to sound system design, free of design vulnerabilities ■ Ensure the new concept is robust for downstream development	DFSS	■ QFD ■ Taguchi methods/robust design ■ TRIZ ■ Axiomatic design ■ DOE ■ Simulation/optimization ■ Reliability-based design

(Continued)

Table 4.4 (Continued)

Product life cycle stages	Six Sigma tasks	Six Sigma strategy	Six Sigma tools
3. Process design/tryout	■ Ensure process can deliver desired functions ■ Ensure process will perform consistently and robustly ■ Validate process for performance and consistency	DFSS	■ Taguchi methods/robust design ■ DOE ■ Simulation/optimization ■ Reliability-based design/testing and estimation ■ Statistical validation
4. Process and routine operations	■ Ensure process will perform consistently	Six Sigma process improvement	■ SPC ■ Troubleshooting and diagnosis ■ Error-proofing
5. Process Improvement	■ Improve to satisfy new requirements	Six Sigma process improvement	■ DMAICT strategy ■ Customer analysis, Kano analysis ■ QFD ■ Statistical measurement system ■ DOE, Shanin methods, multivariate analysis, regression analysis ■ Process analysis, value stream mapping ■ SPC

The major objective of the Six Sigma improvement process is "to do it right and do it right all the time." The major objective of DFSS is "to design it right the first time" to avoid complications during the product life cycle. Most managers who are unable to improve Six Sigma process performances retreat to the design phase to reach Six Sigma process capability. Generally, a bad design results in a bad performance. The sources for bad design are either conceptual vulnerabilities that exist due to violations of design axioms (Yang and andEl-Haik 2003)[6] and principles, or operational vulnerabilities that exist due to lack of robustness in the usage environment.

DFSS aims to tackle both operational vulnerabilities and conceptual vulnerabilities. Conceptual vulnerabilities are generally anticipated by using quality engineering (Taguchi 1986; Taguchi 1987), TRIZ (Altshuller 1984), axiomatic design (Suh 1990, 2001), and theory of probability and statistical modeling. Operational vulnerabilities are generally anticipated by using robust design, DMAIC Six Sigma process improvement, and tolerance design/tolerance analysis. Conceptual vulnerabilities are usually overlooked or underestimated because of the lack of a compatible systematic approach to finding an ideal solution; overcoming the errors of the designer; the pressure of schedule deadlines; and budget limitations.

Traditional quality methods focus on improvement, since the process is already ongoing, and it is therefore easy to measure the costs and benefits. This triggers an endless cycle of design-test-fix-retest, because the quality improvement process is based on the necessity to tackle operational vulnerabilities that have been overlooked at the conceptual phase. Corrective action to improve conceptual vulnerabilities by

---

[6]An axiom is a truth that cannot be derived but for which no counterexamples or exceptions exist.

repairing operational vulnerabilities is not an efficient approach. That is why many of current problem-solving techniques are hard to implement and very costly.

Traditional design is based on both empirical data (experience) and subjectivity (creativity). Nonscientific design produces less-than-optimal solutions for achieving Six Sigma process capability. Axiomatic design is introduced to provide a scientifically based design principle.

Human nature in the business world is typically more reactive than proactive and focuses attention on the later phases of the design cycle. Although DFSS takes more effort at the beginning, it will benefit an organization in the long run by designing Six Sigma quality into products and processes. There are several methodologies for DFSS, such as define, measure, analyze, design, and verify (DMADV) and identify, design, characterize the design, optimize, and validate (ICOV [or IDOV]). DMADV is a popular methodology, and basically follows the DMAICT model (omitting the transfer step). ICOV is a well-known design methodology, especially in the manufacturing world. Thus, DFSS integrates many of the well-known methods, tools, and philosophies for quality and reliability improvement; research, development, and design strategies; and management strategies to build teamwork and collaboration from cradle to grave for products and processes in any organization.

The suggested ICOV DFSS strategy has four phases:

1. Identify (I) the requirements.
2. Characterize (C) the design.
3. Optimize (O) the design.
4. Verify (V) the design.

4.5.1 Identify (I)

The design project can be categorized as design and redesign.

Step 1: Draft Project Charter. This is the same as the DMAIC strategy. However, the draft project charter in the DFSS project is longer because, as argued earlier, the design phase takes longer than the improvement phase. The latter is like patching a hole, while design involves creating something from nothing.

Step 2: Identify Customer Requirements. Since all processes are defined in terms of customer satisfaction, the customer requirements need to be identified before they can be translated and mapped into engineered functional requirements.

QFD and Kano analysis are examples of early-stage tools that can be used to help identify critical-to-quality customer requirements. An algorithmic approach is used to ensure that all the elements of the customer requirements are identified and included. The approach involves the following steps:

1. Identify methods of ascertaining customer needs and wants.
2. Ascertain customer needs and wants.
3. Translate the voice of the customer (VOC) into functional and measurable requirements.
4. Finalize requirements: establish minimum requirements definition; identify and fill gaps in customer-provided requirements; validate application and usage environments.

5. Identify points that are critical to quality (CTQ), critical to delivery (CTD), and critical to cost (CTC). (CTQ, CTD, and CTC can be referred to as CTXs.)
6. Quantify CTXs: establish metrics for CTXs; establish performance levels and operating windows; and perform flow-down of CTXs.

4.5.2 Characterize (C)

The customer's requirements may be too abstract to be meaningful to the product/process engineer. Therefore, the CTQ, CTD, and CTC elements must be translated into product/process functional requirements—that is, those things necessary so that the product can function at the level of customer satisfaction. After these functional requirements (FRs) have been identified, the design parameters and process variables can be determined. Thus, it is a very important step to tackle conceptual vulnerabilities. Some tools used in this phase include: TRIZ, QFD, axiomatic design, robust design, Design for X (X = manufacture and assembly, reliability, maintainability, serviceability, environmentality, life-cycle cost), DFMEA and PFMEA, design review, CAD, simulation, and process management. The characterization phase comprises a few strategic and algorithmic steps to aid designers in good design:

Step 1: Translate CTS into Process-Functional Requirements.

Step 2: *Generate Design Alternatives.* Sometimes, the existing technology is unable to deliver the CTS. It is therefore very important to design alternatives to deliver the CTS. The new design can be creatively begun from scratch or incrementally evolved from the baseline design.

Step 3: *Evaluate Design Alternatives to Select the Best Process-Functional Requirements.*

4.5.3 Optimize (O)

There are many combinations of parametric designs that engineers can use to satisfy functional requirement goals. Optimization aims to identify the best way to tailor the functional requirements and minimize operational vulnerabilities. The objective is to provide a logical and objective basis for setting manufacturing tolerances. Once the optimal parameter design is established, engineers can determine the level of system robustness best suited to the environment. Some tools used in this phase include: design/simulation tools, DOE (design of experiments), the Taguchi method, parameter design, and tolerance design, reliability-based design, and robustness assessment.

4.5.4 Verify (V)

Once the parameters of the design are optimized, a validation or design inspection is performed before the design is launched into mass production and process implementation.

Step 1: *Conduct Pilot Testing and Refining.* Pilot testing can be used to test and evaluate real-life performance.

Step 2: Validate the Results of the Pilot Testing. Confirm that process variables accounted for in the parameter design are mapped to functional requirements and can produce the identified customer attributes (quality functional validation). Confirm that the final process can produce Six Sigma process capability (statistical validation).

Step 3: Roll Out the Product Commercially and Hand It Over to the New Process Owner. The following tools are used in this phase: process capability modeling, DOE (design of experiment), reliability testing, poka-yoke, error-proofing, confidence analysis, process control plan, and training.

A summary of the key steps for DFSS and the applicable Six Sigma tools is given in Table 4.5. Table 4.6 shows the key differences between Design for Six Sigma (DFSS) and the traditional Six Sigma process improvement strategy based on DMAICT.

Six Sigma and other continuous improvement strategies are extremely valuable tools in today's global competition. The ideas presented in this chapter are important in terms of both the research and its applications for the analysis and improvement phases of the DMAICT process. Six Sigma will contribute to the design of many products and processes and also improve the quality and productivity in any organization.

Table 4.5 Summary of key steps in DFSS and the Six Sigma tools available at each step

Key DFSS steps		Six Sigma tools
(I) Identify	■ Draft project charter ■ Identify and document process ■ Identify VOC	■ SIPOC modeling ■ P-diagram ■ Ishikawa diagram ■ Kano Analysis ■ QFD
(C) Characterize	■ Translate CTS to process functional requirements ■ Define design alternatives ■ Map FR to DP (design parameter)	■ TRIZ ■ QFD ■ Axiomatic design ■ Robust design ■ DFMEA ■ PFMEA ■ Design for "X" ■ CAD ■ Simulation
(O) Optimize	■ Parametric design optimization to determine optimal process variable (PV)	■ DOE ■ Taguchi method ■ Tolerance design ■ Reliability-based design ■ Robustness assessment
(V) Verify	■ Validation of experimental data to customer satisfaction attributes	■ Process capability modeling ■ DOE ■ Reliability testing ■ Poka-yoke ■ Confidence analysis ■ Process control plan ■ Training

Table 4.6 The key differences between Six Sigma process improvement and design for Six Sigma

Key aspects	Six Sigma process improvement	Design for Six Sigma
Strategy and approaches	■ DMAIC: define, measure, analyze, improve, control	■ DMADV: define, measure, analyze, design, verify ■ DMADOV: define, measure, analyze, design, optimize, verify
Operating mode	■ Reactive	■ Proactive
Focus	■ Fixing problems in existing process	■ Up-front design of the process to prevent problems from happening
Benefits	■ Easier to quantify in dollars	■ Hard to quantify but tend to be greater long-term

4.6 Summary

Successful industrial, manufacturing, and service organizations are interested in reducing variance in their products and processes. Customers as well judge the quality of a process or product based on the variance in quality that they encounter in their transactions with processes or repeated uses of a product. Six Sigma is a business management process that companies implement to achieve a reduction in variance and continuously improve their products and processes. The ideal goal of Six Sigma is to deliver near-zero defects for every product, process, and transaction within an organization.

There are six steps in the Six Sigma process; these steps can be remembered by the acronym DMAICT. The first step is to define the problem that needs to be solved. The second step is to measure the current capability of the process. The third step is to analyze the root causes of process variability. The fourth step is to improve the process capability. The fifth step is to determine which controls can be put in place to sustain the improvement. The sixth step is technology transfer. The Six Sigma process can be optimized in various ways to fit the specific needs of an organization. In addition to process optimization, which more or less leaves a company's current processes intact, there is also design for Six Sigma (DFSS), which involves changing or redesigning a process at the early stages of the product or process life cycle. Each company must weigh the various costs and benefits of implementing either Six Sigma process optimization or Design for Six Sigma.

Problems

4.1 What is a Six Sigma process? Why was this process developed and who developed it?

4.2 Explain briefly why it is necessary and good strategy to consider variability around the target value for the underlying quality characteristic from the viewpoint of the customer.

4.3 Consider a Three Sigma process where the mean is at the target value. As discussed in this chapter, this gives a probability of meeting the specifications as 0.9973 which corresponds to 2,700 defective parts per million.

(a) If a product consists of an assembly of 100 independent parts and all of these parts must be nondefective for the product to function successfully, what is the probability that any specific unit of the product is nondefective?

(b) If a complex product has 10,000 parts that function independently, what is the probability that any specific unit of the product will be nondefective?

(c) Now suppose that we have a Six Sigma process without mean shift for each part of the complex product that has 10,000 parts. What is the probability that any specific unit of the product will be nondefective?

(d) Now suppose that we have a Six Sigma process with mean shifted for each part of the complex product which has 10,000 parts. What is the probability that any specific unit of the product will be nondefective?

4.4 The upper specification limit for a product is 10.00 and the lower specification limit is 8.00. It is assumed that the target value is 9.00. Based on statistical process control, it was found the process is in control and the mean of the process is 9.26 and the standard deviation is 0.21.

(a) What is the C_{pk} index for this process?
(b) What is the DPMO for this process?

4.5 This chapter covered the DMAIC process and explained the steps for such a process. Some companies have tollgates between each of the major steps in the DMAIC process. Briefly explain the need and importance of these tollgates.

4.6 Suppose an organization for a particular product is operating at 4 level (where the mean is shifted from the target). This will result in 6210 DPMO. The objective is to achieve 6 performance (3.4 DMPO). Suppose the organization quality improvement effort is 25% annual improvement in quality level. How many years will it take to achieve 6 performance?

4.7 Suppose your business is operating at Three Sigma quality level, and the project has an average annual improvement rate of 50%. How many years will it take to achieve Six Sigma quality?

4.8 During the analysis phase, an organizations finds that it has discovered the solution for the underlying problem. Please discuss whether the solution should be immediately implemented and remaining steps of DMAIC process are abandoned.

4.9 Consider any type of service system that you use and are very familiar with it. What are the CTQs for such a system and how will you apply the Six Sigma DMAIC process to this service system?

4.10 What is the difference between Six Sigma DMAIC process and the DFSS?

4.11 Explain various steps for DMADV process for any organization whose products you are familiar with.

4.12 Explain briefly various steps of ICOV, especially for manufacturing, which is used for DFSS for any organization that you are familiar with. Briefly explain some of the tools, methods, and philosophies that you will use during various steps of ICOV.

5 Product Development

To ensure acceptable product reliability, an organization must follow certain practices during the product development process. These practices impact reliability through the selection of parts (materials), product design, manufacturing, assembly, shipping and handling, operation, maintenance, and repair. Best practices for reliability, listed below and described in this book, dictate that the organization should:

- Define realistic product reliability requirements determined by factors including the targeted life cycle application conditions and performance expectations. The product requirements should consider the customer's needs and the manufacturer's capability to meet those needs.
- Define the product life-cycle conditions by assessing relevant manufacturing, assembly, storage, handling, shipping, operating, and maintenance conditions.
- Ensure that the supply-chain participants have the capability to produce the parts (materials) and services necessary to meet the final reliability objectives.
- Select parts (materials) that have sufficient quality and are capable of delivering the expected performance and reliability in the application.
- Identify the potential failure modes, failure sites, and failure mechanisms by which the product can be expected to fail.
- Design to the process capability (i.e., the quality level that can be controlled in manufacturing and assembly), considering the potential failure modes, failure sites, and failure mechanisms, obtained from the physics-of-failure analysis, and the life-cycle profile.
- Qualify the product to verify the reliability of the product in the expected life-cycle conditions. Qualification encompasses all activities that ensure that the nominal design and manufacturing specifications will meet or exceed the reliability goals.

Reliability Engineering, First Edition. Kailash C. Kapur and Michael Pecht.
© 2014 John Wiley & Sons, Inc. Published 2014 by John Wiley & Sons, Inc.

- Ensure that all manufacturing and assembly processes are capable of producing the product within the statistical process window required by the design. Variability in material properties and manufacturing processes will impact the product's reliability, so characteristics of the process must be identified, measured, and monitored.
- Manage the life-cycle usage of the product using closed-loop, root-cause monitoring procedures.

5.1 Product Requirements and Constraints

Various reasons justify the creation, modification, or upgrade of a product. For example, as discussed in Chapter 4, a company may want to address a perceived market need or open new markets. In some cases, a company may need to develop new products to remain competitive in a key market or to maintain market share and customer confidence. In other cases, a company may want to satisfy specific strategic customers, demonstrate experience with a new technology or methodology, or improve the maintainability of an existing product. In addition, product updates are often developed to reduce the life-cycle costs of an existing product.

To make reliable products, suppliers and customers throughout the supply chain must cooperate. The IEEE 1332 (IEEE Std. 1332–1998) addresses this cooperation through the three reliability objectives discussed in the previous chapter. First, the supplier must understand the customer's requirements and product needs in order to generate a comprehensive design specification. Second, the supplier must employ appropriate engineering activities so that the resulting product satisfies the customer's reliability requirements. Finally, the supplier must assure the customer that the reliability requirements and product needs have been satisfied.

Initially, requirements are formulated into a requirements document, where they are prioritized. The specific people involved in prioritization and approval will vary with the organization and the product. For example, for safety-critical products, safety, reliability, and legal representatives may all provide guidance.

As we have noted, once a set of requirements has been completed, the product engineering function creates a response to the requirements in the form of a specification. The specification states the requirements that must be met; the schedule for meeting the requirements; the identification of those who will perform the work; and the identification of potential risks. Differences in the requirements document and the preliminary specification become the topic of trade-off analyses.

After product requirements are defined and the design process begins, there should be an assessment of the product's requirements against the actual product design. As the product's design becomes increasingly detailed, it becomes more important to track the product's characteristics in relation to the original requirements. The rationale for making changes should be documented. The completeness with which requirement tracking is performed can significantly reduce future product redesign costs. Planned redesigns or design refreshes through technology monitoring and use of roadmaps ensure that the company is able to market new products or redesigned versions of old products in a timely, effective manner to retain its customer base and ensure continued profits.

5.2 Product Life Cycle Conditions

The life cycle conditions of the product influence decisions regarding product design and development, materials and parts selection, qualification, product safety, warranty, and product support (maintenance). The phases in a product's life cycle include manufacturing and assembly, testing, rework, storage, transportation and handling, operation[1] (modes of operation, on-off cycles, etc.), and repair and maintenance.

During each phase of its life cycle, a product will experience various environmental and usage loads. These loads may be thermal (steady-state temperature, temperature ranges, temperature cycles, and temperature gradients); mechanical (pressure levels, pressure gradients, vibrations, shock loads, and acoustic levels); chemical (aggressive or inert environments, ozone, pollution humidity levels, contamination, and fuel spills); environmental (radiation, electromagnetic interference, and altitude); electrical loading conditions (power, power surge, current, voltage, and voltage spikes); or the extent and rate of product degradation, among others. Reliability depends upon the nature, magnitude, and duration of exposure to such loads.

Defining and characterizing life-cycle loads is often an uncertain element of the overall design-for-reliability process. The challenge occurs because products can experience completely different application conditions depending on the application location, the product utilization or nonutilization profile, the duration of utilization, and maintenance and servicing conditions. For example, typically all desktop computers are designed for home or office environments. However, the operational profile of each unit may be completely different depending on user behavior. Some users may shut down the computer after it is used each time; others may shut down only once at the end of the day; still others may keep their computers powered all the time. Furthermore, one user may keep the computer by a sunny window, while another may keep the computer near an air conditioner; thus, the temperature profile experienced by each product, and hence its degradation due to thermal loads, would be different.

Four methods are used to estimate product life-cycle loads: market surveys and standards, similarity analysis, field trial and service records, and in situ monitoring. Market surveys and standards provide a very coarse and often inaccurate estimate of the environmental loads possible in various field applications. The environmental profiles available from these sources are typically classified according to industry type, such as military, consumer, telecommunications, automotive, and commercial avionics.

Similarity analysis is a technique for estimating environmental loads when sufficient field histories for similar products are available. Before using data on existing products for proposed designs, the characteristic differences in design and application use for the comparison products need to be reviewed. For example, electronics inside a washing machine in a commercial laundry are expected to experience a wider distribution of loads and use conditions (due to a larger number of users) and higher usage rates than a home washing machine. As another example, it has been found that some Asians use a dishwasher to wash vegetables, in addition to eating utensils. These dishwashers experience higher usage rates than those used only for washing dishes.

---

[1]Operational conditions are sometimes referred to as the life-cycle application conditions.

Field trial records provide estimates of the environmental profiles experienced by the product. The data depend on the durations and conditions of the trials, and can be extrapolated to estimate actual environmental conditions. Service records provide information on the maintenance, replacement, or servicing performed. These data can give an idea of the life-cycle environmental and usage conditions that lead to servicing or failure.

Environmental and usage conditions experienced by the product over its life cycle can be monitored in situ (Vichare et al. 2004). These data are often collected using sensors, either mounted externally or integrated with the product and supported by telemetry systems. Load distributions should be developed from data obtained by monitoring products used by different customers, ideally from various geographical locations where the product is used. The data should be collected over a sufficient period to provide an estimate of the loads and their variation over time. In situ monitoring provides the most accurate account of load histories and is most valuable in design-for-reliability (DFR) and product reliability assessment.

5.3 Reliability Capability

The selection of a supply chain is often based on factors that do not explicitly address reliability, such as technical capabilities, production capacity, geographic location, support facilities, and financial and contractual factors. A selection process that takes into account the ability of suppliers to meet reliability objectives during manufacturing, testing, and support can improve the reliability of the final product throughout its life cycle and provide valuable competitive advantages.

Reliability capability is a measure of the practices within an organization that contribute to the reliability of the final product and the effectiveness of these practices in meeting the reliability requirements of customers. Reliability capability assessment is the act of quantifying the effectiveness of reliability activities, using a metric called reliability capability maturity. From a reliability perspective, maturity indicates whether the key reliability practices employed by an organization are well understood, supported by documentation and training, applied to all products throughout the organization, and continually monitored and improved.

5.4 Parts and Materials Selection

A parts and materials selection and management methodology helps a company to make risk-informed decisions concerning the incorporation of parts and materials into a product. The part assessment process is shown in Figure 5.1. Key elements of part assessment include performance, quality, reliability, and ease of assembly.

The goal of performance assessment is to evaluate the part's ability to meet the performance requirements (structural, mechanical, electrical, thermal, biological, etc.) of the product. In general, there is often a minimum and a maximum limit beyond which the part will not function properly, at least in terms of the datasheet specifications. These limits, or ratings, are often called the recommended operating conditions.

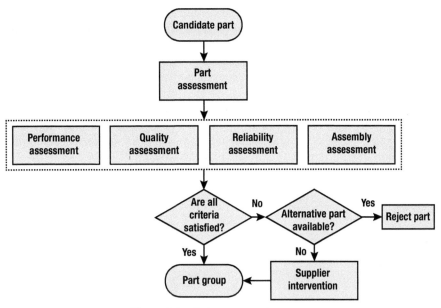

Figure 5.1 Part assessment process.

Quality is evaluated by outgoing quality and process capability metrics. Reliability assessment results provide information about the ability of a part to meet the required performance specifications in its targeted life-cycle application for a specified period of time. Reliability is evaluated through part qualification and reliability test results.

A part is acceptable from an assembly viewpoint if it is compatible with the downstream assembly equipment and processes. Assembly guidelines should be followed to prevent damage and deterioration of the part during the assembly process. Examples include a recommended temperature profile, cleaning agents, adhesives, moisture sensitivity, and electrical protection. As new technologies emerge and products become more complex, assembly guidelines become more necessary to ensuring the targeted quality and reliability of the parts and the product.

5.5 Human Factors and Reliability

All systems are of, by, and for humans. Human factors therefore are critical in the system design process and must be weighed against safety, reliability, maintainability, and other system parameters in order to affect trade-offs that increase system effectiveness. Human interaction with a system includes:

- Design and production of systems
- Operators and repairers of systems
- Operators and repairers as decision elements.

The human machine interface consists of such aspects as allocation of functions (human vs. machine), automation, accessibility, human tasks, stress characteristics, and both the information presented to the operator or repairer and the reliability of interfaces and decisions based on such information. Both human and machine elements of a system can fail, and their failures have varying effects on the system's performance. Some human errors cause total system failure or increase the risk of such failure. Human factors exert a strong influence on the design and ultimate reliability of a system (Kirwan 1994).

Both reliability and human factors are concerned with predicting, measuring, and improving system effectiveness. When the human machine interface is complex, the possibility of human error increases, resulting in an increase in the probability of system failure. An interesting facet of relationship among human factors, reliability, and maintainability is that the system's reliability and maintainability depends on the detection and correction of system malfunctions. This task is generally performed by people. Thus, the system performance can be enhanced or degraded depending on the human response. The quantification of human reliability characteristics and the development of a methodology for quantifying human performance, error prediction, control, and measurement are given in many sources (Gertman and Blackman 1994; Meister 1996).

The reliability of a system is affected by the allocation of system functions to humans, machines, or both. Favorable human characteristics include the ability to:

1. Detect certain forms of energy.
2. Be sensitive to a wide variety of stimuli within a restricted range.
3. Detect signals and patterns in high noise environments.
4. Store large amounts of information for long periods and remember relevant facts.
5. Learn from experience.
6. Use judgment.
7. Improvise and adopt flexible procedures.
8. Arrive at new and completely different solutions to problems;
9. Handle low probability or unexpected events.
10. Perform fine manipulations.
11. Reason instinctively.

Characteristics tending to favor machines are:

1. Computing capacity
2. Performance of routine, repetitive, and precise tasks
3. Quick response to control signals
4. Ability to exert large amounts of force smoothly and precisely
5. Ability to store and recall large amounts of data
6. Ability to reason deductively
7. Insensitivity to extraneous factors
8. Ability to handle highly complex operations that involve doing several things at once.

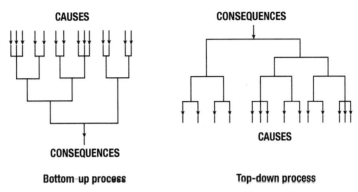

Figure 5.2 Bottom-up versus top-down methods.

5.6 Deductive versus Inductive Methods

Deduction comprises reasoning from the general to the specific. In a deductive system analysis, it is postulated that the system itself has failed in a certain way, and an attempt is made to find out what modes of system or subsystem (component) behavior contribute to this failure. These methods are also called top-down. One of the very popular and useful deductive methods is fault tree analysis (FTA), which is covered in Section 5.8.

Induction involves reasoning from individual cases to a general conclusion. In this case, a particular fault or initiating condition is postulated and an attempt to ascertain the effect of that fault or condition on system operation is made. These methods are also called bottom-up. The reliability block diagram (RBD) is one example of an inductive method that is covered in Chapter 17. Another very popular and useful method is failure modes, effects and criticality analysis (FMECA), which is discussed in the next section. Figure 5.2 shows the difference between backward versus forward methods. The arrows indicate the direction of these tree-like graphs.

In general, both deductive and inductive approaches must be employed to get a complete set of failure/fault/accident sequences. The deductive approach has the benefit of focusing the analysis on the undesired event, while the inductive approach is useful in assuring that the analysis is broad enough to encompass all possible scenarios.

5.7 Failure Modes, Effects, and Criticality Analysis

Failure modes, effects, and criticality analysis (FMECA) is a design evaluation procedure used to identify all conceivable and potential failure modes and to determine the effect of each failure mode on system performance. Criticality analysis in FMECA helps to develop priorities for continuous improvement. This procedure is accomplished by formal documentation, which serves (1) to standardize the procedure, (2) as a means of historical documentation, and (3) as a basis for future improvement.

Correct usage of the FMECA process will result in two improvements:

1. An improvement in the reliability of the product through the anticipation of problems and the institution of corrections prior to going into production.
2. An improvement in the validity of the analytical method itself through strict documentation that illuminates the rationale for every step.

Failure modes and effects analysis is an iterative, systematic, documented process performed to identify basic failure/faults at the part level and determine their effects at higher levels of assembly. Criticality analysis in FMECA helps to develop priorities for continuous improvement. The analysis can be performed utilizing either actual failure modes from field data or hypothesized failure modes derived from design analysis, reliability prediction activities, and experience with how parts fail.

In their most complete form, failure modes are identified at the part level, which is usually the lowest level of direct concern to the designer of the product or process. In addition to providing insight into failure cause-and-effect relationships, FMECA provides a disciplined method for proceeding part by part through the system to assess failure consequences.

Failure modes are analytically induced into each component, and failure effects are evaluated and noted, including the severity and frequency (or probability) of occurrence. As the first mode is listed, the corresponding effect on performance at the next higher level of assembly is determined. The resulting failure effect becomes, in essence, the failure mode that impacts the next higher level. Iteration of this process results in establishing the ultimate effects at the system level.

The analysis of all failure modes usually reveals that each effect or symptom at the system level is caused by several different failure modes at the lowest level. This relationship to the end effect provides the basis for grouping the lower-level failure modes.

Using this approach, probabilities for the occurrence of system failure can be calculated, based on the probability of occurrence of the lower-level failure modes. Based on these probabilities and a severity factor assigned to the various system effects, a criticality number can be calculated. Criticality numerics also provide the basis for corrective action priorities, engineering changes, and resolution of problems in the field.

The procedure consists of a sequence of logical steps, starting with the analysis of lower level subsystems or components. The analysis assumes a failure point of view and identifies all potential modes of failure, along with the causative agent, termed the "failure mechanism." The effect of each failure mode is then traced up to the systems level (MIL_STD_1629 (SHIPS)).

As mentioned before, a criticality rating is developed for each failure mode and its resulting effect. The rating is based on the probability of occurrence, severity, and detectability. For failures scoring a high rating, design changes to reduce criticality are recommended. This procedure is aimed at providing a more reliable design.

A failure mode is the manner in which a failure can occur—that is, the way in which the products fails to perform its intended design function, or performs the function but fails to meet its objectives. For example, failure modes of a cell phone include a button that doesn't cause a number to register, or a microphone that doesn't pick up your voice.

Sometimes, the failure modes are intentionally accentuated so that the user of the product will become aware of the existence of a problem. For example, a bad-smelling substance is sometimes added to natural gas to indicate the existence of a leak. Another example is the grinding noise when the brake pads wear out on a car.

Failure mechanisms are the processes by which a specific combination of physical, electrical, chemical, and mechanical stresses induces failures. For example, fracture, fatigue, and corrosion are failure mechanisms.

The purpose of failure modes, mechanisms, and effects analysis (FMMEA) is to identify potential failure mechanisms and models for all the potential failures modes of a product, and then to prioritize failure mechanisms for efficient product development. FMMEA is based on understanding (1) the relationships between product requirements and the physical characteristics of the product (and their variation in the production process), and (2) the interactions of product materials with loads (stresses under application conditions) and their influence on the product's susceptibility to failure with respect to the use conditions.

5.8 Fault Tree Analysis

FTA is a method for system safety and reliability analysis (*Fault Tree Handbook* 2002). The concept was originated by Bell Telephone Laboratories as a technique to evaluate the safety of the Minuteman Launch Control System. Many reliability techniques are inductive and are concerned primarily with ensuring that hardware will accomplish its intended functions. FTA is a detailed deductive analysis that usually requires considerable information about the system. Concerned with ensuring that all critical aspects of a system are identified and controlled, it is a graphical representation of the Boolean logic associated with the development of a particular system failure (consequence), called the "top event," into basic failures (causes), and called "primary events." These top events can be broad, all encompassing events, such as "the release of radioactivity from a nuclear power plant" or "the inadvertent launch of an ICBM missile," or they can be specific events, such as "failure to insert control rods" or "energizing power available to ordnance ignition line."

FTA is of value for:

- Providing options for qualitative and quantitative reliability analysis
- Helping the analyst to understand system failures deductively
- Pointing out the aspects of a system that are important with Respect to the failure of interest
- Providing the analyst an insight into system behavior.

A fault tree is a model that graphically and logically represents the various combinations of possible events, both fault and normal, that occur in a system and lead to the top event. The term "event" denotes a dynamic change of state that occurs in a system element. A fault event is an abnormal system state. A normal event is an event that is expected to occur. System elements include hardware, software, and human and environmental factors. (Details about the construction of fault trees can be found in the reference mentioned at the beginning of this section.)

FTA is a deductive methodology to determine the potential causes of failures and to estimate the failure probabilities. FTA addresses system design aspects and potential failures, tracks down system failures deductively, describes system functions and behaviors graphically, focuses on one error at a time, and provides qualitative and quantitative reliability analyses. The purpose of a fault tree is to show the sets of events—particularly the primary failures—that will cause the top event in a system.

FTA provides critical information that can be used to prioritize the importance of the contributors to the undesired event. The contributing importances provided by FTA vividly show the causes that are dominant and that should be the focus of any safety or reliability activity.

More formal risk–benefit approaches can also be used to optimally allocate resources to minimize both resource expenditures and the probability of occurrence of the undesired event. These risk–benefit approaches are useful for allocating resource expenditures, such as safety upgrades to complex systems like the Space Shuttle.

FTA can be applied to both an existing system and a system that is being designed. When it is applied to a system being designed for which specific data do not exist, FTA can provide an estimate of the failure probability and the important contributors to failure, using generic data to bracket the design components or concepts. FTA can also be used as an important element in the development of a performance-based design.

When applied to an existing system, FTA can be used to identify weaknesses and to evaluate possible upgrades. It can also be used to monitor and predict behavior. Furthermore, FTA can be used to diagnose causes and potential corrective measures for an observed system failure.

The approaches and tools to obtain this information and apply it in decision-making are important topics. FTA can be simply described as an analytical technique, through which (1) an undesired state of the system is specified (usually a state that is critical from a safety or reliability standpoint), and (2) the system is then analyzed in the context of its environment and operation to find all the realistic ways in which the undesired top event can occur.

The fault tree itself is a graphic model of the various parallel and sequential combinations of faults that will result in the occurrence of the predefined undesired event. The faults can be events associated with component hardware failures, human errors, software errors, or any other pertinent factors that can lead to the undesired event. A fault tree thus depicts the logical interrelationships of basic events that lead to the top event of the fault tree.

A fault tree is composed of a complex of entities known as "gates" that serve to permit or inhibit the passage of fault logic up the tree. The gates show the relationships of events needed for the occurrence of a "higher" event. The "higher" event is the output of the gate; the "lower" events are the "inputs" to the gate. The gate symbol denotes the type of relationship of the input event required for the output event.

The qualitative evaluations basically transform the FT logic into logically equivalent forms that provide more focused information. The principal qualitative results that are obtained are the minimal cut sets (MCSs) of the top event. A cut set is a combination of basic events that can cause the top event. An MCS is the smallest combination of basic events that result in the top event. The basic events are the bottom events of the fault tree. Hence, the MCSs relate the top event directly to the basic event causes. The set of MCSs for the top event represent all the ways that the basic events can cause the top event.

A more descriptive name for a MCS may be "minimal failure set." The set of MCSs can be obtained not only for the top event, but for any of the intermediate events (e.g., gate events) in the FT. A significant amount of information can be obtained from the structure of MCSs. Any MCS with one basic event identifies a single failure or single event that alone can cause the top event to occur. These single failures are often weak links and are the focus of upgrade and prevention actions. Examples of such single failures are a single human error or component failure that can cause a system failure.

An MCS having events with identical characteristics indicates a susceptibility to implicit dependent failure, or a common cause that can negate a redundancy. An example is an MCS of failures of identical valves. A single manufacturing defect or single environmental sensitivity can cause all the valves to simultaneously fail.

Failures can be classified in several ways (e.g., hardware faults or human error, or one of many possible hardware faults: early, random, or aging; primary, secondary or command; passive or active). More information on this classification is given in *Fault Tree Handbook with Aerospace Applications* (2002).

The quantitative evaluations of a FT consist of the determination of top event probabilities and basic event importances. Uncertainties in any quantified result can also be determined. Fault trees are typically quantified by calculating the probability of each MCS and by summing all the cut set probabilities. The cut sets are then sorted by probability. The cut sets that contribute significantly to the top event probability are called the dominant cut sets. While the probability of the top event is a primary focus in the analysis, the probability of any intermediate event in the fault tree can also be determined.

Different types of probabilities can be calculated for different applications. In addition to a constant probability value that is typically calculated, time-related probabilities can be calculated to provide the probability distribution of the time of first occurrence of the top event. Top event frequencies, failure or occurrence rates, and availabilities can also be calculated. These characteristics are particularly applicable if the top event is a system failure.

In addition to the identification of dominant cut sets, the importances of the events in the FT are among the most useful information that can be obtained from FT quantification. Quantified importances allow actions and resources to be prioritized according to the importances of the events causing the top event. The importance of the basic events, the intermediate events, and the MCSs can be determined.

Different importance measures can be calculated for different applications. One measure is the contribution of each event to the top event probability. Another is the decrease in the top event probability if the event were prevented from occurring. A third measure is the increase in the top event probability if the event were assured to occur. These importance measures are used in prioritization, prevention activities, upgrade activities, and maintenance and repair activities. Thus, substantial rich qualitative and quantitative information can be obtained from a FT.

5.8.1 Role of FTA in Decision-Making

FTA has numerous uses in enhancing product reliability:

- To understand the logic leading to the top event
- To prioritize the contributors leading to the top event
- As a proactive tool to prevent the top event
- To monitor the performance of the system
- To minimize and optimize resources

- To assist in designing a system
- As a diagnostic tool to identify and correct causes of the top event.

5.8.2 Steps of Fault Tree Analysis

A successful FTA requires the following steps be carried out:

1. Identify the objective for the FTA.
2. Define the top event of the FT.
3. Define the scope of the FTA.
4. Define the resolution of the FTA.
5. Define ground rules for the FTA.
6. Construct the FT.
7. Evaluate the FT.
8. Interpret and present the results.

5.8.3 Basic Paradigms for the Construction of Fault Trees

The basic paradigm in constructing a fault tree is to "think small," or more accurately, "think myopically." For each event that is analyzed, the *necessary and sufficient immediate events* (i.e., the most closely related events) that result in the event must be identified. The key phrase is "the necessary and sufficient immediate events." The analysis does not jump to the basic causes of the event. Instead, a small step is taken and the immediate events that result in the event are identified. This taking small steps backwards assures that all of the relationships and primary causes will be uncovered. It also provides the analyst with insight into the relationships that are necessary and sufficient for the occurrence of the top event of the fault tree. This backward stepping ends as the basic causes are identified, which constitute the *resolution* of the analysis.

5.8.4 Definition of the Top Event

Some guidelines for the definition of the top event include the following:

1. To define the top event, define the criteria for the occurrence of the event. For a system failure, first define the system success criteria.
2. Assure that the top event is consistent with the problem to be solved and the objectives of the analysis.
3. If unsure of the top event, define alternative definitions that cover the top event and assess the applicability of each one.

5.8.5 Faults versus Failures

A distinction is made here between the rather specific word "failure" and the more general word "fault." As an example of the distinction, consider a relay. If the relay closes properly when a voltage is applied across its terminals, this is a relay "success." If, however, the relay fails to close under these circumstances, this is a relay "failure." Another possibility is that the relay closes at the wrong time due to the improper functioning of some upstream component. This is clearly not a relay failure; however, untimely relay operation may well cause the entire circuit to enter into an unsatisfactory state. An occurrence like this is referred to here as a "fault." Generally speaking, all failures are faults but not all faults are failures. Failures are basic abnormal occurrences, whereas faults are "higher order" or more general events.

There are three phases in FTA. The first step is to develop a logic block diagram or a fault tree using elements of the fault tree. This phase requires complete system definition and understanding of its operation. Every possible cause and effect of each failure condition should be investigated and related to the top event. The second step is to apply Boolean algebra to the logic diagram and develop algebraic relationships between events. If possible, simplify the expressions using Boolean algebra. The third step is to apply probabilistic methods to determine the probabilities of each intermediate event and the top event. The probability of occurrence of each event has to be known; that is, the reliability of each component or subsystem for every possible failure mode has to be considered.

The graphical symbols used to construct the fault tree fall into two categories: gate symbols and event symbols. The basic gate symbols are AND, OR, k-out-of-n voting gate, priority AND, exclusive OR, and inhibit gate. The basic event symbols are basic event, undeveloped event, conditional event, trigger event, resultant event, transfer-in and transfer-out event (Kececioglu 1991; Lewis 1996; Rao 1992). Quantitative evaluation of the fault tree includes calculation of the probability of the occurrence of the top event. This is based on the Boolean expressions for the interaction of the tree events. Figure 5.3 shows the commonly used symbols in creating a fault tree. For the quantitative analysis, the basic Boolean relations are shown in Table 5.1.

In engineering analysis, the symbol for $\cup$ is $+$ and the symbol for $\cap$ is $\bullet$. Using the engineering symbols, for an application of the use of these rules, consider the simplification of the expression

$$(A+B)\bullet(A+C)\bullet(D+B)\bullet(D+C).$$

Applying the distributive law to $(A + B) \bullet (A + C)$ results in

$$(A+B)\bullet(A+C)=A+(B\bullet C).$$

Likewise,

$$(D+B)\bullet(D+C)=D+(B\bullet C).$$

An intermediate result produced is

$$(A+B)\bullet(A+C)\bullet(D+B)\bullet(D+C)=(A+B\bullet C)\bullet(D+B\bullet C).$$

Letting E represent the event B $\bullet$ C results in

$$(A+B\bullet C)\bullet(D+B\bullet C)=(A+E)\bullet(D+E)=(E+A)\bullet(E+D).$$

5 Product Development

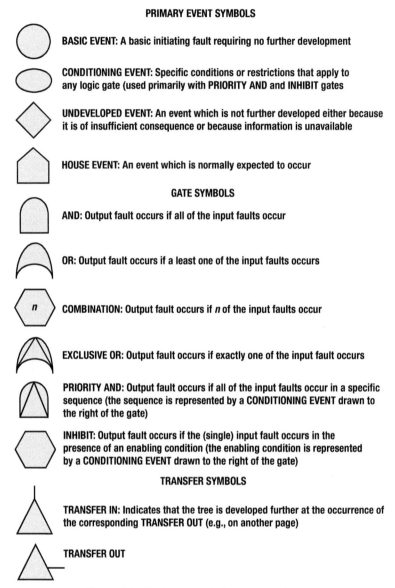

Figure 5.3 Fault tree symbols: events and gates.

Another application of distributive law yields

$$(E+A)\bullet(E+D) = E + A\bullet D = B\bullet C + A\bullet D.$$

Therefore, the final result is

$$(A+B)\bullet(A+C)\bullet(D+B)\bullet(D+C) = B\bullet C + A\bullet D.$$

The original expression has been substantially simplified for purposes of evaluation. This idea can be applied to simplify fault trees.

5.8 Fault Tree Analysis

Table 5.1 Rules of Boolean algebra

Mathematical symbolism	Designation
$X \cup (Y \cap Z) = (X \cup Y) \cap (X \cup Z)$	Distributive law
$X \cap (Y \cup Z) = (X \cap Y) \cup (X \cap Z)$	
$X \cup X = X$	Idempotent law
$X \cap X = X$	
$X \cup (X \cap Y) = X$	Law of absorption
$X \cap (X \cup Y) = X$	
$(X \cap Y)' = X' \cup Y'$	DeMorgan's theorem
$(X \cup Y)' = X' \cap Y'$	
$X \cup (X' \cap Y) = X \cup Y$	Useful result
$X' \cap (X \cup Y') = X' \cap Y'$	

Example 5.1

Reliability Block Diagram for Blackout (see Figure 5.4)
Blackout happens if both the off-site power and the emergency power fail. The emergency power fails if either the voltage monitor or the diesel generator fails. The voltage monitor signals the diesel generator to start when the offsite voltage falls below a threshold level. The fault tree for the blackout event is shown in Figure 5.5.

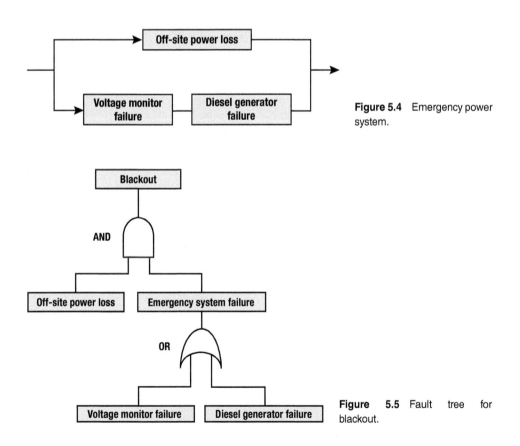

Figure 5.4 Emergency power system.

Figure 5.5 Fault tree for blackout.

5 Product Development

Example 5.2

Analyze the following fault tree:

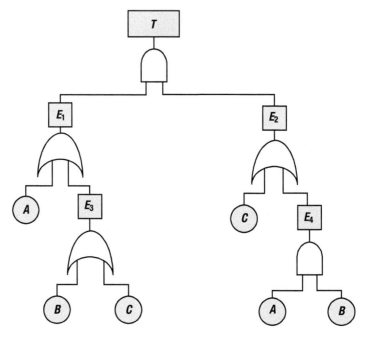

Top-Down Evaluation

1. $T = E_1 \cap E_2$
2. $E_1 = A \cup E_3$; $E_2 = C \cup E_4$
3. $E_3 = B \cup C$; $E_4 = A \cap B$
4. $T = (A \cup E_3) \cap (C \cup E_4) = [A \cup (B \cup C)] \cap [C \cup (A \cap B)]$.

Bottom-Up Evaluation

1. $E_3 = B \cup C$; $E_4 = A \cap B$
2. $E_1 = A \cup E_3$; $E_2 = C \cup E_4$
3. $E_1 = A \cup (B \cup C)$
4. $E_2 = C \cup (A \cap B)$
5. $T = E_1 \cap E_2 = [A \cup (B \cup C)] \cap [C \cup (A \cap B)]$.

Either evaluation direction can be used for FTA.

- Associative law: $A \cup (B \cup C) = (A \cup B) \cup C$
- Commutative law: $(A \cup B) \cup C = C \cup (A \cup B)$
  Thus $T = [C \cup (A \cup B)] \cap [C \cup (A \cap B)]$
- Distributive law: $T = C \cup [(A \cup B) \cap (A \cap B)]$
  $A \cap B = B \cap A$

- Associative law: $T = C \cup [(A \cup B) \cap B \cap A]$
- Absorption law: $(A \cup B) \cap B = B$
- $T = C \cup (B \cap A)$.

Hence, the tree can be reduced to show T occurs only when C or both A and B occur:

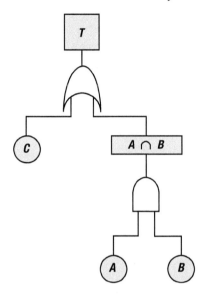

One of the main purposes of representing a fault tree in terms of Boolean equations is that these equations can then be used to determine the fault tree's associated MCSs and minimal path sets. Once the MCSs are obtained, the quantification of the fault tree is more or less straightforward. The minimal path sets are essentially the complements of the MCSs and define the "success modes" by which the top event will not occur. The minimal path sets are often not obtained in a fault tree evaluation; however, they can be useful in particular problems.

5.8.6 Minimal Cut Sets

By definition, a MCS is a combination (intersection) of primary events sufficient for the top event. The combination is a "minimal" combination in that all the failures are needed for the top event to occur; if one of the failures in the cut set does not occur, then the top event will not occur (by this combination).

Any fault tree will consist of a finite number of MCSs that are unique for that top event. One-component MCSs, if there are any, represent those single failures that will cause the top event to occur. Two-component MCSs represent the two failures that together will cause the top event to occur. For an n-component MCS, all n components in the cut set must fail in order for the top event to occur.

The MCS expression for the top event can be written in the general form,

$$T = M_1 + M_2 + \cdots + M_k,$$

where T is the top event, and $M_{i=1,2,\ldots k}$ are the MCSs. Each MCS consists of a combination of specific component failures, and hence the general n-component minimal cut can be expressed as

$$M_i = X_1 \bullet X_2 \bullet \cdots \bullet X_n,$$

where X_1, X_2, and so on, are basic component failures in the tree. An example of a top event expression, as shown in Example 5.2, is

$$T = A + B \bullet C,$$

where A, B, and C are component failures. This top event has a one-component MCS (A) and a two-component MCS ($B \bullet C$). The MCSs are unique for a top event and are independent of the different equivalent forms the same fault tree may have.

To determine the MCSs of a fault tree, the tree is first translated to its equivalent Boolean equations. A variety of algorithms exist to translate the Boolean equations into cut sets. Two of the most common are the "top-down" or "bottom-up" substitution methods to solve for the top event. The methods are straightforward and involve substituting and expanding Boolean expressions. The distributive law and the law of absorption are used to remove the redundancies.

5.9 Physics of Failure

Once the parts (materials), load conditions, and possible failure risks based on the FMMEA have been identified, the design guidelines based on physics-of-failure models aid in making design trade-offs, and can also be used to develop tests, screens, and derating[2] factors. Tests based on physics-of-failure models can be planned to measure specific quantities, to detect the presence of unexpected flaws, and to detect manufacturing or maintenance problems. Screens can be planned to precipitate failures in "weak" products while not deteriorating the design life of the shipped product. Derating or safety factors can be determined to lower the stresses for the dominant failure mechanisms.

5.9.1 Stress Margins

Products should be designed to operate satisfactorily, with margins (the design margins) at the extremes of the stated recommended operating ranges (the specification limits). These ranges must be included in the procurement requirement or specifications.

Figure 5.6 schematically represents the hierarchy of product load (stress) limits and margins. The specification limits are set by the manufacturer to limit the conditions of customer use. The design margins correspond to the load (stress) condition that

---

[2]Derating is the practice of subjecting parts to lower electrical or mechanical stresses than they can withstand to increase the life expectancy of the part.

5.9 Physics of Failure

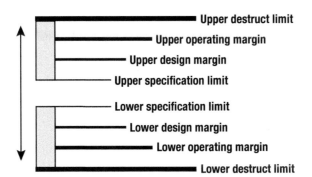

Figure 5.6 Load (stress) limits and margins.

the product is designed to survive without field failures. That is, the operating margin is the expected load (stress) that may lead to a recoverable failure. The destruct margin is the expected load (stress) that may lead to permanent (overstress) failure.

Statistical analysis and worst-case analysis should be used to assess the effects of product parameter variations. In statistical analysis, a functional relationship is established between the output characteristics of the product and its parameters. In worst-case analysis, the effect of the product outputs is evaluated on the basis of end-of-life performance values.

5.9.2 Model Analysis of Failure Mechanisms

Model analysis of failure mechanisms is based on computer-aided simulation. Model analysis can assist in identifying and ranking the dominant failure mechanisms associated with the product under life-cycle loads, determining the acceleration factor for a given set of accelerated test parameters, and determining the time to failure corresponding to the identified failure mechanisms.

Each failure model comprises a load analysis model and a damage assessment model. The output is a ranking of different failure mechanisms, based on the time to failure. The load model captures the product architecture, while the damage model depends on a material's response to the applied loads. Model analysis of failure mechanisms can be used to optimize the product design so that the minimum time to failure of the product is greater than its desired life. Although the data obtained from model analysis of failure mechanisms cannot fully replace those obtained from physical tests, they can increase the efficiency of tests by indicating the potential failure modes and mechanisms that can be expected.

It should be remembered that the accuracy of modality results depends on the accuracy of the process inputs—that is, the product geometry and material properties, the life-cycle loads, the failure models used (e.g., constants in the failure model), the analysis domain, and discretization approach (spatial and temporal). Hence, to obtain a reliable prediction, the variability in the inputs should be specified using distribution functions, and the validity of the failure models should be tested by conducting the appropriate tests.

5.9.3 Derating

To ensure that the product remains within the predetermined margins shown in Figure 5.6, derating can be used. Derating is the practice of limiting loads (e.g., thermal, electrical, or mechanical) to improve reliability. Derating can provide added protection from anomalies unforeseen by the designer (e.g., transient loads or electrical surges). For example, manufacturers of electronic parts often specify limits for supply voltage, output current, power dissipation, junction temperature, and frequency. The product design team may decide to ensure that the operational condition for a particular load, such as temperature, is always below the rated level. The load reduction is expected to extend the useful operating life, when the failure mechanisms under consideration are wearout type. This practice is also expected to provide a safer operating condition by furnishing a margin of safety when the failure mechanisms are of the overstress type.

As inherently suggested by the term "derating," the methodology involves a two-step process: "rated" load values are first determined, and then a reduced value is assigned. The margin of safety that the process of derating provides is the difference between the maximum allowable actual applied load and the demonstrated limits of the product.

In order to be effective, derating must target the appropriate, critical load parameters, based on models of the relevant failure mechanisms. Once the failure models for the critical failure mechanisms have been identified, using, for example, FMMEA, the impact of derating on the effective reliability of the product for a given load can be determined. The goal should be to determine the "safe" operating envelope for the product and then operate within that envelope.

5.9.4 Protective Architectures

The objective of protective architectures is to enable some form of action, after an initial failure or malfunction, to prevent additional or secondary failures. Protective techniques include the use of fuses and circuit breakers, self-sensing structures, and adjustment structures that correct for parametric shifts.

In designs where safety is an issue, it is generally desirable to incorporate some means of preventing a product from failing or from causing further damage when it fails. Fuses and circuit breakers are used to sense excessive current or voltage spikes and disconnect power from the electronic products. Similarly, thermostats can be used to sense critical temperature-limiting conditions, and to power-off the product until the temperature returns to normal. Self-checking circuitry can also be incorporated to sense abnormal conditions and restore them to normal, or to activate circuitry that will compensate for the malfunction.

In some instances, it may be desirable to permit partial operation of the product after a part failure, possibly with degraded performance, rather than completely power off the product. For example, in shutting down a failed circuit whose function is to provide precise trimming adjustment within a deadband of another control product, acceptable performance may be achieved, under emergency conditions, with the deadband control product alone.

Protective architectures must be designed considering the impact of maintenance. For example, if a fuse protecting a circuit is replaced, the following questions need to be answered: What is the impact when the product is reenergized? What protective architectures are appropriate for postrepair operations? What maintenance guidance must be documented and followed when fail-safe protective architectures have or have not been included?

5.9.5 Redundancy

The purpose of redundancy is to enable the product to operate successfully even though one or more of its parts fail. A design team often finds that redundancy is the quickest way to improve product reliability if there is insufficient time to explore alternatives. It can be the most cost-effective solution, or perhaps the only solution, if the reliability requirement is beyond the state of the art.

A redundant design typically adds size, weight, and cost. When not properly implemented, redundancy can also provide a false sense of reliability. If a failure cause can affect all the redundant elements of a product at the same time, then the benefits of redundancy will be lost. Also, failures of sensing and switching circuitry or software can result in failure even in the presence of redundancy.

5.9.6 Prognostics

A product's health is the extent of deviation or degradation from its expected normal physical and performance operating condition (Vichare et al. 2004). Knowledge of a product's health can be used to detect and isolate faults or failures (diagnostics) and to predict an impending failure based on current conditions (prognostics). Thus, by determining the advent of failure based on actual life-cycle conditions, procedures can be developed to mitigate and manage potential failures and maintain the product.

Prognostics can be designed into a product by (1) installing built-in fuses and canary structures that will fail faster than the actual product when subjected to life-cycle conditions (Mishra and Pecht 2002); (2) sensing parameters that are precursors to failure, such as defects or performance degradation (Pecht et al. 2001); (3) sensing the life-cycle environmental and operational loads that influence the system's health, and processing the measured data using physics-of-failure models to estimate remaining useful life (Mishra et al. 2002; Ramakrishnan and Pecht 2003).

5.10 Design Review

The design review, a formal and documented review of a system design, should be conducted by a committee of senior company personnel who are experienced in various pertinent aspects of product design, reliability, manufacturing, materials, stress analysis, human factors, safety, logistics, maintenance, liability, and so on. The design review spans all phases of product development from conception to production and can be extended over the useful life of the product. In each phase, previous work is updated and the review is based on current information.

A mature design requires trade-offs between many conflicting factors, such as performance, manufacturability, reliability, safety and maintainability. These trade-offs depend heavily on experienced judgment and require continuous communication among experienced reviewers. The design review committee approach has been found to be extremely beneficial to this process. The committee adopts the system's point of view and considers all conceivable phases of design and system use, to ensure that the best trade-offs have been made for the particular situation.

A complete design review procedure must be multiphased in order to follow the design cycle until the system is released for production. A typical example of a review

5 Product Development

Table 5.2 Design review committee

Member	Review phase 1	2	3	Responsibility
Chairperson	x	x	x	Ensure that review is conducted efficiently. Issue major reports and monitor follow-up.
Customer rep.	x	x	x	Ensure that the customer's viewpoint is adequately presented (especially at the design trade-off stage).
Design engineer (of this product)	x	x	x	Prepare and present initial design with calculations and supporting data.
Design engineer (not of this product)	x	x	x	Review and verify adequacy of design.
Reliability engineer	x	x	x	Evaluate design for maximum reliability consistent with system goals.
Manufacturing engineer		x	x	Ensure manufacturability at reasonable cost. Check for tooling adequacy and assembly problems.
Materials engineer		x		Ensure optimum material usage considering application and environment.
Stress analyst		x		Review and verify stress calculations.
Quality control engineer		x	x	Review tolerancing problems, manufacturing capability, inspection strategies, and testing problems.
Human factors engineer		x		Ensure adequate consideration of human operator. Identify potential human-induced problems.
Safety engineer		x		Ensure safety of operating and auxiliary personnel.
Maintainability engineer		x	x	Analyze for ease of maintenance repair and field servicing problems.
Logistics engineer		x	x	Evaluate and specify logistical support. Identify logistics problems.

committee, including personnel and their responsibilities, is shown in Table 5.2. Here, the review process has been subdivided into three phases, and each phase is an update of detailed analysis based on the latest knowledge.

Ultimately, the design engineer has the responsibility for investigating and incorporating the ideas and suggestions posed by the design review committee. The committee's chairperson is responsible for adequately reporting all suggestions by way of a formal and documented summary. The design engineer then can accept or reject various points in the summary; however, he or she must formally report back to the committee, stating reasons for the actions taken.

Considerably more thought and detail than the basic philosophy presented here must go into developing the management structure and procedures for conduct in order to have a successful review procedure. The review procedure must consider not only reliability, but all important factors to ensure that a mature design will result from the design effort.

5.11 Qualification

Qualification tests are conducted to identify and assess potential failures that could arise during the use of a product. Qualification tests should be performed during initial product development, and also after any significant design or manufacturing changes to an existing product.

In some cases, the target application, and therefore the use conditions, of the product may not be known. For example, a part or an assembly may be developed for sale to the open market for incorporation into many different types of products. In such cases, standard qualification tests are often employed. However, passing these tests does not mean that the product will be reliable in the actual targeted application. As a result, it is generally not sufficient to rely on qualification tests conducted on the parts (materials) of a product to determine or ensure the reliability of the final product in the targeted application.

Most often, there is insufficient time to test products for their complete targeted application life under actual operating conditions. Therefore, accelerated (qualification) tests are often employed. Accelerated testing is based on the premise that a product will exhibit the same failure mechanisms and modes in a short time under high-load conditions as it would exhibit in a longer time under actual life-cycle load conditions. The purpose is to decrease the total time and cost required to obtain reliability information for the product under study.

Accelerated tests can be divided into two categories: qualitative tests and quantitative tests. Qualitative tests generally overstress the products to determine the load conditions that will cause overstress or early wearout failures. Such tests may target a single load condition, such as shock, temperature extremes, or electrical overstress, or some combination of these. The results of the tests include failure mode information, but qualitative tests are not generally appropriate to estimate time to failure in the application.

Quantitative tests target wearout failure mechanisms, in which failures occur as a result of cumulative load conditions. These tests make analysis possible to quantitatively extrapolate from the accelerated environment to the usage environment with some reasonable degree of assurance.

The easiest form of accelerated life testing is continuous-use acceleration. The objective of this approach is to compress useful life into the shortest time possible. This approach assumes that the product is not used continuously, and that, when the product is not used, there are no loads (stresses) on it. For example, most washing machines are used for 10 hours per week on average. If a washing machine was continuously operated, the acceleration factor[3] would be $(24)(7)/10 = 16.8$. Thus, if the warranty or design life of the product was 5 years, the product should be tested for $5/16.8 = 0.3$ years, or 106 days.

Continuous-use acceleration is not very effective with high-usage products, or with products that have a long expected life. Under such circumstances, accelerated testing is conducted to measure the performance of the product at loads (stresses) that are more severe than would normally be encountered, in order to accelerate the damage accumulation rate in a reduced time period. The goal of such testing is to accelerate time-dependent failure mechanisms and the damage accumulation rate to reduce the time to failure. Based on the data from accelerated tests, the time to failure in the targeted use conditions can be extrapolated.

Accelerated testing begins by identifying all the significant overstress and wearout failure mechanisms from the failure modes, mechanisms, and effects analysis

---

[3]The acceleration factor is defined as the ratio of the life of the product under normal use conditions to that under an accelerated condition.

(FMMEA). The load parameters that cause the failure mechanisms are selected as the acceleration parameters, and are commonly called accelerated loads. Typical accelerated loads include thermal loads, such as temperature, temperature cycling, and rates of temperature change; chemical loads, such as humidity, corrosives, acid, solvents, and salt; electrical loads, such as voltage or power; and mechanical loads, such as vibration, mechanical load cycles, strain cycles, and shock/impulses. Accelerated tests may require a combination of these loads. Interpretation of the results for combined loads requires a quantitative understanding of their relative interactions.

Failure due to a particular mechanism can be induced by several acceleration parameters. For example, corrosion can be accelerated by both temperature and humidity, and creep can be accelerated by both mechanical stress and temperature. Furthermore, a single accelerated load can induce failure by several mechanisms. For example, temperature can accelerate wearout damage accumulation of many failure mechanisms, such as corrosion, electrochemical migration, and creep. Failure mechanisms that dominate under usual operating conditions may lose their dominance as the load is elevated. For example, high-power electronics can generate temperatures that evaporate moisture. Conversely, failure mechanisms that are dormant under normal use conditions may contribute to device failure under accelerated conditions. Thus, accelerated tests require careful planning if they are to accelerate the actual usage environments and operating conditions without introducing extraneous failure mechanisms or nonrepresentative physical or material behaviors.

Once the failure mechanisms are identified, it is necessary to select the appropriate acceleration load; to determine the test procedures and the load levels; to determine the test method, such as constant load acceleration or step-load acceleration; to perform the tests; and to interpret the test data, which includes extrapolating the accelerated test results to normal operating conditions. The test results provide failure information to assess the product reliability, to improve the product design, and to plan warranties and support.

5.12 Manufacture and Assembly

Improper manufacturing and assembly can introduce defects, flaws, and residual stresses that act as potential failure sites or stress enhancers (or raisers) later in the life of the product. The effect of manufacturing variability on time to failure is depicted in Figure 5.7.

A shift in the mean or increase in the standard deviation of key parameters during manufacturing can result in early failure due to a decrease in the strength of the product. Generally, qualification procedures are required to ensure that the normal product is reliable. In some cases, lot-to-lot screening is required to ensure that the variability of assembly and manufacturing-related parameters are within specified tolerances. Here, screening ensures the quality of the product by precipitating latent defects before they reach the final customer.

5.12.1 Manufacturability

The design team must understand material limits and manufacturing process capabilities to construct products that promote produceability and reduce the occurrence of

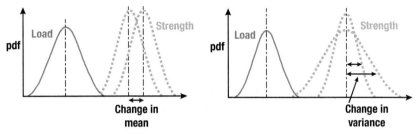

Figure 5.7 Influence of quality on failure probability.

defects. The team must also have clear definitions of the threshold for acceptable quality and of what constitutes nonconformance. Products with quality nonconformances should not be accepted.

A defect is any outcome of a process that impairs or has the potential to impair the performance of the product at any time. A defect may arise during a single process or may be the result of a sequence of processes. The yield of a process is the fraction of products that are acceptable for use in a subsequent process sequence or product life cycle. The cumulative yield of the process is approximately determined by multiplying the individual yields of each of the individual process steps. The source of defects is not always apparent, because defects resulting from a process can go undetected until the product reaches some downstream point in the process.

It is often possible to simplify processes to reduce the probability of workmanship defects. As processes become more sophisticated, however, process monitoring and control are necessary to ensure a defect-free product. The bounds that specify whether the process is within tolerance limits, often referred to as the process window, are defined in terms of the independent variables to be controlled within the process and the effects of the process on the product. The goal is to understand the effect of each process variable on each product parameter to formulate control limits for the process—that is, the condition in which the defect rate begins to have a potential for causing failure. In defining the process window, the upper and lower limits of each process variable beyond which defects might occur must be determined. Manufacturing processes must be contained in the process window by defect testing, analysis of the causes of defects, and elimination of defects by process control, such as using closed-loop corrective action systems. Establishing an effective feedback path to report process-related defect data is critical. Once this is accomplished and the process window is determined, the process window itself becomes a feedback system for the process operator.

Several process parameters may interact to produce a different defect than would have resulted from an individual parameter acting independently. This complex case may require that the interaction of various process parameters be evaluated by a design of experiments.

In some cases, a defect cannot be detected until late in the process sequence. Thus, a defect can cause rejection, rework, or failure of the product after considerable value has been added to it. This cost can reduce return on investment by adding to hidden factory costs. All critical processes require special attention for defect elimination by process control.

5.12.2 Process Verification Testing

Process verification testing is often called screening. Screening involves 100% auditing of all manufactured products to detect or precipitate defects. The aim of this step is to preempt potential quality problems before they reach the field. Thus, screening can aid in reducing warranty returns and increase customer goodwill. In principle, screening should not be required if parts (materials) are selected properly and if processes are well-controlled.

Some products exhibit a multimodal probability density function for failures, with peaks during the early period of their service life due to the use of faulty materials, poorly controlled manufacturing and assembly technologies, or mishandling. This type of early-life failure is often called infant mortality. Properly applied screening techniques can successfully detect or precipitate these failures, eliminating or reducing their occurrence in field use. Screening should only be considered for use during the early stages of production, if at all, and only when products are expected to exhibit infant mortality field failures. Screening will be ineffective and costly if there is only one main peak in the failure probability density function. Further, failures arising due to unanticipated events, such as lightning or earthquakes, may be impossible to cost-effectively screen.

Since screening is conducted on a 100% basis, it is important to develop screens that do not harm good products. The best screens, therefore, are nondestructive evaluation techniques, such as microscopic visual exams, X-rays, acoustic scans, nuclear magnetic resonance, electronic paramagnetic resonance, and so on. Stress screening involves the application of loads, possibly above the rated operational limits. If stress screens are unavoidable, overstress tests are preferred over accelerated wearout tests, since the latter are more likely to consume some useful life of good products. If damage to good products is unavoidable during stress screening, then quantitative estimates of the screening damage, based on failure mechanism models, must be developed to allow the design team to account for this loss of usable life. The appropriate stress levels for screening must be tailored to the specific product. As in qualification testing, quantitative models of failure mechanisms can aid in determining screen parameters.

A stress screen need not necessarily simulate the field environment, or even utilize the same failure mechanism as the one likely to be triggered by this defect in field conditions. Instead, a screen should exploit the most convenient and effective failure mechanism to stimulate the defects that can show up in the field as infant mortality. This requires an awareness of the possible defects that may occur in the product and familiarity with the associated failure mechanisms.

Any commitment to stress screening must include the necessary funding and staff to determine the root cause and appropriate corrective actions for all failed units. The type of stress screening chosen should be determined by the design, manufacturing, and quality teams. Although a stress screen may be necessary during the early stages of production, stress screening carries substantial penalties in capital, operating expense, and cycle time, and its benefits diminish as a product approaches maturity. If many products fail in a properly designed screen test, the design is probably faulty, or a revision of the manufacturing process may be required. If the number of failures in a screen is small, the processes are likely to be within tolerances and the observed faults may be beyond the resources of the design and production process.

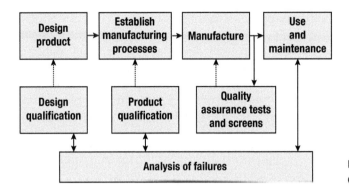

Figure 5.8 Reliability management using a closed-loop process.

5.13 Analysis, Product Failure, and Root Causes

Product reliability needs to be ensured using a closed-loop process that provides feedback to design and manufacturing in each stage of the product life cycle. Data obtained from manufacturing, assembly, storage, shipping periodic maintenance, and use and health monitoring methods can be used to aid future design plans, tests, and perform timely maintenance for sustaining the product and preventing catastrophic failures. Figure 5.8 depicts the closed-loop process for managing the reliability of a product over the complete life cycle.

The objective of closed-loop monitoring is to analyze all failures throughout the product life cycle to identify the root cause of failure. The root cause is the most basic casual factor or factors that, if corrected or removed, will prevent recurrence of the situation. The purpose of determining the root cause(s) is to fix the problem at its most basic source so it does not occur again, even in other products, as opposed to merely fixing a failure symptom.

Correctly identifying root causes during design, manufacturing, and use, followed by taking appropriate corrective actions, results in fewer field returns, major cost savings, and customer goodwill. The lessons learned from each failure analysis need to be documented, and appropriate actions need to be taken to update the design, manufacturing process, and maintenance actions.

After products are developed, resources must be applied for supply chain management, obsolescence assessment, manufacturing and assembly feedback, manufacturer warranties management, and field failure and root-cause analysis. The risks associated with the product fall into two categories:

- *Managed Risks.* Risks that the product development team chooses to proactively manage by creating a management plan and performing a prescribed monitoring regime of the field performance, manufacturer, and manufacturability
- *Unmanaged Risks.* Risks that the product development team chooses not to proactively manage.

If risk management is considered necessary, a plan should be prepared. The plan should contain details about how the product is monitored (data collection), and how the results of the monitoring feed back into various product development processes. The feasibility, effort, and cost involved in management processes must be considered.

5.14 Summary

The development of a reliable product is not a matter of chance; rather, it is a rational consequence of conscious, systematic, and rigorous efforts conducted throughout the entire life cycle of the product. Meeting the targeted product reliability can only be assured through robust product designs, capable processes that are known to be within tolerances, and qualified parts (materials) from vendors whose processes are also capable and within tolerances. Quantitative understanding and modeling of all relevant failure mechanisms can guide design, manufacturing, and the planning of test specifications.

When utilized early in the concept stage of a product's development, reliability analysis serves as an aid to determine feasibility and risk. In the design stage of product development, reliability analysis involves the selection of parts (materials), design trade-offs, design tolerances, manufacturing processes and tolerances, assembly techniques, shipping and handling methods, and maintenance and maintainability guidelines. Engineering concepts such as strength, fatigue, fracture, creep, tolerances, corrosion, and aging play a role in these design analyses. Physics-of-failure concepts, coupled with mechanistic and probabilistic techniques, are used to assess the potential problems and trade-offs and to take corrective actions.

Problems

5.1 Production lots and vendor sources for parts that comprise a design are subject to change, and variability in parts characteristics is likely to occur during the fielded life of a product. How does this affect design decisions that impact reliability?

5.2 Discuss the relationship between manufacturing process control and stress margins. How does this affect qualification? What are the implications for product reliability?

5.3 List five characteristic life-cycle loads for a computer keyboard. Describe how the product design could address these in order to ensure reliability.

5.4 Explain how the globalization of the supply chain could affect the parts selection and management process for a product used for critical military applications.

5.5 Explain the distinction between FMEA and FMMEA and how this is significant for design for reliability. For example, how would an FMMEA affect product qualification testing?

5.6 Explain how the intended application for a product would affect the decision on whether to incorporate redundancy into its design. Include in your answer a discussion of the relevant constraints related to product definition.

5.7 Discuss the concept of design for manufacturability, and how it can lead to improvement of product reliability. Provide a specific example.

5.8 What are the advantages and disadvantages of virtual qualification as compared with accelerated testing? How can these be combined in a qualification program to reduce the overall product design cycle time?

5.9 For a top-level event T, the following MCSs were identified: ABC, BDC, AE, ADF, and BEF. Draw a fault tree for the top event of these MCS.

5.10 Using the rules of Boolean algebra, show that

$$[(A \bullet B)+(A \bullet B')+(A' \bullet B')] = A' \bullet B.$$

5.11 Using the rules of Boolean algebra, show that

$$(A' \bullet B \bullet C') \bullet (A \bullet B' \bullet C')' = C + [(A' \bullet B')+(A+B)].$$

5.12 Using the rules of Boolean algebra, show that

$$[(X \bullet Y)+(A \bullet B \bullet C)] \bullet [(X \bullet Y)+(A'+B'+C')] = X \bullet Y.$$

6 Product Requirements and Constraints

Product development is a process in which the perceived need for a product leads to the definition of requirements that are translated into a design. The definition of requirements is directly derived from the needs of the market and the constraints in producing the product.

6.1 Defining Requirements

One of the first steps in product development is the process of transforming broad goals and vague concepts into realizable, concrete requirements. While the company's core competencies, cultures, goals, and customers all influence the requirements, Figure 6.1 shows that the product definition results from a combination of marketing and business-driven product requirements, design and manufacturing constraints, and various external influences.

Marketing often takes the lead in determining requirements for products such as toys, cell phones, and personal computers. For components of more complex products (e.g., an engine control module for an automobile), the requirements and constraints are often defined by the customer (the manufacturer of the product or system that the component fits into), and the marketing function is less involved.

The development of product specifications begins with an initial set of objectives, which are formulated into a preliminary requirements document. These should be approved by many people, ranging from engineers to corporate management to customers (the actual people involved in the approval depends on the company and the product). Once the requirements are approved, engineering typically prepares a specification indicating the exact set of requirements that are "of value" to implement.

Design decisions are a balance of all the requirements, as per the final specifications for the product. The design may be adjusted to reduce cost or to improve such attributes as ergonomics, safety, performance, quality, and reliability.

Reliability Engineering, First Edition. Kailash C. Kapur and Michael Pecht.
© 2014 John Wiley & Sons, Inc. Published 2014 by John Wiley & Sons, Inc.

6 Product Requirements and Constraints

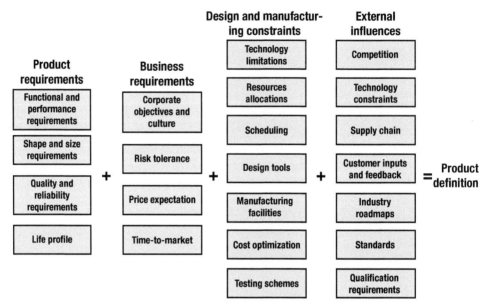

Figure 6.1 Example requirements and constraints in the product definition process.

The cost and ease of product support may also factor into product requirements to improve maintenance and accessibility to spares, support equipment, and personnel who can test and repair the product. A poor definition of requirements will often lead to a poor design, for example, where the air conditioning compressor of a car has to be removed to replace a spark plug or special tools are necessary to replace the oil in a car.

6.2 Responsibilities of the Supply Chain

The IEEE Reliability Program Standard 1332 (IEEE Std. 1332–1998) presents the relationship between the component suppliers and customers in terms of reliability objectives. The standard identifies three such objectives. First, the supplier, working with the customer, must determine and understand the customer's requirements and product needs, so that a comprehensive design specification can be generated. Second, the supplier must structure and follow a series of engineering activities to insure that the resulting product satisfies the customer's requirements and product needs with regard to product reliability. Third, the supplier must include activities that assure the customer that the reliability requirements and product needs have been satisfied.

6.2.1 Multiple-Customer Products

For commercial products, requirements are often defined by technical marketing groups. To define the product requirements, these groups use information from diverse sources, including sales personnel from previous or similar product groups, market focus groups, market research and customer surveys, analysis of the competition, corporate objectives, industry standards and agreements, and product roadmaps and trends. The objective of defining requirements is to maximize profit by appealing to the largest possible customer base.

Business goals must also justify the creation of the product. Business goals that might be considered include creating a product that:

- *Fills a Perceived Market Need.* A new product may be defined and produced because a compelling business case can be made that there is a market for the product that can be tapped.
- *Opens New Markets.* A new product may be defined and produced (possibly at a financial loss) to facilitate the company's entry into a new market that it considers to be strategic and/or believes can be created for the product.
- *Keeps the Company Competitive in a Key Market (Maintaining Market Share).* A new product may be needed to maintain competitiveness in a market that the company considers important. For example, in the cellular phone market, companies must continuously define and create more advanced phones if they want to maintain market share. Skipping a generation of phones may cause a loss of market share that can never be recovered.
- *Fills a Need of Specific Strategic Customers.* Many companies tailor their business decisions to the needs of a few influential customers (or a key market segment). Product requirements may be defined solely to satisfy one customer's needs, either because the customer is viewed as a leader in its marketplace or because that customer has a large enough demand for the product to justify making the product.
- *Improves Maintainability/Extensibility and Cost of an Existing Product.* Redesigning or modifying a product may increase profit margins or competitiveness by reducing the cost of producing, supporting, or extending its life.

6.2.2 Single-Customer Products

Many low volume and specialty applications effectively have only a single customer who predefines the product's requirements. Delivering products with these requirements at the lowest life-cycle cost, and/or strategically exceeding the minimum requirements, gains the customer's business. For single-customer products, the requirements and constraints are determined from customer definitions, analysis of the competition, and technology roadmaps and trends. The business reasons that justify the creation of the product include:

- *Maintaining Market Share and Customer Confidence.* Just as with commercial products, a new product may be needed to remain competitive in a market or with a customer that the company considers important. A new product may also be important to retain the customer's confidence that the company is a long-term partner and a legitimate competitor in a strategic area.
- *Demonstrating Experience with a New Technology or Methodology.* In the single-customer market, customers are often influenced by prior experience with cutting- edge and niche technologies. Continued demonstration of the ability to integrate these techniques maintains customer confidence.

- *Reducing Life-Cycle Costs of an Existing Product.* For single-customer products, the manufacturing cost of the product may not be as important as the costs of product sustainment (support, maintenance, and upgrade). Product redesigns or modifications for reducing the life-cycle costs are often important to the customer.

6.2.3 Custom Products

Some low volume commercial products and "modular" single-customer products do not fit into either of the categories defined earlier. These are products that are designed with a minimum set of "generic" requirements and then customized using each customer's specific requirements. Examples include supercomputers, military equipment (used across multiple Army, Navy, and Air Force platforms), and corporate Intranets. For these types of products, the guidelines defined above are still relevant. Reconfigurability with platform compatibility and upgradeability may be the salient features of these designs.

6.3 The Requirements Document

The actual content of the requirements document will depend on the application; however, the requirements and constraints fall into the general categories shown in Table 6.1. In addition to defining values for the requirements listed in Table 6.1, the requirements document may also assign a priority to each requirement. Table 6.2 provides three grades often used to prioritize requirements. Schedule and cost might not be included in the requirements document for some products if the document is released to other internal or external organizations for bids.

The inclusion of irrelevant requirements can lead to unnecessary expenditures and time for design and testing. Irrelevant or erroneous requirements result from two sources: personnel who do not understand the constraints and opportunities implicit in the product definition, and including requirements for historical reasons. The latter get "cut and pasted" into requirements documents from a previous product. No one knows exactly why the requirement is included, but no one is brave enough to remove it, simply because no one takes the time to see the obvious mistake or to question the norm.

The omission of critical requirements may cause the product not to be functional or may significantly reduce the effectiveness and appeal of the product by not providing the necessary attributes. This may reduce the market size of the product and delay the product launch, shrinking the time window for achieving return on investment.

A single person cannot realistically define all product requirements for a product. To make the requirements realistic and useful, personnel from different disciplines (see Table 6.3) should contribute to, review, and approve the product requirements.

6.4 Specifications

Once a set of requirements has been completed, product engineering creates a response to the requirements in the form of a specification. The specification states:

Table 6.1 Example of product requirements

Requirement and/or constraint	Definition
Physical attributes: ■ Size, dimensions ■ Weight	Describes the physical size, shape, and weight of the final product.
Functionality	Describes what the product does, including the various features.
Performance ■ Electrical, mechanical, chemical, biological performance ■ Speed ■ Noise ■ Power dissipation	Describes the characteristics of operation of the product.
Environmental conditions ■ Temperature, temperature changes ■ Humidity, pressure ■ Vibration, flexure, shock ■ Radiation ■ Electromagnetic interference	Defines the environment conditions within which the product must be constrained to operate properly.
Reliability ■ Useful life ■ Acceptable redundancies ■ Warranty periods	The ability of a product or system to perform as intended (i.e., without failure and within specified performance limits) for a specified time, in its life-cycle conditions.
Cost/quality ■ Procurement costs ■ Assembly costs ■ Testing costs ■ Final product quality	Defines the yield of the resulting product and the cost to field a tested product.
Qualification ■ Cost and time ■ Regulations	Defines how and under what conditions the product's reliability will be assessed.
Schedule ■ Time to market ■ Product volumes	Defines when the first product needs to be delivered to customers and what volume needs to be delivered over the lifetime of the product.
Life cycle ■ Maintainability ■ Upgradeability	Defines the ease with which the product can be maintained and upgraded during its fielded life.
End of life ■ Disassembly costs ■ Recycling ■ Reuse	Defines what happens to the product after the customer is finished using it; also defines whether any end-of-life requirements exist, depending on legislation where the product is sold.
Safety and regulatory	Defines requirements related to customer safety and necessary approvals from legislatures, government agencies, and industry bodies.

■ *The Requirements That Must Be Met.* The fact that a requirement appears in a requirements document does not guarantee that it will be achieved in the final product. Technical marketing does not always understand what can be successfully engineered within a specified time window. Requirements grading defines the priorities during specification development.

6 Product Requirements and Constraints

Table 6.2 Requirements grading

Grade	Definition
Must (shall)	The requirement is essential to the viability of the product.
Should	The requirement is not essential to the viability of the product, but it should be implemented either because it adds great value or because it can be easily implemented, to add value.
Could	The requirement is not essential to the viability of the product, and its development could be delayed either because the requirement is too costly to implement or because it adds only marginal value to the product.

Table 6.3 Example of requirements buy-in for a multiple-customer product

Role	Approval level
Marketing	Approval authority
Engineering manager	Approval authority
Product manager	Approval authority

Role	Buy-in level
Development engineers	Consulted prior to approval
Reliability engineers	Consulted prior to approval
Customer	Consulted prior to approval
Application engineers	Consulted prior to approval
Quality assurance	Informed after approval
Corporate management	Informed after approval

- *The Methods by Which the Requirements Will Be Met.* This requires an outline of the basic process used to meet the requirements. The outline may consist of flowcharts, block diagrams, manufacturing processes, and possible parts lists and physical resources, such as specialized requirements.
- *The Schedule for Meeting the Requirements.* The design, prototype, procurement, and manufacturing schedules for the product are identified. Proposed schedules for contract manufacturers may also be included.
- *An Identification of Those Who Will Perform the Work.* The specific persons who will perform the work may be identified. In addition, groups within the company that perform specific functions necessary to fabricate the product, and all contract manufacturers from outside the company, should also be identified.
- *An Identification of the Potential Risks.* If any specific design, development, or manufacturing risks are known, they should be stated in this document. Possible second sources of components and services and backup plans should also be identified.

6.5 Requirements Tracking

Once product requirements are defined and the design process begins, the process of continuously comparing the product's requirements to the actual product design begins. As the product's design becomes increasingly detailed through selection of components and implementation strategies, it becomes increasingly important to track the product's characteristics (e.g., size, weight, performance, functionality, reliability, and cost) in relation to the original product requirements. The rationale for making changes should be documented and approved.

The completeness with which requirements tracking is performed can significantly reduce future product redesign costs. Planned redesigns or design refreshes through technology monitoring, and use of roadmaps ensure that a company is able to market new products or redesigned versions of old products in a timely, effective manner to retain its customer base and ensure continued profits.

6.6 Summary

Product requirements are usually defined by technical marketing groups and then reviewed and enhanced by other disciplines in the company. The product engineering function creates a response to the product requirements in the form of a preliminary specification that defines what requirements will be implemented, who will perform the work, how the work will be performed, and a schedule. Requirements are tracked to ensure that the product remains in compliance as it is developed.

Two prevalent risks in requirements and constraints definition are the inclusion of irrelevant requirements and the omission of relevant requirements. The inclusion of irrelevant requirements can involve unnecessary design and testing time as well as money. Irrelevant or erroneous requirements generally result from two sources: requirements created by personnel who do not understand the constraints and opportunities implicit in the product definition, and including requirements for historical reasons. The omission of critical requirements can significantly reduce the effectiveness of the product.

Problems

6.1 Using the example of a cellular phone, develop a list of requirements for a business application.

6.2 Classify the list of requirements in question 1 to "must," "could," and "should" categories. Follow the template in Table 6.2. You can add additional rows with justification.

6.3 Give three examples each of (a) multiple-customer products, (b) single-customer products, and (c) custom products.

6.4 Provide an example of an electronic product for which compatibility with another manufacturer's product, such as an Apple iPod, is an essential requirement. Explain how the product requirements could be specified to differentiate your offering from those of similar, competing products. Describe the constraints imposed by the compatibility requirement.

List the appropriate sources of input for this product's requirements and explain the value of each source.

7 Life-Cycle Conditions

The actual loading conditions on a product are often assumed based on engineering specifications or conjecture. This approach can lead to costly overdesign or hazardous underdesign, and consequently, increased investment. Hence, a formal method is necessary to capture the life-cycle load conditions of a product.

This chapter discusses a systematic methodology for developing a life-cycle profile (LCP) for a product. The LCP can thus be used for design and test for reliability assurance. The life-cycle conditions should be collected on actual products if possible.

7.1 Defining the Life-Cycle Profile

The life cycle of a product includes manufacturing and assembly, testing, rework, storage, transportation and handling, operational modes, repair, and maintenance. The life-cycle loads include thermal (steady-state temperature, temperature ranges, temperature cycles, and temperature gradients), mechanical (pressure levels, pressure gradients, vibrations, shock loads, and acoustic levels), chemical (aggressive or inert environments, ozone, pollution humidity levels, contamination, and fuel spills), physical (radiation, electromagnetic interference, and altitude), and/or operational loading conditions (power, power surge, heat dissipation, current, and voltage spikes). The extent and rate of product degradation depend upon the nature, magnitude, and duration of exposure to loads.

Defining and characterizing the life-cycle conditions can be the most difficult part of the overall reliability planning process, because products can be used or not used in different ways, for different amounts of time and with different care, maintenance and servicing. For example, typically all desktop computers are designed for office or home environments. However, the operational profile of each unit will depend on user behavior. Some users may shutdown the computer every time after it is used, others may shut it down only once at the end of the day, while other users may keep their computers powered on all the time. Thus, the temperature profile experienced by each product and hence its degradation due to thermal loads will be different.

Reliability Engineering, First Edition. Kailash C. Kapur and Michael Pecht.
© 2014 John Wiley & Sons, Inc. Published 2014 by John Wiley & Sons, Inc.

A life-cycle profile (LCP) is a time history of events and conditions associated with a product from its release from manufacturing to its removal from service. The life cycle should include the various phases that the product will encounter in its life. In some cases, the environmental factors experienced by constituents of the product begin before manufacturing—for example, storage of parts (material) in advance of their use in manufacturing.

An LCP helps to identify the possible load combinations so that the loads acting on the product can be identified and their effects can be accounted for in the product's design, test, and qualification process. The reliability of a product depends on the magnitude of the stresses, rate of change of stresses, and spatial variation of the stresses that are generated by the loads acting during its life cycle.

Three key steps are given for the development of LCP:

- The first step is to describe expected events for a product from manufacture through end of life, which involves identifying the different events, which the product will pass through. Typical events include testing and qualification, storage at the test facility, transportation to the place of installation, storage at the place of installation, transportation to the specific site of installation, installation, operation, and field service during scheduled maintenance. It also involves identifying product requirements, such as who will use the product, what platform will carry it, and the operational requirements, deployment, and transportation concepts.

- The second step is to identify significant natural and induced environmental conditions or their combinations for each expected event. This involves identifying the load conditions that act in each of the identified events. The natural environment is the product's natural ambient conditions, for example, temperature, pressure, and humidity. The induced environment is the product's environmental conditions related to the specific functionality of the product. For example, electronics on a drilling tool experience mechanical vibration during the drilling process. Electronics used to control an aircraft engine will include high steady-state temperature dwells, temperature cycling, low pressures, and random vibrations.

- The third step is to describe load conditions to which the product will be subjected during the life cycle, which involves the quantification of load conditions identified as a result of the previous two steps. Data should be determined from real-time measurements but may be estimated by simulation and laboratory tests. For example, the vibrations experienced by a product during shipping could be identified by a mock shipping experiment wherein sensors are kept with the product to record vibration data. The loads should be quantified in a statistical manner to identify the range and variability of the load.

7.2 Life-Cycle Events

Since the LCP is application and product dependent, a thorough analysis of the possible load conditions in each of the events is necessary during the design of any product. Typical loads in most events include temperature, vibration, shock, pressure, humidity, and the induced environments. However, load conditions such as radiation, fungi/microorganisms, fog, freezing rain, snow, hail, sand and dust, salt spray, and wind should not be overlooked. In consumer products, drink spills and food may also be important loads to consider.

7.2.1 Manufacturing and Assembly

Assembly of a product also involves load conditions. For example in assembly of electronic products, soldering operations can lead to significant thermal stresses in the components being assembled, as well as surrounding components. The mechanical handling, placement, and assembly procedures can also induce vibration, shock, and loads. Other load conditions which might be critical are radiation, chemical and ionic contamination (plasma machining or welding), humidity, and pressure, depending on the assembly process.

7.2.2 Testing and Screening

The load conditions a product is subjected to during testing and screening should not affect the product if it is to be subsequently placed in the market. However, tests and screens, such as high-temperature bake, high temperature operating life, vibration, shock, temperature humidity bias (THB), highly accelerated life testing (HALT), and highly accelerated stress testing (HAST), will impact the product reliability and remaining useful life to some extent.

7.2.3 Storage

Storage typically has temperature (diurnal cycles) and humidity as the prime load conditions. However, depending on the quality of the storage facility, load conditions such as rain, snow, fungi, sand and dust, and radiation might also come into the picture. Chemical gases are also an issue when the product is stored in chemically aggressive environments.

7.2.4 Transportation

Transportation is often characterized by high vibration, shock, and temperature loads. Transportation by road can cause shock and vibration due to rocky and uneven paths, internal vibrations, and accidents. The product can also be subjected to diurnal temperature cycles, as well as to heat generated by the operation of the vehicle. Transportation by air can subject the product to vibrations while taking off and landing, as well as to temperature cycling due to the differences in ground and airborne temperatures. Apart from these specific loads, the product can also experience sand and dust, gases, humidity, and radiation.

7.2.5 Installation

The installation process is typically characterized by vibration and shock loads. In deployment of permanent monitoring equipment for oil wells, the equipment suddenly encounters very high temperatures when it comes in contact with the hot oil inside the tubes during deployment.

7.2.6 Operation

The load conditions during operation are specific to the application. For example, an electronic product in the under hood of a car encounters temperature cycling and vibration, whereas the electronics inside a desktop computer has limited vibrations. Humidity, on the other hand, might be a consideration in both of these applications.

7.2.7 Maintenance

Maintenance, in some cases, can subject the product to loads due to handling and mishandling of the product. Shock and vibration are typical loads associated with maintenance procedures. For electronic products, electrostatic discharge can be an issue when proper care is not taken during maintenance.

7.3 Loads and Their Effects

Table 7.1 provides some of the load conditions and their possible effects on products. Some of these conditions are discussed in more detail in the following sections.

7.3.1 Temperature

Temperature can influence the electrical, mechanical, chemical, and physical deterioration of materials for two main reasons: many of the properties of materials can be altered by changes in temperature, and the rate of a chemical reaction between two or more reactants is generally dependent on the temperature of the reactants. Some of the adverse effects of temperature include the expansion or contraction of materials due to temperature changes, causing problems with fit between product interfaces and couplings, outgassing of corrosive volatile products due to application of heat, local stress concentrations due to nonuniform temperature, and the collapse of metal

Table 7.1 Load conditions and their possible effects

Load conditions		Principal effects	Possible Failures
Temperature (natural/induced)	High	■ Thermal aging ■ Oxidation ■ Structural change ■ Chemical change ■ Softening and melting ■ Viscosity reduction/ evaporation ■ Physical expansion	■ Insulation failure because of melting ■ Alteration of electrical properties due to changes in resistance ■ Unequal expansion between coupled parts leading to fatigue or fracture ■ Ionic contamination ■ Surface degradation
	Low	■ Physical contraction ■ Brittleness	■ Alteration of electrical properties due to changes in resistance ■ Unequal expansion between coupled parts, leading to fatigue or fracture ■ Increased brittleness of metals

Table 7.1 (Continued)

Load conditions		Principal effects	Possible Failures
Relative humidity/ moisture (natural/induced)	High	■ Moisture absorption ■ Chemical reaction ■ Corrosion ■ Electrolysis	■ Corrosion ■ Electrical shorting ■ Loss of electrical properties owing to corrosion and chemical reactions ■ Cracking of materials due to moisture absorption ■ Reduction in electrical resistance because of conduction through moisture
	Low	■ Desiccation ■ Embrittlement ■ Granulation	■ Loss of mechanical strength ■ Structural collapse ■ Alteration of electrical properties
Pressure (natural/induced)	High	■ Compression	■ Structural collapse ■ Penetration of seals ■ Interference with function
	Low	■ Expansion ■ Outgassing	■ Explosive expansion of parts ■ Alteration of electrical properties ■ Loss of mechanical strength ■ Insulation breakdown
Wind (natural)		■ Force application ■ Deposition of materials ■ Heat loss (low velocity) ■ Heat gain (high velocity)	■ Structural collapse ■ Interference with function ■ Loss of mechanical strength ■ Mechanical interference and clogging ■ Accelerated abrasion ■ Removal of protective coatings ■ Surface deterioration
Salt spray (natural)		■ Chemical reactions ■ Corrosion ■ Electrolysis	■ Increased wear ■ Alteration of electrical properties ■ Interference with function ■ Surface deterioration ■ Increased conductivity
Sand and dust (natural)		■ Abrasion ■ Clogging	■ Increased wear ■ Interference with function ■ Alteration of electrical properties ■ Removal of protective coatings ■ Surface deterioration
Rain (natural)		■ Physical stress ■ Water absorption and immersion ■ Erosion ■ Corrosion	■ Structural collapse ■ Increase in weight ■ Electrical failure ■ Structural weakening ■ Removal of protective coatings ■ Surface deterioration ■ Enhanced chemical reactions like corrosion
Ionized gases (natural)		■ Chemical reactions ■ Corrosion ■ Change in conductivity	■ Change in electrical properties ■ Deterioration in material properties

(Continued)

Table 7.1 (Continued)

Load conditions	Principal effects		Possible Failures
Air pollution (natural)	■ Chemical reactions ■ Clogging		■ Interference in functionality ■ Deterioration in material properties owing to chemical reactions ■ Corrosion
Freezing rain/ frost/snow (natural)	■ Low temperature ■ Moisture ingress ■ Corrosion/chemical reactions ■ Clogging		■ Mechanical stress caused by expansion mismatch between structural components ■ Increase in weight ■ Change in electrical properties due to change in resistance/conductivity ■ Delamination of materials ■ Material deterioration ■ Corrosion
Fungi (natural)	■ Clogging		■ Change in electrical characteristics due to shorts and alteration in electrical resistance ■ Oxidation of structural elements
■ Static electricity ■ Electrostatic discharge (natural/induced)	■ Change in electrical response ■ Electrical overstress		■ Interference in function due to changes in electrical properties (resistance, voltage) ■ Shorts or opens in circuit
Chemicals (induced)	■ Chemical reactions ■ Reduced dielectric strength		■ Alteration of physical and electrical properties ■ Insulation breakdown ■ Corrosion
Explosion (induced)	■ Severe mechanical stress		■ Rupture and cracking ■ Structural collapse of the product
Shock (induced)	Thermal	■ Mechanical stress	■ Unequal expansion between coupled materials of the product leading to fatigue or fracture ■ Surface degradation ■ Melting
	Mechanical	■ Mechanical stress ■ Fatigue	■ Loss of mechanical strength ■ Interference with function ■ Increased wear ■ Fatigue ■ Structural collapse of products
Vibration (induced)	■ Vibration/ acceleration	■ Mechanical stress ■ Fatigue	■ Loss of mechanical strength ■ Interference with function ■ Increased wear ■ Fatigue ■ Structural collapse of product
	■ Rotation	■ Mechanical stress ■ Torsional acceleration	■ Twisting of parts ■ Loss of mechanical strength ■ Deformation
	■ Bending	■ Mechanical stress ■ Fatigue	■ Bending failure ■ Cracking ■ Deformation

structures when subjected to cyclic heating and cooling due to induced stresses and fatigue caused by repeated flexing.

7.3.2 Humidity

Water vapor in the air or any other gas is called "humidity"; water in solids or absorbed in liquids is usually designated "moisture." Relative humidity (RH) is the ratio of actual vapor pressure to saturation vapor pressure at the prevailing temperature. It is usually expressed as a percentage. Absolute humidity is the mass of water vapor per unit mass of dry air in the sample volume at the prevailing temperature. Vapor pressure is the part of the total pressure contributed by the water vapor. Dew point temperature is the temperature to which a gas must be cooled at constant pressure to achieve saturation.

Humidity or moisture can play a major role in accelerating failures in products. Failure mechanisms, such as corrosion, contamination, and swelling of polymer-based structural elements or potting, are all adversely impacted by the presence of moisture. Moisture can cause mated parts in a product to lock together, especially when water condenses on them and then freezes. Many materials that are normally pliable at low temperatures can become hard and brittle due to absorption of water, which subsequently freezes at low temperatures. The volume increase due to freezing of water can also separate parts, materials, or connections.

Moisture can also act as a medium for the interaction between several otherwise relatively inert materials. For example, chlorine will be released by polyvinyl chloride (PVC), and form hydrochloric acid when combined with moisture. Moisture with certain ionic materials can cause shorts or leak paths between metal traces or adjacent conductors on printed circuit boards in electronic products.

Although the presence of moisture may cause deterioration, the absence of moisture can also cause reliability problems. Many nonmetallic materials become brittle and crack when they are very dry. The properties of these materials depend upon an optimum level of moisture. For example, fabrics wear out at an increasing rate as moisture levels are lowered, and fibers become dry and brittle. Environmental dust, which is usually held in suspension by moisture, can cause increased wear and friction on moving parts. Freed dust can clog filters due to the absence of moisture.

Design techniques can be used to counteract the effects of moisture. For example, moisture traps can be eliminated by providing drainage or air circulation, using desiccant systems to remove moisture when air circulation or drainage is not possible, applying protective coatings, providing rounded edges to allow uniform coating of protective material. Using materials resistant to fungi, corrosion, and other moisture-related effects and hermetically sealing components by using gaskets and other sealing products can also prevent degradation due to moisture. Other design techniques include impregnating or encapsulating materials in moisture-resistant waxes, plastics, or varnishes, separating dissimilar metals or materials that might combine or react in the presence of moisture or of components, which might damage protective coatings.

The design team must consider possible adverse effects caused by specific methods of protection. Hermetic sealing, gaskets, and protective coatings may, for example, increase moisture by sealing moisture inside or contributing to condensation. Gasket materials must be evaluated carefully for outgassing of volatile vapors or for incompatibility with adjoining surfaces or protective coatings.

7.3.3 Vibration and Shock

Vibrations result from dynamic forces that set up a series of motions within a product. The forced motions may be linear, angular (torsion), or a combination of both. A vibratory system includes, in general, a means for storing potential energy (spring or elasticity), a means for storing kinetic energy (mass or inertia), and a means by which energy is gradually lost (damping or resistance).

Fatigue, which is the tendency of a material to yield and fracture under cyclic stress loads considerably below its tensile strength, is a failure mechanism that may result from vibrations. Fatigue failures include high cycle fatigue, acoustic fatigue, and fatigue under combined stresses such as temperature extremes, temperature fluctuations, and corrosion.

Some of the common faults that may be caused by vibration include bent shafts, damaged or misaligned drives and bearings, fretting corrosion, onset of cavitations, and worn gears. Vibration and shock can harmfully flex electrical leads and interconnects, cause parts to strike the housing, dislodge parts from their positions, cause acoustical and electrical noise, and lead to structural instabilities.

Protective measures against vibration and shock are generally determined by an analysis of the deflections and mechanical stresses produced by these load conditions. This involves the determination of natural frequencies and evaluation of the mechanical stresses within components and materials produced by the shock and vibration environment. If the mechanical stresses are below the acceptable safe working stress levels of the materials involved, no direct protection methods are required. If the stresses exceed the safe levels, corrective measures such as stiffening, reduction of inertia and bending moment effects, and incorporation of further support members, as well as possible uses of isolators, may be required. If such approaches do not reduce the stresses below the acceptable safe levels, further reduction is usually possible by the use of shock-absorbing mounts.

In addition to using proper materials and configurations, it is necessary to control the amount of shock and vibration experienced by the product. Damping systems are used to reduce peak oscillations and special stabilizers can be employed when unstable configurations are involved. Typical examples of dampers are viscous hysteresis, friction, and air damping. Vibration isolators are commonly identified by their construction and material used for resilient elements like rubber, coiled spring, and woven metal mesh. Shock isolators differ from vibration isolators in that shock requires a stiffer spring and a higher natural frequency for the resilient element. Isolation mounting systems are of the type installed underneath, the over-and-under type, and inclined isolators. In some cases, however, even though a product is properly insulated and isolated against shock and vibration damage, repetitive forces may loosen the fastening systems. If the fastening systems loosen enough to permit movement, the product will be subjected to increased forces and may fail. Many specialized self-locking fasteners are available to counter this occurrence.

7.3.4 Solar Radiation

Solar radiation contributes several types of loads to the life-cycle environment. The solar flux provides radiant heating, ionizing radiation, including ultraviolet exposure, and visible wavelengths that can interfere with optics.

The maximum solar load outside the atmosphere occurs on January 2 of each year when the Earth is closest to the sun. The solar flux is taken at an average of 1367 W/m^2, with a January peak of 1413 W/m^2 and the July 4 minimum at 1332 W/m^2. The sun can be modeled as a black body radiator at 6000 K. Therefore, the sun emits ultraviolet (UV) radiation. Objects in orbit receive this flux projected onto their area unless shadowed by the Earth. The Earth's atmosphere attenuates and scatters much of the incident solar energy. The solar radiation on objects on the surface of the Earth is the sum of the projected area normal to the Earth–sun line flux, a function of the time of day and location, energy incident by a scattered path, and energy reflected off other surface objects.

The primary effect of sunlight is heating. Surface temperatures in space are directly dependent upon the ratio of solar absorbtivity to infrared emissivity. This ratio is important on the surface of the Earth, but convection also plays a dominant role in determining surface temperatures.

The sun's light also provides damaging UV radiation on products. For example, organics used in plastics and paints, wiring, cables, and connectors are especially vulnerable to damage by UV radiation. Optical components such as security cameras are vulnerable to damage by heat, direct solar exposure, thermal loading, and functional interference by glint or overexposure.

7.3.5 Electromagnetic Radiation

Products stored near nuclear reactors, isotropic nuclear sources, accelerators, or nuclear detonations must be designed to tolerate the effects of nuclear irradiation. For example, time-dependent wearout failures can cause an embrittlement phenomenon that increases the hardness and decreases the ductility of metals. Another failure mechanism is random overstress when a single radiation particle interacts with the electronic circuitry.

In general, metals are quite resistant to radiation damage in the space environment. Semiconductor devices may be affected by gamma rays, which increase leakage currents. The lattice structure of semiconductors can be damaged by high energy electrons, protons, and fast neutrons, which cause permanent effects through atomic displacement and damage to the lattice structure. Organic materials are particularly susceptible to physical changes in cross-linking and scission of molecular bonds. Radiation-induced formation of gas, decreased elasticity, and changes in hardness and elongation are some of the predominant changes in plastics which have been subjected to radiation of the type encountered in the space environment.

Protection against the effects of electromagnetic radiation has become an engineering field by itself: electromagnetic compatibility design. The most direct approach to protection is to entirely avoid the region in which high radiation levels are found. When exposure cannot be avoided, shielding and filtering are the protective measures used. In other cases, material design changes or operating procedural changes must be instituted in order to provide protection or to minimize the effects on normal operation of the product.

7.3.6 Pressure

Pressure is defined as the normal force per unit area exerted by a fluid (either a liquid or a gas) on any surface. The surface is typically a solid boundary in contact with the fluid. Finding the component of the force normal to the surface is sufficient for determining the pressure. Pressure can be expressed in four ways:

- *Absolute Pressure.* The same as the definition given above. It represents the pressure difference between the point of measurement and a perfect vacuum where the pressure is zero.
- *Gage Pressure.* The pressure difference between the point of measurement and the ambient pressure.
- *Differential Pressure.* The pressure difference between two points, one of which is chosen to be the reference.
- *Stagnation Pressure.* The pressure due to fluid flow.

In high vacuum conditions (such as space), materials having a high vapor pressure will sublimate or evaporate rapidly, particularly at elevated temperatures. In some plastics, the loss of the plasticizing agent by evaporation will cause cracking, shrinking, or increased brittleness. Inorganic coatings with low vapor pressures can be used to protect metals such as magnesium, which would normally evaporate rapidly.

In a high vacuum, adjoining solid surfaces can become cold-welded after losing adsorbed gases from their surfaces. Some form of lubrication is therefore necessary. Conventional oils and greases evaporate quickly. Graphite becomes unsatisfactory and actually behaves as an abrasive because of the loss of absorbed water. However, thin films of soft metals, such as lead, silver, or gold, are effective lubricants in a high vacuum. Thin films of molybdenum disulfide are often sprayed over chrome or nickel plating, forming easily sheared layers.

7.3.7 Chemicals

The Earth's environment contains numerous chemically active elements, such as sulfur, phosphorus, chlorine, nitrogen, snow, ice, sand, dust, saltwater spray, and organic matter, which have the ability to corrode and deteriorate materials.

A material or structure can undergo a chemical change in a number of ways. Among these are interactions with other materials, such as corrosion, metal migration and diffusion, and modifications in the material itself, such as recrystallization, stress relaxation, and phase change. In addition to the deterioration problems associated with the external environments to which products are subjected, adhesives, batteries, and certain types of capacitors are susceptible to chemical aging and biological growths due to biochemical reactions.

Materials widely separated in the electrochemical series are subject to galvanic action, which occurs when two chemically dissimilar metals are in contact in an electrolytic liquid medium. The more active metal dissolves, and an electric current flows from one metal to the other. Coatings of zinc are often applied to iron so that the zinc, which is more active, will dissolve and protect the iron. This process is commonly known as "galvanization." Galvanic action is also known to occur within the same piece of metal if one portion of the metal is under stress and has a higher free-energy level than the other. The part under stress will dissolve if a suitable liquid medium is present.

Stress-corrosion cracking occurs in certain magnesium alloys, stainless steels, brass, and aluminum alloys. It has also been found that a given metal will corrode much more rapidly under conditions of repeated stress than when no stress is applied.

Proper design of a product therefore requires trade-offs in selecting corrosion-resistant materials, specifying protective coatings, use of dissimilar metallic contacts, controlling metallurgical factors to prevent undue internal life-cycle conditions, preventing water entrapment, using high temperature resistance coatings when necessary, regulating the environment through dehydration, rust inhibition, and electrolytic and galvanic protective techniques.

7.3.8 Sand and Dust

In relatively dry environments, such as deserts, fine particles of dust and sand can readily be agitated into suspension in the air, where they may persist for many hours, sometimes reaching heights of several thousand feet. Thus, even though there is virtually no wind present, the speed of vehicles that may be housing an electronic product and moving through these dust clouds can also cause surface abrasion by impact.

Although dust commonly is considered to be fine, dry particles of earth, it also may include minute particles of metals, combustion products, and solid chemical contaminants. These other forms may cause direct corrosion or fungal effects on products, because this dust may be alkaline, acidic, or microbiological. Dust accumulations have an affinity for moisture, and this may accelerate corrosion and biological growth.

Dust reduction methods are mainly of two types: active and passive. Active methods include installation of fans to increase the flow of air and use of filters and shelters. Passive methods include measures such as planting trees, and improving pollution standards. Dust protection must be planned in conjunction with protective measures against other environmental factors. For example, specifying a protective coating against moisture, if sand and dust are present, is useless unless the coating is carefully chosen to resist abrasion and erosion. When products require air circulation for cooling or for removing moisture, the issue is not whether to allow dust to enter, but rather to control the size of the dust particles. The problem becomes one of filtering the air to remove dust particles above a specific nominal size. For a given working filter area, these filters decrease the flow of air or other cooling fluids through the filter, while the ability of the filter to stop smaller and smaller dust particles is increased. Therefore, there must be a trade-off between the filter surface and the decrease of flow of the fluid through the filter or the allowable particle size.

7.3.9 Voltage

Voltage load in the form of input voltage, feedback voltage, voltage drops, and transient spikes can affect functionality and trigger several failure mechanisms in electronic products. Over voltage can cause electrical overstress (EOS). High voltages may also result in gate oxide breakdown. The high voltage (100 V to 20 KV) associated with electrostatic discharge (ESD) can cause damage to thin dielectrics, such as the gate oxides in CMOS processes, and the high energy can result in thermal damage in both bipolar and CMOS devices. Low voltage electrostatic pulse can cause damage to the gate oxides of MOS transistors if no protection circuit is present. Drain-source shorts are the most severe form of damage observed.

7.3.10 Current

Current loads manifests as steady-state high level of current, current variations, excessive leakage current (such as supply leakage, gate leakage, and drain-source leakage), and transient current spikes. This load condition is particularly important for electronic products. Supply current monitoring is routinely performed for testing of CMOS ICs. This method is based upon the notion that defective circuits produce an abnormal or at least significantly different amount of current than the current produced by fault-free circuits. This excess current can be sensed to detect faults. The high power supply quiescent current has been reported as a precursor for defects such as bridging, opens, and parasitic transistor defects.

Overcurrent can cause electrical overstress (EOS). As the semiconductor junctions get hotter, more current flows in the hot regions and a thermal runaway condition is reached. Eventually, the device is driven into a second breakdown as the temperature approaches the melting point of silicon. Failures may be due to silicon melting, causing the junctions to short circuit, or the metallization to melt and open circuit.

7.3.11 Human Factors

Humans can directly induce load conditions to a product. Humans are active participants in the operation of most systems, and this interaction must be weighed against safety, reliability, maintainability, and other product parameters to assess product reliability, maintainability, time performance, safety analyses, and specific human engineering design criteria. Humans by virtue of the way they handle a product can contribute to failures and affect the reliable operation of the product.

7.4 Considerations and Recommendations for LCP Development

The following are recommendations for obtaining data on the load conditions.

7.4.1 Extreme Specifications-Based Design (Global and Local Environments)

Extreme environmental conditions in the location of deployment of the product are often used for design.[1] Extreme conditions are unlikely to be encountered by the product in its lifetime. Moreover, the duration of maximum conditions is typically short. Further, the environment in the vicinity of the product can be modified by its functionality (local environment). Hence, the use of extreme-based specifications for the design of a product can lead to overdesign or underdesign (due to a change in the local environment).

The part's local environment, that is, the environment in the immediate vicinity of the part, often varies from the overall product's global environment, that is, the environment in the larger vicinity of the part For example, the local environment of certain parts in a desktop computer, given the heat generated from the power dissipation of the parts on the board, will be significantly hotter than the regulated office

---

[1]The highest temperature recorded on Earth is 57.77°C in Al Aziziyah, Libya, in September 1922. Death Valley, California, recorded 56.77°C in July 1913. The place that has the world's highest average temperature is Dakol, Ethiopia, in the Danakil Depression, with a mean temperature of 34.44°C. Places in Pakistan (e.g., Pad Idan) have recorded temperatures up to 50.55°C. The lowest recorded temperature on Earth to date is −89.44°C in Vostok, Antarctica.

7.4 Considerations and Recommendations for LCP Development

environment.[2] The variation between the global environment and the local environment may be a function of the part's isolation from the global environment, the existence of cooling systems within the product, the heat generated by nearby parts, and insulating air between the part and the product environment.

For example, the lowest recorded ambient temperature in Greenland is −70°C. To meet needed performance and reliability objectives, the local environment of parts in products located in Greenland must be thermally insulated or regulated (i.e., through the use of heaters). The design procedure should incorporate extreme specifications as a baseline, along with their probability of occurrence, and should modify it according to the expected local environments.

7.4.2 Standards-Based Profiles

Standards-based environmental data can be found in standards including MIL-STD-210 (United States Department of Defense 1987), MIL-STD-810 (United States Department of Defense 1989), and IPC-SM-785 (1992). MIL-STD-210 is a database of regional and worldwide climatic data. The data are divided into three groups—worldwide surface environment, regional surface environment, and worldwide air environment—and include details about basic regional types: hot regions, cold regions, severe cold regions, and coastal/ocean regions. The load conditions discussed for each of the groups include temperature, humidity, pressure rainfall rate, wind speed, blowing snow, snow load, ice accretion, hail size, ozone concentration, sand and dust, and freeze-thaw cycles in terms of extreme values, nominal (average) values, and frequency of occurrences. In spite of the details provided in MIL-STD-210, climatic data derived from this standard should not be used directly for design criteria. Rather, they should be used to derive design criteria for each product based on the response of the product to both the natural environment and the forcing functions induced by the platform on or within which the product is located (local environments).

MIL-STD-810 provides guidelines for conducting environmental engineering tasks to tailor environmental tests to end-item product applications. It contains test methods for determining the effects of natural and induced environments on product performance and is mainly focused on system-level design. The conditions and procedures described in MIL-STD-810 can be used for deriving the LCP for electronic products. Other standards, like those of the EIA, IPC, and SAE, also provide environmental data, which can be used to derive the LCP.

7.4.3 Combined Load Conditions

Combined loads (incorporating two or more environmental factors) may affect product reliability differently than a single environmental factor. If the combined effect of the environmental factors proves to be more harmful than that of a single environmental condition, then the product must be designed for failures arising from the combined effects. Some examples of the possible effects of pairs of environmental factors appear in Table 7.2.

An increase in one environmental factor can lead to an increase in another, thereby intensifying the net effect. For example, high temperatures accelerate the growth of

---

[2] For example, the local environment of certain parts in a desktop computer, given the heat generated from the power dissipation of the parts on the board, will be significantly hotter than the regulated office environment.

Table 7.2 Examples of generic effects of combined loads on products

Combined loads	Classification of effects	Possible effects
High temperature and salt spray	Intensified deterioration	High temperature tends to increase the rate of corrosion caused by salt spray and thereby increase the net effect.
High temperature and high relative humidity	Intensified deterioration	High temperature increases the rate of moisture penetration and the rate of corrosion. Thus the combination can aggravate failures caused by humidity (e.g., corrosion).
High temperature and high pressure	Intensified deterioration	Each of these environmental factors leads to deterioration in the strength of the material and can cause structural failure in electronic assemblies.
High temperature and fungi	Intensified deterioration	High temperatures provide a congenial environment for growth of fungi and microorganisms. Thus high temperatures aggravate failures caused by fungal growth.
High temperature and acceleration	Intensified deterioration/ weakened net effect	Both acceleration and high temperature affect material properties. The combination, however, can reduce failure caused by fatigue/ fracture because the material stress relaxes at high temperatures and the material becomes more pliable. In the case of brittle materials, however, this combination can lead to early failures because the material becomes weak at high temperatures and can easily fracture. In electronic products, failures caused by solder joint fatigue and cracking are diminished by the combination.
High temperature, sand, and dust	Intensified deterioration/ weakened net effect	The erosion caused by sand may be accelerated by high temperature, which can cause wear of structural parts due to abrasion. High temperature also reduces the penetration of sand and dust, thereby decreasing failures that occur from dust penetration.
High temperature, shock, and vibration	Intensified deterioration/ weakened net effect	Vibration, shock, and high temperature affect material properties and cause deterioration of mechanical properties. The combination, however, reduces failure caused by fatigue/fracture, because the material stress relaxes at high temperatures and the material becomes more pliable. Failures caused by solder joint fatigue and cracking are diminished by the combination. In case of brittle materials, however, this combination can lead to early failures because the material becomes weak at high temperatures and can easily fracture.
Low temperature and humidity	Intensified deterioration	Relative humidity increases as temperature decreases (especially in moist conditions), and lower temperature may induce moisture condensation. If the temperature is low enough, frost or ice may result. Hence, low temperatures can aggravate failures caused by humidity, frost, or ice (e.g., corrosion).
Low temperature and high pressure	Intensified deterioration	The combination can cause structural failure, such as leakage through seals and airtight enclosures.
Low temperature and salt spray	Weakened net effect	Low temperature reduces the corrosion caused by salt spray; the combination causes weakening.
Low temperature, sand, and dust	Intensified deterioration	Low temperature increases dust penetration and can aggravate failures caused by wear of assemblies and alteration of electrical properties.

7.4 Considerations and Recommendations for LCP Development

Table 7.2 (Continued)

Combined loads	Classification of effects	Possible effects
Low temperature and fungi	Weakened effect	Low temperature reduces fungus growth. At subzero temperatures, fungi remain in suspended action, thereby weakening the net effect.
Low temperature, shock, and vibration	Intensified deterioration	Low temperature tends to intensify the effects of shock and vibration, because certain materials (such as aluminum) tend to become brittle at lower temperatures. However, this is a consideration only at very low temperatures.
Low temperature and acceleration	Intensified deterioration	Acceleration produces shock, vibration, or both. Hence, low temperature and acceleration intensify the effects of acceleration because of brittleness at low temperatures.
Humidity and high pressure	Intensified deterioration	The effect of this combination varies with the temperature. High temperature can aggravate the deleterious effects caused by humidity and high pressure, indirectly increasing the net effect on a product.
Humidity and salt spray	Intensified deterioration	High humidity may dilute the salt concentration and could affect the corrosive action of the salt by increasing its mobility and spread, thereby increasing the conductivity. Corrosion failures are typically aggravated.
Humidity and fungi	Intensified deterioration	Humidity helps the growth of fungus and microorganisms but adds nothing to their effects.
Humidity, sand and dust	Intensified deterioration	Sand and dust have a natural affinity for water, and this combination increases deterioration by corrosion.
Humidity and vibration	Intensified deterioration	This combination tends to increase the rate of breakdown of material and connections.
Humidity, shock, and acceleration	Intensified deterioration	The periods of shock and acceleration, if prolonged, aggravate the effects of humidity, because humidity tends to cause deterioration of material properties. The combination can lead to early structural failure.
High pressure and vibration	Intensified deterioration	This combination intensifies structural failures in a product.
High pressure, shock, and acceleration	Intensified deterioration	This combination intensifies structural failures in a product.
Salt spray and dust	Intensified deterioration	Sand and dust have a natural affinity for water, and this combination increases deterioration by corrosion.
Salt spray, shock, or acceleration	Coexistence without any synergistic effects on deterioration of the product	These combinations produce no added effect.
Salt spray and vibration	Intensified deterioration	This combination tends to increase the rate of breakdown of material and connections.
Salt spray and explosive atmosphere	Incompatible	This is considered an incompatible combination.
Sand, dust, and vibration	Intensified deterioration	Vibration increases the wearing effects of sand and dust.

(Continued)

Table 7.2 (*Continued*)

Combined loads	Classification of effects	Possible effects
Shock and vibration	Coexistence without any synergistic effects on deterioration of the product	Since shock is a form of vibration, this combination does not produce any added effects.
Vibration and acceleration	Intensified deterioration	This combination produces increased effects when encountered with high temperatures and low pressures (typically in applications such as oil suction).
High temperature and low pressure	Intensified deterioration	As pressure decreases, outgassing of constituents of materials increases. As temperature increases, outgassing increases. Hence, each tends to intensify the effects of the other.
High temperature and explosive atmosphere	Coexistence without any synergistic effects on deterioration of the product	Temperature has minimal effect on the ignition of an explosive atmosphere but does affect the air–vapor ratio, which is an important consideration.
High pressure and explosive atmosphere	Intensified deterioration	High pressure aggravates the effects of explosion and thereby enhances the hazards of an explosive atmosphere.
Low temperature and low pressure	Intensified deterioration	This combination can accelerate leakage through seals and airtight regions. It can cause material deterioration and loss of functionality in hermetic parts.
Low temperature and explosive atmosphere	Coexistence without any synergistic effects on deterioration of the product	Temperature has minimal effect on the ignition of an explosive atmosphere but does affect the air–vapor ratio, which is an important consideration.
Humidity and explosive atmosphere	Weakened net effect	Humidity has no effect on the ignition of an explosive atmosphere, but high humidity will reduce the pressure of an explosion.
Low pressure and salt spray	Intensified deterioration	This combination can lead to increased penetration of moisture into the product and thus enhance the rate of material deterioration and corrosion-related failure mechanisms.
Low pressure and fungi	Coexistence without any synergistic effects on deterioration of the product	This combination does not add to overall effects.
Low pressure and explosive atmosphere	Intensified deterioration	At low pressures, an electrical discharge is easier to develop, but the explosive atmosphere is harder to ignite.

some fungi and microorganisms. With a small amount of humidity present, microorganisms can grow on electronic assemblies and the organic processes can cause chemical changes and contamination, resulting in loss of performance.

In some cases, two load conditions may act independently on a product and do not influence each other's effect. For example, acoustic vibrations to which electronic product might be subjected do not have any significant additive effect on the potential hazards caused by fungal activity in the vicinity of electronic parts.

In some cases, two load conditions may diminish the effect of each other. For example, high temperature can increase outgassing of constituents of the structural material of electronic parts, while high pressure generally decreases it. Permanently installed downhole gauges typically experience high temperature and high pressure conditions.

The increase in one load condition can also lead to the reduction of another; consequently, the net effect is reduced. For example, low temperature generally retards growth of fungi; therefore, the effects of the presence of fungi are reduced with low temperature.

7.4.4 Change in Magnitude and Rate of Change of Magnitude

Failure mechanisms in a product can be caused by steady-state loads or changes in the magnitude of the load (absolute change or rate of change). Therefore, the nature of the application of the loads (steady state or dynamic) should be determined. For example, in electronic products, functional failures caused by reduced propagation of signals is often caused by high temperature conditions, while failures in electrical interconnections often depend more on the rate of temperature change (Lall et al. 1997).

7.5 Methods for Estimating Life-Cycle Loads

The life-cycle loads need to be quantified in terms of range of possible values and expected variability of these values. Ideally, the design team should know the distribution of the loads experienced by the product. Several methods to quantify the life-cycle loads are discussed in the next section. Methods such as conducting in situ monitoring provide the most accurate information. Designs that are based on life-cycle loads obtained from market studies and field trials are usually much less accurate.

7.5.1 Market Studies and Standards Based Profiles as Sources of Data

Market surveys and reports generated independently by agencies[3] or conducted by industries as a part of their design process are often used as the basis for load conditions characterization. These kinds of data are derived most often from a similar kind of load conditions and give a very coarse estimate of the actual load conditions that

---

[3] These agencies include focus groups in organizations and standards committees like those that develop military standards. For example, IPC SM-785 specifies the use and extreme temperature conditions for electronic products categorized under different industry sectors, such as telecommunication, commercial, and military.

the targeted product will experience. The use of standard-based profiles such as military standards and IPC was discussed in the previous section. These methods should only be applied after similarity analysis shows considerable agreement with the new product.

7.5.2 In Situ Monitoring of Load Conditions

Environmental and usage loads experienced by the product in its life cycle can be monitored in-situ. These data are often collected using sensors, either mounted externally or integrated with the product and supported by telemetry systems. Devices such as health and usage monitoring systems (HUMS) are popular in aircraft and helicopters for in situ monitoring of usage and environmental loads.

Load distributions should be developed from data obtained by monitoring products used by different customers, ideally from various geographical locations where the product is used. The data should be collected over a sufficient period to provide an accurate estimate of the loads and their variation over time. In situ monitoring has the potential to provide the most accurate account of load history for use in design and test of future products.

7.5.3 Field Trial Records, Service Records, and Failure Records

Field trial records are sometimes used to get estimates on the load profiles. Field trial records provide estimates on the load conditions experienced by the product. The data depend on the durations and conditions of the trials, and can be extrapolated to get an estimate of actual load conditions.

Service records and failure records usually document the causes for scheduled or unscheduled maintenance and the nature of failure in the product, which might have been due to certain load or usage conditions. These data are sometimes used to estimate the kinds of load conditions the product might be subjected to.

7.5.4 Data on Load Histories of Similar Parts, Assemblies, or Products

Similarity analysis is a technique for estimating loads when sufficient field histories for similar products are available. Before using data on existing products for proposed designs, the characteristic differences in design and application for the two products need to be reviewed. Changes and discrepancies in the conditions of the two products should be critically analyzed to ensure the applicability of available loading data for the new product. For example, electronics inside a washing machine in a commercial laundry is expected to experience a wider distribution of loads and use conditions (due to several users) and higher usage rates compared with a home washing machine. These differences should be considered during similarity analysis.

7.6 Summary

To design a reliable product, it is necessary to understand the events and loads that a product will experience throughout its life cycle. Many times, the life-cycle events and loads of a product are merely assumed based on engineering specifications or conjecture. But this approach can lead to costly overdesign or hazardous underdesign, resulting in increased investment and risk. Hence, a formal method is needed to capture the life-cycle events and loads that a product will experience. Such a method involves determining systematically the life-cycle profile of a product based on data collected from actual products, if at all possible.

A life-cycle profile is a time history of events and conditions associated with a product from its release from manufacturing to its removal from service. There are three steps in the development of a life-cycle profile. The first step is to describe expected events for a product from manufacture through end of life, which involves identifying the different events that the product will pass through. The second step is to identify significant natural and induced environmental conditions or their combinations for each expected event. The third step is to describe load conditions to which product will be subjected during the life cycle, which involves the quantification of load conditions identified as a result of the previous two steps. Beyond these three steps, there are certain considerations that companies should take into account when developing life-cycle profiles for their products. These considerations include life-cycle profile recommendations from standards, the impact of combined loads on their products, and changes in the magnitude of loads.

Recommended methods for estimating life-cycle loads include conducting market studies and researching standards-based profiles to collect data. Various sensors and prognostic and diagnostic techniques can be used for in situ monitoring of load conditions in a product. Data can also be culled from field trial records, service records, and failure records, as well as from load histories from similar parts, assemblies, or products. As these data are collected and analyzed, companies will be able to predict with increasing accuracy the life-cycle events that their products will experience, thus enabling them to produce products in a way that minimizes cost and maximizes reliability.

Problems

7.1 Poor manufacturing, handling, assembly, storage, and transportation are often found as early failure causes. Can the life-cycle profile for a product help in reduction of early failures? Explain using the example of incandescent (tungsten filament) light bulbs.

7.2 Prepare a life-cycle profile for a bicycle for three different usage conditions: commuting from apartment or dormitory to school, courier service in a city and recreational mountain biking.

7.3 Prepare a life-cycle profile for a military heavy-lift helicopter. What the operational and environmental conditions that the helicopter will be subjected too?

7.4 What are the different combined loads that a car can experience? How can these combined loads affect the reliability of the car? What are the failure mechanisms that are induced by the combined loads?

8 Reliability Capability

The last decade of the twentieth century and the first decade of the twenty-first have witnessed the rapid globalization of many industries. Competitive and regulatory pressures have driven companies to low-cost manufacturing and to the evolution of a worldwide supply chain. Today, external sourcing of components and contract manufacturing is widespread. Companies are dependent upon worldwide suppliers who provide them with materials, parts, and subassemblies. Therefore, for any product design, it is essential that the reliability requirements be applied to all the incoming subcontracted elements so that reliability can be managed across all the tiers of the supply chain. The ultimate goal is that each supplier's reliability practices will be adequate to satisfy the end-product requirements of their customers.

System integrators, who are at the top of the supply chain, generally set the requirements for system reliability. Parts and manufacturing processes purchased on the market as commodities are selected based on information provided by suppliers. However, system integrators cannot wait until they receive purchased parts or subassemblies to assess whether they are reliable. This would lead to an expensive iterative process of part delivery, product assembly, and reliability testing followed by part respecification. An upfront evaluation of suppliers based on their ability to meet reliability requirements can provide a valuable competitive advantage. A company's capability to design for reliability and to implement a reliable design through manufacturing and testing can yield important information about the likelihood that the company will provide a reliable product.

8.1 Capability Maturity Models

The maturity approach to determining organizational abilities has roots in quality management. Crosby's quality management maturity grid (Crosby 1996) describes the typical behavior of a company, which evolves through five phases (uncertainty, regression, awakening, enlightenment, and certainty) in its ascent to quality management excellence. Consequently, maturity models have been proposed for a wide range of activities, including software development (Bamberger 1997; Bollinger and McGowan

Reliability Engineering, First Edition. Kailash C. Kapur and Michael Pecht.
© 2014 John Wiley & Sons, Inc. Published 2014 by John Wiley & Sons, Inc.

1991; Paulk et al. 1993), supplier relationships (Macbeth and Fergusson 1994), research and development effectiveness (Szakonyi 1994a, 1994b), product development (McGrath 1996), innovation (Chiesa et al. 1996), collaboration (Fraser and Gregory 2002), product design (Fraser et al. 2001; Strutt 2001; Williams et al. 2003), and reliability information flows (Boersma et al. 2004; Brombacher 1999; Sander and Brombacher 1999, 2000). This leads to the following metric for reliability capability:

> A reliability capability maturity metric is a measure of the practices within an organization that contribute to the reliability of the final product, and the effectiveness of these practices in meeting the reliability requirements of customers.

8.2 Key Reliability Practices

The IEEE Reliability Program Standard 1332 (IEEE Standard 1332–1998; Pecht and Ramakrishnan 2000) defines broad guidelines for the development of a reliability program, based on three objectives:

1. The supplier, working with the customer, should determine and understand the customer's requirements and product needs so that a comprehensive design specification can be generated.
2. The supplier should structure and follow a series of engineering activities so that the resulting product satisfies the customer's requirements and product needs with regard to product reliability.
3. The supplier should include activities that adequately verify that the customer's reliability requirements and product needs have been satisfied.

For each of these reliability objectives, key practices for evaluating reliability capability can be assigned. Figure 8.1 presents eight key practices identified from a study of reliability standards from the industry and reliability literature. Each of the eight key reliability practices is described in the following sections.

8.2.1 Reliability Requirements and Planning

During product development, the customer's needs and operational conditions for all phases of the product life cycle must be understood to arrive at a set of customer reliability requirements. The different considerations for establishing reliability requirements for a product include the design and operational specifications (information about the manner in which the product will be used), regulatory and mandatory requirements, definition of failure, expected field life, criticality of application, cost and schedule limitations, and business constraints, such as potential market size.

Establishing reliability requirements and planning early incorporates activities needed to understand customers' requirements, generates reliability goals for products, and plans reliability activities to meet those goals. The inputs for generating reliability requirements for products include customer needs, reliability data specifications for competitive products, and lessons learned from the reliability experience with previous products, including test and field failure data.

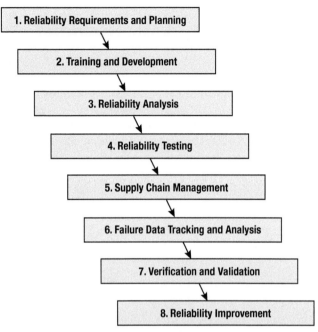

Figure 8.1 Key reliability practices.

Reliability planning is needed to establish and maintain plans that define reliability activities and manage the defined activities. The planning activity starts with identifying available resources, such as materials, human resources, and equipment, and determining the need for additional resources. Reliability analysis and testing needed for the product and the logistics to obtain feedback on the implementation of these activities can be identified.

The output from this key practice is a reliability plan. The reliability plan identifies and ties together all the reliability activities. The plan should allocate resources and responsibilities and include a schedule to follow. Decision criteria for altering reliability plans can also be included.

8.2.2 Training and Development

Training and development enhances the specialized skills and knowledge of people so that they can perform their roles in the development of a reliable product effectively and efficiently. The aim is to ensure that employees understand the reliability plans and goals for products, and have sufficient expertise in the methods required to achieve those goals. This includes the development of innovative technologies or methods to support business objectives.

Training and educating employees enhances the possibility of obtaining a better, more reliable product. Reliability managers must be aware of how specific reliability activities can impact or improve reliability, and business managers should appreciate the importance of reliability to ensure successful implementation of reliability practices within a company. The implementation of regular training programs indicates the willingness of business managers to spend time, effort, and money on training employees.

Effective training requires assessment of needs, planning, instructional design, and appropriate training media. The main components of employee training include a training development program with documented plans and means for measuring its effectiveness. The main activity for this key practice is the development of a training plan, including training needs for individual personnel, with a schedule. The implementation of the plan requires the procurement of a training infrastructure, including training instructors and training material.

The different modes of imparting training include in-class training, mentoring, web-based training, guided self-study, or a formal on-the-job training program. Employees must be trained in the life-cycle reliability management of products, including specific areas such as failure analysis, root cause analysis, and corrective action system. The training should incorporate an understanding of reliability concepts and statistical methods.

8.2.3 Reliability Analysis

Reliability analysis incorporates activities to identify potential failure modes and mechanisms, to make reliability predictions, and to quantify risks for critical components in order to optimize the life-cycle costs for a product. Prior experience and history can be helpful in this analysis. The data used to make reliability predictions may be historical, from previous testing of similar products, or from the reported field failures of similar products.

Reliability analysis activities include conducting failure modes, mechanisms, and effects analysis (FMMEA) to identify potential single points of failure, failure modes, and failure mechanisms for a product. The next step is to identify the criticality of these failure modes and mechanisms. Criticality may be based on complexity, application of emerging technologies, demand for maintenance and logistics support and, most importantly, the impact of potential failure on overall product success. Reliability analysis also includes identification of reliability logic for products as a system, and creating reliability models at the component and product levels in order to make reliability predictions. Assessing adherence to design rules, including derating, electrical, mechanical, and other guidelines, is also a part of reliability analysis.

The outputs from this analysis include an assessment of the reliability of the product, expected failure modes, and identification of design weaknesses to determine the suitability of the existing design for avoiding early-life failures and the product's susceptibility to wear-out failures. The information from reliability analysis can be used to create a list of reliability critical materials, parts, subassemblies, or processes, and to design reliability tests. Predictions regarding expected warranty costs and logistics support, including spares provisioning, can also be made.

8.2.4 Reliability Testing

Reliability testing can be used to explore the limits of a product, to screen products for design flaws, and to demonstrate (or qualify) the reliability of products. The tests may be conducted according to some industry standards or to required customer specifications. The reliability testing procedures may be generic—that is, common for all products—or the tests may be custom designed for specific products. The tests may or may not be used for the verification of known failure modes and mechanisms.

Detailed reliability testing plans can include the sample size for tests and the corresponding confidence level specifications.

Important considerations for any type of reliability testing include establishing the nature of the test (failure or time terminated), the definition of failure, the correct interpretation of the test results, and correlating the test results with the reliability requirements for the product. The information required for designing product-specific reliability tests includes the expected life-cycle conditions, the reliability plans and goals for a product, and the failure modes and mechanisms identified during reliability analysis. The different types of reliability tests that can be conducted include tests for design marginality, destruct limits determination, design verification testing before mass production, ongoing reliability testing, and accelerated testing.

The output from this key practice is the knowledge obtained from different types of tests. Test data analysis can be used as a basis for design changes prior to mass production, for identifying the failure models and model parameters, and for modification of reliability predictions for the product. Test data can also be used to create guidelines for manufacturing tests, including screens, and to create test requirements for materials, parts, and subassemblies obtained from suppliers.

8.2.5 Supply-Chain Management

Supply-chain management activities include monitoring a list of potential suppliers, conducting supplier assessment and audits, and selecting vendors or subcontractors for parts or processes. Other activities include part and process qualification through review of process, quality, reliability testing, or accelerated test data from the suppliers. Activities such as tracking product change notices, changes in the part traceability markings, and management of part obsolescence are also included under this key practice. These activities are essential for sustaining product reliability throughout its life cycle.

The decision criteria for supplier selection include their ability to provide reliable components effectively and their demonstrated ability to control their own supply chain. Possible control over the supplier's reliability practices through exchange of technological expertise and sharing of information also increases the possibility of achieving and maintaining product reliability. In some cases, multisourcing of parts may be necessary due to considerations of product manufacturing schedules, supplier capacity, or anticipated supply fluctuations.

Key outputs from this key practice are a list of preferred/qualified/approved parts, vendors and subcontractors, and a system for supplier rating. Other outputs include component qualification reports, supplier audit reports, and development of supply contracts that include contractual quality and reliability requirements.

8.2.6 Failure Data Tracking and Analysis

Failure tracking activities are used to collect manufacturing, test, and field-failed components, as well as related failure information. Failures must then be analyzed to identify the root causes of manufacturing defects and test or field failures and to generate failure analysis reports. These records can include the date and lot code of the returned product, the failure point (quality testing, reliability testing, or field), the return date, the failure site, the failure mode and mechanism, and recommendations for avoiding the failure mode in existing and future products. For each product category, a Pareto chart of failure causes can be created and continually updated.

The failure sources that initiate failure analysis of a product include manufacturing, production testing, reliability testing, pre- and postwarranty field returns, and customer complaints. Failure analysis includes statistical analyses of failure data and analysis of the cause of failure at various levels down to the identification of the root cause of failure.

The outputs for this key practice are a failure summary report arranged in groups of similar functional failures, forward and backward traceability of failed components through date and lot code information, actual times to failure of components based on time-specific part returns, and a documented summary of corrective action implementation and effectiveness. All the lessons learned from failure analysis can be included in a corrective actions database for future reference. This database can help save the considerable cost in fault isolation and rework associated with problems that may be encountered.

8.2.7 Verification and Validation

Verification and validation through an internal review/audit of reliability planning, testing and analysis activities helps to ensure that planned reliability activities are implemented so that the product fulfills the specified reliability requirements. Benchmarking can be used to study the best internal practices that produce superior reliability performance and for ensuring that noncompliance is addressed. Part of the process is to understand how some practices are better than others and to find ways to improve others by pushing for improved facilities, equipment, and methodologies.

The inputs for this key practice are the outputs from previous practices like planning, analysis, testing, and failure data tracking. The inputs include reliability plans and goals for products, potential failure modes and mechanisms identified during reliability analysis, information on failure mechanisms from reliability testing, specific reliability test plans and specifications, and the corrective actions database.

Verification and validation activities include comparison of identified potential problems against those experienced in the field. This includes comparison of expected and field failure modes and mechanisms and of reliability prediction models for a product against field failure distributions.

The outputs from this key practice include an updated failure modes and mechanisms database, modification of reliability predictions and failure models for a product, and modification of warranty costs and spares provisioning. Reliability test conditions may also be modified based on field information on products.

8.2.8 Reliability Improvement

Reliability improvement is concerned with applying lessons learned from testing, reported field failures, technological improvements, and any additional information from previous tests or experiences. This key practice primarily involves implementing corrective actions based on failure analysis. It also involves initiating design changes in products or processes due to changes in reliability requirements or in life-cycle application conditions (operating and nonoperating).

Reliability improvements can be affected either by making design changes in products or by using alternative parts, processes, or suppliers. Design changes can include an improved design using an established technology, or implementing developing technologies within an older design. New modeling and analysis techniques and trends that could improve reliability can also be used.

The inputs required to initiate reliability improvement also come from previous key practices. Such information includes Pareto charts for field failure modes and mechanisms, recommendations from the corrective actions database, and documented anomalies from verification and validation. Other factors that can initiate a reliability improvement process are changes in life-cycle usage conditions for a product or changes in reliability requirements due to business or other considerations.

The outputs from this practice include methods to prevent the recurrence of identified failures and implementation of corrective actions stemming from failure analysis. Corrective actions can be implemented by issuing engineering change notices, or through modifications in manufacturing and design guidelines for future products.

8.3 Summary

In the last 20 years, competitive and regulatory pressures have driven many companies to develop low-cost manufacturing processes and a worldwide supply chain. Since reliability represents a risk factor associated with profit-making, it is essential that reliability is managed across all the tiers of the supply chain. System integrators need an upfront evaluation of suppliers' ability to meet reliability requirements to effectively manage reliability and provide competitive advantage.

Reliability capability maturity assesses the effectiveness of the organizational practices that contribute to the reliability of the final product. Eight key reliability practices are essential to a strategy for reliability management and form the basis for reliability capability evaluation. These practices underlie the development of a reliability capability maturity model that can help companies assess their potential suppliers or help suppliers to assess themselves. Reliability tasks under each key practice can be used as evaluation items to assign maturity scores that provide a quantitative metric for grading companies.

The reliability capability maturity model can also help to establish reliability management practices for designers, suppliers, customers, and independent authorities. It can produce increased customer satisfaction, provide competitive opportunities, and shorten the product development cycle. It is expected that this model can also be used to identify shortcomings in a company's reliability program, which can then be overcome by improvement actions.

Problems

8.1 What are the three reliability objectives established for suppliers and customers by IEEE Reliability Program Standard 1332? How do they relate to the key reliability practices in reliability capability evaluation?

8.2 A reliability capability assessment is useful to the organization being evaluated because it can help identify the blind spots where a company can make improvements. List three possible blind spots for an emergency dispatch service (like police or ambulance), a gas station, or a manufacturer of pipes that can be identified by a reliability capability assessment. Relate each one of those possible blind spots to a unique key reliability practice and justify why each blind spot correlates to that key practice.

8.3 Discuss the benefits of conducting a reliability capability evaluation for an organization. What is the outcome of the evaluation, and can it be used to ensure delivery of reliable products?

8.4 List the eight key reliability practices in order and provide a brief description of each.

9 Parts Selection and Management

To produce a product, there is usually a complex supply chain of companies that are involved directly and indirectly in producing and parts (materials) for the final product. Thus, to produce a reliable product, it is necessary to select the parts that have sufficient quality and are capable of delivering the expected performance for the targeted life-cycle conditions.

This chapter discusses parts selection and management. The key elements to a practical selection process are presented. Then, the practices necessary to ensure continued acceptability over the product life cycle are discussed.

9.1 Part Assessment Process

The parts (materials)[1] selection and management process is usually carried out by a multidisciplinary team, which develops part assessment criteria and acceptability levels to guide part selection. A part is selected if it conforms to the targeted requirements, is cost-effective and available[2] to meet the schedule requirements. If there are problems, the parts management team also helps to identify alternative sources of parts or ways to help the supplier produce a better part.

Many product design teams maintain a list of preferred parts of proven performance and reliability. A "preferred part" is typically mature, and has a history of successful manufacturing, assembly, and field operation, so it is usually the conservative approach to parts selection. Thus, in some cases, new technologies, processes, markets, materials, and price pressures make a mature part undesirable or obsolete. When a new product is being developed or a mature product improved, a new part may be required.

---

[1] In this book, the materials that comprise the product are also considered parts. This can include everything from structural materials, to added material ingredients, such as flame retardants.

[2] The availability of a part is a measure of the ease with which the part can be procured. Availability is assessed by determining the amount of inventory at hand, the number of parts required over the life of the product production, the economic order quantity for the part, the lead time between placing an order for the part and receiving the part, production schedules and deadlines, and part discontinuation plans.

Reliability Engineering, First Edition. Kailash C. Kapur and Michael Pecht.
© 2014 John Wiley & Sons, Inc. Published 2014 by John Wiley & Sons, Inc.

9 Parts Selection and Management

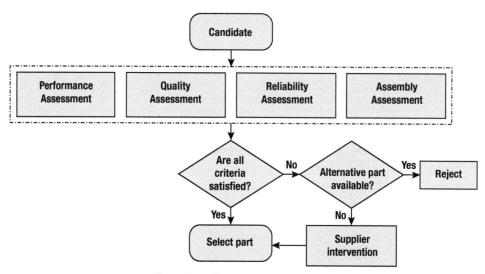

Figure 9.1 Part assessment process.

Because a company might make parts for many different customers, who have different target applications, it is not always possible for a customer of parts to dictate requirements. An "eyes-on, hands-off" approach may be needed to select parts with the required attributes. In some cases, the selection of a proper part might require an evaluation of the part, beyond that conducted by the part manufacturer.

Key elements of part assessment (see Figure 9.1) include performance quality, reliability, and ease of assembly. Performance is evaluated by functional assessment against the datasheet specifications. Quality is evaluated by process capability and outgoing quality metrics. Reliability is evaluated through part qualification and reliability test results. A part is acceptable from an assembly viewpoint if it is compatible with the downstream assembly equipment and processes.

Part assessment results may not remain valid if ingredients and process changes are made to the part. Even parts that are deemed acceptable may need to be reassessed periodically to ensure their continued acceptability.

If the part is not acceptable, then the assessment team must determine if an acceptable alternative is available. When no alternative is available, the team may have to pursue intervention techniques (e.g., work with the part manufacturer and conduct special screens) to mitigate the possible risks.

9.1.1 Performance Assessment

The goal of performance assessment is to evaluate the part's ability to meet the functional requirements (e.g., structural, mechanical, electrical, thermal, and biological) of the product. In general, there is often a minimum and a maximum limit beyond which the part may not function properly according to the datasheet specifications. These limits, or ratings, are often called the recommended operating conditions.

Manufacturers also typically set reliability limits for their parts, called absolute maximum rating. Companies who integrate parts into their products need to adapt their design so that the parts do not experience conditions beyond their absolute maximum ratings, even under the worst possible operating conditions (e.g., in

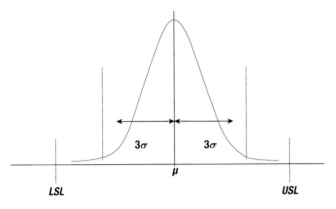

Figure 9.2 Distribution with 3σ process limits and upper and lower specification limits.

electrical products this would include supply voltage variations, load variations, and signal variations).[3]

9.1.2 Quality Assessment

Quality is associated with the workmanship of the product. Quality defects can result in premature (prior to that designed-for) failures of the product. To ensure designed-for reliability, it is necessary that the selected parts have acceptable quality. This is assessed by examining the control of the processes used to make the parts and the outgoing quality of the parts.

9.1.3 Process Capability Index

Statistical control of a process is arrived at by eliminating special causes of excessive variation one by one. Process capability is determined by the total variation that comes from common causes. Therefore, a process must first be brought into statistical control and then its capability to meet specifications can be assessed.

The process capability to meet specifications is usually measured by process capability indices that link process parameters to product design specifications. Using a single number, process capability indices measure the degree to which the stable process can meet the specifications (Kane 1986; Kotz and Lovelace 1998). If we denote the lower specification limit as *LSL* and the upper specification limit as *USL*, and if σ is the true value of the process, then the process capability index C_p is defined as:

$$C_p = \frac{USL - LSL}{6\sigma}, \qquad (9.1)$$

which measures the potential process capability. It is obvious that Equation 9.1 does not have the process mean or expected value μ as well as any information about the target value *T* for the underlying quality characteristic. Figure 9.2 shows the normal

---

[3]Because the part might be used in a manner or in conditions in which it was not intended, a special process, called uprating, was developed. The term uprating was coined by Michael Pecht to signify the use of a part outside its recommended operating conditions (per the datasheet). The interested reader is referred to the book, Das, D., Pecht, M., and Pendse, N., *Rating and Uprating of Electronic Products*, CALCE EPSC Press, University of Maryland, College Park, MD, 2004.

distribution of a process, the 3σ limits from the mean, and the lower and upper specifications. In this figure, the process is centered between the specifications.

To measure the actual process capability for a noncentered process, we use C_{pk} that is defined as:

$$C_{pk} = \min\left(\frac{USL-\mu}{3\sigma}, \frac{\mu-LSL}{3\sigma}\right). \tag{9.2}$$

The measure of C_{pk} takes the process centering into account, by choosing the one side C_p for the specification limit closest to the process mean. The estimation of C_p and C_{pk} are obtained by replacing μ and σ using the estimates $\hat{\mu}$ and $\hat{\sigma}$ from the statistical control charts. To consider the variability in terms of both standard deviation and mean, another process capability index C_{pm} is defined as

$$\hat{C}_{pm} = \frac{USL-LSL}{6\hat{\tau}}, \tag{9.3}$$

where $\hat{\tau}$ is an estimator of the expected square deviation from the target, T, and is given by

$$\tau^2 = E\left[(X-T)^2\right] = \sigma^2 + (\mu-T)^2. \tag{9.4}$$

Therefore, if we know the estimate of C_p, we can estimate C_{pm} as

$$\hat{C}_{pm} = \frac{\hat{C}_p}{\sqrt{1+\left(\frac{\hat{\mu}-T}{\hat{\sigma}}\right)^2}}. \tag{9.5}$$

In addition to process capability indices, capability can also be described in terms of the distance of the process mean from the specification limits in standard deviation units, Z, that is

$$Z_U = \frac{USL-\hat{\mu}}{\hat{\sigma}}, \quad \text{and} \quad Z_L = \frac{\hat{\mu}-LSL}{\hat{\sigma}}. \tag{9.6}$$

Z-values can be used from a table of standard normal distribution to estimate the proportion of process nonconforming units for a normally distributed and statistically controlled process. The Z-value can also be converted to the capability index, C_{pk},

$$C_{pk} = \frac{Z_{\min}}{3} = \frac{1}{3}\min(Z_U, Z_L). \tag{9.7}$$

A process with $Z_{\min} = 3$, which could be described as having $\hat{\mu} \pm 3\hat{\sigma}$ capability, would have $C_{pk} = 1.00$. If $Z_{\min} = 4$, the process would have $\hat{\mu} \pm 4\hat{\sigma}$ capability and $C_{pk} = 1.33$.

9.1 Part Assessment Process

Example 9.1

Find the fraction nonconforming items and the C_{pk} for a process with

$$\hat{\mu} = 0.738, \quad \hat{\sigma} = 0.0725, \quad USL = 0.9, \quad \text{and} \quad LSL = 0.5.$$

Also discuss various improvement strategies.

Solution:
Since the process have two-sided specification limits,

$$Z_{min} = \min\left(\frac{USL - \hat{\mu}}{\hat{\sigma}}, \frac{\hat{\mu} - LSL}{\hat{\sigma}}\right) = \min\left(\frac{0.9 - 0.738}{0.0725}, \frac{0.738 - 0.5}{0.0725}\right)$$
$$= \min(2.23, 3.28) = 2.23.$$

The fraction nonconforming p can be calculated as

$$p = 1 - \Phi(2.23) + \Phi(-3.28) = 0.0129 + 0.0005 = 0.0134.$$

The process capability index using Equation 9.7 is

$$C_{pk} = \frac{Z_{min}}{3} = 0.74.$$

If the process could be adjusted toward the center of the specification, the proportion of process nonconforming can be reduced, even with no reduction in σ, because we have

$$Z_{min} = \min\left(\frac{USL - \hat{\mu}}{\hat{\sigma}}, \frac{\hat{\mu} - LSL}{\hat{\sigma}}\right) = \min\left(\frac{0.9 - 0.7}{0.0725}, \frac{0.7 - 0.5}{0.0725}\right) = 2.76,$$

and the proportion of process fallout would be:

$$p = 2\Phi(-2.76) = 0.0058.$$

The process capability index increases to

$$C_{pk} = \frac{Z_{min}}{3} = 0.92.$$

To improve the actual process performance in a long run, the variation from common causes must be reduced. If the capability criterion is $\hat{\mu} \pm 4\hat{\sigma}$, ($Z_{min} \geq 4$), the process standard deviation for a centered process would be:

$$\sigma_{new} = \frac{USL - \hat{\mu}}{Z_{min}} = \frac{0.9 - 0.7}{4} = 0.05.$$

Therefore, actions should be taken to reduce the process standard deviation from 0.0725 to 0.05, or about 31%.

At this point, the process has been brought into statistical control and its capability has been described in terms of process capability index or Z_{min}. The next step is to evaluate the process capability in terms of meeting customer requirements. The fundamental goal is never-ending improvement in process performance. In the near term, however, priorities must be set as to which processes should receive attention first. This is essentially an economic decision. The circumstances vary from case to case, depending on the nature of the particular process in question. While each such decision could be resolved individually, it is often helpful to use broader guidelines to set priorities and promote consistency of improvement efforts. For instance, certain procedures require $C_{pk} > 1.33$, and further specify $C_{pk} = 1.50$ for new processes. These requirements are intended to assure a minimum performance level that is consistent among characteristics, products, and manufacturing sources.

Whether in response to a capability criterion that has not been met, or to the continuing need for improvement of cost and quality performance even beyond minimum capability requirement, the action required is the same: improve the process performance by reducing the variation that comes from common causes. This means taking management action to improve the system.

9.1.4 Average Outgoing Quality

Average outgoing quality (AOQ) is defined as the average nonconforming part lot fraction from a series of lots, based on sample testing. It represents the total number of parts that are outside specification limits, as determined from sample tests conducted during the final quality control inspection. This number reflects the estimated number of defective parts that will be received by the customer. AOQ is usually reported in parts per million (ppm).

AOQ reflects the effectiveness (or lack of it) of the part manufacturer's quality management system. An effective quality management system will minimize the total number of nonconformities produced, as well as the number that are shipped. High values of AOQ represent a high defective count, implying poor quality management. Low values reflect high part quality.

If all parts are tested prior to shipping, then theoretically, the AOQ should always be zero because all nonconformities should be removed. However, if a large volume of parts is produced, it is usually impractical to test all parts. Instead, a sample is tested, and an estimation of the AOQ is calculated from it.

The parts management team should establish threshold AOQ requirements to determine part acceptability. Limits should be defined to differentiate acceptable and unacceptable parts. Some factors to be considered include application, testability and diagnosability, production volume, reworkability, and the target cost.

9.1.5 Reliability Assessment

Reliability assessment is a means to obtain information about the ability of a part to meet the required performance specifications in its targeted life-cycle application for a specified period of time. If the parametric and functional requirements of the product cannot be met, then a different part may have to be used, or some product changes will be necessary, such as reducing the loads acting on the part, adding redundancy, or implementing maintenance.

Reliability assessment includes the evaluation of supplier qualification data, reliability life testing, and reliability monitoring tests. In some cases, failure models will be needed to assess the results of the tests against the target performance and reliability objectives. Once all the parts are selected and incorporated into a product, there will be additional tests to determine the reliability of the product as a whole. Product tests are necessary to assess the parts under the targeted conditions, and include the effects of assembly, the interaction of parts and the loads that the parts generate.

Most often, the suppliers of parts will not know, and may not want to know, the life-cycle usage profiles, load conditions, and reliability targets of the final product. As a result, suppliers generally follow guidelines and standards when they assess the reliability of their parts, to provide some standard level of assurance to potential customers, and to baseline the reliability of a part against that of previous parts and other parts accepted by the industry. The test results are generally published in a document or on the Internet. Most manufacturers provide test results in terms of the number of failures and the sample size tested.

Qualification tests are conducted prior to volume manufacturing. The purpose is to ensure that the nominal product will meet some standard level. The sample size may range from one to hundreds of parts, depending on the cost, volume to be sold, and risk. For example, some precision sensor parts are so expensive and only a few are ever sold, that only a few are normally tested. On the other hand, many hundreds of parts are qualified when producing a new semiconductor device.

Qualification tests usually employ a wide range of load conditions to assess reliability. These load conditions may include high and low temperature, temperature cycling, thermal shock, vibrations, mechanical shock, various humidity conditions, dust, and contamination. These conditions may be much more severe and exhaustive than those seen in the field. Generally, the tests are conducted for a specified time, number of cycles, step loading, and so on. In some cases, a company may wish to test to failure (called life tests) to determine the limits of the part, but again this depends on the time consumed, the costs, and the perceived risks.

Qualification tests should be repeated if any changes to the materials, processes, structure and operation of the part, could possibly change the failure mechanisms. If the changes are not significant, then reliability monitoring tests are used to assess the ongoing reliability level.

Reliability monitoring tests are conducted at intervals in the manufacturing process, to ensure that the nominal product continues to meet some standard level, especially if there are continuous improvements being made to the materials, processes, structure, and operation of the part. For electronic parts, companies often conduct reliability monitoring tests at least twice per year, and sometimes quarterly.

Reliability monitoring tests are typically some subset of the qualification tests, which have been determined to be the most applicable to the specific part, in terms of the potential failure mechanisms that could be precipitated. However, the test conditions are still usually accelerated and will often be more severe and exhaustive than those seen in the field. Similar to qualification, the sample size may range from one to hundreds, typically depending on the cost, volume to be sold, and risk.

It is the responsibility of the company that selects the parts for use in the final product, to determine if the magnitude and duration of the life-cycle conditions are less severe than those of the reliability tests, and if the test sample size and results are acceptable, the part reliability should be acceptable. If the reliability test data are

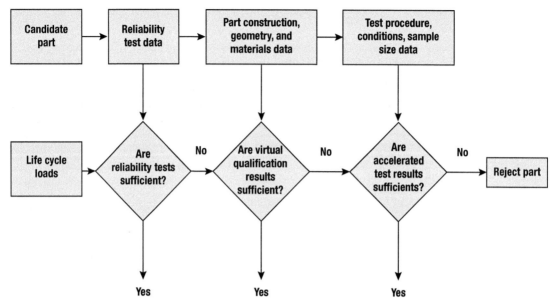

Figure 9.3 Decision process for part reliability assessment.

insufficient to validate part reliability in the application, then additional tests and virtual qualification should be considered. The process flow is show in Figure 9.3. Reliability monitor tests are also called life tests, continuous monitor tests, and environmental tests.

Results should include test types, test conditions, duration, sample size, and number of failures. Root cause analysis should be performed for all failures. If any failures are reported, parts manufactured during the same period as those that the data represent must be cautiously evaluated, and possibly rejected for use. In addition, the same kind of defect may exist in other nontested parts.

Virtual qualification is a simulation-based methodology used to identify the dominant failure mechanisms associated with the part under the life-cycle conditions, to determine the acceleration factor for a given set of accelerated test parameters, and to determine the time-to-failures corresponding to the identified failure mechanisms. Virtual qualification allows the operator to optimize the part parameters (e.g., dimensions and materials) so that the minimum time-to-failure of any part is greater than the expected product life.

Whether integrity test data, virtual qualification results, accelerated test results, or a combination thereof are used, each applicable failure mechanism to which the part is susceptible must be addressed. If part reliability is not ensured through the reliability assessment process, the equipment supplier must consider an alternate part or product redesign. If redesign is not considered a viable option, the part should be rejected, and an alternate part must be selected. If the part must be used in the application, redesign options may include load (e.g., thermal, electrical, and mechanical) management techniques, vibration and shock, damping, and modifying assembly parameters. If product design changes are made, part reliability must be reassessed.

Reliability assessment results provide information about the ability of a part to meet the required performance specifications in its targeted life-cycle application for a specified period of time. If the parametric and functional requirements of the system cannot be met, then the load acting on the part may have to be lessened, or a different part may have to be used.

9.1.6 Assembly Assessment

Assembly guidelines are recommendations by the part manufacturer to prevent damage (e.g., defects) and deterioration of the part during the assembly process. Examples of assembly can include recommended temperature assembly profiles, cleaning agents, adhesives, moisture sensitivity, and electrical protection. Assembly guidelines could also include information regarding part compatibility with equipment or dependent technologies (e.g., heat sinks and paints). As new technologies emerge and products become more complex, assembly guidelines become more important to ensure the quality and integrity of parts used within the product.

9.2 Parts Management

After a part is accepted, resources must be applied to manage the life cycle of the parts used in the product. This typically includes supply chain management, obsolescence assessment, manufacturing and assembly feedback, manufacturer warranties management, and field failure and root-cause analysis.

9.2.1 Supply Chain Management

In supply chain management, one of the key risks is associated with change. Changes occur for many reasons. For example, there may be shifts in consumer demand, new market challenges, advances in technology, and evolution in regulatory requirements and standards. All these changes affect supply-chain interactions.

Changes of concern to product reliability include a change in the companies that comprise the supply chain, a change in any of the materials and processes used to make the part and control quality, a change in any of the processes in which the part is assembled into the product, and a change in any other assembly process (not directly associated with the part) that could affect the reliability of the part.

A change in a company or a company's practices must be considered a risk in the production of a product. In general, no two companies make a product exactly the same way. Furthermore, a common company name does not ensure the same quality system and policies at each of the manufacturer's different locations. A part can be fabricated and assembled at multiple locations around the world and subjected to different company quality policies. In fact, companies may have different quality certifications from site to site. Different certifications can cover different areas of a quality system, and certification audits may examine different criteria (see Table 9.1). Because a part may be manufactured at different sites, and these sites may have different quality policies and certifications, the actual manufacturer of a candidate part should be identified and assessed.

In the manufacturer's supply chain assessment, the manufacturer's quality policies are assessed with respect to five assessment categories: process control; handling, storage, and shipping controls; corrective and preventive actions; product traceability; and change notification (Figure 9.4). These categories contain the minimum set of criteria necessary to monitor the supply chain.

Table 9.1 Example of Fairchild semiconductor corporation certifications[a]

Site	Certificate type	Certifying body
Cebu, the Philippines	ISO-9001 and QS-9000	DNV
	Qualification of Transistors & Diodes for Delco Electronics	Delco Electronics
Penang, Malaysia	ISO-9001 and QS-9000	DNV
	Stack Level 2 Supplier Certification	Stack International
	AEC-A100	AEC
	QSA Semiconductor Edition	
Puchon, South Korea	ISO-9001 and QS-9000	BSI
South Portland, Maine	ISO-9001 and QS-9000	DNV
	Stack Level 1 Supplier Certification	Stack International
Wuxi, Jiangsu, China	ISO-9001 and QS-9000	TÜV Cert

[a] Fairchild Semiconductor, "Fairchild Semiconductor Quality Certificates," South Portland, ME, http://www.fairchildsemi.com/company/quality.html, accessed November 16, 2013.

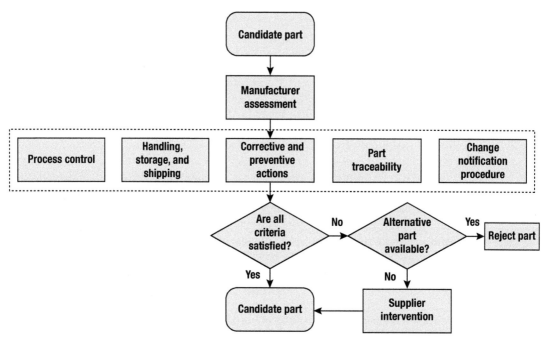

Figure 9.4 Manufacturing assessment process flowchart.

9.2.2 Part Change Management

Changes to parts are made throughout the life cycles of the parts. These changes are usually managed by a manufacturer's change control board. The policies of these boards generally vary from one manufacturer to another.

The types of changes that are made to parts, as well as the motivations for making changes, depend on the life-cycle stage of the parts. For example, a typical semiconductor part goes through sequential phases of introduction, growth, maturity, decline, and obsolescence (see Figure 9.5).

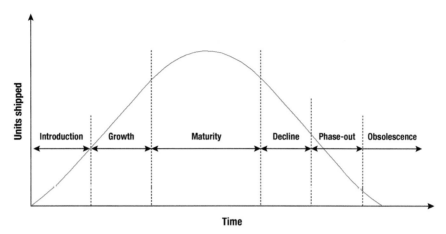

Figure 9.5 Typical life-cycle of an electronic part.

During the introduction stage, the changes implemented are mostly design improvements and manufacturing process adjustments. The part may be continuously modified so that it can meet datasheet requirements, achieve economic yields, and meet reliability and quality requirements.

During the growth and maturity stages, a part is in high volume production. Changes are implemented both to enhance the part and to minimize costs. Feature enhancements may be made to maintain competitiveness and generate new interest in the part. Material, fabrication, and assembly and testing locations may change to reflect changing business needs and capacity. Changes to improve yields and minimize costs may be necessary to maintain competitiveness in the marketplace.

During the decline stage of the part, sales levels start dropping, and manufacturers try to transition customers to newer parts and technologies. Part discontinuance usually occurs when the volume of sales for a part drops to the point where the part can no longer be profitably manufactured. However, it could also occur when a semiconductor company transitions its facilities to a new manufacturing technology.

After the part has been discontinued, it is in the obsolescence stage. Parts are no longer available for purchase, and costumer of parts must utilize stockpiled parts, obtain parts from an aftermarket source, find an equivalent substitute parts, or redesign their products.

9.2.3 Industry Change Control Policies

For most part manufacturers, the change process starts with the submission of a proposal to a change control board, sometimes called an engineering control board This board is usually composed of people from all major divisions within the company, including marketing, manufacturing, product engineering, and reliability engineering. Any division within the company can propose a change to the board.

Upon receipt of the change proposal, the board classifies the change according to some internal classification process. This classification involves deciding how significantly the form, fit, or function of the part would be affected by the change. Part characterization and reliability test results, contractual agreements with customers, and the number of parts affected by the change are also considered. If the board determines that the benefits of the change outweigh the risks, the change is approved.

Change classification systems and change control policies vary widely from one company to the next. Many companies have policies detailing the amount of testing that needs to be conducted to propose a change to a change control board. Many companies also have policies on how quickly the changes are phased into production. The change control process for IBM Microelectronics is illustrated in Figure 9.6. When, all changes go through a single standardized review process, regardless of the type of change.

Change is an inevitable aspect of part manufacturing. The development of new technologies and manufacturing processes, constantly changing business forces, and the emergence of new environmental regulations all necessitate change for a manufacturer to remain competitive. The manner in which a manufacturer manages change can have a large impact on economic success and customer satisfaction. If changes are not implemented in a controlled manner, changes that adversely affect part reliability are more likely to be inadvertently made, damaging the reputation of a manufacturer and increasing the risk of liability. If changes are made frequently or if insufficient notice or reason is provided for changes, manufacturers can also receive negative reactions from customers.

Effective change notification requires manufacturers to communicate with their customers frequently and openly, so that a bond of understanding can develop. The complete effects of changes are often unknown, and the distinction between major and minor changes is often fuzzy. Change control is therefore not only a science but also an art.

For original equipment manufacturers (OEMs), change tracking is becoming increasingly complicated. As captive parts suppliers are divested, the amount of control OEMs have over the change control process has diminished. An increasing number of companies are also purchasing through distributors and contract manufacturers, increasing the number of paths for the flow of change notification information through the supply chain. OEMs must therefore take an active role in the change tracking process and establish contractual agreements with the manufacturers, distributors, and contract manufacturers from which they purchase parts to ensure that they receive the change notifications they need. Larger OEMs that have the benefit of being able to work directly with part manufacturers should clarify what types of changes result in notifications and make special arrangements to address any omissions from this list that may affect their products. A request to be included on advance notification lists allows the most advance warning of impending changes to be received as soon as possible, often early enough so that feedback to the part manufacturer that may influence the implementation of the change can be provided.

9.3 Risk Management

The risks associated with incorporating a part into a product fall into two categories:

- *Managed Risks.* Risks that the product development team chooses to proactively manage by creating a management plan and performing a prescribed

9.3 Risk Management

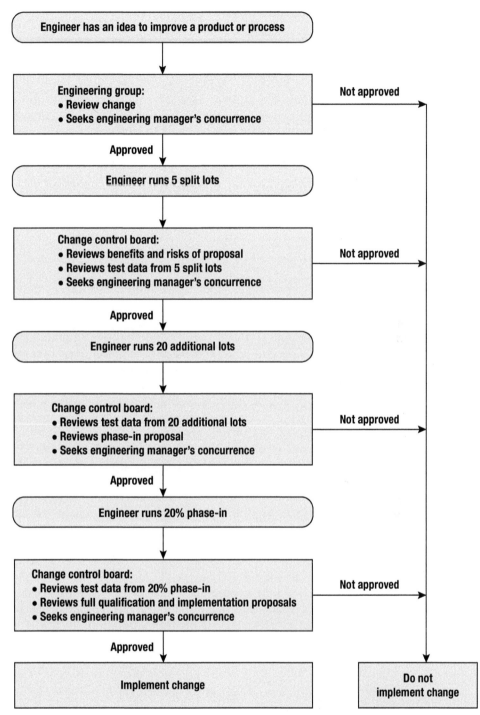

Figure 9.6 Change control process at IBM Microelectronics.

regimen of monitoring the manufacturer of the part, and the part's fabrication and field performance.
- *Unmanaged Risks.* Risks that the product development team chooses not to proactively manage.

If risk management is considered necessary, a plan should be prepared. The plan should contain guidance on how the part is monitored (data collection), and how the results of the monitoring feed back into various manufacturers and parts selection and management processes. The feasibility, effort, and cost involved in management processes should be considered prior to the final decision to select the part. In addition, feedback regarding the part's assembly performance, field performance, and sales history is also essential to ascertain the reliability risks.

9.4 Summary

For many products, there is a complex supply chain involved in producing parts for the final product. To produce a reliable product, it is necessary to select quality parts capable of reliable performance under the life-cycle conditions. The parts selection and management process is usually carried out by a product development team that develops part assessment criteria to guide part selection. Based on these criteria, a part is selected if it conforms to the targeted requirements, is cost-effective, and is available to meet the schedule requirements.

Key elements of part assessment include performance quality, reliability, and ease of assembly. Performance is evaluated by functional assessment against the datasheet specifications. Quality is evaluated by process capability and outgoing quality metrics. Reliability is evaluated through part qualification and reliability test results. A part is acceptable from an assembly viewpoint if it is compatible with the downstream assembly equipment and processes.

In supply chain management, one of the key risks is associated with change. Changes of concern to product reliability include a change in the companies that comprise the supply chain, a change in any of the materials and processes used to make the part and control quality, a change in any of the processes in which the part is assembled into the product, and a change in any other assembly process that could affect the reliability of the part.

The risks associated with incorporating a part into a product fall into two categories: managed risks, which are risks that the product development team chooses to proactively manage, including monitoring the manufacturer of the part; and unmanaged risks, which are risks that the product development team chooses not to proactively manage. A risk management plan should also be prepared. The plan should contain guidance on how a particular part is to be monitored, and how the results of the monitoring will feed back into various parts selection and management processes.

Ensuring the quality of the supply chain is essential for ensuring the quality of a manufactured product. Companies must be proactive in part assessment, managing changes, and risk assessment in order to ensure that their final products are reliable in the field.

Problems

9.1 Discuss the different test data that can be used to assess reliability of parts. Which of these data are most appropriate in making reliability assessments?

9.2 Consider the supply-chain assessment in parts selection and management.

(a) How does the supply-chain assessment help in developing reliable products?
(b) Why is the identification of each company in the supply-chain of value in the design of reliable products?
(c) If you have already assessed a manufacturer, and then the company changes its name, is it necessary to reassess the manufacturer? Explain the various possible circumstances.

9.3 Identify the following as a quality issue or a reliability issue, or both. Explain why.

(a) Two out of every 10 products made have a critical part put in backward that causes malfunction in test.
(b) Devices show failure after a 1000-hour long, 125°C high temperature operating life test during qualification testing.
(c) One out of every five shafts produced is out of tolerance.
(d) One out of every five devices has a joint that weakens in operation after 3–5 years in a 10-year application, causing electrical signal noise to increase over the allowed performance limit.

9.4 Find C_{pk} for $\mu = 3$, $\sigma = 0.45$, $USL = 4.3$, and $LSL = 1.9$. Comment on the result.

10 Failure Modes, Mechanisms, and Effects Analysis

This chapter presents a methodology called failure modes, mechanisms, and effects analysis (FMMEA), used to identify potential failures modes, mechanisms, and their effects. FMMEA enhances the value of failure modes and effects analysis (FMEA) and failure modes, effects, and criticality analysis (FMECA) by identifying the "high priority failure mechanisms" to help create an action plan to mitigate their effects. The knowledge about the cause and consequences of mechanisms found through FMMEA helps in efficient and cost-effective product development. The application of FMMEA for an electronic circuit board assembly is described in the chapter.

10.1 Development of FMMEA

The competitive market places demands on manufacturers to look for economic ways to improve the product development process. In particular, the industry has been interested in an efficient approach to understand potential product failures that might affect product performance over time. Some organizations are either using or requiring the use of a technique called FMEA to achieve this goal, but most of these companies are not completely satisfied with this methodology.

FMEA was developed as a formal methodology in the 1950s at Grumman Aircraft Corporation, where it was used to analyze the safety of flight control systems for naval aircrafts. From the 1970s through the 1990s, various military and professional society standards and procedures were written to define and improve the FMEA methodology (Bowles 2003; Guidelines for Failure Mode and Effects Analysis 2003; Kara-Zaitri et al. 1992).

In 1971, the Electronic Industries Association (EIA) G-41 committee on reliability published "Failure Mode and Effects Analysis." In 1974, the U.S. Department of Defense published MIL-STD 1629 "Procedures for Performing a Failure Mode, Effects and Criticality Analysis," which through several revisions became the basic

---

Reliability Engineering, First Edition. Kailash C. Kapur and Michael Pecht.
© 2014 John Wiley & Sons, Inc. Published 2014 by John Wiley & Sons, Inc.

approach for analyzing systems. In 1985, the International Electrotechnical Commission (IEC) introduced IEC 812 "Analysis Techniques for System Reliability—Procedure for Failure Modes and Effects Analysis." In the late 1980s, the automotive industry adopted the FMEA practice. In 1993, the Supplier Quality Requirements Task Force comprised of representatives from Chrysler, Ford, and GM, introduced FMEA into the quality manuals through the QS 9000 process. In 1994, the Society of Automotive Engineers (SAE) published SAE J-1739 "Potential Failure Modes and Effects Analysis in Design and Potential Failure Modes and Effects Analysis in Manufacturing and Assembly Processes" reference manual that provided general guidelines in preparing an FMEA. In 1999, Daimler Chrysler, Ford, and GM, as part of the International Automotive Task Force, agreed to recognize the new international standard "ISO/TS 16949" that included FMEA and would eventually replace QS 9000 in 2006.

FMEA is used across many industries as one of the Six Sigma tools. FMEA may be applied to various applications, such as System FMEA, Design FMEA, Process FMEA, Machinery FMEA, Functional FMEA, Interface FMEA, and Detailed FMEA. Although the purpose and terminology can vary according to type and the industry, the principle objectives of the different FMEA processes are to anticipate problems early in the development process and either prevent the problems or minimize their consequences (SAE Standard SAE J1739 2002).

An extension of FMEA, called FMECA was developed to include techniques to assess the probability of occurrence and criticality of potential failure modes. Today, the terms FMEA and FMECA are used interchangeably (Bowles 2003; Bowles and Bonnell 1998). FMEA is also one of the Six Sigma tools (Franceschini and Galetto 2001), and is utilized by the Six Sigma organizations in some form. The FMEA methodology is based on a hierarchical approach to determine how potential failure modes affect a product. This involves inputs from a cross-functional team having the ability to analyze the whole product life cycle. A typical design FMEA worksheet is shown in Figure 10.1.

Failure mechanisms are the processes by which specific combinations of physical, electrical, chemical, and mechanical stresses induce failure (Hu et al. 1993). Neither FMEA nor FMECA identify the failure mechanisms and models in the analysis and reporting process. In order to understand and prevent failures, failure mechanisms must be identified with respect to the predominant stresses (mechanical, thermal, electrical, chemical, and radiation) that precipitate these failures. Understanding the cause and consequences of failure mechanisms aid the design and development of a product, including virtual qualification, accelerated testing, root-cause analysis, and life consumption monitoring.

System ____________ | **Potential Failure Modes and Effects Analysis (Design FMEA)** | FMEA Number ____________
Subsystem ____________ | | Prepared by ____________
Component ____________ | | FMEA Date ____________
Design Lead ____________ | Key Date ____________ | Revision Date ____________
Core Team ____________ | | Page ______ of ______

Item/ Function	Potential Failure Mode(s)	Potential Effect(s) of Failure	Sev	Potential Cause(s) of Failure	Prob	Current Design Controls	Det	RPN	Recommended Action(s)	Responsibility & Target Completion Date	Action Results				
											Actions Taken	New Sev	New Occ	New Det	New RPN

Figure 10.1 FMEA worksheet (Guidelines for Failure Mode and Effects Analysis 2003).

In virtual qualification, failure models are used to analytically estimate the times to failure distributions for products. Without knowledge of the relevant dominant failure mechanisms and the operating conditions, virtual qualification for a product cannot be meaningful. For accelerated testing design, one needs to know the failure mechanisms that are likely to be relevant in the operating condition. Only with the knowledge of the failure mechanism, one can design appropriate tests (stress levels, physical architecture, and durations) that will precipitate the failures by the relevant mechanism without resulting in spurious failures.

All the root-cause analysis techniques, including cause and effect diagram and fault tree analysis, require that we know how the conditions during an incident may have an impact on the failure. The hypothesis development and verification processes are also affected by the failure mechanisms analysis. Knowledge of failure mechanisms and the stresses that influence these mechanisms is an important issue for life consumption monitoring of a product. The limitations on physical space and interfaces available for data collection and transmission put a limit on the number of sensors that can be implemented in a product in a realistic manner. To make sure that the appropriate data are collected and utilized for the remaining life assessment during health monitoring, the prioritized list of failure mechanisms are essential.

The traditional FMEA and FMECA do not address the key issue of failure mechanisms to analyze failures in products. To overcome this, a FMMEA methodology has been developed. The FMMEA process merges the systematic nature of the FMEA template with the "design for reliability" philosophy and knowledge. In addition to the information gathered and used for FMEA, FMMEA uses application conditions and the duration of the intended application with knowledge of active stresses and potential failure mechanisms. The potential failure mechanisms are considered individually and are assessed using appropriate models for design and qualification of the product for the intended application. The following sections describe the FMMEA methodology in detail.

10.2 Failure Modes, Mechanisms, and Effects Analysis

FMMEA is a systematic approach to identify and prioritize failure mechanisms and models for all potential failures modes. High priority failure mechanisms determine the operational stresses and the environmental and operational parameters that need to be controlled or accounted for in the design.

FMMEA is based on understanding the relationships between product requirements and the physical characteristics of the product (and their variation in the production process), the interactions of product materials with loads (stresses at application conditions), and their influence on product failure susceptibility with respect to the use conditions. This involves finding the failure mechanisms and the reliability models to quantitatively evaluate failure susceptibility. The steps in conducting an FMMEA are illustrated in Figure 10.2. The individual steps are described in greater detail in the following subsections.

10.2.1 System Definition, Elements, and Functions

The FMMEA process begins by defining the system to be analyzed. A system is a composite of subsystems or levels that are integrated to achieve a specific objective.

10 Failure Modes, Mechanisms, and Effects Analysis

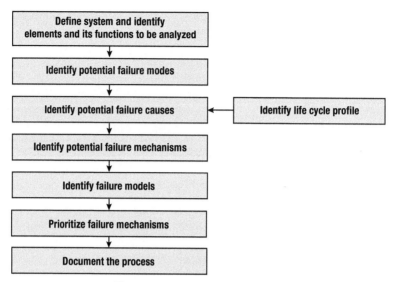

Figure 10.2 FMMEA methodology.

The system is divided into various subsystems or levels. These subsystems may comprise of further divisions or may have multiple parts that make up this subsystem. The parts are "components" that form the basic structure of the product.

Based on convenience or needs of the team conducting the analysis, the system breakdown can be either by function (i.e., according to what the system elements "do"), or by location (i.e., according to where the system elements "are"), or both (i.e., functional within the location based, or vice versa). For example, an automobile is considered a system, a functional breakdown of which would involve the cooling system, braking system, and propulsion system. A location breakdown would involve the engine compartment, passenger compartment, and dashboard or control panel. In a printed circuit board system, a location breakdown would include the package, plated though hole (PTH), metallization, and the board itself. Further analysis is conducted on each element thus identified.

10.2.2 Potential Failure Modes

A failure mode is the effect by which a failure is observed to occur (SAE Standard SAE J1739 2002). It can also be defined as the way in which a component, subsystem, or system could fail to meet or deliver the intended function.

For all the elements that have been identified, all possible failure modes for each given element are listed. For example, in a solder joint, the potential failure modes are open or intermittent change in resistance, which can hamper its functioning as an interconnect. In cases where information on possible failure modes that may occur is not available, potential failure modes may be identified using numerical stress analysis, accelerated tests to failure (e.g., HALT), past experience, and engineering judgment. A potential failure mode may be the cause of a failure mode in a higher level subsystem, or system, or be the effect of one in a lower level component.

10.2.3 Potential Failure Causes

A failure cause is defined as the circumstances during design, manufacture, or use that lead to a failure mode (IEEE Standard 1413.1-2002 2003). For each failure mode, the possible ways a failure can result are listed. Failure causes are identified by finding the basic reason that may lead to a failure during design, manufacturing, storage, transportation, or use condition. Knowledge of potential failure causes can help identify the underlying failure mechanisms driving the failure modes for a given element. For example, consider a failed solder joint of an electronic component on a printed circuit board in an automotive underhood environment. The solder joint failure modes, such as open and intermittent change in resistance, can potentially be caused due to fatigue under conditions such as temperature cycling, random vibration and/or shock impact.

10.2.4 Potential Failure Mechanisms

Failure mechanisms are the processes by which specific combination of physical, electrical, chemical, and mechanical stresses induce failure (Hu et al. 1993). Failure mechanisms are determined based on combination of potential failure mode and cause of failure (JEDEC Publication JEP 148 2004) and selection of appropriate available mechanisms corresponding to the failure mode and cause. Studies on electronic material failure mechanisms, and the application of physics-based damage models to the design of reliable electronic products comprising all relevant wearout and overstress failures in electronics are available in literature (Dasgupta and Pecht 1991; JEDEC Publication JEP 122-B 2003).

Failure mechanisms thus identified are categorized as either overstress or wearout mechanisms. Overstress failures involve a failure that arises as a result of a single load (stress) condition. Wearout failure on the other hand involves a failure that arises as a result of cumulative load (stress) conditions (IEEE Standard 1413.1-2002 2003). For example, in the case of a solder joint, the potential failure mechanisms driving the opens and shorts caused by temperature, vibration, and shock impact are fatigue and overstress shock. Further analyses of the failure mechanisms depend on the type of mechanism.

10.2.5 Failure Models

Failure models use appropriate stress and damage analysis methods to evaluate susceptibility of failure. Failure susceptibility is evaluated by assessing the time-to-failure or likelihood of a failure for a given geometry, material construction, environmental, and operational condition. For example, in case of solder joint fatigue, Dasgupta et al. (1992) and Coffin-Manson (Foucher et al. 2002) failure models are used for stress and damage analysis for temperature cycling.

Failure models of overstress mechanisms use stress analysis to estimate the likelihood of a failure based on a single exposure to a defined stress condition. The simplest formulation for an overstress model is the comparison of an induced stress versus the strength of the material that must sustain that stress. Wearout mechanisms are analyzed using both stress and damage analysis to calculate the time required to induce failure based on a defined stress condition. In the case of wearout failures, damage is accumulated over a period until the item is no longer able to withstand the applied load. Therefore, an appropriate method for combining multiple conditions must be determined for assessing the time to failure. Sometimes, the damage due to the individual loading conditions may be analyzed separately, and the failure assessment results may be combined in a cumulative manner (Guidelines for Failure Mode and Effects Analysis 2003).

Failure models may be limited by the availability and accuracy of models for quantifying the time to failure of the system. It may also be limited by the ability to combine the results of multiple failure models for a single failure site and the ability to combine results of the same model for multiple stress conditions (IEEE Standard 1413.1-2002 2003). If no failure models are available, the appropriate parameter(s) to monitor can be selected based on an empirical model developed from prior field failure data or models derived from accelerated testing.

10.2.6 Life-Cycle Profile

Life-cycle profiles include environmental conditions such as temperature, humidity, pressure, vibration or shock, chemical environments, radiation, contaminants, and loads due to operating conditions, such as current, voltage, and power (Society of Automotive Engineers 1978). The life-cycle environment of a product consists of assembly, storage, handling, and usage conditions of the product, including the severity and duration of these conditions. Information on life-cycle conditions can be used for eliminating failure modes that may not occur under the given application conditions.

In the absence of field data, information on the product usage conditions can be obtained from environmental handbooks or data monitored in similar environments. Ideally, such data should be obtained and processed during actual application. Recorded data from the life-cycle stages for the same or similar products can serve as input towards the FMMEA process. Some organizations collect, record, and publish data in the form of handbooks that provide guidelines for designers and engineers developing products for market sectors of their interest. Such handbooks can provide first approximations for environmental conditions that a product is expected to undergo during operation. These handbooks typically provide an aggregate value of environmental variables and do not cover all the life-cycle conditions. For example, for general automotive applications, life-cycle environment and operating conditions can be obtained from the SAE handbook (Society of Automotive Engineers 1978), but for specific applications more detailed information of the particular application conditions need to be obtained.

10.2.7 Failure Mechanism Prioritization

Ideally, all failure mechanisms and their interactions will be considered for product design and analysis. In the life cycle of a product, several failure mechanisms may be activated by different environmental and operational parameters acting at various stress levels, but only a few operational and environmental parameters, and failure mechanisms are in general responsible for the majority of the failures. High priority mechanisms are those select failure mechanisms that may cause the product to fail earlier than the product's intended life duration. These mechanisms occur during the normal operational and environmental conditions of the products application. High priority failure mechanisms provide effective utilization of resources and are identified

10.2 Failure Modes, Mechanisms, and Effects Analysis

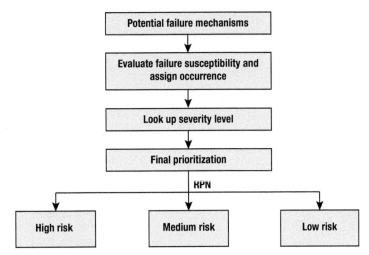

Figure 10.3 Failure mechanism prioritization.

through prioritization of all the potential failure mechanisms. The methodology for failure mechanism prioritization is shown in Figure 10.3.

Environmental and operating conditions are set up for initial prioritization of all potential failure mechanisms. If the load levels generated by certain operational and environmental conditions are nonexistent or negligible, the failure mechanisms that are exclusively dependent on those environmental and operating conditions are assigned a "low" risk level and are eliminated from further consideration.

For all the failure mechanisms remaining after the initial prioritization, the susceptibility to failure by those mechanisms is evaluated using the previously identified failure models when such models are available. For the overstress mechanisms, failure susceptibility is evaluated by conducting a stress analysis to determine if failure is precipitated under the given environmental and operating conditions. For the wearout mechanisms, failure susceptibility is evaluated by determining the time-to-failure under the given environmental and operating conditions. To determine the combined effect of all wearout failures, the overall time-to-failure is also evaluated with all wearout mechanisms acting simultaneously. In cases where no failure models are available, the evaluation is based on past experience, manufacturer data, or handbooks.

After evaluation of failure susceptibility, occurrence ratings under environmental and operating conditions applicable to the system are assigned to the failure mechanisms. For the overstress failure mechanisms that precipitate failure, the highest occurrence rating ,"frequent," is assigned. In case no overstress failures are precipitated, the lowest occurrence rating, "extremely unlikely," is assigned. For the wearout failure mechanisms, the ratings are assigned based on benchmarking the individual time-to-failure for a given wearout mechanism, with overall time-to-failure, expected product life, past experience and engineering judgment. Table 10.1 shows the occurrence ratings.

A "frequent" occurrence rating involves failure mechanisms with very low time-to-failure (TTF) and overstress failures that are almost inevitable in the use condition. A "reasonably probable" rating involves cases that involve failure mechanisms with low TTF. An "occasional" involves failures with moderate TTF. A "remote" rating

Table 10.1 Occurrence ratings

Rating	Criteria
Frequent	Overstress failure or very low TTF
Reasonably probable	Low TTF
Occasional	Moderate TTF
Remote	High TTF
Extremely unlikely	No overstress failure or very high TTF

Table 10.2 Severity ratings

Rating	Criteria
Very high or catastrophic	System failure or safety-related catastrophic failures
High	Loss of function
Moderate or significant	Gradual performance degradation
Low or minor	System operable at reduced performance
Very low or none	Minor nuisance

involves failure mechanisms that have a high TTF. An extremely unlikely rating is assigned to failures with very high TTF or overstress failure mechanisms that do not produce any failure.

To provide a qualitative measure of the failure effect, each failure mechanism is assigned a severity rating. The failure effect is assessed first at the level being analyzed, then the next higher level, the subsystem level, and so on to the system level (SAE Standard SAE J1739 2002). Safety issues and impact of a failure mechanism on the end system are used as the primary criterion for assigning the severity ratings. In the severity rating, possible worst case consequence is assumed for the failure mechanism being analyzed. Past experience and engineering judgment may also be used in assigning severity ratings. The severity ratings shown in Table 10.2 are defined later in the chapter.

A "very high or catastrophic" severity rating indicates that there may be loss of life of the user or un-repairable damage to the product. A "high" severity rating indicates that failure might cause a severe injury to the user or a loss of function of the product. A "moderate or significant" rating indicates that the failure may cause minor injury to the user or show gradual degradation in performance over time through loss of availability. A "low or minor" rating indicates that failure may not cause any injury to the user or result in the product operating at reduced performance. A "very low or none" rating does not cause any injury and has no impact on the product or at the best may be a minor nuisance.

The final prioritization step involves classification of the failure mechanisms into three risk levels. This can be achieved by using the risk matrix as shown in Table 10.3. The classifications may vary based on the product type, use condition, and business objectives of the user/manufacturer.

10.2.8 Documentation

The FMMEA process involves documentation. FMMEA documentation includes the actions considered and taken based on the FMMEA. For products already

Table 10.3 Risk matrix

Severity		Occurrence				
		Frequent	Reasonably probable	Occasional	Remote	Extremely unlikely
	Very high or catastrophic	High risk	High risk	High risk	High risk	Moderate risk
	High	High risk	High risk	High risk	Moderate risk	Low risk
	Moderate or significant	High risk	High risk	Moderate risk	Low risk	Low risk
	Low or minor	High risk	Moderate risk	Low risk	Low risk	Low risk
	Very low or none	Moderate risk	Low risk	Low risk	Low risk	Low risk

manufactured, documentation may exist in the form of records of root-cause analysis conducted for the failures that occur during product development and testing. The history and lessons learned contained within the documentation provide a framework for future product FMMEA. It is also necessary to maintain and update documentation about the FMMEA after the corrective actions so as to generate a new list of high priority failure mechanisms for future analysis.

10.3 Case Study

A simple printed circuit board (PCB) assembly used in an automotive application was selected to demonstrate the FMMEA process. The PCB assembly was mounted at all four corners in the engine compartment of a 1997 Toyota 4Runner. The assembly consisted of an FR-4 PCB with copper metallizations, plated through-hole (PTH) and eight surface mount inductors soldered into the pads using 63Sn-37Pb solder. The inductors were connected to the PTH through the PCB metallization. The PTHs were solder filled and an event detector circuit was connected in series with all the inductors through the PTHs to assess failure. Assembly failure was defined as one that would result in breakdown, or no current passage in the event detector circuit.

For all the elements listed, the corresponding functions and the potential failure modes were identified. Table 10.4 lists the physical location of all possible failure modes for the elements. For example, for the solder joint, the potential failure modes are open and intermittent change in resistance.

For sake of simplicity and demonstration purposes, it was assumed that the test setup, the board, and its components were defect free. This assumption can be valid if proper screening was conducted after manufacture. In addition, it must be assumed that there was no damage to the assembly after manufacture. Potential failure causes were then identified for the failure modes and are shown in Table 10.4. For example, for the solder joint, the potential failure causes for open and intermittent change in resistance are temperature cycling, random vibration, or sudden shock impact caused by vehicle collision.

Based on the potential failure causes that were assigned to the failure modes, the corresponding failure mechanisms were identified. Table 10.4 lists the failure mechanisms for the failure causes that were identified. For example, for the open and

Table 10.4 FMMEA worksheet for the case study

Element	Potential failure mode	Potential failure cause	Potential failure mechanism	Mechanism type	Failure model	Failure susceptibility	Occurrence	Severity	Risk
PTH	Electrical open in PTH	Temperature cycling	Fatigue	Wearout	CALCE PTH barrel thermal fatigue (Bhandarkar et al. 1992)	>10 years	Remote	Very low	Low
Metallization	Electrical short/open, change in resistance in the metallization traces	High temperature	Electromigration	Wearout	Black (1983)	>10 years	Remote	Very high	Moderate
		High relative humidity	Corrosion	Wearout	Howard (1981)	>10 years	Remote	Very high	Moderate
		Ionic contamination							
Component (Inductors)	Short/open between windings and the core	High temperature	Wearout of winding insulation	Wearout	No Model		Remote[a]	Very high	Moderate
Interconnect	Open/intermittent change in electrical resistance	Temperature cycling	Fatigue	Wearout	Coffin–Manson (Foucher et al. 2002)	170 days	Frequent	Very high	**High**
		Random vibration		Wearout	Steinberg (1988)	43 days	Frequent	Very high	**High**
		Sudden impact	Shock	Overstress	Steinberg (1988)	No failure	Extremely unlikely	Very high	Moderate

PCB	Electrical short between PTHs	High relative humidity	CFF	Wearout	Rudra and Pecht (Rudra et al. 1995)	4.6 years	Occasional	Very low	Low
	Crack / Fracture	Random vibration	Fatigue	Wearout	Basquin (Steinberg 1988)	>10 years	Remote	Very high	Moderate
		Sudden impact	Shock	Overstress	Steinberg (1988)	No failure	Extremely unlikely	Very high	Moderate
	Loss of polymer strength	High temperature	Glass transition	Overstress	No model	No failure	Extremely unlikely	Very high	Moderate
	Open	Discharge of high voltage through dielectric material	EOS/ESD	Overstress	No model	Eliminated in first level prioritization			Low
	Excessive noise	Proximity to high current or magnetic source	EMI	Overstress	No model	Eliminated in first level prioritization			Low
Pad	Lift /crack	Temperature cycling/ random vibration	Fatigue	Wearout	No model		Remote	Very high	Moderate
		Sudden impact	Shock	Overstress			Extremely unlikely	Very high	Moderate

[a]Based on failure rate data of inductors in Telcordia (Telcordia Technologies 2001).

intermittent change in resistance in solder joint, the mechanisms driving the failure were solder joint fatigue and fracture.

For each of the failure mechanisms listed, the appropriate failure models were then identified from the literature. Information about product dimensions and geometry were obtained from design specification, board layout drawing, and component manufacturer datasheets. Table 10.4 provides all the failure models for the failure mechanisms that were listed. For example, in case of solder joint fatigue, a Coffin-Manson (Steinberg 1988) failure model was used for stress and damage analysis for temperature cycling.

The assembly was powered by a 3-V battery source independent of the automobile electrical system. There were no high current, voltage, magnetic, or radiation sources that were identified to have an effect on the assembly. For the temperature, vibration, and humidity conditions prevalent in the automotive underhood environment, data were obtained first from the Society of Automotive Engineers (SAE) environmental handbook (Society of Automotive Engineers 1978) as no manufacturer field data were available for the automotive underhood environment for the Washington, DC area. The maximum temperature in the automotive underhood environment was listed as 121°C (Society of Automotive Engineers 1978). The car was assumed to operate on average 3 hours per day in two equal trips in the Washington, DC area. The maximum shock level was assumed to be 45G for 3ms. The maximum relative humidity in the underhood environment was 98% at 38°C (Society of Automotive Engineers 1978). The average daily maximum and minimum temperature in the Washington DC area for the period the study was conducted were 27°C and 16°C, respectively.

After all potential failure modes, causes, mechanisms, and models were identified for each element; an initial prioritization was made based on the life-cycle environmental and operating conditions. In automotive underhood environment for the given test setup, failures driven by electrical overstress (EOS) and electrostatic discharge (ESD) were ruled out because of the absence of active devices, and the low voltage source of the batteries. Electromagnetic interference (EMI) was also not anticipated because the circuit function was not susceptible to transients. Hence, EOS, ESD, and EMI were each assigned a "low" risk level.

The time to failure for the wearout failure mechanisms was calculated using calcePWA.[1] Occurrence ratings were assigned based on comparing the time-to-failure for a given wearout mechanism with the overall time-to-failure with all wearout mechanisms acting together. For the inductors, the occurrence rating was assigned based on failure rate data obtained from Telcordia (Telcordia Technologies 2001). From prior knowledge regarding wearout associated with the pads, it was assigned a "remote" occurrence rating.

An assessment of a shock level of 45G for 3 ms using calcePWA produced no failure for interconnects and the board. Hence it was assigned an "extremely unlikely" occurrence rating. Since no overstress shock failure was expected on the board and the interconnects, it was assumed there would also be no failure on the pads. Hence overstress shock failure on pads was also assigned an "extremely unlikely" rating. The glass transition temperature for the board was 150°C. Since the maximum temperature in the underhood environment was only 121°C (Society of Automotive Engineers

---

[1] A physics-of-failure-based virtual reliability assessment tool developed by CALCE, University of Maryland.

1978), no glass transition was expected to occur, and it was assigned an "extremely unlikely" rating.

A short or open PTH would not have had any impact on the functioning of circuit, as it was used only as terminations for the inductors. Hence, it was assigned a "very low" severity rating. For all other elements, any given failure mode of the element would have led to the disruption in the functioning of circuit. Hence, all other elements were assigned a "very high" severity rating.

Final prioritization and risk assessment for the failure mechanisms is shown in Table 10.4. Out of all the failure mechanisms that were analyzed, fatigue due to thermal cycling and vibration at the solder joint interconnect were the only failure mechanisms that had a high risk. Being a high risk failure mechanism, they were identified as high priority.

An FMEA on the assembly would have identified all the elements, their functions, potential failure modes, and failure causes as in FMMEA. FMEA would then have identified the effect of failure for each failure mode. For example, in the case of a solder joint interconnect, the failure effect of the open joint would have involved no current passage in the test set up. Next, the FMEA would have identified the severity, occurrence, and detection probabilities associated with each failure mode. For example, in case of a solder joint open failure mode, based on past experience and use of engineering judgment, each of the metrics, severity, occurrence and detection would have received a rating on a scale of ten. The product of severity, occurrence, and detection would then have been used to calculate RPN. The RPNs for other failure modes would have been calculated in a similar manner, and then all the failure modes would have been prioritized based on the RPN values. This is unlike FMMEA, which used failure mechanisms and models and used the combined effect of all failure mechanism to quantitatively evaluate the occurrence. The occurrence rating in conjunction with severity was then used to assign a risk level to each failure mechanisms for prioritization.

10.4 Summary

FMMEA allows the design team to take into account the available scientific knowledge of failure mechanisms and merge them with the systematic features of the FMEA template with the intent of "design for reliability" philosophy and knowledge. The idea of prioritization embedded in the FMEA process is also utilized in FMMEA to identify the mechanisms that are likely to cause failures during the product life cycle.

FMMEA differs from FMEA in a few respects. In FMEA, potential failure modes are examined individually and the combined effects of coexisting failures causes are not considered. FMMEA, on the other hand, considers the impact of failure mechanisms acting simultaneously. FMEA involves precipitation and detection of failure for updating and calculating the RPN, and cannot be applied in cases that involve a continuous monitoring of performance degradation over time. FMMEA on the contrary does not require the failure to be precipitated and detected, and the uncertainties associated with the detection estimation are not present. The use of environmental and operating conditions is not made at a quantitative level in FMEA. At best, they are used to eliminate certain failure modes. FMMEA prioritizes the failure mechanisms using the information on stress levels of environmental and operating conditions to identify high priority mechanisms that must be accounted for in the design or be controlled. This prioritization in FMMEA overcomes the shortcomings of RPN prioritization used in FMEA, which provide a false sense of granularity. Thus, the use of FMMEA provides additional quantitative information regarding product reliability and opportunities for improvement than FMEA, as it takes into account specific failure mechanisms and the stress levels of environmental and operating conditions into the analysis process.

There are several benefits to organizations that use FMMEA. It provides specific information on stress conditions so that that the acceptance and qualification tests yield useable result. Use of the failure models at the development stage of a product also allows for appropriate "what-if" analysis on proposed technology upgrades. FMMEA can also be used to aid several design and development steps considered to be the best practices, which can only be performed or enhanced by the utilization of the knowledge of failure mechanisms and models. These steps include virtual qualification, accelerated testing, root-cause analysis, life consumption monitoring, and prognostics. All the technological and economic benefits provided by these practices are realized better through the adoption of FMMEA.

Problems

10.1 How are failure mechanisms identified? Explain with realistic examples.

10.2 What are the differences between overstress mechanisms and wearout mechanisms?

10.3 Give an example of the life-cycle profile for an electronic product.

10.4 The steps in FMMEA are listed in random order.

- Prioritize failure mechanisms.
- Define system and identify elements and functions to be analyzed.
- Identify failure models.
- Identify potential failure modes.
- Identify potential failure causes.
- Identify life-cycle profile.
- Identify potential failure mechanisms.

(a) Arrange the steps listed in their proper order.
(b) Suggest another step that could be added to this list to make the process more useful. Explain and provide a realistic example.

11 Probabilistic Design for Reliability and the Factor of Safety

To ensure the production of a reliable product, reliability activities must start early in the product development cycle. In order to analyze the reliability of a product, we have to first understand how to analyze the reliability of its components. In order to achieve the desirable reliability level, various reliability methodologies and tools can be used throughout the life cycle of the product—from the early planning stages through design, development, production, field testing, and customer use.

This chapter covers basic models and principles to quantify and evaluate reliability during the design stage. It presents the probabilistic design approach and relationships between reliability and safety factor. The relationship between tolerances on the characteristics of the parts and reliability is also discussed. Probabilistic design requires analysis of the functions of random variables.

11.1 Design for Reliability

Reliability is a design parameter and must be incorporated into a product at the design stage. One way to quantify reliability during design and to design for reliability is the probabilistic approach to design (Haugen 1968; Kececioglu 1991; Kececioglu and Cormier 1968). The design variables and parameters are random variables, and hence the design methodology must consider them as random variables.

The basic premise in reliability analysis from the viewpoint of probabilistic design methodology is that a given component has a certain strength which, if exceeded, will result in the failure of the component. The factors that determine the strength of the component are random variables, as are the factors that determine the stresses or load acting on the component. Stress is used to indicate any agency that tends to induce failure, whereas strength indicates any agency resisting failure. Failure is taken to mean failure to function as intended; it occurs when the actual stress exceeds the actual strength for the first time.

Reliability Engineering, First Edition. Kailash C. Kapur and Michael Pecht.
© 2014 John Wiley & Sons, Inc. Published 2014 by John Wiley & Sons, Inc.

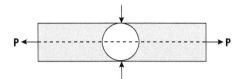

Figure 11.1 Design of a tension element.

11.2 Design of a Tension Element

Let us consider the design of a tension element for a tensile load of $P = 4{,}000$ units of load, as shown in Figure 11.1. The design engineer is considering the failure mode due to tensile fracture. Based on the material and its manufacturing processes, the designer finds the value of the ultimate tensile strength to be 10,000 units of load per square inch. This value typically is some average or mean value for the strength.

The classical approach to design uses the equation (where A is the cross-sectional area of the element)

$$\text{Mean Strength} \geq \text{Mean Stress} \times \text{Factor of Safety}$$

$$\text{or } 100{,}000 \geq (4{,}000 / A) \times 2$$

$$\text{or } A \geq 0.08 \text{ in}^2.$$

If we consider that the element has a circular cross-section with diameter D, then we can calculate that $D = 0.3192$ in.

Thus, it is clear that this approach does not consider the concept of reliability. We cannot answer the following questions:

- How reliable is this design? The answer is not provided by the above design approach and analysis.
- If a certain level of reliability is specified for a given mode of failure, what should be the value of the design variable (the diameter) of the tension element?

We do know that

- The load is a random variable due to varying conditions of customer usage and environmental factors.
- The ultimate tensile strength is a random variable due to material variation and manufacturing processes.
- The diameter of the element is a random variable due to manufacturing variability and is typically dealt with by introducing tolerances.

Thus we want to know what effect all types of variability have on the reliability.

The concept of design by probability, or probabilistic design, recognizes the reality that loads or stresses, and the strength of products subjected to these stresses, cannot be identified as specific values but have ranges of values with a probability of occurrence associated with each value in the range. Figure 11.2 shows $f(x)$ as the probability density function (pdf) for the stress random variable X, and $g(y)$ as the pdf for the strength random variable Y.

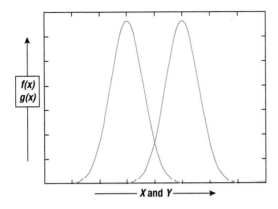

Figure 11.2 Stress and strength distributions.

The words *stress* and *strength* are used here in a broad sense applicable in a variety of situations well beyond traditional mechanical or structural systems. As mentioned, stress is used to indicate any agency that tends to induce failure, while strength indicates any agency resisting failure. Formulaically, we can say that

$$\text{Reliability} = R = P[\text{Strength} > \text{Stress}].$$

The reliability of the component is the probability that the strength of the component will be greater than the stress to which it will be subjected. The factor of safety, represented by number n, is the ratio of strength (Y) and the stress (X). Since both Y and X are random variable, one definition of the factor of safety is

$$n = \frac{\mu_Y}{\mu_X}. \tag{11.1}$$

There are four basic ways in which the designer can increase reliability:

1. *Increase Mean Strength.* This is achieved by increasing size or weight of materials, using stronger materials, and so on.
2. *Decrease Average Stress.* This can be done by controlling loads or using higher dimensions.
3. *Decrease Stress Variations.* This variation is harder to control, but can be effectively truncated by putting limitations on use conditions.
4. *Decrease Strength Variation.* The inherent part-to-part variation can be reduced by improving the basic process, controlling the process, and utilizing tests to eliminate less desirable parts.

11.3 Reliability Models for Probabilistic Design

For a certain mode of failure, let $f(x)$ and $g(y)$ be the probability density functions for the stress random variable X and the strength random variable Y, respectively.

Also, let $F(x)$ and $G(y)$ be the cumulative distribution functions for the random variables X and Y, respectively. Then the reliability, R, of the product for a failure mode under consideration, with the assumption that the stress and the strength are independent random variables, is given by

$$\begin{aligned}R = P\{Y > X\} \\ = \int_{-\infty}^{\infty} g(y) \left\{ \int_{-\infty}^{y} f(x) dx \right\} dy \\ = \int_{-\infty}^{\infty} g(y) F(y) dy \\ = \int_{-\infty}^{\infty} f(x) \left\{ \int_{x}^{\infty} g(y) dy \right\} dx \\ = \int_{-\infty}^{\infty} f(x) \{1 - G(x)\} dx.\end{aligned} \quad (11.2)$$

Consider a product where the stress and strength are normally distributed. Specifically, stress random variable X is normally distributed, with mean μ_X and with the standard deviation as σ_X. Similarly, the strength random variable Y is normally distributed, with mean μ_Y and standard deviation σ_Y. The reliability, R, for this mode of failure can be derived by:

$$R = P[Y > X] = P[(Y - X) > 0]. \quad (11.3)$$

It is known that $U = Y - X$ is also normally distributed with

$$\begin{aligned}\mu_U = \mu_Y - \mu_X \\ \sigma_U^2 = \sigma_Y^2 + \sigma_X^2.\end{aligned} \quad (11.4)$$

Hence,

$$R = P[U > 0] = \Phi\left[Z > \left(\frac{-\mu_U}{\sigma_U}\right)\right] = \Phi\left[Z < \left(\frac{\mu_U}{\sigma_U}\right)\right], \quad (11.5)$$

or

$$R = \Phi\left[\frac{\mu_Y - \mu_X}{\sqrt{\sigma_Y^2 + \sigma_X^2}}\right], \quad (11.6)$$

where $\Phi(\cdot)$ is the cumulative distribution function for the standard normal variable.

The reliability computations for other distributions, such as exponential, lognormal, gamma, Weibull, and several extreme value distributions, have been developed (Kapur and Lamberson 1977). In addition, the reliability analysis has been generalized when the stress and strength variables follow a known stochastic process.

11.4 Example of Probabilistic Design and Design for a Reliability Target

Example 11.1

Suppose we have the following data for strength Y and stress X for a particular failure mode:

$$\mu_Y = 40,000 \quad \sigma_Y = 4000$$

$$\mu_X = 30,000 \quad \sigma_X = 3000.$$

One definition for the factor of safety as mentioned before is that it is the ratio of mean value of strength divided by the mean value of stress. Hence, for this problem,

$$\text{Factor of safety} = 40,000/30,000 = 1.33.$$

Using Equation 11.6, the reliability for this failure mode is

$$R = \Phi\left[\frac{40,000 - 30,000}{\sqrt{(4000)^2 + (3000)^2}}\right] = \Phi(2.0) = 0.97725.$$

From the failure perspective, there would be 2275 failures per 100,000 use conditions. The above reliability calculation is for a factor of safety of 1.33. We can increase the factor of safety by changing the design (such as increasing dimensions). This change makes μ_X equal to 20,000, increasing the factor of safety to 2. Thus, higher reliability can be expected, given as

$$R = \Phi\left[\frac{40,000 - 20,000}{\sqrt{(4,000)^2 + (3,000)^2}}\right] = \Phi(4.0) = 0.99997.$$

Now there are only three failures per 100,000 use conditions, which is a tremendous decrease from the previous situation. Reliability could also have been increased by decreasing the stress and strength variation.

11.4 Example of Probabilistic Design and Design for a Reliability Target

Let us now illustrate how to solve the design of the tension element given in Section 11.2 by using the probabilistic approach, which considers variability in the load conditions, ultimate tensile strength, and the diameter of the element. Suppose the variability of the load is quantified by its standard deviation as 100 units, and the ultimate tensile strength has a standard deviation of 5×10^3 units of strength. Thus we have

Load: $\mu_P = 4 \times 10^3, \quad \sigma_P = 100.$

Ultimate tensile strength: $\mu_Y = 100 \times 10^3, \quad \sigma_Y = 5 \times 10^3.$

Now we want to design the element for a specified reliability $= R = 0.99990$, with tensile fracture as the failure mode. Suppose the standard deviation of the diameter

Table 11.1 Relationship between dimensional tolerances and reliability

% Tolerances on D	Reliability
0	0.999915
1.0	0.999908
1.5	0.999900
3.0	0.999847
7.0	0.999032

Table 11.2 Relationship between strength variability and reliability

Standard deviation for strength	Reliability
2×10^3	0.99999
4×10^3	0.99996
5×10^3	0.99990
8×10^3	0.99157
10×10^3	0.97381

based on manufacturing processes is 0.5% of the diameter. The standard deviation can be converted to tolerances based on the idea of using $\pm k\sigma$ tolerances, where k is typically chosen as 3. If k is equal to 3 and the underlying random variable has normal distribution, then 99.73% of the values of the variable will fall within $\pm 3\sigma$ (it must be emphasized that it can be any other number depending on the requirements). Thus, for our example, $\pm 3\sigma$ tolerances will be $\pm 1.5\%$ of the diameter of the element. Then we can design (Kapur and Lamberson 1977) the tension element using a probabilistic approach, and the mean value of the diameter will be $\bar{D} = 0.2527$, compared with 0.3192 (calculated in Section 11.2).

We can also do a sensitivity analysis of reliability with respect to all the design variables. For example, consider the effect of tolerances or the standard deviation of the diameter of the design element. Table 11.1 shows the effect on the diameter of tolerances, based on the nature of the manufacturing processes, as a percent of the dimension.

Similarly, we can study the sensitivity of R with respect to the standard deviation of the strength, which also may be a reflection of the design and manufacturing processes. This is given in Table 11.2.

11.5 Relationship between Reliability, Factor of Safety, and Variability

When stress and strength are normally distributed,

$$R = \Phi\left[\frac{\mu_Y - \mu_X}{\sqrt{\sigma_Y^2 + \sigma_X^2}}\right]. \tag{11.7}$$

The factor of safety is some ratio of the strength and stress variables. Since both are random variables, the question arises of which measure of the strength or the stress

11.5 Relationship between Reliability, Factor of Safety, and Variability

should be used in the computation of the factor of safety. One definition is based on using the mean values of the strength and the stress variables; then, the factor of safety, n, is defined as

$$n = \frac{\mu_Y}{\mu_X} = \text{factor of safety}. \tag{11.8}$$

The variability of any random variable can be quantified by its coefficient of variation, which is the ratio of the standard deviation and the mean value. Thus, the coefficient of variation is a dimensionless quantity. If it is 0.05, we can say that the standard deviation is 5% of the mean value, and if we use $\pm 3\sigma$ tolerances, we can say that the tolerances are $\pm 15\%$ of the mean value of the underlying variable. Thus,

$$CV_Y = \text{coefficient of variation for strength random variable}$$
$$= \frac{\sigma_Y}{\mu_Y} \tag{11.9}$$

$$CV_X = \text{coefficient of variation for stress random variable}$$
$$= \frac{\sigma_X}{\mu_X}. \tag{11.10}$$

Then, Equation 11.7 can be rewritten as (by dividing both the numerator and the denominator by μ_Y):

$$R = \Phi\left[\frac{n-1}{\sqrt{CV_Y^2 n^2 + CV_X^2}}\right]. \tag{11.11}$$

Thus, the above relation can be used to relate reliability, factor of safety, coefficient of variation for the stress random variable, and coefficient of variation for the strength random variable.

Example 11.2

The stress (X) and the strength (Y) random variables for a given failure mode of a component follow the normal distributions with the following parameters:

$$\mu_X = 10{,}000 \quad \mu_X = 2400$$

$$\mu_Y = 15{,}000 \quad \mu_Y = 2000.$$

(a) Find the reliability for the component for this failure mode.

Solution:

$$R = \Phi\left(\frac{\mu_Y - \mu_X}{\sqrt{\sigma_X^2 + \sigma_Y^2}}\right) = \Phi\left(\frac{5000}{3124.1}\right) = \Phi(1.60) = 0.9452.$$

(b) The customer wants a reliability of 0.9990. The only thing that the designer can change for this failure mode is the mean value for the strength random variable Y (thus increasing the factor of safety). Find the new target for μ_Y to achieve the reliability goal.

Solution:

$$R = \Phi\left(\frac{\mu_Y - \mu_X}{\sqrt{\sigma_X^2 + \sigma_Y^2}}\right) = 0.9990$$

$$3.10 = \frac{\mu_Y - \mu_X}{\sqrt{\sigma_X^2 + \sigma_Y^2}} = \frac{\mu_Y - 10,000}{3124.1}$$

$$\mu_Y = 3.1 \times 3124.2 + 10,000 = 19,684.7.$$

Thus, the new target for the mean value of the strength should be 19,684.7 units.

Example 11.3

The stress (X) and the strength (Y) for a given failure mode of a component follow a normal distribution with the following information about their coefficient of variation, CV:

$$CV_X = 0.25 \quad CV_Y = 0.17.$$

The customer wants a reliability of 0.99990 for this failure mode. What is the safety factor that the designer must use to meet the requirements of the customer?

Solution:

$$R = 0.99990 = \Phi\left[\frac{n-1}{\sqrt{n^2 CV_Y^2 + CV_X^2}}\right] = \Phi(3.715)$$

$$3.715 = \frac{n-1}{\sqrt{n^2 0.17^2 + 0.25^2}}$$

$$(n-1)^2 = (0.0289n^2 + 0.0625) \times 13.8012$$

$$0.601145n^2 - 2n + 0.137423 = 0$$

$$n = \frac{-b \pm \sqrt{b^2 - 4ac}}{2a}$$

$$= \frac{2 \pm \sqrt{4 - 4 \times 0.601145 \times 0.137423}}{2 \times 0.601145}$$

$$= \frac{2 \pm 1.91561}{1.20229} = 3.26 \quad \text{or} \quad 0.0702.$$

Choose $n = 3.26$. Note that the other root of the quadratic equation gives us the value of the unreliability.

11.6 Functions of Random Variables

For any design problem, there is one design variable that is a function of several other design variables. For example, for the tension element considered in Section 11.2, the stress random variable, X, for the circular cross-section tension element is given by

$$X = \frac{4P}{\pi D^2}, \quad (11.12)$$

where P and D are both random variables. Generally, it is very difficult to find the probability density function for one variable that is a nonlinear function of several other variables. In such cases, for engineering analysis, knowledge of the first and the second moments of the transformed variable is quite useful. This knowledge can be used for probabilistic design. Consider the following general model:

$$Y = f(X_1, X_2, \ldots, X_n), \quad (11.13)$$

where Y is a function of n other variables, represented by vector $X = (X_1, X_2, \ldots, X_n)$.

This represents a general equation where a design random variable, Y, is a function of other design random variables. Information about the mean and variance of these n random variables is given as

$$\begin{aligned} E[X_i] &= \mu_i, \quad i = 1, 2, \ldots, n \\ V[X_i] &= \sigma_i^2, \quad i = 1, 2, \ldots, n. \end{aligned} \quad (11.14)$$

Then, we can find the approximate values for μ_Y and σ_Y^2 using Taylor's series approximations as follows:

$$\mu_Y \cong f(\mu_1, \mu_2, \ldots, \mu_n) + \frac{1}{2} \sum_{i=1}^{n} \left. \frac{\partial^2 f(X)}{\partial X_i^2} \right|_{X = \mu} V(X_i)$$

$$V_Y = \sigma_Y^2 \cong \sum_{i=1}^{n} \left\{ \left. \frac{\partial f(X)}{\partial X_i} \right|_{X = \mu} \right\}^2 V(X_i), \quad (11.15)$$

where

$$X = (X_1, X_2, \ldots, X_n) \quad \text{and} \quad \mu = (\mu_1, \mu_2, \ldots, \mu_n). \quad (11.16)$$

For approximate analysis, the second derivative terms for the mean are typically ignored.

For engineering analysis, designers think in terms of tolerances. Let the tolerances about the mean value be denoted by

$$X = \mu_X \pm t_X. \tag{11.17}$$

If X is normally distributed with a mean of μ_X and variance of σ_X^2, and we use 3σ limits—that is, $t_X = 3\sigma_X$, then for symmetrical tolerances about the mean, 0.27% of the items will be outside the 3σ limits.

If we use 4σ limits, then $t_X = 4\sigma_X$ and 99.993666% of the values of X are within 4σ limits, or 0.006334% will be outside 4σ limits.

Example 11.4

Consider the probabilistic analysis of part of an electrical circuit that has two resistances in parallel. The terminal resistance, R_T, as a function of the two other resistances, R_1 and R_2, is given by

$$R_T = f(R_1, R_2) = \frac{R_1 R_2}{R_1 + R_2}.$$

Suppose 3σ tolerances on R_1 are 100 ± 30, and 3σ tolerances on R_2 are 200 ± 45. Thus,

$$\mu_{R_1} = 100\ \Omega \quad \sigma_{R_1} = 10\ \Omega$$
$$\mu_{R_2} = 200\ \Omega \quad \sigma_{R_2} = 15\ \Omega.$$

Then, using Equation 11.15,

$$E[R_T] = f(100, 200) = \frac{100 \times 200}{100 + 200} = 66.67\ \Omega$$

$$\frac{\partial f}{\partial R_1} = \frac{R_2^2}{(R_1 + R_2)^2} \quad \frac{\partial f}{\partial R_2} = \frac{R_1^2}{(R_1 + R_2)^2}$$

$$\left.\frac{\partial f}{\partial R_1}\right|_{R=\mu} = 0.444 \quad \left.\frac{\partial f}{\partial R_2}\right|_{R=\mu} = 0.111$$

$$\sigma_{R_T}^2 = 0.444^2 \times 10^2 + 0.111^2 \times 15^2 = 22.4858 \quad \text{or} \quad \sigma_{R_T} = 4.74\ \Omega.$$

Thus, the three tolerances on R_T are 66.67 ± 14.22. These statistical tolerances are much tighter than the tolerances based on worst-case analysis, which may consider minimum and maximum values of the resistances based on 3σ tolerances for the two resistances.

Example 11.5

Determine the tolerance for the volume of a cylinder having the following 3σ tolerances for the diameter, D, and its length, L

$$D = 2.5 \pm 0.002\ \text{m}, \quad L = 4.0 \pm 0.005\ \text{m}.$$

Also assume that D and L are probabilistically independent.

11.6 Functions of Random Variables

Solution:
The volume of the cylinder is given by

$$V = \frac{\pi D^2 L}{4}.$$

so the mean volume is approximately given by

$$\mu_V \approx f(\mu_D, \mu_L) = \pi \frac{\mu_D^2 \mu_L}{4} = \pi \frac{2.5^2 \times 4.0}{4} = 19.635.$$

The partial derivatives are:

$$\frac{\partial V}{\partial D} = \frac{\pi DL}{2} \quad \frac{\partial V}{\partial L} = \frac{\pi D^2}{4},$$

so,

$$\sigma_V^2 = \left\{\frac{\partial V}{\partial D}\bigg|_{\mu_D,\mu_L}\right\}^2 \sigma_D^2 + \left\{\frac{\partial V}{\partial L}\bigg|_{\mu_D,\mu_L}\right\}^2 \sigma_L^2$$

$$= \left(\frac{\pi \mu_D \mu_L}{2}\right)^2 \sigma_D^2 + \left(\frac{\pi \mu_D^2}{4}\right)^2 \sigma_L^2.$$

If we let the tolerance on V as $T_V = 3\sigma_V$, then the above equation can be written in terms of tolerances as,

$$T_V^2 = = \left\{\frac{\partial V}{\partial D}\bigg|_{\mu_D,\mu_L}\right\}^2 T_D^2 + \left\{\frac{\partial V}{\partial L}\bigg|_{\mu_D,\mu_L}\right\}^2 T_L^2$$

$$= \left(\frac{\pi \mu_D \mu_L}{2}\right)^2 T_D^2 + \left(\frac{\pi \mu_D^2}{4}\right)^2 T_L^2 = 0.001589.$$

Hence the tolerances on $V = \mu_V \pm T_V = 19.635 \pm 0.0399$ m^3.

Example 11.6

A random variable, Y, for a product is a function of three other random variables, X_1, X_2, X_3, and is given by

$$Y = \frac{2\sqrt{X_1}}{X_2 X_3} \quad \text{or} \quad Y = \frac{2 X_1^{1/2}}{X_2 X_3}.$$

(a) Find the expected value and the standard deviation of the random variable Y, given the following information on the three-sigma tolerances for the variables using Taylor's series approximation:

$$X_1 = 4.00 \pm 1.20$$

$$X_2 = 2.00 \pm 0.60$$

11 Probabilistic Design for Reliability and the Factor of Safety

$$X_3 = 1.00 \pm 0.30$$

$$\mu_Y \cong \frac{2\sqrt{\mu_{X_1}}}{\mu_{X_2}\mu_{X_3}} = \frac{2 \times \sqrt{4}}{2 \times 1} = 2.$$

Now, the derivatives are:

$$\frac{\partial Y}{\partial X_1} = \frac{1}{\sqrt{X_1} X_2 X_3}; \quad \frac{\partial Y}{\partial X_2} = -\frac{2\sqrt{X_1}}{X_2^2 X_3};$$

$$\frac{\partial Y}{\partial X_3} = -\frac{2\sqrt{X_1}}{X_2 X_3^2}.$$

Evaluate the derivatives at μ:

$$\left.\frac{\partial Y}{\partial X_1}\right|_\mu = \frac{1}{\sqrt{4} \times 2 \times 1} = \frac{1}{4}; \quad \left.\frac{\partial Y}{\partial X_2}\right|_\mu = -\frac{2\sqrt{4}}{2^2 \times 1} = -1$$

$$\left.\frac{\partial Y}{\partial X_3}\right|_\mu = -\frac{2\sqrt{4}}{2 \times 1^2} = -2.$$

To find the variance, we have:

$$\sigma_Y^2 \cong \left(\left.\frac{\partial Y}{\partial X_1}\right|_\mu\right)\sigma_{X_1}^2 + \left(\left.\frac{\partial Y}{\partial X_1}\right|_\mu\right)\sigma_{X_2}^2 + \left(\left.\frac{\partial Y}{\partial X_1}\right|_\mu\right)\sigma_{X_3}^2$$

$$= \left(\frac{1}{4}\right)^2 (0.4)^2 + (-1)^2 (0.2)^2 + (-2)^2 (0.1)^2$$

$$= 0.01 + 0.04 + 0.04 = 0.09.$$

Hence, $\sigma_Y = \sqrt{0.09} = 0.3$.

(b) The upper specification limit on Y is 3.00, and the lower specification limit is 1.50. What percent of the products produced by this process will be out of specifications?

$$= \left[1 - \Phi\left(\frac{3-2}{0.3}\right)\right] + \left[1 - \Phi\left(\frac{2-1.5}{0.3}\right)\right]$$

$$= [1 - \Phi(3.33)] + [1 - \Phi(1.67)]$$

$$= [1 - 0.9995658] + [1 - 0.95254]$$

$$= 0.04789.$$

11.7 Steps for Probabilistic Design

Considering the total design-for-reliability program, the steps related to the probabilistic approach may be summarized as follows:

- Define the design problem. Develop system functional and reliability block diagrams to the lowest level of detail.
- Identify the design variables and parameters needed to meet customer's requirements for each component. Focus on understanding the physics/chemistry/biology of failure.
- Conduct a failure modes, mechanisms, and effects analysis (FMMEA). Focus on understanding failure mechanisms.
- Select and verify the significant design parameters.
- Formulate the relationship between the critical parameters and the failure-governing criteria related to the underlying components.
- Determine the stress function governing failure.
- Determine the failure governing stress distribution.
- Determine the failure governing strength function.
- Determine the failure governing strength distribution.
- Calculate the reliability associated with these failure-governing distributions for each critical failure mode.
- Iterate the design to obtain the design reliability goal.
- Optimize the design in terms of other qualities, such as performance, cost, and weight.
- Repeat optimization for each critical component.
- Calculate system reliability.
- Iterate to optimize system reliability.

11.8 Summary

Producing a reliable product requires planning for reliability from the earliest stages of product design. There are models and principles that can be used to quantify and evaluate reliability in the design stage. One approach is known as probabilistic design for reliability. The basic premise of probabilistic design for reliability is that a given component has a certain strength which, if exceeded, will result in failure. The factors that determine the strength of the component are random variables, as are the factors that determine the stresses or load acting on the component. Stress is used to indicate any agency that tends to induce failure, whereas strength indicates any agency resisting failure. The factor of safety is some ratio of the strength and stress variables. Since both are random variables, the engineers designing a product must determine which measures of strength or the stress should be used in the computation of the factor of safety based on probability. Following the steps for probabilistic design provided in

11 Probabilistic Design for Reliability and the Factor of Safety

this chapter can result in the production of a product that will achieve the desired level of reliability in its application environment.

Problems

11.1 The stress (X) and the strength (Y) random variables for a given failure mode of a product follow the normal distributions with the following parameters:

$$\mu_X = 11{,}000 \quad \sigma_X = 2400$$
$$\mu_Y = 15{,}000 \quad \sigma_Y = 1500.$$

(a) Find the system reliability for this failure mode.

(b) The customer wants a reliability of 0.99990. The only thing that the designer can change for this failure mode is the mean value for the strength random variable Y (thus increasing the factor of safety). Find the new target for μ_Y to achieve the reliability goal.

11.2 The stress (X) and the strength (Y) are random variables for a given failure mode and follow the normal distribution with the following information (CV is the coefficient of variation):

$$CV_X = 0.20 \quad CV_Y = 0.15.$$

(a) The customer wants a reliability of 0.9990. What is the safety factor that the designer must use to meet the requirements of the customer?

(b) The customer wants a reliability of 0.990. What is the safety factor that the designer must use to meet the requirements of the customer?

(c) Another customer wants a reliability of 0.9990. The design team does not want to increase the safety factor any more. The only thing the team can easily change is the variation of the strength variable. What should be the value of CV_Y to meet the requirements of the customer?

11.3 A beam with a tubular cross-section, shown in the figure below, is to be used in an automobile assembly.

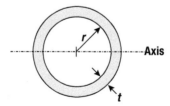

To compute the stresses, the moment of inertia (I) of the beam about the neural axis is calculated as

$$I = \pi r^3 t.$$

The mean radius and thickness of the tubular cross-section have the following dimensions with 3σ tolerances:

$$r = 2.00 \pm 0.06$$
$$t = 0.11 \pm 0.015.$$

Find the mean value of the moment of inertia and its standard deviation.

11.4 A random variable, Y, for a product is a function of three random variables, X_1, X_2, X_3, and is given by

$$Y = \frac{4\sqrt{X_1}}{X_2 X_3^2} \quad \text{or} \quad Y = \frac{4X_1^{1/2}}{X_2 X_3^2}.$$

(a) Find the expected value and the standard deviation of the random variable Y, given the following information on the 3σ tolerances for the variables using Taylor's series approximation:

$$X_1 = 4.00 \pm 0.60$$
$$X_2 = 2.00 \pm 0.40$$
$$X_3 = 1.00 \pm 0.15.$$

(b) The upper specification limit on Y is 5.00 and the lower specification limit is 3.00. What percent of the products produced by this process will be out of specifications?

(c) The design team thinks that the percent of nonconforming products, as calculated in part (b), is relatively high. If the tolerances can be reduced on only one random variable, which variable would you pick? Let us say that we can decrease the tolerances on the chosen random variable by half. What percent of the products will be out of specification with the new process?

11.5 A random variable, Y, which determines the function of a system, is a function of three other random variables, X_1, X_2, X_3, and is given by

$$Y = \frac{3X_1^{\frac{1}{2}}}{X_2^2 X_3}.$$

Find the expected value and the standard deviation of the random variable Y, given the following information using the first-order Taylor's series approximation:

$$\mu_{X_1} = 9.00 \quad \sigma_{X_1} = 0.60$$
$$\mu_{X_2} = 2.00 \quad \sigma_{X_2} = 0.20.$$
$$\mu_{X_3} = 1.50 \quad \sigma_{X_3} = 0.15$$

Also develop the 3σ tolerance limits for Y.

11 Probabilistic Design for Reliability and the Factor of Safety

11.6 Suppose a mechanism is made of three components with dimensions X_1, X_2, X_3. The response of this mechanism, Y, is related to X_is by: $Y = 2X_1 + X_2 - 3X_3$.

The 3σ tolerances on the dimensions are as follows:

$$X_1 = 5.00 \pm 0.18$$
$$X_2 = 4.00 \pm 0.12$$
$$X_3 = 2.00 \pm 0.15.$$

All the dimensions, X_i, $i = 1, 2, 3$, follow the normal distribution.

(a) Find the mean and variance for the random variable Y, and specify its 3σ tolerance limits.

(b) If the specification limits on the response, Y, of this mechanism are 8.00 ± 0.50, what percentage of the mechanism will not meet these specifications?

(c) Another response, Z, of the mechanism is given by:

$$Z = \frac{X_1^2 X_2}{X_3}.$$

Find the mean and variance of the random variable Z and specify its 3σ tolerance limits.

12

12 Derating and Uprating

Derating is the practice of limiting thermal, electrical, and mechanical "stresses" to levels below the manufacturer's specified ratings, to improve reliability. Derating allows added protection from anomalies unforeseen by the designer (e.g., transient loads and electrical surge).

12.1 Part Ratings

Ratings set by manufacturers of parts and subsystems on their environmental and operational limits affect the decision-making by the part users and equipment manufacturers. This section explains the ratings with the examples of electronic parts.

Part datasheets provide two types of ratings: *absolute maximum* ratings and *recommended operating conditions*. In general:

- *Absolute maximum ratings* (AMR) are provided as a limit for the "reliable" use of parts.
- *Recommended operating conditions* (ROC) are the conditions within which electrical functionality and specifications given with the part datasheet are guaranteed.

Intel (1995) considers the difference between absolute and maximum ratings as guidance to users on to how much variation from the recommended ratings can be tolerated without damage to the part. Motorola (*Boyle* v. *United Technologies Corp.* 1988) states that, when a part is operated between the ROC and AMR, it is not guaranteed to meet any electrical specifications on the datasheet, but the physical failure or adverse effects on reliability are not expected. Motorola notes margins of safety are added to the absolute maximum ratings to ensure the recommended operating conditions (*Boyle* v. *United Technologies Corp.* 1988).

Reliability Engineering, First Edition. Kailash C. Kapur and Michael Pecht.
© 2014 John Wiley & Sons, Inc. Published 2014 by John Wiley & Sons, Inc.

12 Derating and Uprating

12.1.1 Absolute Maximum Ratings

The absolute maximum ratings section in the datasheet includes limits on operational and environmental conditions, including power, power derating, supply and input voltages, operating temperature (e.g., ambient, case, and junction), and storage temperature. The IEC (IEC/PAS 62240 2001) defines absolute maximum ratings as "limiting values of operating and environmental conditions applicable to any electronic device of a specific type as defined by its published data, which should not be exceeded under the worst possible conditions. These values are chosen by the device manufacturer to provide acceptable serviceability of the device taking no responsibility for equipment variations, and the effects of changes in operating conditions due to variations in the characteristics of the device under consideration and all other electronic devices in the equipment."

The IEC (IEC/PAS 62240 2001) also states, "The equipment manufacturer should design so that, initially and throughout life, no absolute-maximum value for the intended service is exceeded for any device under the worst probable operating conditions with respect to supply voltage variation, equipment component variation, equipment control adjustment, load variations, signal variation, environmental conditions, and variation in characteristics of the device under consideration and of all other electronic devices in the equipment." In summary, companies that integrate electronic parts into products and systems are responsible for assuring that the AMR conditions are not exceeded.

Part manufacturers generally state that below the AMR but above the recommended conditions, the performance of the part is not guaranteed, but the useful life of the part will not be affected. That is, there are no reliability concerns below the AMR. Some manufacturers (e.g., Motorola) suggest that operating parameters within the recommended operating range are not guaranteed at or near the AMR, and there may be reliability concerns over the long term[1] (Lieberman 1998; Pfaffenberger and Patterson 1987; United States Department of Defense 1996). Motorola (Lycoudes 1995) also states that noise can push the environment beyond "destruct" limits when parts are operated near the absolute maximum ratings.

Philips notes, "The 'RATINGS' table (limiting values in accordance with the Absolute Maximum System—IEC 134) lists the maximum limits to which the device can be subjected without damage. This doesn't imply that the device will function at these extreme conditions, only that when these conditions are removed and the device operated within the recommended operating conditions, it will still be functional and its useful life won't have been shortened (Philips 1988).

12.1.2 Recommended Operating Conditions

Recommended operating conditions provided by part manufacturers include parameters such as voltage, temperature ranges, and input rise and fall time. Part manufacturers guarantee the electrical parameters (e.g., typical, minimum, and maximum) of the parts only when the parts are used within the recommended operating conditions.

---

[1]Some EIA/JEDEC documents refer to absolute maximum ratings as absolute maximum "continuous" ratings. In those documents, transient conditions under which these ratings may be exceeded are defined. For example, the JEDEC standard for description of low voltage TTL-Compatible CMOS logic devices [53], states that "Under transient conditions these rating [AMR] may be exceeded as defined in this specification."

Philips notes, "The recommended operating conditions table [in the Philips datasheet] lists the operating ambient temperature and the conditions under which the limits in the DC characteristics and AC characteristics will be met" (Philips 1988). Philips also states that "The table (of recommended operating conditions) should not be seen as a set of limits guaranteed by the manufacturer, but the conditions used to test the devices and guarantee that they will then meet the limits in the DC and AC characteristics table" (Solomon et al. 2000).

12.1.3 Factors Used to Determine Ratings

Factors used to determine the AMR and recommended operating conditions include (Rofail and Elmasry 1993):

- Margins determined through electrical testing and procedures and methods used to set specifications from device test characterization data
- Competitors' second source advantages and limits set to maintain parity with competitors' products
- Design rule limitations: physical dimensions of device elements
- Semiconductor fabrication process: manufacturing processes and conditions that affect temperature sensitivity of parameters.

12.2 Derating

The stress limits on the products are often determined through a combination of manufacturer[2]-specified stress limits and some "derating" criteria. Derating is the practice of using an electronic part in a narrower environmental and operating envelope than its manufacturer designated limits. The purpose of derating is to lower the (i.e., electrical, thermal, and mechanical) stresses acting on part. These lower stresses are expected to extend useful operating life where the failure mechanisms under consideration are wear out type. This practice is also expected to provide a safer operating condition by furnishing a "margin of safety" when the failure mechanisms are of overstress type.

The concept of derating is schematically shown in the Figure 12.1, Figure 12.2, and Figure 12.3. These concepts are based on the load strength interference relationship. Figure 12.1 shows the load distribution on a component and the strength distribution of the component in the same scale. There are potential reliability issues only if there is an intersection between the load and strength distribution.

12.2.1 How Is Derating Practiced?

The techniques for derating of electronic parts customarily comprise of the steps described later in the list. The exact methodology varies between organizations and

---

[2]The terms "manufacturer," "user," "OEM," and the like are used to identify different levels of entities in the electronics industry supply chain. It should be noted that under different circumstances, the same company can play the role of more than one such entity. Also (particularly in vertically integrated companies), the different division of the same organization can play the role of the separate entities.

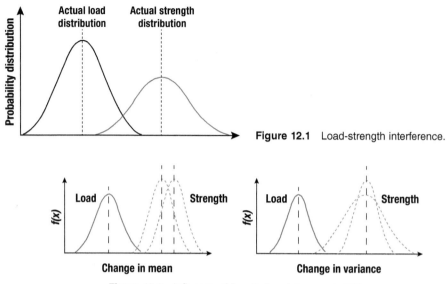

Figure 12.1 Load-strength interference.

Figure 12.2 Influence of "quality" on failure probability.

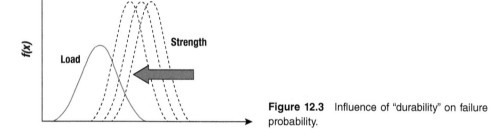

Figure 12.3 Influence of "durability" on failure probability.

the types of products under consideration. Sometimes, these procedures are presented as standards by acquisition agencies of dominant users. Larger original equipment and system manufacturers often develop and use own derating standards.

- The equipment or systems in which the parts are used are classified into categories according to their importance and criticality in the reliable functioning of the whole system. In military and space applications, this classification follows the criticality of the missions in which the systems are to be deployed. Rome Laboratories (Eskin et al. 1984) uses three derating levels, and so does Boeing Corporation (1996) and Naval Air Systems Command, Department of the Navy (1976).
- Sets of parts are identified as groups, which are believed to have similar responses to different environmental and operational stresses. It can be based on the material (Si, GaAs, and SiC), functional type of part (analog, digital, logic, and passive), technology (bipolar, field effect), types of packaging, and other considerations.
- For all such groups, stresses that possibly affect reliability of that group of parts are enumerated. Electrical and thermal stresses are used most often.

- For each type of part (or part class), derating guidelines are developed for the categories of equipments in which they are used for each such stress. These guidelines are usually one of the following types. Sometimes, a combination of these limits or coupled limits between a number of different parameters are used:
  - A percentage or fraction of the manufacturer specified limits
  - A maximum (or minimum) value
  - An absolute difference from the manufacturer limits.

The term "derating" by definition suggests a two-step process; first a "rated" stress value is determined from a part manufacturer's databook and then some reduced value is assigned. The "margin of safety," supposed to be provided by derating is the difference between the maximum allowable actual applied stress and the "demonstrated limits" of the part capabilities. The part capabilities as given by manufacturer specifications are taken as the "demonstrated limits." Sometimes, an upper limit on certain stress values is used irrespective of the actual manufacturer limit (some derated value of the manufacturer limit is used if that is lower than the derating limit). The propensity for the system design team inclines toward using conservative stresses at the expense of overall productivity. There are reasons to believe that the part manufacturers already provide safety margin while choosing the operating limits. When those values are derated by the users, it is effectively adding a second level of "safety margin."

The military standard, MIL-STD-975, was issued in 1976 and it remains the baseline for most derating guidelines in use by military and civilian users and manufacturers. Those guidelines have remained largely unchanged up to current versions, except for the addition of new part types and families. Although the intended purpose of derating was to provide a margin of safety, in reality, this has become a perceived guard against inadequate qualification and acceptance criteria. It is believed that the principal benefit of derating is not to extend the life of reliable parts but to protect against the presence of "maverick" parts and substandard lots. The design teams assume that the incoming parts will not fall within a narrow range of quality/performance characteristics. It is also assumed that these lower operational parameters will protect against random failures. Considering the arbitrary nature of such assumptions, a NASA Goddard Space Flight Center engineer recently noted that "It would be nice to be able to say the guidelines are all based on sound scientific and engineering principals and calculations but this does not seem to be so. Rather most are based on experience, best engineering estimate and a conservative philosophy. This should come as no surprise considering that many of the requirements are for simple quantities such as 50%."

12.2.1.1 Resistors The following is the typical resistor derating methodology.

- Maximum ambient temperature for use at full power (T_s)
- Maximum allowed operating temperature without power (T_{max})
- Generate "absolute maximum rating" curve (see Figure 12.4) 100% wattage and horizontal until T_s
  - Linearly connects T_s, 100% to T_{max}, 0%

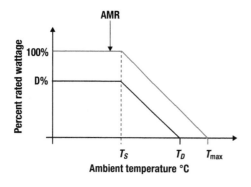

Figure 12.4 Generic resistor derating procedure.

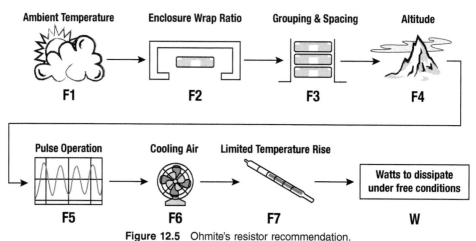

Figure 12.5 Ohmite's resistor recommendation.

- Generate "Derating Requirement" curve
  - Horizontal at % derated wattage until T_s
  - Linearly derated until maximum derated temperature, T_d

Selected resistor should have I2R rating of: W * F1 * F2 * F3 * F4 * F5 * F6 * F7 (see Figure 12.5). Note that the "F" values are multipliers, and some can be less than 1.

The maximum permissible operating temperature is a set amount. Any increase in the ambient temperature subtracts from the permissible temperature rise and therefore reduces the permissible power dissipation (Figure 12.6).

Enclosure limits the removal of heat by convection currents in the air and by radiation. The walls of the enclosure also introduce a thermal barrier between the air contacting the resistor and the outside cooling air. Hence, size, shape, orientation, amount of ventilating openings, wall thickness, material, and finish all affect the temperature rise of the enclosed resistor.

Figure 12.7 indicates for a particular set of conditions how the temperatures varied with the size of enclosure for a moderate size power resistor.

The temperature rise of a component is affected by the nearby presence of other heat-producing units, such as resistors, and electronic tubes. The curves (Figure 12.8) show the power rating for groups of resistors with various spacing between the closest points of the resistors, assuming operation at maximum permissible hot spot temperature.

12.2 Derating

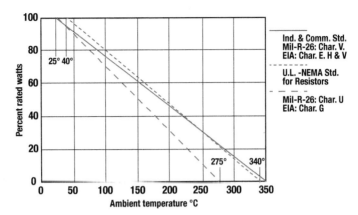

Figure 12.6 Derating of resistors for high ambient temperatures (Ohmite 2002).

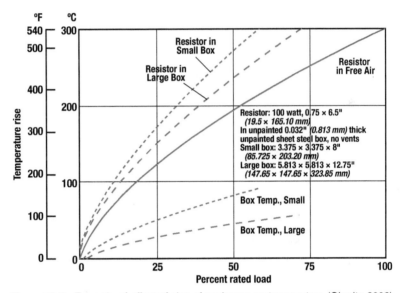

Figure 12.7 Example of effect of size of enclosure on temperature (Ohmite 2002).

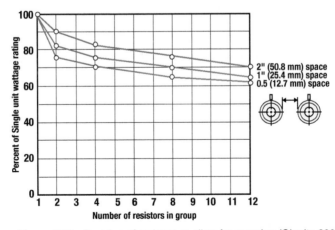

Figure 12.8 Derating of resistors to allow for grouping (Ohmite 2002).

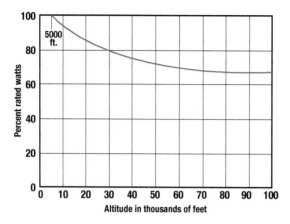

Figure 12.9 Altitude (Ohmite 2002).

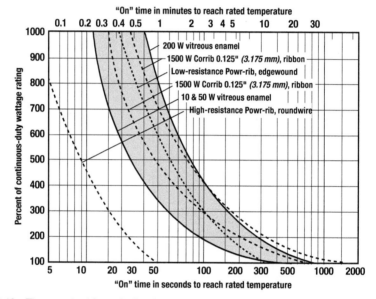

Figure 12.10 Time required for typical resistors to reach rated operating temperatures at various watt loads (Ohmite 2002).

The amount of heat which air will absorb varies with the density, and therefore with the altitude above sea level. At altitudes above 100,000 feet, the air is so rare that the resistor loses heat practically only by radiation (Figure 12.9).

Unlike the environmental factors, which result in reduction of the watt rating, pulse operation may permit higher power in the pulses than the continuous duty rating (Figure 12.10).

Resistors can be operated at higher-than-rated wattage when cooled. Forced circulation of air over a resistor removes more heat per unit time than natural convection does and therefore permits an increased watt dissipation. Liquid cooling and special conduction mountings also can increase the rating (Figure 12.11).

It is sometimes desirable to operate a resistor at a fraction of the Free Air Watt Rating in order to keep the temperature rise low. When it is desired to operate a

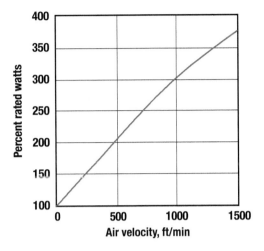

Figure 12.11 Cooling air.

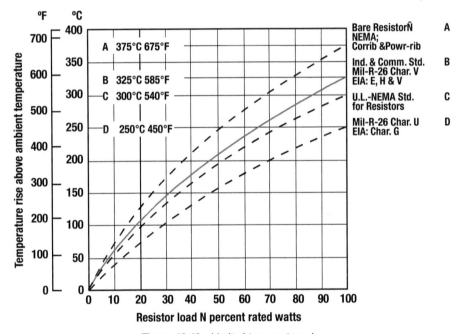

Figure 12.12 Limited temperature rise.

resistor at less than maximum temperature rise, the percent watts for a given rise can be read from the curve (Figure 12.12).

12.2.2 Limitations of the Derating Methodology

Thermal derating for semiconductor parts typically involves controlling the maximum steady-state temperature at some location in the system or part. For example, Naval Air Systems Command uses a set of thermal derating guidelines for electronic components, in which the junction temperature of TTL, CMOS, and linear amplifiers must be maintained 20°, 25°, or 30°C below of the manufacturers' rated junction temperature depending on the criticality of the system. Besides trying to control the stresses directly resulting from them, limits on power, current, and voltages are also often used with a view to reduce the joule heating of the part. For example, NASA (Jet Propulsion Laboratory 1996) guidelines require an iterative process of alternately calculating junction temperature and reduction in power output until the junction temperature goes below its derated value. In fact, many military (Army, Air Force) and government (NASA) agencies, as well as industrial (Boeing) and consumer products (Philips, Siemens) manufacturers, either use or require use of some variations of these derating criteria. Unfortunately, there are several major shortcomings with this approach to derating which can render this process ineffective in achieving its purported goals.

12.2.2.1 Emphasis on Steady-State Temperature The use of temperature as the most common derating guideline follows the misplaced belief that steady-state temperature is the dominant cause of part failure, or deviation from rated performance. This outlook has been shaped by a belief system that dictates that "cooler is better" for electronics performance and reliability. There are serious doubts raised about this philosophy by many researchers (Evans 1988; Hakim 1990; Lall et al. 1997; Pecht 1996a; Pecht and Ramappan 1992a; Wong 1989). These articles clearly demonstrate the lack of faith in the steady-state temperature-based reliability prediction methodology. Although discredited by reliability community, statements such as "The reliability of a silicon chip is decreased by about 10% for every 2°C temperature rise" are still made by practitioners in the electronic packaging discipline (Yeh 1995).

Today's failure analysis reports rarely find equipment failures driven by part failure mechanisms dependent solely on steady-state temperature (Dasgupta et al. 1995; Hakim 1990; Lasance 1993). A significant portion of the failure mechanisms at the package-level is driven by thermal mismatch at mating surfaces of bimaterial interfaces (e.g., die fracture and wire bond fatigue) (Lall et al. 1995, 1997). Damage at such interfaces is actuated by thermal excursions due to environmental or power cycling. Just raising the steady-state temperature does not accelerate rate of failure at those interfaces, so lowering temperature is not going to increase reliability.

Temperature often manifests itself as a second order effect at bimaterial interfaces—as diffusion and intermetallic formation. Typically, diffusion is critical to formation of bimaterial joints at interfaces—at the same time, too much intermetallic often degrades the strength of an interface. The rate of intermetallic formation can be controlled by choosing appropriate mating metals, controlling surface roughness, and using barrier metals—such that interface strength after degradation is still greater than required to ensure field life.

Several chip-level failure mechanisms are also driven by localized temperature gradients. Temperature gradients can exist in the chip metallization, chip, substrate, and package case, due to variations in the conductivities of materials and defects such as voids or nonuniformities in metallizations. The locations of maximum temperature gradients in chip metallization are sites for mass transfer mechanisms, including electromigration. Current thermal derating practices do not provide any limits on these types of thermal stresses. Further, Pecht and Ramappan (1992a) found that the majority of electronic hardware failures over the past decade were not component failures at all, but were attributable to interconnects and connectors, system design, excessive environments, and improper user handling (Table 12.1). No amount of thermal derating is going to reduce the number of occurrences of these types of failures. Thus,

12.2 Derating

Table 12.1 Common failure mechanisms for microelectronics (Pecht and Ramappan 1992a)

Source of data	Year	The dominant causes of failure
Failure analysis for failure rate prediction methodology (Manno 1983)	1983	Metallization (52.8%); oxide/dielectric (16.7%)
Westinghouse failure analysis memos (Westinghouse 1989)	1984–1987	Electrical overstress (40.3%)
Failure analysis based on failures experienced by end-user (Bloomer 1989)	1984–1988	Electrical overstress and electrostatic discharge (59%); wirebonds (15%)
Failure analysis based on Delco data (Dolco 1988)	1988	Wirebonds (40.7%)
Failure analysis by power products division (Taylor 1990)	1988–1989	Electrical overstress damage (30.2%)
Failure analysis on CMOS (private correspondence)	1990	Package defects (22%)
Failure in vendor parts screened per MIL-STD-883	1990	Wire bonds (28%); test errors (19%)
Pareto ranking of failure causes per Texas Instruments study (Ghate 1991)	1991	Electrical overstress and electrostatic discharge (20%)

derating of individual parts without regard to the complete system can be unproductive and costly.

Performance improvement with lower temperature is another assumption made when one decides to lower the operating temperature for microelectronics. It is indeed true that certain performance parameters show significant improvement at lower temperature, but the temperature ranges at which those effects become prominent are generally well below even the derated temperature. In the normal operating range, it is well known that the performance parameters vary with temperature. For the system designer, it is important to understand the nature and extent of such variations. Deviations in electrical parameters from the required range can occur at both ends of the rated operating temperature range and using only lower-than-rated temperature is not going to provide improvement in performance. In addition, it was found by Hitachi that peak performance metrics of computers show a downward turn when there exists large temperature gradient within the system, even when the maximum temperature was maintained within rated limits.

It should be acknowledged that lower maximum temperature might also result in lower values of thermal gradients or lower temperature excursion, which might be beneficial for the system. Even in those situations, it is more advantageous and scientifically sound to directly find the limits on gradients and cycling rather than using lower temperature believing that to be a panacea. One can also employ better design techniques, which can ensure control on those thermal parameters even at relatively higher maximum temperature.

12.2.2.2 Is the "Right" Temperature Derated? There are many choices of the temperature, which might be controlled through derating in an electronic system. The logical choice of the temperature being derated should be the one at the location where temperature influences failure mechanisms and/or performance parameters. The electrical parameter variations are calculated based on the junction temperature of semiconductor parts. For passive parts, temperatures at the winding, dielectric, coils, or other parts depending on the type of the part can influence the electrical characteristics. Depending on the packaging and attachment process of the passive parts to boards, temperatures at certain points can influence its reliability and/or electrical parameter. In the derating factor choices, the temperature most often derated is the junction temperature for the semiconductor parts; for other electrical parts, specification of operating temperature is also common. At close examination, it becomes difficult to identify the logic behind the choice of those temperatures.

The most common thermally driven (not necessarily by steady state temperature) driven reliability problems on the part or package are not at the junction; it is not even on the semiconductor. These problems occur at various parts of the package, including the traces, etallizations, and bond pads. As discussed in the previous section, other thermal stresses acting on areas other than the die affect the reliability more strongly. Lall et al. (1995) provide a thorough and detailed analysis of the temperature effects on part reliability with examples. They also show that how design modifications at different levels (from die to system) can suppress those failure mechanisms regardless of temperature rise. Attempts to improve the long-term reliability of a part through reduction of "junction" temperature cannot be justified from scientific standpoint.

This practice of derating of junction temperature would be defensible from an engineering standpoint if the temperatures within a part or package were constant or temperature differences were negligible. If there exists any significant temperature difference within a package, then it is important to identify the temperature pattern within such a package and to choose the valid temperature to derate. Junction temperature appears to be a reasonable compromise to use the junction temperature assuming that the temperature throughout the die, its metallization, and passivation is constant. Recent studies involving simultaneous solving of the electrical and thermal field equations (Fushinobu et al. 1995; Goodson et al. 1995) show that there are non-negligible difference in temperature between the different portions of the part. Fushinobu et al. (1995) found a hot spot in GaAs MESFET devices at the drain side of the channel. The temperature difference between the source and the drain can be in the same order of magnitude as the level of the derating of the junction temperature. In this situation, it will be futile to expect the derating to have any significant effect on the reliability.

The temperature discrepancies within the die, metallization, passivation, and other parts in close proximity of the die become more prominent in the Silicon-on-Insulator (SOI) devices. In SOI devices, the thermal resistance of the silicon dioxide is more than the bulk silicon in other devices. That makes the channel to substrate thermal resistance more than 10 times higher as compared with bulk devices. The value of that thermal resistance can be comparable to the thermal resistance of the package itself. The manifestation of this thermal resistance on the thermal map of the package is more complex than just a temperature difference. Due to all these variations related to junction temperature, Philips Semiconductors (1993) proposed that instead of the so-called junction temperature, a virtual temperature derived from the reliability data be used. This temperature would be analogous to electron temperature, which is used to describe the energy level of electrons.

As smaller and smaller (deep submicron level) devices are becoming more prevalent, another inadequacy of the emphasis on junction temperature usage becomes prominent. The transmission delay coming from the interconnects has become a major contributor to the total delay time of a device (Nakayama 1993). The current derating practice does not take into account the effects of temperature changes on the electrical properties of the interconnects. Also, as many new specially developed and processed materials are being used in the parts, it becomes more important to identify the effects of temperature change on the electrical properties of those materials. Naem et al. (1994) describe the thermal effect on the resistivity of polycrystalline silicon titanium salicide, which is used in advanced CMOS parts to reduce sheet resistance. That study demonstrates the importance of taking into account the other seemingly nonthermal parameters in determining the thermal coefficient of resistivity (TCR). The effect of current density on TCR was very pronounced at higher current and temperature. The effects of manufacturing defects on the changes in resistivity also become more conspicuous at those conditions.

Another imprecisely defined and inadequately measured variable makes the calculation of junction temperature unreliable. That variable is the package thermal resistance; both case-to-junction and ambient-to-junction. The values used for these thermal resistances are critical when one attempts to control the junction temperature through derating. Unless this value is determined and used properly, the calculation for the required environmental temperature becomes ineffectual. Lasance (1993) provides a detailed account of the sources and levels of variability in the computation of junction to case and ambient thermal resistances. The nonlinear relationship of thermal resistance with power dissipation and temperature, and the dependence of this value to possible manufacturing defects makes it imprudent to use manufacturer-listed thermal resistance values under all circumstances.

We find that there are many variabilities associated with elusive junction temperature, in the methods of measuring and predicting that temperature, and the published data of package thermal resistances. With the target temperature at the junction being predicted imprecisely, it is difficult to determine what, if any, effect derating of that temperature is having on the reliability and performance of the part.

12.2.2.3 All Parts Are Not Equal A common but unfounded practice is to derate temperature for a technology (e.g., TTL and CMOS) or a packaging class (e.g., hybrids), based on the assumption that reliability of all such parts has similar dependence on temperature. While there are some 67 "major" semiconductor part groups (Ng 1996), each exhibiting fundamentally different characteristics, many derating guidelines use very similar (often identical) derating factors for a broad group of parts. For example, NASA sets an absolute limit on junction temperature for all digital microcircuits, and Naval Air Systems Command, Department of the Navy (1976) guidelines suggest identical thermal derating for all digital CMOS parts. However, there are many variations in the CMOS technology with respect to their temperature tolerance. For example, twin-tub CMOS technology is much more tolerant to latchup breakdown than bulk type CMOS parts; changes in doping profile can and do make many MOS parts resistant to hot carrier aging. Some newer derating guidelines (Boeing 1996) use more subgroups of parts while prescribing temperature limits. Still, due to the fast changes in part technology, it is difficult to derive deterministic derating guidelines for all new types of parts.

Even when the parts and packages are of similar technology in terms of material, scaling, and functionality, their performance parameters can be rather different over the same range of junction temperatures. A high speed benchmarking study (Maniwa and Jain 1996) found differences in the rate of change of propagation delay with temperature for similar ASICs made by various major manufacturers. That fact, coupled with the variations in voltage and process derating, shows significant performance difference between manufacturers. If one is to apply derating to maintain performance parameters within the same level, that parameter should be different for different manufacturers. The current derating guidelines do not grant such options.

12.2.2.4 Derating Is Not without Cost Thermal derating of parts is not without cost. The reduction of temperature can require expensive, large, and heavy thermal structures or the addition of a cooling system. When active cooling is provided for a box as a reliability enhancement measure, that box will then be dependent on the reliable operation of the cooling system (Leonard 1991a). This reference provides a thorough critique of the present practice of using the temperature–reliability relationship. It also provides the readers the broad picture of the role of temperature in overall electronic part management. Reduction in junction temperature without active cooling can require expensive and time-consuming redesign with increase in size and weight of the system.

Possibly the most important cost of derating is in the sacrificed productivity of electronic system for the express purpose of reduction in junction temperature. One of the most common examples is the reduction of power dissipation in electronics parts, which in turn requires the reduction of speed of operation for MOS devices. If the reduction in temperature does not serve any positive purpose, then this loss of productivity is unjustifiable.

12.2.2.5 The Potential for Reduced Reliability and Performance Thermal derating guidelines have the potential for doing more damage than good, in that they give the design team a false sense of security about achieving increased reliability at lower temperatures. For example, lower temperatures may not necessarily increase reliability, since some failure mechanisms (e.g., hot electron) are inversely dependent on temperature. Other device failure mechanisms that show inverse dependence of temperature are ionic contamination when the devices are not operational, and reverse second breakdown (Lall et al. 1995).

Even when the low temperature itself is not the culprit causing failure, the process of achieving that lower temperature can have a serendipitous negative effect on the system. One such unintended consequence of actively lowering the operating temperature is the possible introduction of thermal cycling during the process of startup and shutdown of the system. This is observed in systems with unrealistically low constraints on the junction temperature—imposed by system designers with the perceived notion of improving reliability by lowering temperature. Boeing had to upwardly revise the maximum junction temperature for the Comanche light helicopter, because the initial low temperature limits were causing unique problems, such water deposit through condensation, besides initiating large thermal cycles at startup (Pecht 1996b).

It is generally assumed that the functionality of electronic devices improves at lower temperature. Although it is true for many parameters, there are situations in which performance parameters degrade at lower temperature. Huang et al. (1993) found larger degradation of current gain at lower temperatures while stress testing bipolar transistors. Some other phenomena, such as a large increase in leakage current in Poly-SOI MOSFET (Bhattacharya et al. 1994), logic swing loss in BiCMOS circuits (Rofail and Elmasry 1993), kink, and hysteresis (Kasley et al. 1993) occur at cryogenic temperatures. It should also be noted that for many performance parameters, there are no good or bad values, only a range of acceptable values. If reduction in temperature pushes those parameters beyond acceptable range, then that reduction is detrimental. Threshold or starting voltage for MOS devices is a good example of such a parameter.

12.2.2.6 Interaction of Thermal and Nonthermal Stresses Thermal and nonthermal stresses do not act independent of each other in precipitating failure as is implicitly assumed in many derating guidelines. For example, temperature and current density accelerate electromigration. Lowering the current density reduces the dependence of equipment reliability on steady-state temperature—that is, benefitting in improved life due to lowered temperature in spite of the temperature dependence is often beyond the designed-for field life and is thus of little consequence. In case of electromigration, the site of failure is also dependent on both current density and temperature gradient. In general, part reliability, for mechanisms with a dominant dependence on more than one operating stress (temperature and nontemperature), complicated by dependence on magnitudes of manufacturing defects, can often be maximized more economically through methods other than lowering temperature.

The interaction of various thermal and nonthermal stresses modifies the dominant dependence of the failure mechanisms on one or more of the stresses. For example, temperature transients generated by the On/Off duty cycle (often expressed as a ratio of the on time to the total time ON/(ON+OFF)) modify the dependence of the metallization corrosion on steady-state temperature. At low duty cycle values, metallization corrosion has a dominant dependence on steady-state temperature due to the exponential acceleration of the corrosion chemical reaction rate. However, at higher values (in the neighborhood of 1.0), accompanied by the evaporation of the electrolyte, metallization corrosion has a dominant dependence on duty cycle and a mild dependence on steady-state temperature. Brombacher (1992) lists sets of stresses, which work in combination in affecting different failure mechanisms. Often, optimal sets of these parameters can be found that does not involve lowering of temperature.

12.2.2.7 Technology Improvements Are Not Reflected Recent developments of thermally tolerant active and passive part technologies are often not reflected even in the newer derating guidelines (Boeing 1996). McCluskey (McCluskey 1996; McCluskey et al. 1996) quotes other sources (Pecht 1994), which shows that the common thin-film resistors can operate at temperatures higher than 200°C. The temperature limits for resistors listed in the guidelines are much lower than that limit. McCluskey also reports development of capacitors that are reliable at temperatures much higher than the limits in most derating guidelines.

All semiconductor devices are derated for very similar maximum junction temperatures in derating guidelines. There are two basic flaws in adopting this type of guidelines. The first problem lies in the fact that these guidelines do not take into account the varied response of different semiconductor materials to temperature. The intrinsic temperature of silicon carbide is 1000°C as compared with 400°C for silicon. The second problem with this guideline is its conservative outlook, which does not take into account current developments in high temperature electronics (Bromstead and Baumann 1991; Dreike et al. 1994; McCluskey et al. 1996). These blanket limits imposed on all semiconductors by some of the derating guidelines is applicable to all types of devices irrespective of their individual technology and the architecture in which those are used. This approach precludes the use of new technologies if the guidelines are strictly followed.

There remains another related issue, which needs to be clarified in this context. The main problem with the current derating methodology is in the approach taken, not with the exact numbers. The criticism of unjustified conservative values is not an attack on the exact values. Even if the capacities of the newer materials are reflected in the derating guidelines, unless the approach is changed, one can potentially continue to underutilize the capacities of the newer materials.

12.2.2.8 Summary of the Limitations of Thermal Derating Considering the preceding examples of various types of the shortcomings of the thermal derating process, we can conclude that this practice is deeply flawed. The process is not scientifically sound; besides, it can be costly, unproductive, and even harmful. The process also tends to put the users and manufacturers in a position of adversarial relationship instead of cooperation. In the worst case, it can become a blame allocation methodology. The process itself appears to be an isolated step in the design and use of microelectronics. The basic flaws in the process can briefly be summarized as the following:

- Overemphasizes steady-state temperature.
- The temperature the process tends to derate may not be the most important factor in determining performance and reliability.
- Does not recognize recent advances in semiconductor design and production methodology.
- Does not take into account the physical models developed to describe the failure mechanisms and performance parameters.
- Tries to safeguard individual parts without regard to their design, production and the actual use in circuitry.
- Groups parts together for similar derating in an arbitrary manner.
- Does not utilize the expertise of the manufacturers in determining the use profile of parts.

The process of component-by-component derating will make the system operate at a level that is presumed to be safe for the weakest link in the system. It is definitely a process where productivity is sacrificed in the quest of perceived safety. Even if the process does not harm reliability or performance, it is not wise to employ derating if it adds no value, reduces the functionality of systems, or increases the cost of operation.

12.2.3 How to Determine These Limits

We have seen in the previous sections that it might be necessary to find or verify the thermal limits on a device in certain conditions. Before any such task is undertaken, the user should verify through simulation/experimentation or both that the extended operational conditions or stricter performance requirements are truly necessary. Military electronic history is replete with cases of specifying unrealistically harsher (Fink 1997) environment than necessary. As this task of determining new limits at the end of the user for complex circuitry can be expensive, this should be done only after verifying that there is reason to go beyond the manufacturer specified limits.

When one needs to set such limits, that methodology has to be based on scientific grounds. The salient features of this process are given in the next section.

The inputs required for deriving these stress limits are more comprehensive than the apparent simple choice of device type, and, in some cases, operating environment in the current derating methodologies. One needs to enumerate the desired mission life, device and system architecture (including material, geometry, and layout), required performance parameters, and the worst-case manufacturing defects for the particular manufacturer and device type. It also requires closer cooperation and open sharing of information between the manufacturers and users of devices.

The limiting values of steady-state temperature, temperature cycle magnitude, temperature gradient, and time-dependent temperature change, including nontemperature operating stresses are determined for a desired device mission life. Physics-of-failure are used to relate allowable operating stresses to design strengths through quantitative models for failure mechanisms. Failure models will be used to assess the impact of stress levels on the effective reliability of the component for a given load. The quantitative correlations outlined between stress levels and reliability will enable design teams and users to tailor the margin of safety more effectively to the level of criticality of the component, leading to better and more cost-effective utilization of the functional capacity of the component.

Theoretical, numerical, and experimental results on the effect of temperature on the electrical performance of devices are used to determine the thermal limits for maintaining electrical performance. Inclusion of the effect of temperature on the packaging and interconnect electrical properties will allow more accurate determination of such limits.

12.3 Uprating

Uprating is a process to assess the capability of an electronic part to meet the functional and performance requirements of an application in which the part is used outside the manufacturers' specified operating conditions. In an ideal world, there would not be a need for a book on uprating. The part manufacturers would supply parts with appropriate ratings, for all products and systems. One would not have to be concerned about using any part beyond its ratings. Also performance, cost, assembly, test, and obsolescence would not be factors of concern. However, the ideal world does not exist.

Electronic parts are most often commodity items. The profitability in this highly competitive market comes from economies of scale. Electronic manufacturers do not generally benefit from the creation of boutique parts, unless alternatives do not exist and a significant price premium can be charged. This has been exemplified by the decline of the U.S. military's qualified part list (QPL) and qualified manufacturer list (QML) procurement program; showing that niche parts with specific ratings, tests, screens, and documentation cannot be kept available by mandate.

12 Derating and Uprating

This chapter provides the rationale for uprating and the role of uprating in the part selection and management process. Options are then given for situations where the ratings of a part are narrower than the application requirements.

Uprating is a process to assess the ability of a part to meet the functionality and performance requirements of the application in which the part is used outside the manufacturers' recommended operating range.[3]

Today's semiconductor parts are most often specified for use in the "commercial" 0–70°C, and to a lesser extent in the "industrial" −40 to 85°C, operating temperature range, thus satisfying the demands of the computer, telecommunications, and consumer electronics and their markets. There is also demand for parts rated beyond the "industrial" temperature range, primarily from the aerospace, military, oil and gas exploration, and automotive industries. However, the demand is often not large enough to attract and retain the interest of major semiconductor part manufacturers to make extended temperature range parts.

It is becoming increasingly difficult to procure parts that meet the engineering, economic, logistical, and technical integration requirements of product manufacturers, and are also rated for temperature ranges (Solomon et al. 2000). Yet there are products and applications that do require parts that can operate at temperatures beyond the industrial temperature range. It is desired that parts for these products incorporate technological advancements in the electronics industry in terms of performance, cost, size, and packaging styles.

For electronic parts, there is a limit of voltage, current, temperature, and power dissipation, called the absolute maximum ratings, beyond which the part may not be reliable. Thus, to operate in a reliable manner, the part must be operated within the absolute maximum rating.

There are also operational limits for parts, within which the part will satisfy the electrical functional and performance specifications given in the part datasheet. These ratings are generally narrower (within) than the absolute maximum ratings. A part may be used beyond the recommended operating rating but never beyond the absolute maximum rating.

Product manufacturers who perform system integration need to adapt their design so that the parts do not experience conditions beyond their absolute maximum ratings, even under the worst possible operating conditions (e.g., supply voltage variations, load variations, and signal variations) (IEC Standard 60134 1961). It is the responsibility of the parts selection and management team to establish that the electrical, mechanical, and functional performance of the part is suitable for the application.

Uprating is possible because there is often very little difference between parts having different recommended operating conditions in the datasheet. For example, Motorola notes (Lycoudes 1995) that "There is no manufacturing difference between PEMs (plastic encapsulated microcircuits) certified from 0 to 70°C and those certified from −55 to 125°C. The same devices, the same interconnects, and the same encapsulants are used. The only difference is the temperature at which the final electrical testing is done." In fact, many electronic parts manufacturers have used the same die for various

---

[3]Thermal uprating is a process to assess the ability of a part to meet the functionality and performance requirements of the application in which the part is used beyond the manufacturer-specified recommended operating temperature range. Upscreening is a term used to describe the practice of attempting to create a part equivalent to a higher quality level by additional screening of a part (e.g., screening a JANTXV part to JAN S requirements).

"temperature grades" of parts (commercial, industrial, automotive, and military). For example, Intel[4] (Intel 1990) stated in their military product data book: "there is no distinction between commercial product and military product in the wafer fabrication process. Thus, in this most important part of the VLSI manufacturing process, Intel's military products have the advantages of stability and control which derive from the larger volumes produced for commercial market. In the assembly, test and finish operations, Intel's military product flow differs slightly from the commercial process flow, mainly in additional inspection, test and finish operations."

Parts may also be uprateable for temperature because part manufacturers generally provide a margin between the recommended operating temperature specification of a part and the actual temperature range over which the part will operate. This margin helps maximize part yields, reduce or eliminate outgoing tests, and optimize sample testing and statistical process control (SPC). Sometimes, this margin can be exploited, and thus the part can be uprated.

12.3.1 Parts Selection and Management Process

Equipment manufacturers must have procedures in place for the selection and management of electronic parts used in their products. When uprating electronic parts, it is necessary to follow documented, controlled, and repeatable processes, which are integrated with the parts selection and management plans. The parts selection and management plan ensures the "right" parts for the application, taking into account performance requirements, assemblability, quality, reliability, and part obsolescence. The maintenance and support of some existing products require replacement parts to be available over the product life cycle. In the case of avionics, this period can be more than 10 years (Jackson et al. 1999a; Solomon et al. 2000). When companies stop producing avionics parts with wide recommended operating ranges, replacement parts become obsolete. One option is to use a "commercial" or "industrial" temperature range part as a substitute.

The performance assessment step of the parts selection management process assesses whether a part "will work" in its intended application. If the recommended operating condition in the datasheet of the part is outside the actual environment in the application, then options to mitigate this problem must be addressed.

12.3.2 Assessment for Uprateability

Uprating of parts can be expensive and time consuming, if there is no analysis of "promising" parts prior to the actual uprating process (Pecht 1996b). In other words, candidate parts should be accessed for their uprateability prior to conducting any uprating tests.

The best way to see if a part is uprateable is to obtain the simulation and characterization data from the part manufacturers. The data include product objective specifications, product and packaging roadmaps, device electrical simulation models, and temperature characterization data. Depending on the part manufacturer, some

---

[4]The Intel statement on the military and commercial parts shows that the practice of using the same die for various temperature ranges is common among manufacturers. In the mid-1990s, Intel stopped producing military temperature grade parts for business reasons.

of these data are available freely, while other may be available upon request, or in some cases, by signing a nondisclosure agreement.

Some datasheets or associated documents include electrical parameter data beyond the recommended operating temperature limits. These data may be useful in preassessing if a part can be uprated.

The effects of temperature (and other factors such as voltage and frequency) on different electrical parameters can be estimated using models available from part manufacturers. Often, the device electrical simulation models are made available to the public, although the models are often "sanitized" so that any proprietary information is masked (Micron Semiconductor 1998). Simulation models of devices can be used to calculate the effects of temperature variation on device parameters[5] (e.g., the BSIM3 model for short channel MOSFETs) (Foty 1997). Device simulations therefore can be used to estimate if the part will be uprateable, and what parameter changes may be expected at application operating conditions.

The cost of offering the models are minimal since they are developed during the design process. Circuit level models can be prepared in such a way that they do not reveal details of physical design. SPICE models are available from many companies, including Analog Devices, TI, and National Semiconductor. IBIS, VHDL, and VERILOG models are available from some companies such as Cypress Semiconductor and Intel. The model parameters can be examined to assess the effects of different factors on part electrical parameters over the target application conditions.

12.3.3 Methods of Uprating

Uprating is carried out after the part, the part manufacturer, and the distributors have been assessed (Jackson et al. 1999a, 1999b; Maniwa and Jain 1996) based on datasheets, application notes, and any other published data. Three methods for part uprating (see Figure 12.13) are overviewed in this chapter. The International Electrotechnical Commission and the Electronics Industry Association (IEC/PAS 62240 2001) accept these methods as industry best practices. Publications for the U.S. Department of Defense acknowledge these methods as effective and rigorous (Lasance 1993).

12.3.3.1 Parameter Conformance Parameter conformance is a process of uprating in which the part is tested to assess if its functionality and electrical parameters meet the manufacturer's recommended operating conditions over the target temperature range. Electrical testing is performed with the semiconductor manufacturer-specified test setups to assess compliance within the semiconductor manufacturer-specified parameter limits. The tests are of "go/no-go" type, and are generally performed at the upper and lower ends of the target application conditions. A margin may be added to the test, either in a range wider than the target application conditions or tighter electrical parameter limits for the test. The electrical parameter specifications in the datasheet are not modified by this method.

12.3.3.2 Parameter Recharacterization Parameter recharacterization is a process of uprating in which the part functionality is assessed and the electrical parameters are

---

[5]Different models provide different levels of details on the parameter estimates. Some examples of SPICE models are: Level 3, HSPICE, and BSIM3. Some SPICE versions allow the user to select the model to be used for transistor level analysis.

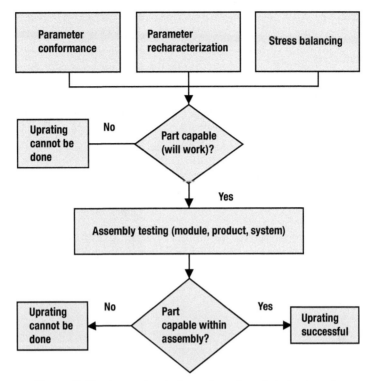

Figure 12.13 Approaches to thermal uprating of electronic parts.

characterized over the target application conditions, leading to a possible respecification of the manufacturer-specified datasheet parameter limits. The parameter recharacterization method of uprating seeks to mimic the part manufacturer's characterization process. The electrical parameter limits of parts rated for multiple temperature ranges are often obtained using the concept of parameter recharacterization (Pecht 1996b; Pendsé and Pecht 2000) and is shown in Figure 12.13. Electrical testing is followed by data analysis and margin estimation.

In parameter recharacterization, part electrical parameters are tested at several points in the target application conditions, the parameter values are recorded, and the parameter distributions are plotted. Figure 12.14 exemplifies the process. Here, propagation delay is on the horizontal axis and the population distribution on the vertical axis. Curve "1" is the distribution of the parameter at the manufacturer's specified temperature limit, and curve "2" is the distribution of the same parameter for the target application temperature limit.

The margin at the manufacturer-specified temperature range is the difference between the limit[6] of the distribution "1" and the specification limit on the parameter, PSPEC.[7] From distribution "2," at the target temperature, a new limit can be defined

---

[6]The limit may be chosen by the designers (e.g., 6-σ limit) as per the application yield and risk tolerances.

[7]Several factors influence the margin on a parameter, including the test equipment accuracy and confidence interval for the sample size. From distribution "2," at the target temperature, a new limit can be defined after adding a margin to it, and the modified parameter limit PNew can be obtained. One may chose to not modify the parameter limit, if the margin is still acceptable at the target temperature.

12 Derating and Uprating

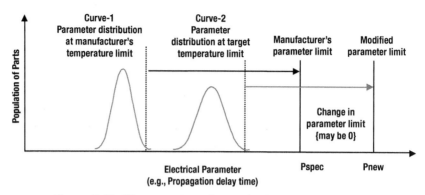

Figure 12.14 The parameter recharacterization method of uprating.

after adding a margin to it, and the modified parameter limit PNew can be obtained. One may chose to not modify the parameter limit, if the margin is still acceptable at the target temperature.

12.3.3.3 Stress Balancing

Stress balancing is a process of thermal uprating in which at least one of the part's electrical parameters is kept below its maximum allowable limit to reduce heat generation, thereby allowing operation at a higher ambient temperature than that specified by the semiconductor part manufacturer (McCluskey 1996). The process assesses the possibility that the application may not need to use the full performance capability of the device, and that a power versus operating temperature trade-off for the part may be possible. For active electronic parts, the power temperature relation is:

$$T_J = T_A + P \cdot \theta_{JA}, \qquad (12.1)$$

where T_J is the junction temperature, T_A is the ambient temperature, P is the power dissipation, and θ_{JA} is the junction-to-ambient thermal resistance. The performance of the part will generally depend upon the junction temperature. If the junction temperature is kept constant, then the temperature-dependent performance of the part should not change.

For a constant junction temperature, Equation 12.1 shows that higher ambient temperatures can be allowed if the power dissipation is reduced. However, the power dissipation of the part is often a function of some electrical parameters (e.g., operating voltage, and frequency), which will have to be changed. Thus, a trade-off can be made between increased ambient temperature and a change in some electrical parameter(s).[8]

The first step in stress balancing is to assess the electrical parameters, which can be used to change the power dissipation. The second step is to calculate the reduction in power dissipation required at the application temperature, by using the relationship given in Equation 12.1. The third step is to determine the changes in electrical parameters necessary to achieve the reduction in power dissipation. The fourth step is to conduct electrical tests to ensure the capability of the part to operate in the application environment with changed electrical parameters.

---

[8] Another option is to reduce the thermal resistance θ_{JA} of the part in the application, which may be achieved using heat sinks or providing cooling systems.

12.3.4 Continued Assurance

Part manufacturers provide product change notices (PCNs) for form, fit, and functional changes in their parts. However, change notices provided by the manufacturer do not necessarily reflect the changes in electrical performance that may occur beyond the recommended operating conditions in the datasheet. Thus, all changes need to be assessed by the part selection and management team for their possible effects on the use of parts beyond their manufacturer-specified recommending operating conditions in the datasheet. The changes in the parts that generally warrant a new uprating assessment include:

- Change in the temperature rating(s) of the part
- Change in power dissipation
- Changes in the thermal characteristics of the part, caused by changes in the package type, size or footprint, die size change, and materials
- Changes in the electrical specifications of the parts.

Semiconductor process changes (e.g., a die shrink, a new package, or an improvement in a wafer process) may or may not affect the part datasheet, but may affect the part performance beyond the recommended operating conditions. The changes in production sites may also result in changes in the uprateability of a part.

Specific protocols for reassessment of uprateability should be included in all uprating documents. For example, one may perform go/no-go tests on a sample of all incoming lots. Any failure of the part performance or deviation from the original lot (on which uprating was performed) will mean that uprating needs to be repeated.

Changes in the application of the part may also warrant reconsideration of the uprateability of the part. Changes in the environment, target temperature range, and system thermal design are factors that must be monitored. Electrical margin changes or part replacements (uprated or not uprated) may result in changes in system-level thermal interactions, and additional testing at the system level may be necessary.

12.4 Summary

Uprating is often possible because of the way electronic parts are designed and manufactured. The methods of uprating take into consideration the issues related to part testing and specification margins to assess the ability of the parts to operate over their target temperature range.

The uprating assessment of a part determines the electrical functional capability of parts in their target application conditions. This determines whether a part "can work" in a given environment. However, to determine if a part "won't fail" in the application environment, the reliability of the part needs to be determined for the application. The methods of determination of reliability can vary and may include assessment of manufacturers' qualification test results, additional tests performed by the equipment manufacturers, and virtual qualification.

12 Derating and Uprating

Problems

12.1 Select any nonelectronic item of everyday use and identify the recommended operating conditions and absolute maximum ratings from its documentation (e.g., datasheet, product specification, and web literature). List both the ratings clearly with source.

12.2 List all the environmental and operational parameters for which you may be able to uprate a computer. You will need to refer to the specifications and the ratings.

12.3 Select a mechanical design item (e.g., gear, beam, shaft, and engine).

(a) List three stresses related to that items which can be derated to improve the reliability of the item.
(b) For each listed stress level listed in (a), discuss how you would derate those stresses to improve reliability of the mechanical item.
(c) Relate the stress level listed in (b) to the definition of reliability and explain what part of reliability definition is addressed in the derating plan.
   (i) Within specified performance limits: A product must function within certain tolerances in order to be reliable.
   (ii) For a specified period of time: A product has a useful life during which it is expected to function within specifications.
   (iii) Under the life-cycle application conditions: Reliability is dependent on the products life-cycle operational and environmental conditions.

12.4 Can you utilize the derating factors provided by a part manufacturer to design an accelerated test? Under what conditions can you make use of that information? Use an example from today's lecture notes to explain.

12.5 Find a datasheet on a nonelectrical product specifying AMR.

12.6 Find a datasheet on a nonelectrical product specifying the recommended operating conditions. Why are these not AMR?

13 Reliability Estimation Techniques

Reliability is defined as "the ability of a product to function properly within specified performance limits for a specified period of time, under the life-cycle application conditions." Reliability estimation techniques include methods to evaluate system reliability throughout the product life cycle. The major components of reliability estimation techniques are the test program and the analysis of data from the tests. Test programs are developed throughout the life cycle, that is, design, development, production, and service, to ensure that reliability goals are met at different stages in the product life cycle. Data from test programs during each stage are acquired and processed to evaluate the reliability of a product at each stage in the life cycle.

The purpose of reliability demonstration testing is to determine whether the product has met a certain reliability requirement with a stated confidence level prior to shipping. Tests should be designed to obtain maximum information from the minimum number of tests in the shortest time. To achieve this, various statistical techniques are employed. A major problem in the design of adequate tests is simulating the real-world environment. During its lifetime, a product is subjected to many environmental factors, such as temperature, vibration, shock, and rough handling. These stresses may be encountered singly, simultaneously, or sequentially, and there are many other random factors.

13.1 Tests during the Product Life Cycle

Various tests are carried out at different stages in a product's life cycle to ensure that the product is reliable and robust. The different stages in a product's life cycle and tests suggested to be carried out in each stage are listed in Table 13.1.

13.1.1 Concept Design and Prototype

The purpose of the tests in the concept design and prototype stage is to verify breadboard design and functionality. Tests are conducted to determine the need for parts,

Reliability Engineering, First Edition. Kailash C. Kapur and Michael Pecht.
© 2014 John Wiley & Sons, Inc. Published 2014 by John Wiley & Sons, Inc.

Table 13.1 Stages in a product's life cycle and suitable tests in each stage

Phase description	Suitable tests
Concept design and prototype	Engineering verification test
Performance validation to design specification	Design verification test
Design maturity validation	Highly accelerated life test, accelerated stress test, stress plus life test (STRIFE), accelerated life test
Design and manufacturing process validation	Design maturity test, firmware maturity test, process maturity test
Preproduction low volume manufacturing	Burn-in, reliability demonstration test
High volume production	Highly accelerated stress screen, environmental stress screen, reliability demonstration test, ongoing reliability test, accelerated life test
Feedback from field data	Fleet/field monitoring/surveillance

materials, and component evaluation or qualification to meet system performance and other reliability design criteria.

13.1.2 Performance Validation to Design Specification

In this stage, tests are carried out to verify the functional adequacy of the design and the product performance. Tests are used to corroborate preliminary predictions. Failure modes and effects analysis is carried out to disclose high-risk areas and reliability problems in the proposed design.

13.1.3 Design Maturity Validation

Tests are carried to evaluate a design under environmental conditions, to verify the compatibility of subsystem interfaces, and to review the design. The design margins and robustness of the product are tested and quantified at this stage. The results from these tests will assist in developing a better design with minimal reliability issues.

13.1.4 Design and Manufacturing Process Validation

Design acceptance tests are used to demonstrate that the design meets the required levels of reliability. A reliability demonstration test is considered mandatory for design acceptance. Tests conducted at this stage include the design maturity test (DMT), firmware maturity test (FMT), and process maturity test (PMT). The purpose of DMT is to show that the product design is mature and frozen and is ready for production. The integration of hardware and software components is tested in the FMT. The process control is demonstrated in PMTs.

13.1.5 Preproduction Low Volume Manufacturing

Product screening—the process of separating products with defects from those without defects—is carried out at the preproduction stage to reduce infancy defects due to process, manufacturing and workmanship. Screening tests in the preproduction stage assist in reducing the infant mortality of a product. Burn-in is one of the mostly commonly used product screening tests. A reliability demonstration test may also be carried out to demonstrate the reliability of low volume manufacturing.

13.1.6 High Volume Production

Tests are carried out to determine the acceptability of individual products in order to ensure production control and critical interfaces, parts, and material quality. Screening at this stage does not improve the overall reliability of the product or change the yield; rather, it allows companies to ship products without defects to customers. Commonly conducted screens include the highly accelerated stress screen and environmental stress screen. To demonstrate the long-term reliability at high volume production, reliability demonstration tests are also carried out.

13.1.7 Feedback from Field Data

Tests and evaluation programs are carried out during field use of the product for continued assessment of reliability and quality. The data from field use are utilized to improve the design and reliability of the next version or generation of a product.

13.2 Reliability Estimation

Product reliability can be estimated from the test data using parametric or nonparametric techniques. In parametric estimation, the distribution of the test data should be known or assumed. Parametric techniques provide an inaccurate estimation if the assumptions are incorrect. The parameters are the constants that describe the distribution. Nonparametric estimates do not assume that the data belong to a given probability distribution. Generally, nonparametric estimates make fewer assumptions than parametric estimates. Nonparametric estimates apply only to a specific test interval and cannot be extrapolated. In this chapter, parametric estimates with underlying binomial, exponential, and Weibull distributions are described. The frequently used parametric estimates include: (1) point estimate: a single valued estimate of a parameter/reliability measure; (2) interval estimate: an estimate of an interval that is believed to contain the true value of the parameter; and (3) distribution estimate: an estimate of the parameters of a reliability distribution.

Data are often collected from a sample that is representative of the population to estimate the parameters of the entire population. For instance, the time to failure of light bulbs manufactured in a lot may be assessed to estimate the longevity of all the light bulbs manufactured by the company. Another example is the periodic sampling of manufactured goods to estimate the defect rate of the total population. Sample acceptance testing can also be conducted at the receipt of goods in order to assess and estimate the ability of the entire lot to meet specifications. The confidence interval is a measure of the uncertainty associated with making a generalization about the population based on a sample. These concepts are given later on in this chapter.

13.3 Product Qualification and Testing

A successful accelerated stress test (AST) program meets customer requirements, lowers life-cycle costs, and reduces the time to market for a product. A physics-of-failure (PoF)-based qualification methodology has been developed. The fundamental concepts of the PoF approach and a set of guidelines to design, plan, conduct, and implement a successful accelerated stress test are discussed below.

The inputs to the qualification methodology include hardware configuration and life-cycle environment. The output from the methodology is a *durability assessment*, where the accelerated stress test results are correlated to field life estimates through quantitative damage metrics and acceleration transforms. The PoF-based qualification methodology is a five-step approach, as shown in Figure 13.1. Step 1, *virtual qualification*, identifies the potential failures under the life-cycle loading inputs and their respective failure sites, modes, and mechanisms. Virtual qualification is a PoF-based process to assess the life expectancy under anticipated life-cycle loading conditions. The tasks in step 2, *accelerated test planning and development*, are to design test specimens, set up the experiment, determine accelerating loads, collect data using sensors, monitor data responses, and devise data postprocessing schemes. In step 3, *specimen characterization*, the test specimens' responses to test loads are determined, and the overstress limits of the test specimen are identified to precipitate the failure mechanisms of interest. *Accelerated life testing* (ALT) in step 4 evaluates the intrinsic product vulnerability to applied loads due to wear-out mechanisms. *Virtual testing* is conducted in step 5 on a limited sample size using test load levels scaled back from the destruct level profiles. The steps in the qualification methodology are explained in detail in the following sections.

13.3.1 Input to PoF Qualification Methodology

The inputs to the PoF qualification methodology may be broadly classified as hardware configuration and life-cycle loads. The following sections provide a description of the input parameters.

13.3.1.1 Hardware Configuration The product architecture and material properties must be identified and documented. First, the architecture of the product and the primary performance features of the product are studied. A database of a product's configuration (including physical dimensions, functionality, and constitutive elements) and layout (including electrical traces, routing, and connectivity) will assist in the development of effective analysis and verification procedures. A PoF-based approach cannot be applied to a black box configuration. On the contrary, specific information and detailed understanding of the hardware is required. A comprehensive documentation of the parts list, part dimensions, and primary performance features (electrical, mechanical, etc.) will be useful for PoF analysis.

Second, PoF methodology accentuates the understanding of material behavior and therefore requires that all the intrinsic material properties be documented. Material models must be characterized over the loading conditions experienced by the product. For instance, the fracture toughness of a material must be characterized over a range of temperatures or loading rates, depending on the loading conditions. The fracture toughness of a material to be used in a thermal management system must be

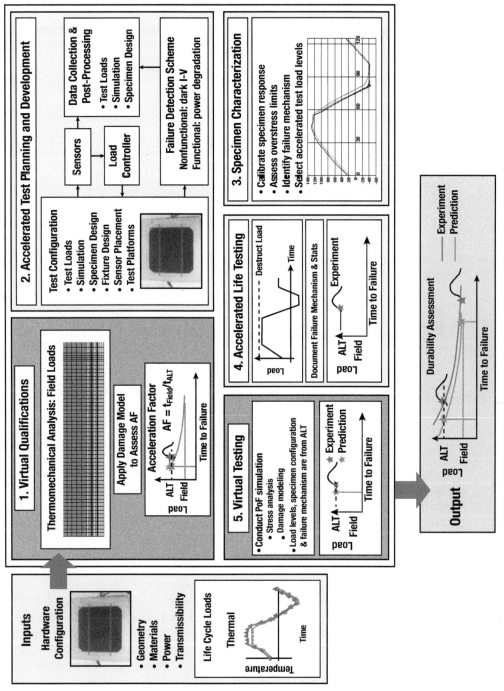

Figure 13.1 Example flowchart of PoF methodology.

13 Reliability Estimation Techniques

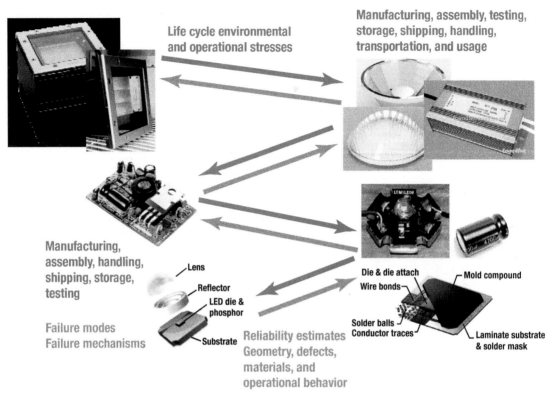

Figure 13.2 PoF requires information about and understanding of the entire supply chain.

characterized over a range of temperatures. On the other hand, if the same material were to be used in a moving system, the fracture toughness of the material must be characterized over a range of loading rates. The properties of common materials are obtained from material handbooks or literature. If the material properties are not readily available, then coupons of the materials are constructed to measure the appropriate material property. Additionally, the stress limits of the materials used in the hardware will assist in determining the design limits of the product. The PoF methodology is a detail-oriented process that requires detailed inputs for the materials used and how they are put together. All the materials and material properties that go into each and every single part of a product down to every IC must be available. An illustration of the breakdown of materials used in an LED is shown in Figure 13.2. Understanding the degradation mode and root-cause failure mechanisms that will be triggered under life-cycle stresses enables engineers to select appropriate materials to be used in the hardware. Product life can be calculated using analytical and PoF models and, based on the estimated product life, appropriate materials can be selected.

An LED device is an optoelectronic device that consists of various electronics and optical materials packaged in a mold material. Although the LED manufacturer builds the final product, the materials are provided by various suppliers in the supply chain. Hence, the manufacturer has to rely on the suppliers for information on failure modes and mechanisms. The manufacturers rely on engineers at each level in the supply chain to conduct physics-of-failure modeling. Each engineer in that supply chain contributes their own piece to the puzzle. One of the major benefits of the PoF approach is that the material properties can be cataloged long before a particular product enters the design cycle and development cycle. Therefore, designers and engineers do not have to wait to build a product and test it to determine the reliability model constants. In other words, before building any product, engineers can estimate the expected life using existing PoF models. However, the results from PoF models have an associated uncertainty due to the uncertainty in the inputs. This associated uncertainty can be orders of magnitude different from the actual life. If a company estimates that a product will last 5 years, it could really be half of that, or it could be year 10. Unfortunately, this is all within the band of uncertainty, and it is very difficult to design for finite life in this region. As a result, companies try to overdesign in this region. For example, if a company has to design for vibration for 5 years, that company would design the product to last 25 years to eliminate the possibility of failures within 5 years.

Electronics manufacturers must take into account the variability introduced into the material properties by the manufacturing and assembly processes. For instance, to measure the properties of the copper used in electronics, fatigue tests are carried out on copper coupons in a laboratory. These material properties are tested without building any electronics. Unfortunately, the building of electronics affects those properties depending on the assembly process of the copper in the printed circuit card. The damage accumulated due to manufacturing processes and manufacturing defects can cause variability in these model constants.

Appropriate PoF models should be selected to estimate the damage accumulated during use and predict the anticipated life under applied loading conditions. The selection of PoF models is based on the type of material used and the loading conditions. For instance, since solder is a viscoplastic material solder, it has to be modeled with a viscoplastic model, which is a combination of elastic deformation, plastic deformation, and time-dependent or rate-dependent creep deformation. Figure 13.3 shows the evolution of solder microstructure under fatigue loading until crack formation. The amount of damage accumulation and the corresponding microstructure evolution are functions of the temperature cycling range and rate.

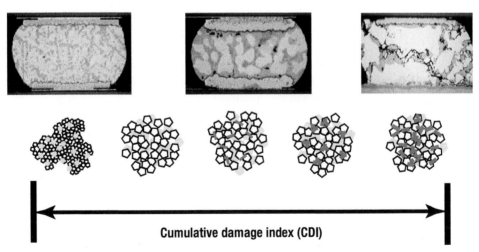

Figure 13.3 Physics-of-failure assessment: failure mechanism identification.

In order to qualify different materials in a product, the engineer has to go through the entire physics of failure process. The engineer needs the dimensions and material properties of the chip, solder, board, underfill, and so on. Finite element simulation is carried out at different loading conditions to find the stress distributions in the critical solder joints.

Life-Cycle Environment   The second part of the input is to identify the operational use environments and document field failures history.

Operational Use Environment   Environmental environment defines the loading conditions experienced by the product during use. There is a distinction between a load on a system and a stress. Loads are the boundary conditions on a system and refer to what the environment is doing to the system. The temperature, electrical usage, voltage, current, temperature, humidity, and mechanical vibration that a product is subjected to are all examples of loads. Information on loads does not necessarily provide the failure mechanism at a failure site. For that, understanding of the stresses causing the failure at the failure site is required. Stress is the intensity of an applied load at a failure site. Intensity will be defined differently depending on which stress type and which failure mechanism is in view. In order to monitor degradation, many parameters must be measured. One must know the usage pattern, because if the usage pattern changes over the life of the product, then the degradation rate also changes. Thus, the environmental conditions must be monitored, including humidity, temperature, mechanical acceleration, cyclical change of temperature, and rate of change of temperature. To relate these functional parameters to the environment, pattern recognition and correlation estimation are carried out. It is advantageous to know which environmental conditions have the highest impact on degradation. In addition to the applied loading conditions, diurnal environmental changes can also cause stresses. For instance, temperature changes from day to night, or from indoors to outdoors, result in stresses in portable electronics.

A product experiences loads even before field use. These loads include stresses during manufacturing, testing, transportation, handling, and storage. Stresses prior to field use can result in product failure before reaching the customer, known as "dead on arrival." The rate of change and duration of exposure of the loads prior to field use are important determinants of the stress magnitudes induced in a product. In addition to the operational loading conditions, a designer should adequately understand the storage conditions of the product. The shelf life, storage and transportation environments, and rework criteria should be understood clearly to adequately design the product for long-term storage. A well-designed, rugged product should survive the loads applied during operation, handling, and storage. Any accelerated life durability tests should take into account all the environmental conditions, rework environments, and workmanship factors that a product is expected to encounter.

Field Failures History   Understanding the history of prior field failures and preliminary failure analysis results is useful for identifying the dominant failure sites, modes, and mechanisms. Every time a product is changed in favor of a new technology, the company must utilize relevant information from the previous product's history to identify potential failure mechanisms. From a PoF perspective, if a company effectively utilizes the previous product's failure history, the more likely it is that the new product will be reliable. The knowledge and model constants of the previous product

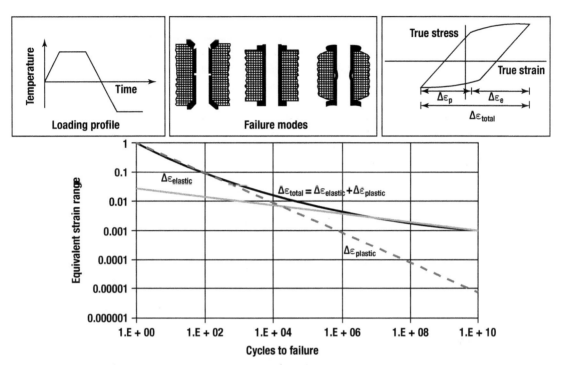

Figure 13.4 Plated through hole (PTH) low cycle fatigue in printed wiring boards (PWBs).

can be extrapolated for the new product. The field life is related to the test results by acceleration factors. Based on the observed modes in the field failure history, the test program can be appropriately tailored to identify and precipitate the weak links. For example, if the field failure history reports interconnect failures from creep-fatigue interactions as the predominant failure mechanism, temperature and vibration loads may be applied during testing to precipitate fatigue failures at the interconnects.

Fatigue-induced fracture due to cyclic loading is one of the most common failure mechanisms observed in electronic products. Figure 13.4 is a thermal cycling loading profile applied to a printed circuit assembly and the corresponding stress-strain history. The hysteresis loop for cyclic loading consists of elastic loading, which causes low cycle fatigue, and plastic loading, which causes high cycle fatigue. In the strain range versus cycles to failure plot, the steeper slope is the low cycle fatigue region, whereas the shallower slope represents the high cycle fatigue region. The actual fatigue data follow the superposition of these two regions in a log-log scale, represented by the solid black curve. If thermal cycling were conducted at different strain levels, the fatigue data would fall along the black curve with some scatter around it.

13.3.2 Accelerated Stress Test Planning and Development

The design of test loads, choice of test vehicle, and identification of issues related to test setup must be carried out prior to the accelerated testing of a product. This constitutes step 2 of the PoF approach to accelerated product qualification.

13.3.2.1 Test Loads and Test Matrix The first step in accelerated stress test planning is to determine the accelerated stress test matrix based on the dominant environmental loads and failure mechanisms. For example, if creep and mechanical fatigue are identified as the dominant failure mechanisms, the potential test loads are thermal and vibrational loads, respectively. Designing the test matrix depends on the dominant test loads and program objectives. For example, in a case study where the primary interest is to explore the interaction effects between thermal and vibrational loads, the test matrix can consist of several load cases involving simultaneous and sequential applications of repetitive shock vibration (RSV) and temperature cycling (TC). To conduct physics-of-failure modeling, quantitative information about the design of the product and its hardware configuration is imperative. The geometries, materials, and overall life-cycle loading of the product are also required. Overall life-cycle loading refers to the combination of stresses experienced by the product over its entire life cycle, including diurnal environmental loading.

13.3.2.2 Test Vehicle An appropriate test specimen should be designed to reduce the test time. For example, if the focus of a test is to understand surface-mount interconnect failures, a nonfunctional circuit card assembly is an appropriate test vehicle instead of testing an entire functional electronic assembly. A nonfunctional test specimen enables acceleration of the stresses beyond the operating limit of the product and is limited by the intrinsic material destruct limits. Accelerating the applied stresses results in considerable test time reduction.

13.3.2.3 Test Setup Test fixtures should be designed to have appropriate transmissibility. The main components of a test setup are selecting test platforms, identifying sensor monitoring schemes, designing fixtures, identifying failure detection, monitoring procedures, and postprocessing.

13.3.2.4 Selecting the Test Platform Selection of the test platform is driven by the test loads. For example, electro dynamic shakers or repetitive shock chambers can be used for vibration loading. On the other hand, a repetitive shock chamber is more appropriate for simultaneous application of multiple loading, such as temperature and vibration.

13.3.2.5 Stress Monitoring Scheme
Sensor Type The selection of sensor type is dependent on the type of stresses being monitored. For example, accelerometers and strain gauges are used to monitor and control vibrational loads, thermocouples are used to monitor temperatures, and capacitive gauges are used to sense moisture.

Optimal Sensor Location An engineering analysis has to be performed to determine the optimal quantity and location of the sensors. For example, the number of vibration sensors and their strategic placement on the test vehicle is based on preliminary modal analysis of electronic assemblies.

Fixturing Issues The test fixture should be designed and built such that the applied loads (thermal, mechanical, electrical, and chemical) are transmitted to the test vehicle with minimum loss. In addition, the applied loads should be effectively transmitted to potential weak links of the product.

Failure Monitoring and Detection Schemes If the test specimens are functional, dedicated electrical monitoring is required for failure detection, and the stress limit for the applied loads can be the operating limit or the destruct limit. If the test specimens are nonfunctional (daisy-chained), event detectors can be used to detect transient electrical opens, and the maximum stresses applied are limited by the destruct limit. For the verification of electrical failures, spectrum analyzers, oscilloscopes, circuit tracers, and signal conditioners may be used.

Data Acquisition and Postprocessing Schemes The chosen test platform needs to be adequately supported by control systems and data acquisition systems. For example, a test setup for combined temperature and vibration testing of daisy-chained components at the circuit card assembly level requires a test chamber capable of applying the stresses simultaneously. In addition, sensors to monitor, control, and collect data, event detectors for failure detection, and spectral, cycle-counting, and wavelet algorithms to postprocess the data are also required. Based on the stress-monitoring schemes, commercially available or custom-made software may be used for postprocessing schemes. For example, for vibration testing, commercially available software is equipped with time-domain, frequency-domain, and fatigue analysis tools. Testing of circuit card assemblies under a random vibration environment requires collection of data that, upon further postprocessing, can be used to compute the associated damage.

13.3.3 Specimen Characterization

Specimen characterization includes overstress tests to determine destruct limits of the specimen, tests to characterize the response of the specimen to the entire range of loads anticipated in the accelerated test, failure analysis, design of accelerated test profiles, and PoF assessment of expected time to failure under accelerated test loads.

13.3.3.1 Overstress Tests The objective of an accelerated test is to reduce test time by accelerating the degradation that occurs during field conditions. However, the accelerated testing should not precipitate failure mechanisms that may not occur in field or at unintended locations by stressing the product beyond its stress limit. Therefore, to efficiently design an accelerated test profile, the stress limits of a product should be known. Overstress tests are conducted to determine the stress limits, in particular the destruct limits, of a product. For example, to stimulate interconnect fatigue failures in flip-chip packages, the maximum temperature to be applied in the accelerated test may be limited by the thermal properties of the underfill material. The stress limits (specification limits, design limits, and operating limits) obtained from step 1 can be used in load profile design to determine the destruct limits for the product. The general criterion for profile design is to precipitate failures by overstress mechanisms.

13.3.3.2 Specimen Response Characterization under Anticipated Accelerated Test Loads The specimen response should be calibrated over the entire accelerated test load range. The characterization should also include the interaction effects between the applied environmental loads that otherwise would have been overlooked. Furthermore, results from the characterization serve as a verification of the PoF stress analysis conducted in step 1. Note: If the load profiles of overstress tests and ALT are significantly different, additional specimen characterization is required. Otherwise, the data collected from destruct tests can be used to characterize the specimen response over the entire accelerated load range.

13.3.3.3 *Failure Analysis* Failure analysis is conducted to verify that the dominant failure modes observed in overstress tests are indeed due to overstress and not wearout. Quantifying the overstress data enables verification of preliminary PoF models. Overstress tests, in conjunction with accelerated tests, are useful for determining the spatial location of acceleration transforms estimated in step 1.

13.3.3.4 *Accelerated Life Test Profiles* Accelerated test profiles should be designed such that wearout failure mechanisms encountered during field use are precipitated. Failure data from overstress tests are used to compute the necessary stress levels required to excite wearout failure mechanisms during ALT. Time-scaling techniques based on the PoF model of the relevant failure mechanism are utlized. For example, in vibration testing of assemblies, time-scaling techniques based on Steinberg's criteria can be used.

13.3.3.5 *PoF Assessment of Expected Failures under Accelerated Test Loads* Physics of failure is a multistep process. First, the stresses at various key sites of the product are determined from stress analysis. Mechanical, electrical, or chemical stress analyses are generally employed. Second, a finite element analysis (FEA) or some simple closed-form model is used, or a prototype may be built to measure the impact of these stresses on the product under study.

Third, PoF models are employed to assess the reliability under accelerated loading conditions. The quantitative parameters of the loading stresses are the inputs to the PoF failure models. PoF assessment is carried out for all the potential failure mechanisms. For each potential failure mechanism, the corresponding stresses are provided as inputs to the appropriate physics of failure model to obtain life data. The life results from each PoF model are then analyzed to determine the dominant failure mechanisms likely to occur the earliest. To determine the reliability of the system as a whole requires much more complex calculation.

PoF methods generally identify dominant failure mechanisms and treat each failure mechanism individually. However, failures can occur due to interactions between different degradation modes. To aggregate all the degradation mode information to the system level requires reliability tools, such as plot diagrams. The input for each block in a plot diagram comes from the individual physics of failure models. The plot diagram provides information about the system-level reliability.

Sensitivity studies are then carried out to determine the outcome when the environment or the design of the product is altered. Sensitivity studies enable reliability engineers to understand the behavior of a product under different environments, thereby making the product more robust. Pareto ranking of potential sources of failure can be used to determine the most dominant failure modes. To improve the reliability of a product, the stress and life margins may be increased. A quicker and more efficient method would be risk mitigation. To mitigate risks, a product can be ruggedized or the stresses can be managed by auxiliary systems. For example, if there is excessive vibration in a system, shock absorbers may be added. If there are excessive thermal stresses, active cooling devices may be implemented. Risk mitigation solutions can be implemented based on the margins.

Based on the initial stress analysis and damage modeling, preliminary estimates of in-service life are obtained. This involves conducting stress analysis, assessing the failure modes, and estimating the acceleration transforms. Accelerated stress tests in conjunction with overstress tests are used to determine the spatial location of the acceleration transform whose functional form has been predetermined from preliminary PoF assessment.

13.3.4 Accelerated Life Tests

Accelerated life tests evaluate the intrinsic vulnerability of a product to applied loads due to wear-out failure mechanisms. Successful implementation of accelerated wear-out test strategies requires: failure mechanisms generated in ALT to be the same as those observed and identified in the preliminary PoF assessment; and extrapolation of results from accelerated tests to field life conditions using acceleration transforms as reliability predictors to enable proactive product design. The primary tasks in this step include implementation of ALT and verification of observed failure modes.

13.3.4.1 ALT Implementation, Data Acquisition, and Failure Analysis Figure 13.5 shows how the time to failure changes with the change in stress level. The time to failure and stress level are plotted on the *x*-axis and *z*-axis, respectively. The time to failure is depicted as a distribution, as the ALT results are based on a set of tested samples. As the stress level increases, the mean value of the time to failure decreases, thereby shortening the test duration.

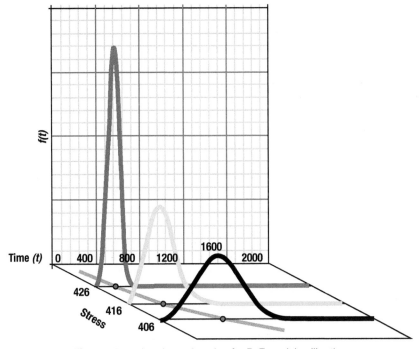

Figure 13.5 Accelerated testing for PoF model calibration.

The shape of the distribution changes at different stress levels. As the width of the distribution decreases, the peak goes up, since the area under the curve is constant. Each curve represents the probability density function at that particular stress level. The concept of accelerated testing is that if a product has to last in the field for 5 years, then, based on this graph, the reliability engineer can estimate the stress levels to precipitate failures in a shorter duration. Thus, with the help of acceleration factors, it can be verified in a short time period that a product indeed has a 5-year life for that failure mechanism. Acceleration transformations need to be developed for each failure mechanism. Typically, accelerated testing is a five-step process, two of which involve physical testing. The rest require PoF modeling, because without PoF modeling, the outcome of the test cannot be quantified. There should be two sets of inputs: the hardware configuration and the life-cycle usage conditions (loading). The output is a time-to-failure assessment or reliability assessment.

The reliability of a product must be designed based on the application conditions. The anticipated savings in the life-cycle and replacement costs as a result of having a reliable product should be the primary motivation for designing a reliable product. For example, if a thousand light bulbs in an auditorium that need to be replaced once per year were replaced with long-life LEDs, the cost savings would be significant. Understanding the life-cycle loading conditions is necessary for designing a reliable product for a specified period of time. For instance, a customer may request an LED manufacturer for a product that lasts for 50,000 hours. To ensure that the product will last 50,000 hours, the LED manufacturer should design its hardware configuration such that accelerated testing is carried out based on the life-cycle stresses experienced by the product.

If a company is in the middle of the life-cycle chain, then it must process information in both directions. The company in the middle of the chain affects the product through the stresses it puts the product through in manufacturing, testing, shipping, handling, storage, and distribution. The LED manufacturer thus needs to know not only what the customer will do with the product, but what the company in the middle of the chain is going to do with it.

13.3.5 Virtual Testing

Virtual testing (VT) is similar to virtual qualification (VQ). However, VQ is for the field configuration (determined in the input phase), whereas VT is for the accelerated test configuration. If we obtain the life-cycle loading conditions, understand hardware configuration, and implement a physics of failure model, then we can develop a virtual assessment methodology of the system's behavior, degradation, and failure. In real time, prognostic and health management (PHM) techniques can track the health of the system and predict anomalies and failures ahead of time. If PHM technologies are unavailable, we can keep updating our assessment by continuous life-cycle monitoring. The goal is to have an instantaneous assessment that continuously updates the remaining useful life of the product.

Physics of failure models update the assessment in real time to determine the remaining useful life. VT assesses time to failure under accelerated life-cycle testing (ALT) loads for the potential failure mechanisms by simulation. Based on a comparison with VQ results, acceleration factors are assessed. PoF methods are used to assess anomalies. PoF methods are also be used to assess the fraction of life used at any

13.3 Product Qualification and Testing

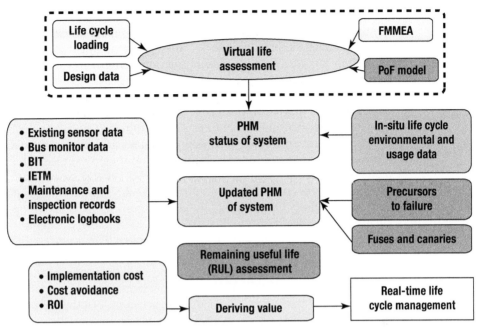

Figure 13.6 PHM hybrid approach.

point of time. However, these are assessments, not measured values. They are results from simulation. The purpose of PHM-based VT is to estimate the health of a system in real time and determine the remaining life (see Figure 13.6). VQ consists of two steps: stress and damage analyses.

13.3.5.1 Stress Analysis Simulations of all dominant load cases during the accelerated test conditions of the specimen are conducted—for example, thermal, thermomechanical, vibration/shock, hygromechanical, diffusion, radiation, and electromagnetic. Potential failure sites are identified through simulations. Physics of failure models are used to determine the prognostic distances. At every stage, by monitoring the actual usage and feeding that back into the PoF model, the virtual life assessment is periodically updated. One can also use PoF to design fuses and identify the best precursors to failure.

13.3.5.2 Damage Analysis After identification of potential failure sites, PoF models for each stress type are applied.

13.3.6 Virtual Qualification

Depending on the specific case, the VQ step is conducted first or last. VQ is similar to VT, except that VQ is for field configuration, while VT is for accelerated test configuration.

The goal of VQ is to determine the reliability risks under combined life-cycle loads by identifying the potential weak links and dominant failure mechanisms. The output of VQ is a ranking of the potential weak links under the expected life-cycle load combinations. PoF is used to identify the potential failures under life-cycle loads. The stress and damage analysis steps in VQ are similar to that in VT, except that the life-cycle loading experienced during operation is simulated.

It is necessary to identify and prioritize the failure mechanisms that could potentially be activated in the product during its life-cycle operation. Depending on the relative severity of the identified failure mechanisms, the predominant stresses of the respective failure mechanisms can be used as parameters for the AST. Accelerated tests should represent actual operating conditions without introducing extraneous failure mechanisms or nonrepresentative physical and/or material behavior. For example, temperature, a parameter that accelerates corrosion-induced failures, may also accelerate ionic contamination failures. The AST should be designed such that there is no possibility of the failure mechanism shifting from corrosion to ionic contamination during the test. Based on the initial stress analysis and damage modeling, preliminary estimates of in-service life are obtained. This involves conducting stress analysis, assessing the failure modes, and estimating the acceleration transforms. Preliminary PoF assessment determines the functional form of a product's acceleration transform for varying qualities that result from manufacturing variabilities. The exact spatial location of the acceleration transform is determined through systematic implementation of exploratory and accelerated stress tests.

13.3.7 Output

An acceleration factor (AF) is estimated by using the results from virtual qualification (step 1) and virtual testing (step 5). The durability of the product is then estimated by extrapolating ALT results using the acceleration factors estimated from simulation. The assumption is that the virtual test and virtual qualification results have the same interrelationship as the product durability under accelerated conditions and life-cycle load conditions. Assuming the relationship holds true, the acceleration factor is estimated as:

$$AF = t_{field}/t_{ALT},$$

where t_{field} is the predicted time to failure under life-cycle conditions (result from virtual qualification), and t_{ALT} is the predicted time to failure under accelerated test conditions (result from virtual testing).

PoF models are periodically verified and updated with the results from accelerated testing.

Life assessment provides scientific methods to interpret and extrapolate ALT results to field life estimates through quantitative acceleration factors. The primary features of life assessment techniques include: the ability to correlate, verify, and update PoF model predictions with the observed ALT results; and the ability to forecast product reliability under field conditions and thereby make design trade-offs to enhance reliability. The time-to-failure under life-cycle loads is obtained by multiplying the acceleration factor with the experimental time-to-failure. Life assessment techniques shift the emphasis from end-of-line testing to PoF-based test methodologies, thereby improving product life-cycle with proactive design and process techniques.

13.4 Case Study: System-in-Package Drop Test Qualification

To demonstrate the PoF qualification methodology, a case study to quantify the reliability of a land grid array (LGA), laminate-based RF (LBRF) system-in-package (SiP) component under drop loading is presented. The physics-of-failure approach (Figure 13.7), which uses the results of accelerated tests and simulations of field and accelerated loadings to estimate the expected lifetime under life-cycle loading, is shown.

The PoF approach to qualification of SiP with accelerated stress testing (AST) under drop test loading is shown in Figure 13.7. As explained in the previous section, the PoF approach consists of five phases. The input information depends on the specific SiP package that is selected for PoF qualification. The life-cycle loading depends on the end use and application. However, it is difficult for the SiP manufacturer to identify all possible use environments. The manufacturer often chooses to qualify a product for a set of standard environments and provides acceleration transforms that different end users can then use to extrapolate the results to their respective end-use environments. The types of loads expected for a SiP are temperature, humidity, vibration, and drop.

13.4.1 Step 1: Accelerated Test Planning and Development

The design and fabrication of a test specimen and experimental setup to subject the SiP package to accelerated stress environments is carried out in this step. The test board is a standard Nokia mechanical shock test board with 12 LGA SiP components laid out in rows of two each (Figure 13.9). The connector harness for daisy-chain monitoring is on the edge of the test board. Each component has a daisy-chain network connecting all of the perimeter bumps of the LGA such that, if one interconnect fails, the network reports failure. Five of the 10 boards are populated on one side, and five on the other side. The test board is fixed to the drop platform from the four mounting holes at each corner of the board.

The accelerated loads to be used in the qualification program are then selected based on the loading expected in use environments. The types of accelerated tests planned for SiP packages based on the input loads are thermal cycling, thermal shock, temperature/humidity/bias, moisture cycling, vibration, and drop. In this case study, the qualification of a SiP package under drop loads is demonstrated.

The accelerated test plan for drop testing of LGA SiP specifies dropping the test vehicle to an acceleration of 1500 G, which corresponds to just over 1 m height. The drop is guided in the out-of-plane direction with the component side down.

The center four components (rows 3 and 4) are expected to experience the maximum loading under drop loading (see Figure 13.8). The loading experienced is dependent on the stiffness of the board material. However, even if the stiffness at the boundaries were infinite, the center of the board would experience the maximum loading. Rows equidistant from the centerline of the board are expected to experience the same loading conditions due to the symmetry of the board. (It will be shown that this is not, in fact, a valid assumption.) The continuity is monitored in real time with high-speed DAQ.

Hardware and software are required to monitor the response of test vehicles to the applied loading and for data acquisition and postprocessing. Hardware to monitor

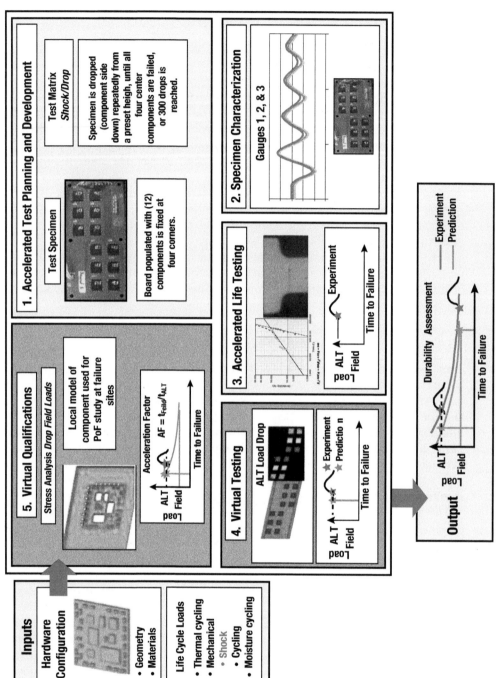

Figure 13.7 Physics-of-failure flowchart for SiP qualification.

13.4 Case Study: System-in-Package Drop Test Qualification

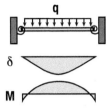

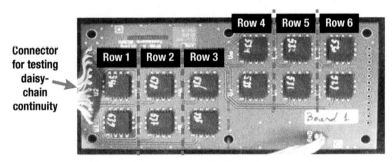

Figure 13.8 Loading, deflection, and moment diagram (2D) for the test vehicle.

Figure 13.9 Drop test vehicle showing the six classifications of component. The rows of components that are expected to see similar loading conditions are 1 and 6; 2 and 5; and 3 and 4.

response includes sensors such as thermocouples, moisture sensors, accelerometers, strain gages, and electrical parametric sensors.

Hardware designed and implemented for real-time failure detection during accelerated stress testing is also used. Examples include built-in tests for real-time functional checks, event detectors for checking the status of interconnects in daisy-chained mechanical specimens, and data loggers that detect increases in resistance due to interconnect degradation. Methods and algorithms are identified for data postprocessing to quantify the specimen response and damage.

13.4.2 Step 2: Specimen Characterization

The purpose of specimen characterization is to understand how the specimen is physically responding to accelerated loading and to collect data to calibrate simulation-assisted models for acceleration factors. The test specimen's response to test loads is explored by sensor placement. Examples include modal characterization for vibration or drop tests, temperature mapping for thermal tests, or moisture absorption mapping for humidity tests.

To characterize the specimen's response to mechanical drop, specimens attached with a strain gauge and an accelerometer were dropped from a drop tower. This step documented actual strain and acceleration histories at strategic locations on the board during a drop event. The gauge locations can be seen in Figure 13.10. Gauges 4 and 5 are oriented in the direction of the shorter edge of the board, whereas all other gauges are in the direction of the longer edge of the board.

Readings were recorded at four different drop heights, from 0.25 to 1 m. The strain histories for gauges 1, 2, and 3 are shown in Figure 13.11. The acceleration histories of the test vehicle, for each drop height, are shown in Figure 13.12. Gauges 2 and 3, which correspond to components U5 and U8, respectively, gave the most consistent data and were used to extrapolate a majority of the durability data. As seen in Figure

13 Reliability Estimation Techniques

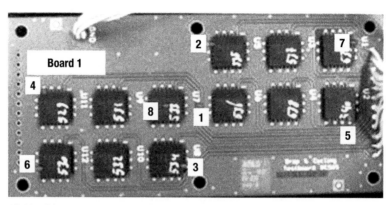

Figure 13.10 Gauge placement: gauges 1–7 are on the reverse side of the board shown here.

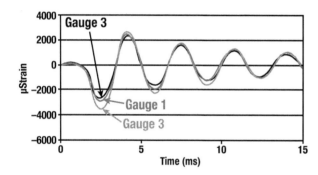

Figure 13.11 Examples of strain gauge time histories: 0.75-m drop. Gauge 3 output is larger than gauge 2 output. Gauge 1 is not the mean of 2 and 3.

13.13, gauges 2 and 3 did not experience the same loading profile, as expected. The ratio of the two strain ranges is listed in Figure 13.13 and is used later when the relations between drops-to-failure and various experimental data are calculated (strain range, acceleration, drop height, etc.). The strain range ratio between gauges 2 and 3 is $\Delta\varepsilon_3/\Delta\varepsilon_2 = 6567/5220 = 1.26$ (in microstrain).

The purpose of AST is to accelerate relevant wear-out failure mechanisms to quickly qualify the product for its life cycle. The overstress limits of the test specimen are identified at this stage to avoid the possibility of inadvertently accelerating the loading levels beyond the overstress limits of the test vehicle. These limits can often be assessed by examining material datasheets. For instance, temperature limits can be identified by looking at the phase transition limits (e.g., glass transition temperatures, recrystallization temperatures, and melting temperatures) of the materials used in the test article. When the datasheets cannot provide relevant data, step-stress testing techniques such as HALT™ can be used to systematically explore the overstress limits of the product. The safe load limits for the accelerated wearout test can then be selected by suitably scaling back from the overstress limits.

13.4.3 Step 3: Accelerated Life Testing

In this step, the accelerated stress regimes designed in step 2 are applied long enough to cause accelerated wear-out failures in the test specimen. Tests are carried out on a statistically significant sample size, and the time to failure of each sample is recorded. Failure data are grouped by failure mechanisms, and the statistics of failure

13.4 Case Study: System-in-Package Drop Test Qualification

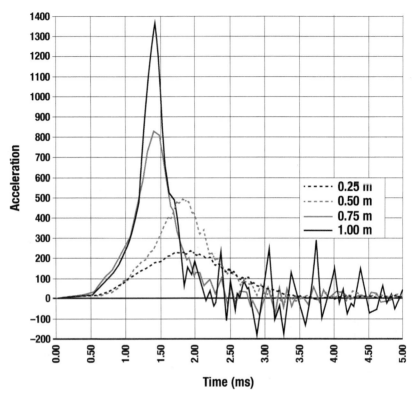

Figure 13.12 Filtered acceleration history data for each of the four drop heights.

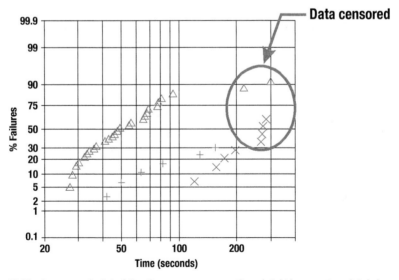

Figure 13.13 Lognormal plot of the three groups: rows 3 and 4 (△), rows 1 and 6 (+), and rows 2 and 5 (X). The data that seems to follow a separate failure mechanism is censored.

distributions are assessed for similar failure mechanisms. These failure distributions are later extrapolated using PoF simulations to field loading conditions. Failure analysis is conducted on the failed specimens to determine the failure sites and mechanisms.

Resource limitations may sometimes force early termination of the test after a fixed period, even if no failures are generated. In this case, PoF simulations for potential failure mechanisms can be used to assess the acceleration factors for each mechanism. Thus, the test duration can be extrapolated to the equivalent field duration for each potential failure mechanism.

Drop testing was conducted for 10 test boards, each with 12 components. For the first and last drop tests, the board was dropped until one specimen registered failure. The remaining eight boards were dropped until either all center four components registered failures, or 300 drops, whichever occurred first. The raw data from drop test is reported in Table 13.2.

In Table 13.3, it appears that the high end of the data follow a different pattern and therefore belong to a different failure mechanism. These data are censored from this qualification.

Table 13.4 shows the ratios of drops to failure for the two components that correspond to strain gauges 2 and 3 from the specimen characterization section. These data will be used later to calculate the durability data.

Failure analysis is conducted after termination of the test to identify the failure sites and mechanisms. Nondestructive failure analysis techniques should be employed initially to extract as much information about the failure sites and mechanisms as possible, followed by destructive failure analysis techniques. Visual, optical microscopy, and X-ray inspection methods are commonly employed nondestructive techniques. In the case of SiP, X-ray spectroscopy was utilized to identify the potential failure sites, as shown in the left image in Figure 13.14. Although extensive voiding was observed in the X-ray images, no cracks were seen.

After X-ray inspection, the SiP specimens were subjected to a "dye and pry" technique. In a standard dye and pry process, a failed component attached to the board is soaked in an indelible dye. In the case of a mechanically shocked/cycled specimen, the board is also flexed while soaking. The dye is expected to seep into cracks that exist on the interconnects. After the dye dries, the component is pried off the board. The presence of dye on the interconnects of the pried sample confirms the presence of an existing crack due to drop testing. In this study, the dye and pry technique was modified by milling up through the board close to the component but stopping just before reaching it. The purpose of this modified approach was to compensate for an LGA's extremely low stand-off height, which makes it very difficult to pry off without damaging the perimeter.

Typical results from a dye and pry analysis are shown on the right side of Figure 13.14. The presence of dye shows that there are edge delaminations in the large center thermal pad, but no fractures in the peripheral solder joints. Five interconnects had fracture through the solder, and the rest had copper traces pulled off near the component during the prying process. However, since no dye was observed in these fractures, the fracture is assumed to be an artifact of the prying.

Another commonly used destructive failure analysis technique is to cross-section the sample in order to identify the failure sites under optical or scanning electron microscopy. However, no fracture sites that would result in electrical failure were observed in the failed SiP.

Table 13.2 Overall drop test data of 10 boards

Side	U1–U12	U13–U25	U13–U25	U1–U12	U13–U25	U1–U12	U13–U25	U1–U12	U13–U25	U1–U12
Board No.	1	2	3	4	5	6	7	8	9	10
Device No.										
1				42		63		53		
2				0	82					
3		0	196	284		0				
4			219	0		274		271		
5		68	69		78	41	56	300	33	
6	45	49	32	43	28	66	30	48	41	46
7		32	29	55	37	77	28	65	29	
8		36		93	27		34	81	27	
9			172					267		
10						120		156		
11										
12				128		154				

Boards 2–9 were dropped until failure of the center devices (U5–U8). The first and last boards were dropped until the first failure. The data of the three symmetrical groups (rows 3 and 4, 2 and 5, and 1 and 6) are plotted in a lognormal plot of percent failed versus drops to failure.

Table 13.3 Reliability data for the three groups: number of drops until 50% fail (N50), 95% confidence level (95% CL), and shape parameter (σ)

Location	N50	95% CL	σ
Row 1, 6	270	183–414	0.806
Row 2, 5	394	250–656	0.806
Row 3, 4	56	43–73	0.806

Table 13.4 Drops-to-failure ratios for the components in the U5 and U8 positions for side 1 and side 2

	U1–U12	U13–U25
N_{f5}/N_{f8}	300/83.67 = 3.6	90.8/30.6 = 3.0

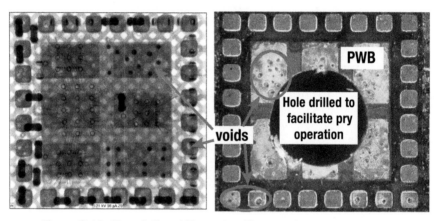

Figure 13.14 X-ray (left) and "dye and pry" images of a typical component.

The failure site and mechanism were identified through a process of desoldering the interconnects, polishing the surface of the board, and examining the copper traces on the board. All desoldered components were found to have failures: in the copper traces at the neck after transition from the pad area, and on the side of the components that is parallel to the short edge of the board (shown in Figure 13.15). Figure 13.16 is an elevation view of the crack that confirms that the cracks were induced by fatigue due to the bending of the board.

13.4.4 Step 4: Virtual Testing

Virtual testing is conducted with PoF simulations to identify the potential failures under the accelerated stress testing (AST). First, the stresses are assessed at the failure sites observed in step 3 under the applied accelerated loads. A combination of experimental measurements on the test vehicle and/or modeling of the test vehicle response to the accelerated loading are carried out.

Second, damage accumulation rates are assessed using appropriate damage models for each wearout failure mechanism. The output of this step is a prediction of mean time to failure for accelerated test loading. The test results from step 3 are used to calibrate the model, if necessary.

13.4 Case Study: System-in-Package Drop Test Qualification

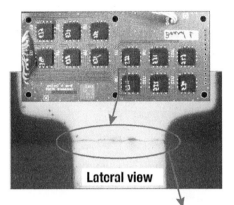

Lateral view

Trace cracks on PWB bond pads

Figure 13.15 Failures were finally found after desoldering the components from the board and polishing the surface of the board. A faint line can be seen in the FR4 on either side of the copper crack.

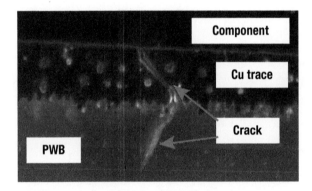

Figure 13.16 Elevation view of the crack, confirming that the failure mode was bending as opposed to pad lift-off.

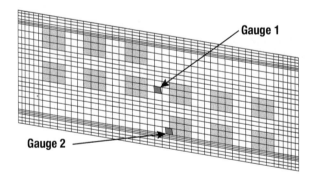

Figure 13.17 Modal analysis to correlate flexural strain to failure data 2D four-node "shell" elements. PWA is modeled with uniform thickness. Footprints of SiPs are given different properties: density and stiffness are increased to represent PWB + SiP; boundary conditions: four corners fixed.

13.4.5 Global FEA

A modal analysis using a 2D representative model using shell elements was carried out in finite element modeling (FEM), as shown in Figure 13.17. Boundary conditions, geometries of the board/components, and material properties of the board/components were inputs to the model.

Using the specimen characterization data, the FEA model can be correlated to drop test conditions. Data can also be collected from the areas in the model where the strain gauges were located and compared.

13 Reliability Estimation Techniques

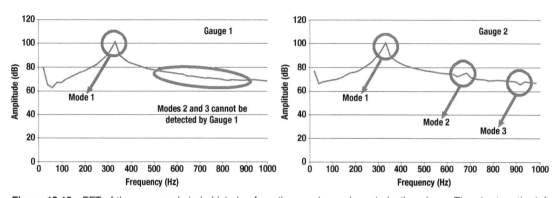

Figure 13.18 FFT of the measured strain histories from the specimen characterization phase. The chart on the left is from gauge 1 and clearly shows influence from only mode 1. The chart on the right is from gauge 2 and shows influence from the first three modes.

Table 13.5 Comparison chart of the measured modal frequencies versus those calculated by the finite element model

Mode frequency comparison	Measured (Hz)	FEA (Hz)	% Difference
Mode 1	332	350	5
Mode 2	703	730	4
Mode 3	918	1000	9

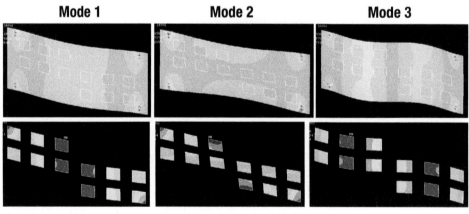

Figure 13.19 Bending strain contours of the first three mode shapes of the entire model (top row) and just the component footprints (bottom row).

Figure 13.18 shows fast Fourier transforms (FFTs) of the strain histories gathered in the specimen characterization phase. Table 13.5 shows the comparison of the first three modes observed in the characterization and calculated by the FEM.

13.4.6 Strain Distributions Due to Modal Contributions

FEM analysis shows why gauge 1 and gauge 2 provided different data. Figure 13.19 shows that mode 2 is a twisting mode. Since the major contributions to the

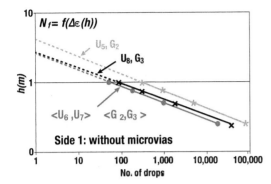

Figure 13.20 Number of drops to failure as a function of drop height for U8, U5, and the average of U6 and U7.

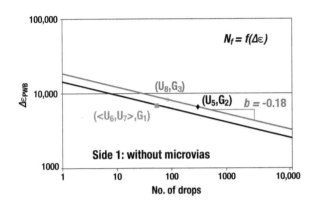

Figure 13.21 Number of drops to failure as a function of the strain-range measured on the PWB for gauges 2 and 3 (U8 and U5).

deformation seem to be from modes 1 and 2, by superimposing mode 2 on mode 1, the deformations are subtracted at gauge 1 and added at gauge 2.

13.4.7 Acceleration Curves

Acceleration curves are created by combining the strain gauge data from the gauges placed near the critical solder joints of the critical components. Acceleration curves are created as functions of parameters such as drop height (Figure 13.20) and PWB strain range (Figure 13.21). The results from acceleration curves can be extrapolated to different drop heights or different strain ranges for this specimen.

13.4.8 Local FEA

A local model of one component and the surrounding PWB was developed in FEA (Figure 13.22). The purpose of a local FEA model was to better understand the stress and strain fields in the component during bending and to correlate the observed PWB strain at the gauges to the solder strain experienced in the critical solder joint. Even though the failure site was observed in the copper trace, this FEA transform can used to predict the strain at the solder interconnects.

The local model was constrained at the bent edge, in the plane of the board, to simulate the stretch in the PWB that also happens during the drop event. The strain field predicted in the FEA (seen in Figure 13.23) was used to develop the strain transfer function shown in Figure 13.24. Using the strain transfer function and the PWB

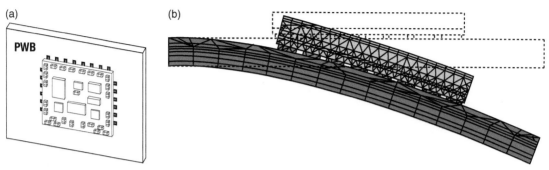

Figure 13.22 3D local FEA model of U8 component: (a) top view, decapsulated; and (b) side view, deflected.

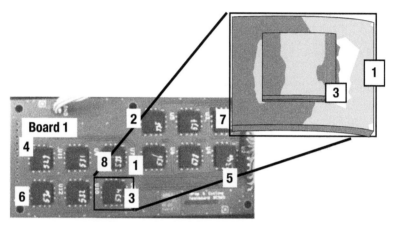

Figure 13.23 Schematic of the location of the local model with respect to the test board and the corresponding strain fields developed in component U8 during drop.

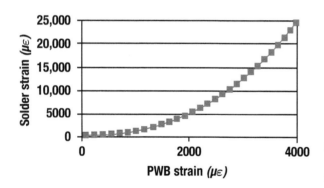

Figure 13.24 Strain transform from measured PWB strain to solder strain experienced by the critical solder joint, as predicted by FEA.

strain measured from the strain gauge, the bending strain experienced by the critical solder joint is estimated.

13.4.9 Step 5: Virtual Qualification

In this step, the time to failure is assessed for the same failure mechanisms addressed in step 4, but now under life-cycle loading conditions. The ratio of the estimated time

13.4 Case Study: System-in-Package Drop Test Qualification

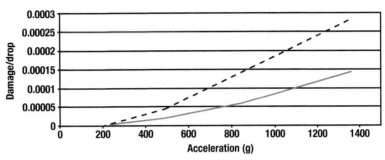

Figure 13.25 Damage curves for components U5 and U8 as a function of the drop acceleration.

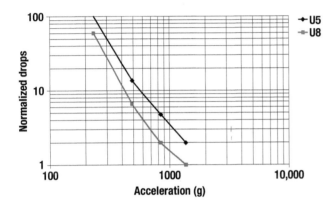

Figure 13.26 Acceleration curves for solder failure: normalized number of drops to failure versus drop acceleration.

to failure in steps 4 and 5 provides the acceleration factor between the test environment and the field environment.

The simulations to obtain acceleration factors for the SiP include stress analysis: thermal (steady-state/transient), mechanical (static/dynamic), thermomechanical, moisture absorption, and hygromechanical; and damage models, including fatigue, fracture, creep, corrosion. The final output of the process is the durability assessment of the product, which is obtained by extrapolating the accelerated test results to field conditions through the use of acceleration factors.

The durability data can be transformed to a function of drop acceleration, as shown in Figure 13.25. This is a more general metric to use, as the actual acceleration experienced by the impact event can change, for any given height, based on the drop conditions. The total strain range experienced by the gauge for each of the first four pulses during the impact was measured. This includes initial impact and clatter afterwards. The damage for each pulse was calculated and related to the acceleration that caused the strain range.

13.4.10 PoF Acceleration Curves

Finally, the acceleration curves can be calculated, as shown in Figure 13.26.

13.4.11 Summary of the Methodology for Qualification

The five-step PoF approach was illustrated for the qualification of a SiP-type package in a drop-loading environment. Although failure did not occur at the solder interconnects, valuable insights were gained from the PoF study. It was observed that the increased robustness of the package type resulted in the transfer of weak points to the copper traces in the test board. The insights from this PoF study will allow for board redesign.

It is good practice to state the results of all engineering analyses with the degree of certainty (or uncertainty) associated with it, for example, confidence intervals. Similar to the way confidence intervals around estimates can be used to estimate unknown distribution parameters, confidence intervals around a regression line can be used to estimate the uncertainties associated with regression relationships.

Specifics about the type of confidence interval used are also imperative while reporting the failure data. The confidence level, one- or two-sided interval, sample size, how the samples were chosen, and the methods of analysis are some information to be included.

Under some circumstances, estimation and visualization of the confidence interval may not be possible. For example, a very small sample size is likely to produce a very wide confidence interval that has no practical use. In such cases, data visualization techniques are used to display the complete results without making any statistical claim to facilitate making judgments on the data.

13.5 Basic Statistical Concepts

A population is a set of data collected from all the members of a group. A sample is a set of data collected from a portion of the population. Since it is not possible or even advisable to measure the whole population (e.g., the act of measurement could damage the samples and make them unusable), data obtained from a sample are used to make estimates about the population. Figure 13.27 describes a schematic of estimating population parameters from a sample. To obtain the population parameters from a sample, the population and sample must be created from the same process.

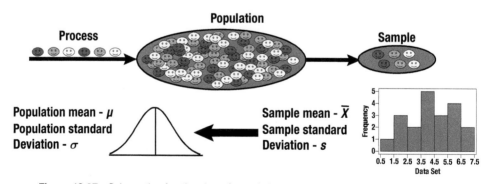

Figure 13.27 Schematic of estimation of population parameters from sample parameters.

13.5 Basic Statistical Concepts

An underlying random variable of interest is denoted by X. The variables $X_1, X_2, \ldots, X_n$ are random samples of size n from the population represented by X, if they are all independent and have the same probability distribution based on the random variable X. The observed data, $X_1, X_2, \ldots, X_n$, is also referred to as a random sample. A statistic is a point estimate derived from the observed data and is defined as $\hat{\Theta} = g(X_1, X_2, \ldots, X_n)$. Some examples of a statistic are mean

$$\bar{X} = \sum_{i=1}^{n} X_i$$

and variance

$$S^2 = \frac{1}{n-1} \sum_{i=1}^{n} (X_i - \bar{X})^2.$$

13.5.1 Confidence Interval

A confidence interval is an interval estimate computed from a given data sample that includes the actual value of the parameter with a degree of certainty. The width of the confidence interval is an indication of the uncertainty about the actual parameter. The confidence interval puts a boundary around these point estimates and provides the likelihood that the population parameters are within those boundaries.

Inferential statistics is used to draw inferences about a population from a sample. Statistics from a sample include measures of location, such as mean, median, and mode, and measures of variability, such as variance, standard deviation, range, or interquartile range.

Standard deviation of a set of measurements is not the same as confidence interval. Standard deviation is a measure of the dispersion of a measurement. In general, the greater the standard deviation is, the wider is the confidence interval on the mean value of that measurement. However, there is more to the statistics of a set of measurements than standard deviation.

When the probability of θ being in the interval between l and u is given by $P(l \leq \theta \leq u) = 1 - \alpha$, where $0 \leq \alpha \leq 1$, the interval $l \leq \theta \leq u$ is called a $100 \times (1 - \alpha)$ percent confidence interval. In this definition, l is the lower confidence limit, u is the upper confidence limit, and $(1 - \alpha)$ is called the confidence level, usually given as a percentage.

A confidence interval can be either one or two sided. A two-sided (or two-tailed) confidence interval specifies both a lower and upper bound on the interval estimate of the parameter. A one-sided (or one-tailed) confidence interval specifies only a lower or upper bound on the interval estimate of the parameter. A lower one-sided $100(1 - \alpha)$ percent confidence interval is given by $l \leq \theta$, where l is chosen so that $P(l \leq \theta) = 1 - \alpha$. Conversely, an upper one-sided $100(1 - \alpha)$ percent confidence interval is given by $\theta \leq u$, where u is chosen so that $P(\theta \leq u) = 1 - \alpha$.

13.5.2 Interpretation of the Confidence Level

The common perception is that the confidence level is the probability of a parameter being within the confidence interval. Although this assumption is intuitive and gives

13 Reliability Estimation Techniques

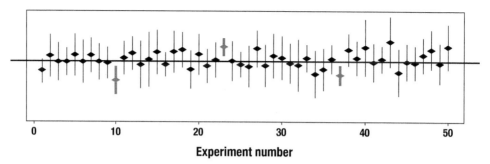

Figure 13.28 Conceptualization of confidence interval.

a measure of understanding, the conceptual definition of confidence interval is more subtle. One engineering statistics textbook (Montgomery and Runger 2007, p. 262) states the nuance in the following way: "in practice, we obtain only one random sample and calculate one confidence interval. Since this interval either will or will not contain the true value of θ, it is not reasonable to attach a probability level to this specific event. The appropriate statement would be that the observed interval [*l, u*] brackets the true value of θ with confidence level 100(1 − α). This statement has a frequency implication; that is, we don't know if the statement is true for a specific sample, but the method used to obtain the interval [*l, u*] yields correct statements 100(1 − α) percent of times."

Figure 13.28 shows fifty confidence intervals on the mean computed from samples taken from a population at a confidence level of 95%. The solid line represents the true mean calculated from the whole population. We expect that 95% of all possible samples taken from the population would produce a confidence interval that includes the true value of the parameter being estimated, and only 5% of all samples would yield a confidence interval that would not include the true value of the parameter. The simulated case shows that three (approximately 5%) of the confidence intervals do not contain the true mean.

With a fixed sample size, the higher the confidence level is, the larger the width of the interval will be. A confidence interval estimated at a 100% confidence level will always contain the actual value of the unknown parameter, but the interval will stretch from −∞ to +∞. However, such a large confidence interval provides little insight. For example, we can say with a very high confidence level that the age of all students in a reliability class is between 1 and 150 years, but that does not provide any useful information.

Selection of the confidence level is part of the engineering risk analysis process. For example, with a confidence interval analysis, the expected worst cases on warranty returns over a period can be estimated. An estimate can then be made of the spare parts to stock based on the point estimate of a 95% or 99% confidence level (or any other chosen value) of the expected warranty return. The decision will depend on the balance between the cost of storing the spares versus the cost of delay in repair time due to the unavailability of spares. In many engineering situations, the industry practices or customer contracts may require the use of a specific confidence level—frequently, values of 90% or 95% are quoted.

13.5.3 Relationship between Confidence Interval and Sample Size

The value of the confidence intervals depends on the measurements for each sample. As long as the measurements made on the samples are from the same population, an increase in sample size will reduce the width of the confidence interval, provided that the confidence level is kept constant. However, when conducting an experiment or gathering data from the field, data may come from multiple populations; in those cases, a large sample size may actually increase the confidence interval. For example, in the manufacturing of baseball bats, the hardness values of samples taken from the production line can be recorded. If the production parameters are all under control, then increasing the number of samples that come from the same population will narrow the confidence interval. However, if for some period the production parameters are out of control, the hardness values for samples taken during those times will differ. Therefore, increasing the sample size by including samples from the "out of control" population will increase the confidence interval.

13.6 Confidence Interval for Normal Distribution

Concepts of the confidence interval are often illustrated using the normal distribution, partly because it is a symmetric distribution described by two parameters. In a population with normal distribution, there is a direct relation between confidence interval and sample size.

This section describes the calculation of confidence intervals for three cases: confidence interval on an unknown mean with known variance, confidence interval on an unknown mean with an unknown variance, and confidence interval on differences between two population means with a known variance.

13.6.1 Unknown Mean with a Known Variance for Normal Distribution

Consider a population with an unknown mean, μ, and a known variance, σ^2. The variance may be known from past experience or prior data, such as physical processes that create the population or the control charts. For this population, random samples of size n yield a sample mean of $\bar{X}$. The $100(1 - \alpha)$ percent confidence interval for the population mean is given by:

$$\bar{X} - \frac{Z_{\alpha/2}\sigma}{\sqrt{n}} \leq \mu \leq \bar{X} + \frac{Z_{\alpha/2}\sigma}{\sqrt{n}}, \qquad (13.1)$$

where $Z_{\alpha/2}$ is the upper $\alpha/2$ percentage point of the standard normal distribution. Correspondingly, to obtain the one-sided confidence intervals, Z_α replaces $Z_{\alpha/2}$; setting $l = -\infty$, and $u = +\infty$, in the two cases, respectively, the one-sided confidence intervals are given by:

$$\mu \leq u = \bar{X} + \frac{Z_\alpha \sigma}{\sqrt{n}} \qquad (13.2)$$

and

$$\bar{X} - \frac{Z_\alpha \sigma}{\sqrt{n}} = l \le \mu. \qquad (13.3)$$

When using a sample mean, $\bar{X}$, to estimate the actual but unknown mean, μ, the "error" is $E = |X - \mu|$. With a confidence level of $100(1 - \alpha)$, for a two-sided interval, the error is within the precision of estimation given by:

$$E \le \frac{Z_{\alpha/2} \sigma}{\sqrt{n}}. \qquad (13.4)$$

Therefore, we can choose a sample size, n, that allows $100(1 - \alpha)$ percent confidence that an error will not exceed a specified amount, E.

$$n = \left[\frac{Z_{\alpha/2} \sigma}{E} \right]^2, \qquad (13.5)$$

where n is rounded up to the next integer.

Example 13.1

Consider measuring the propagation delay of a digital electronic part. You want to have a 99.0% confidence level that the measured mean propagation delay is within 0.15 ns of the real mean propagation delay. What sample size do you need to choose, knowing that the standard deviation of the propagation delay is 0.35 ns?

Solution:
Using Equation 13.5, the value of n is found to be 37.

$$n = \left(\frac{Z_{\alpha/2} \sigma}{E} \right)^2 = \left(\frac{Z_{.005} \times 0.35}{0.15} \right)^2 = \left(\frac{2.58 \times 0.35}{0.15} \right)^2 \approx 37.$$

In this application, α is 0.01 and $\alpha/2$ is 0.005. From the standard normal table, $Z_{0.005} = 2.58$.

13.6.2 Unknown Mean with an Unknown Variance for Normal Distribution

The t-distribution is used to develop the confidence interval in this case. Assuming the population to be normal, the sample variance, S^2, is used to estimate the population variance, σ^2, which is not known. Then,

$$T = \frac{\bar{X} - \mu}{S/\sqrt{n}}, \qquad (13.6)$$

has t-distribution with $n - 1$ degrees of freedom.

Suppose a population has an unknown variance, σ^2. A random sample of size n yields a sample mean, $\bar{X}$, a sample variance, S^2, and as an upper $\alpha/2$ percentage point of the t-distribution with $(n-1)$ degrees of freedom. The two-sided $100(1-\alpha)$ percent confidence interval in this case is given by:

$$\bar{X} - \frac{t_{\alpha/2, n-1} S}{\sqrt{n}} \leq \mu \leq \bar{X} + \frac{t_{\alpha/2, n-1} S}{\sqrt{n}}. \tag{13.7}$$

Example 13.2

The tensile strength of a synthetic fiber used to manufacture seatbelts is an important characteristic in predicting the reliability of the product. From past experience, the tensile strength can be assumed to be normally distributed. Sixteen samples were randomly selected and tested from a batch of fibers. The sample's mean tensile strength was found to be 49.86 psi, and the sample's standard deviation was found to be 1.66 psi. Determine an appropriate interval to estimate the batch mean tensile strength.

Solution:
Since we are only concerned with tensile strengths that are too low, a one-sided confidence interval on the batch mean, μ, is appropriate. Since the population (batch) variance is unknown and the sample size fairly small, a confidence interval based on the t-distribution is necessary. A one-sided, 99% confidence interval for the batch mean μ is:

$$\bar{X} - \frac{t_{\alpha, n-1} S}{\sqrt{n}} \leq \mu \Rightarrow 49.86 - \frac{(1.753)1.66}{\sqrt{16}} \leq \mu \Rightarrow 49.13 \leq \mu.$$

13.6.3 Differences in Two Population Means with Variances Known

A confidence interval for the difference between means of two normal distributions specifies a range of values within which the difference between the means of the two populations $(\mu_1 - \mu_2)$ may lie. A random sample, n_1, from the first population, with a known standard deviation of σ_1, yields a sample mean of X_1. Similarly, a random sample, n_2, from the second population, with a known standard deviation of σ_2, yields a sample mean of X_2. Then, a two-sided $100(1-\alpha)$ percent confidence interval for the difference between the means is given by:

$$\bar{X}_1 - \bar{X}_2 - Z_{\alpha/2} \sqrt{\frac{\sigma_1^2}{n_1} + \frac{\sigma_2^2}{n_2}} \leq (\mu_1 - \mu_2) \leq \bar{X}_1 - \bar{X}_2 + Z_{\alpha/2} \sqrt{\frac{\sigma_1^2}{n_1} + \frac{\sigma_2^2}{n_2}}, \tag{13.8}$$

where $Z_{\alpha/2}$ is the upper $\alpha/2$ percentage point of the standard normal distribution.

Example 13.3

Tensile strength tests are performed on two different types of aluminum wires used for wire bonding power electronic devices. The results of the tests are given in the following table:

Type	Sample size, n	Sample mean tensile strength (kg/mm²)	Known population standard deviation (kg/mm²)
1	15	86.5	1.1
2	18	79.6	1.4

What are the limits on the 90% confidence interval on the difference in mean strength ($\mu_1 - \mu_2$) of the two aluminum wires?

Solution:

$$l = \bar{X}_1 - \bar{X}_2 - Z_{\alpha/2}\sqrt{\frac{\sigma_1^2}{n_1} + \frac{\sigma_2^2}{n_2}}$$

$$= 86.5 - 79.6 - 1.645\sqrt{\frac{(1.1)^2}{15} + \frac{(1.4)^2}{18}} = (6.9 - 0.716)$$

$$= 6.184 \text{ kg/mm}^2.$$

Also

$$u = \bar{X}_1 - \bar{X}_2 + Z_{\alpha/2}\sqrt{\frac{\sigma_1^2}{n_1} + \frac{\sigma_2^2}{n_2}}$$

$$= 86.5 - 79.6 + 1.645\sqrt{\frac{(1.1)^2}{15} + \frac{(1.4)^2}{18}} = (6.9 + 0.716)$$

$$= 7.616 \text{ kg/mm}^2.$$

13.7 Confidence Intervals for Proportions

In engineering applications, the outgoing quality of a product is often estimated based on testing a sample of the parts. If $\hat{p}$ is the proportion of observations in a random sample of size n that belongs to a class of interest (e.g., defects), then an approximate $100(1 - \alpha)$ percent confidence interval on the proportion, p, of the population that belongs to this class is:

$$\hat{p} - z_{\alpha/2}\sqrt{\frac{\hat{p}(1-\hat{p})}{n}} \leq p \leq \hat{p} + z_{\alpha/2}\sqrt{\frac{\hat{p}(1-\hat{p})}{n}}, \tag{13.9}$$

where $z_{\alpha/2}$ is the upper $\alpha/2$ percentage point of a standard normal distribution. This relationship holds true when the proportion is not too close to either 0 or 1 and the sample size n is large.

Example 13.4

An inspector randomly selects 200 boards from the process line and finds 5 defective boards. Calculate the 90% confidence interval for the proportion of good boards from the process line.

Solution:
Use Equation 13.9:

$$\hat{p} - z_{\alpha/2}\sqrt{\frac{\hat{p}(1-\hat{p})}{n}} \leq p \leq \hat{p} + z_{\alpha/2}\sqrt{\frac{\hat{p}(1-\hat{p})}{n}}$$

$$\frac{195}{200} - 1.64\sqrt{\frac{0.975(0.025)}{200}} \leq p \leq \frac{195}{200} + 1.64\sqrt{\frac{0.975(0.025)}{200}}$$

$$0.957 \leq p \leq 0.993.$$

The result implies that the total population is likely (90% probability) to have a proportion of good boards between 0.997 and 0.993. Note that no assumption is made regarding what the total population is.

13.8 Reliability Estimation and Confidence Limits for Success–Failure Testing

Success–failure testing describes a situation where a product (component, subsystem) is subjected to a test for a specified length of time, T_0 (or cycles, stress reversals, miles, etc.). The product either survives to time T_0 (i.e., it survives the test) or fails prior to time T_0.

Testing of this type can frequently be found in engineering laboratories where a test "bogy" has been established and new designs are tested against this bogy. The bogy will specify a set number of cycles in a certain test environment and at predetermined stress levels.

The probability model for this testing situation is the following binomial distribution, which gives the probability that the number of successes is y out of n items tested:

$$P(y) = \binom{n}{y} R^y (1-R)^{n-y}, \quad y = 0, 1, \ldots, n, \qquad (13.10)$$

where

n = the number of items tested
R = the probability of surviving the test for the product
y = the number of survivors out of n,

and

$$\binom{n}{y} = \frac{n!}{y!(n-y)!}, \quad y = 0, 1, \ldots, n. \qquad (13.11)$$

The value R is the reliability, which is the probability of surviving the test. The minimum variance unbiased estimator of R is

$$\hat{R} = \frac{y}{n}. \qquad (13.12)$$

The $100(1 - \alpha)$ percent lower confidence limit on the reliability R is calculated by

$$R_L = \frac{y}{y + (n - y + 1) F_{\alpha, 2(n-y+1), 2y}}, \qquad (13.13)$$

where $F_{\alpha, 2(n-y+1), 2y}$ is obtained from the F tables. Here again, n is the number of items tested and y is the number of survivors.

The $100(1 - \alpha)$ percent upper confidence limit on R is given by

$$R_U = \frac{(y+1) \times F_{\alpha, 2(y+1), 2(n-y)}}{(n-y) + (y+1) F_{\alpha, 2(y+1), 2(n-y)}}. \qquad (13.14)$$

The F tables that are usually available are somewhat limited in terms of degrees of freedom. Therefore, it is convenient to have an approximation for the lower confidence limit that uses the standard normal distribution. The lower confidence on reliability can be approximated by:

$$R_L = \frac{y-1}{n + Z_\alpha \sqrt{\frac{n \times (n-y+1)}{(y-2)}}}, \qquad (13.15)$$

where

Z_α = the standard normal variable, as given in Table 13.6
y = the number of successes
n = the sample size.

It should be noted that Z is the standard normal variable. Values given in Table 13.6 can be read from cumulative distribution tables for standard normal variables given in Appendix C.

Table 13.6 Standard normal variables

Confidence level (1 − α)	Z_α
95	1.645
90	1.281
80	0.841
75	0.678
70	0.525
50	0.000

13.8 Reliability Estimation and Confidence Limits for Success–Failure Testing

Example 13.5

A weapon system has completed a test schedule. The test is equivalent to 60 missions. Dividing the test schedule up into 60 missions results in seven failed missions. Estimate the mission reliability.

Solution:
In this case, the number of successes (y) is $y = 60 - 7 = 53$ successful missions out of $n = 60$ missions. Then the point estimate for mission reliability is

$$\hat{R}_m = \frac{53}{60} = 0.883.$$

Let us now find a 75% lower confidence limit. The exact lower 75% limit is found by using an F value of

$$F_{0.25, 16, 106} = 1.24.$$

Substituting this into the confidence limit equation gives

$$R_L = \frac{53}{53 + (8 \times 1.24)} = 0.842.$$

The 75% lower confidence limit on mission reliability is $0.842 \leq R_m$. If the normal approximation was used, the lower limit's value would be

$$R_L = \frac{52}{60 + 0.675 \times \sqrt{\frac{60 \times (60 - 53 + 1)}{51}}} = 0.838.$$

As can be seen, this approximation provides limits that are reasonably close to the exact values.

Example 13.6

Gas turbine engines are subjected to a 10-hour burn-in test after assembly. Out of 30 engines produced in a month, one engine failed to pass the test.

(a) Find a 95% lower confidence limit on engine reliability relative to this test using the exact equations with the F-distribution.

Solution:

$$\alpha = 0.05, n = 30, y = 29$$

$$R_L = \frac{y}{y + (n - y + 1) F_{\alpha, 2(n-y+1), 2y}}$$

$$= \frac{29}{29 + 2 \times F_{0.05, 4, 58}} = \frac{29}{29 + 2 \times 2.538} = 0.851039.$$

(b) Find a 95% upper confidence limit on engine reliability using the above test results.

Solution:

$$\alpha = 0.05, \quad n = 30, \quad y = 29$$

$$R_U = \frac{(y+1) \times F_{\alpha, 2(y+1), 2(n-y)}}{(n-y) + (y+1) F_{\alpha, 2(y+1), 2(n-y)}}$$

$$= \frac{30 F_{0.05, 60, 2}}{1 + 30 \times F_{0.05, 60, 2}} = \frac{30 \times 19.48}{1 + 30 \times 19.48} = 0.998292.$$

Example 13.7

The customer wants to demonstrate a reliability of 98% relative to this test with 96% confidence using success testing. What sample size should the test engineer use (with no failures) to demonstrate the customer's reliability requirements?

Solution:

$$n = \frac{\ln(1-C)}{\ln R} = \frac{\ln 0.04}{\ln 0.98} = 159.3289 \approx 160.$$

13.8.1 Success Testing

Sometimes in receiving inspection and engineering test labs a no-failure ($r = 0$ or $y = n$) test is specified. The goal is usually to ensure that a reliability level has been achieved at a specified confidence level. A special adaptation of the confidence limit formula can be derived for this situation. For the special case where $r = 0$ (i.e., no failures), the lower 100 $(1 - \alpha)$ percent confidence limit on the reliability is given by:

$$R_L = \alpha^{1/n} = (1-C)^{1/n}. \tag{13.16}$$

where α is the level of significance and n is the sample size (i.e. number of units placed on test).

If $C = (1 - \alpha)$, the desired confidence level (0.80, 0.90, etc.), then the necessary sample size to demonstrate a desired lower limit on reliability level, R_L, is

$$n = \frac{\ln(1-C)}{\ln R_L}. \tag{13.17}$$

For example, if $R_L = 0.80$ is to be demonstrated with a 90 percent confidence level,

$$n = \frac{\ln(0.10)}{\ln(0.80)} = 11. \tag{13.18}$$

Thus, 11 items must be tested, with no failures. This is frequently referred to as success testing.

13.9 Reliability Estimation and Confidence Limits for Exponential Distribution

Two types of tests are typically considered:

1. *Type 1 Censored Test.* The items are tested for a specified time, T, and then the testing is stopped.
2. *Type 2 Censored Test.* The test time is not specified, but the testing is stopped when a desired number of items fail.

Let us consider the situation when n items are being tested and the test is stopped as soon as r failures are observed ($r \leq n$). This is type 2 censoring, with nonreplacement of items. Let the observed failure times be, in order of magnitude,

$$0 = t_0 = < t_1 < t_2 < \cdots < t_{r-1} < t_r. \tag{13.19}$$

Then, making the transformation,

$$u_i = \begin{cases} nt_1, & \text{when } i = 0 \\ (n-i)(t_{i-1} - t_i) & \text{when } i = 1, 2, \ldots, r-1. \end{cases} \tag{13.20}$$

The (u_i, $i = 0, \ldots, r-1$) are independently and identically distributed with the common density function,

$$\left(\frac{1}{\theta}\right) e^{-u/\theta}. \tag{13.21}$$

The total time on test is given by

$$V(t_r) = \text{total time on test}$$

$$= \sum_{i=0}^{r-1} u_i \tag{13.22}$$

$$= \sum_{i=0}^{r-1} t_i + (n-r) t_r.$$

Then

$$\hat{\theta} = \frac{V(t_r)}{r} = \frac{1}{r}\left[\sum_{i=1}^{r} t_i + (n-r)t_r\right], \tag{13.23}$$

is the minimum variance unbiased estimator of θ. Since

$$V(t_r) = \sum_{i=0}^{r-1} u_i, \tag{13.24}$$

and the $\{u_i\}$ are independently distributed with a common exponential density function, it follows that $V(t_r)$ has a gamma distribution with parameters (θ, r). Hence,

$$2V(t_r)/\theta = 2\hat{\theta}r/\theta, \qquad (13.25)$$

is distributed as χ^2_{2r}.

The $100(1-\alpha)\%$ confidence limits on θ are given by:

$$P\left[\chi^2_{1-(\alpha/2),2r} \leq \frac{2\hat{\theta}r}{\theta} < \chi^2_{\alpha/2,2r}\right] = 1-\alpha \qquad (13.26)$$

or

$$\frac{2\hat{\theta}r}{\chi^2_{\alpha/2,2r}} \leq \theta \leq \frac{2\hat{\theta}r}{\chi^2_{1-(\alpha/2),2r}}. \qquad (13.27)$$

Life testing procedures are often used in a quality control context to detect the deviations of θ below some desired levels, such as θ_0. For a significance level of α, the probability of accepting H_0 is

$$P_\alpha = P\left(\frac{2r\hat{\theta}}{\theta_0} \leq \chi^2_{\alpha,2r} \Big| \theta = \theta_0\right) = 1-\alpha. \qquad (13.28)$$

The expected time to complete the test is given by

$$E(t_r) = \theta \sum_{i=1}^{r} \frac{1}{n-i+1}. \qquad (13.29)$$

Let

$\theta_0 =$ desired reliability goal for mean time between failures (MTBF)
$1-\alpha =$ probability of accepting items with a true MTBF of θ_0
$\theta_1 =$ alternative MTBF $(\theta_1 < \theta_0)$
$\beta =$ probability of accepting items with a true MTBF of θ_1.

With this information, reliability testing consists of putting n items on test and stopping the test when the number of failures is given by the smallest integer satisfying:

$$\frac{2T}{\chi^2_{\alpha/2,2(r+1)}} \leq \theta \leq \frac{2T}{\chi^2_{1-(\alpha/2),2r}}. \qquad (13.30)$$

Thus, when we know θ_0, θ_1, α, and β, we can compute the necessary value for r.

13.9 Reliability Estimation and Confidence Limits for Exponential Distribution

For the *type 1 censored test*, where r failures are observed on an interval of total test time, $V(t_r) = T$, the $100(1 - \alpha)$ percent confidence limits on θ are given by a modification of Equation 13.30:

$$\frac{2T}{\chi^2_{\alpha/2,2(r+1)}} \leq \theta \leq \frac{2T}{\chi^2_{1-(\alpha/2),2r}}. \tag{13.31}$$

Example 13.8

Sixteen thousand device-hours (total time on test) are accumulated in a failure-terminated test, with four failures.

(a) What are the upper and lower one-sided 90% confidence limits on MTBF?
(b) What are the one-sided 90% confidence limits on reliability for a 100-hour period?

Solution:
For this problem, we have

$$T = 16{,}000 \text{ hours}$$
$$C = 1 - \alpha = 0.90; \quad \alpha = 0.10; \quad \alpha/2 = 0.05; \quad 1 - \alpha/2 = 0.95$$
$$r = 4.$$

Therefore,

$$\text{MTBF}(l) = \frac{2(16{,}000)}{\chi^2_{0.10;8}} = \frac{32{,}000}{13.362} = 2395 \text{ hours}$$

$$\text{MTBF}(u) = \frac{2(16{,}000)}{\chi^2_{0.90;8}} = \frac{32{,}000}{3.490} = 9195 \text{ hours.}$$

If the lower and upper 0.90 confidence limits on the MTBF for the item are 2395 and 9195 hours, the lower and upper 0.90 confidence limits on its reliability for any 100-hour interval are:

$$R(l) = e^{\frac{-100}{2395}} = e^{-0.0417} = 0.9591$$

$$R(u) = e^{\frac{-100}{9195}} = e^{-0.0109} = 0.9891.$$

Example 13.9

Twenty-one thousand device-hours (total time on test) are accumulated in a time-terminated test, with seven failures. What are the upper and lower one-sided limits on MTBF with 0.99 confidence?

13 Reliability Estimation Techniques

Solution:
Here

$$T = 21{,}000 \text{ hours}$$
$$C = 1 - \alpha = 0.99; \quad \alpha = 0.01; \quad \alpha/2 = 0.005; \quad 1 - \alpha/2 = 0.995$$
$$r = 7.$$

Therefore,

$$\text{MTBF}(l) = \frac{2(21{,}000)}{\chi^2_{0.01;16}} = \frac{42{,}000}{32.000} = 1313 \text{ hours}$$

$$\text{MTBF}(u) = \frac{2(21{,}000)}{\chi^2_{0.99;14}} = \frac{42{,}000}{4.660} = 9013 \text{ hours}.$$

Example 13.10

Ten automotive air conditioning switches were cycled and observed for failure. Testing was suspended when the fourth failure occurred. Failed switches were not replaced. The failures occurred at the following cycles: 8900, 11,500, 19,200, and 29,300.

The assumption for this problem is that time to failure for the switch follows an exponential distribution with parameter θ.

Solution:

(a) Find the point estimator for the mean life (θ) of the switches.

This is a failure-truncated test. Hence the point estimator for θ is found by using Equation 13.23 and is

$$\hat{\theta} = \frac{8900 + 11{,}500 + 19{,}200 + 29{,}300 + (10 - 4)29{,}300}{4}$$

$$= \frac{244{,}700}{4} = 61{,}175.$$

(b) Find the 90% two-sided confidence limits on θ.

$$\chi^2_{0.05,8} = 15.507 \quad \chi^2_{0.95,8} = 2.733$$
$$\frac{2 \times 244{,}700}{15.507} \leq \theta \leq \frac{2 \times 244{,}700}{2.733}$$
$$31{,}559.9 \leq \theta \leq 179{,}071.$$

(c) The warranty for these switches is for 3000 cycles. Find the 95% one-sided upper confidence limit on the percent failures during the warranty period.

$$R_L(3000) = e^{-\frac{3000}{31{,}559.9}} = 0.9093$$

13.9 Reliability Estimation and Confidence Limits for Exponential Distribution

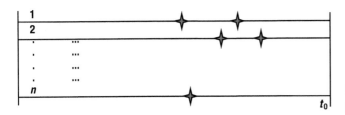

Figure 13.27 Time-truncated testing. Failure points are denoted by ✦.

Table 13.7 Values of parameter γ and dF for confidence limit calculations on MTBF

Type of Test	MTBF (*l*)		MTBF (*u*)	
	γ	dF	γ	dF
Two-sided failure terminated	$\alpha/2$	$2r$	$1 - \alpha/2$	$2r$
One-sided failure terminated	α	$2r$	$1 - \alpha$	$2r$
Two-sided time terminated	$\alpha/2$	$2r + 2$	$1 - \alpha/2$	$2r$
One-sided time terminated	α	$2r + 2$	$1 - \alpha$	$2r$
No failures observed	α	2	–	–

Note: *r* is the number of failures observed.

or

$$F_U(3000) = 0.0907.$$

We are 95% confident that for 3000 cycles, the percent of failures is less than 9.07%.

Figure 13.27 illustrates a time-truncated situation in which there are n test stands. Items are replaced by new items on the test stand when they fail and testing is stopped on every test stand at time t_0. Thus the total time on test is nt_0.

The confidence limits for MTBF assuming an exponential distribution can be summarized by:

$$\text{MTBF} = \frac{2T}{\chi^2_{r;dF}}, \qquad (13.32)$$

where T is the total time on test, and the values for the parameter γ and dF (degrees of freedom) for the χ^2 distribution can be obtained for different testing conditions from Table 13.7.

A common situation occurs when an estimate of the MTBF and the confidence interval around it is of interest, but no failures have occurred. You can still calculate a lower one-sided confidence limit, which is a conservative value for MTBF. Of course, there is no upper confidence limit. The lower confidence limit on MTBF is given by

$$\frac{2T}{\chi^2_{\alpha,2}} \leq \theta. \qquad (13.33)$$

291

13.10 Summary

The purpose of reliability estimation, demonstration, and testing is to determine whether a product has met certain reliability requirements with a stated statistical confidence level. Various tests are done throughout the life cycle of the product and are discussed. A five-step physics of failure approach was presented for the qualification of a SiP-type package in a drop-loading environment. Basic statistical concepts for estimation and confidence intervals are covered. Confidence intervals for both normal and binomial distributions are presented with examples. Finally, reliability estimation and confidence limits when the time to failure follows exponential distribution are discussed.

Problems

13.1 To get a 95% confidence interval on mean thermal conductivity, with an error less than 0.10 Btu/hr-ft-°F, what is the desired sample size? Assume $\sigma = 0.30$ Btu/hr-ft-°F at 100°F and 550 W.

13.2 An inspector found 10 defective keyboards from a sample of 300. Calculate the 95% confidence interval for the proportion of good units. What would be the 95% confidence interval for the proportion of bad units?

13.3 Gas turbine engines are subjected to a 10-hour burn-in test after assembly. Out of 40 engines produced in a month, three engines failed to pass the test. Develop a 95% two-sided symmetrical (both lower and upper) confidence limits on the engine reliability relative to this test using the F distribution.

13.4 The following data represent kilometers to failure for a set of vehicles:

43,000	27,200	10,600	12,400
27,000	4,100	200,000	18,200
68,000	40,500	109,000	14,200
46,000	2600	2400	24,500

(a) Estimate the MTBF.
(b) Set a 90% lower confidence limit on the 10% failure kilometer or B10 life.
(c) With 90% confidence, find the 2400 km lower limit on reliability.

13.5 For a test vehicle, major electrical failures occurred at the following kilometers:

63	17,393	23,128
114	18,707	24,145
14,820	19,179	33,832
16,105	22,642	34,345

The vehicle was driven a total of 36,000 kilometers.

(a) Estimate the MTBF.
(b) Determine the 90 percent two-sided confidence interval for the MTBF.
(c) Estimate the reliability function.
(d) Determine the 95 percent lower confidence limit for the 1,200 kilometer reliability.
(e) With 90 percent confidence estimate the kilometer at which 10 percent of the population will fail.

13.6 Twelve disk drives for computers were cycled and observed for failure. Testing was suspended when the third failure occurred. Failed disk drives were not replaced. The failures occurred at the following hours: 791; 909; 1522. The assumption for this problem is that time to failure for the disk drives follows an exponential distribution with parameter θ.

(a) Find the point estimate for the mean life θ of the disk drives.
(b) Find the 80% two-sided confidence limits on θ.
(c) The warranty for these disk drives is for 5000 hours. Find the 90% one-sided upper confidence limit on the percent failure during the warranty period.

13.7 What is accelerated testing? What is the purpose of doing accelerated testing? Explain with examples.

13.8 What is a qualification test? Can qualification tests reduce the over-stress failure of products?

13.9 What is HALT? Can HALT results be used to predict product reliability? Explain.

13.10 Explain with examples the steps in determining qualification testing conditions.

13.11 Describe how accelerated testing conditions and the accelerating factor are determined.

13.12 Discuss the different test data that can be used to assess reliability of parts. Which of these types of data is most appropriate for making reliability assessment?

14 Process Control and Process Capability

One traditional approach to manufacturing and addressing quality is to depend on production to make the product and on quality control to inspect the final product, screening out the items that do not meet the requirements of the customer. This detection strategy using after-the-fact inspection is highly uneconomical, since the rejected products have already been produced. A better strategy is to avoid waste by not producing unacceptable output in the first place, focusing on prevention rather than screening. Statistical process control (SPC) is an effective prevention strategy to manufacture products that will meet the requirements of the customer (Duncan 1986; Montgomery 2005; Shewhart 1931).

This chapter covers process control systems, the different types of variation and how they affect the process output, and control charts and their use. It also covers how control charts and statistical methods identify whether a problem is due to special or common causes and the benefits that can be expected from using the control charts. It also covers what is meant by a process being in statistical control and process capability and its various indices and their applications.

14.1 Process Control System

A process control system (see Figure 14.1) is a kind of feedback system. Four elements of that system are important to the discussions that will follow:

1. *The Process.* The process means the whole combination of people, equipment, input materials, methods, and environment that work together to produce output. The total performance of the process—the quality of its output and its productive efficiency—depends on the way the process has been designed and built and on the way it is operated. The rest of the process control system is useful only if it contributes to improved performance of the process.

Reliability Engineering, First Edition. Kailash C. Kapur and Michael Pecht.
© 2014 John Wiley & Sons, Inc. Published 2014 by John Wiley & Sons, Inc.

14 Process Control and Process Capability

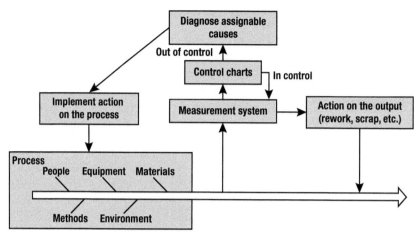

Figure 14.1 Process control system.

2. *Information about Performance.* Much information about the actual performance of the process can be learned by studying the process output. In a broad sense, process output includes not only the products that are produced, but also any intermediate outputs that describe the operating state of the process, such as temperatures, cycle times, and so on. If this information is gathered and interpreted correctly, it can indicate whether action is necessary to correct the process or the product. If timely and appropriate actions are not taken, however, any information-gathering effort is wasted.

3. *Action on the Process.* Action on the process is future oriented, because it is taken when necessary to prevent the production of nonconforming products. This action might consist of changes in the operation (e.g., operator training and changes to the incoming materials) or in the more basic elements of the process itself (e.g., the equipment, which may need rehabilitation, or the design of the process as a whole, which may be vulnerable to changes in shop temperature or humidity).

4. *Action on the Output.* Action on the output is past oriented, because it involves detecting out-of-specification output already produced. Unfortunately, if current output does not consistently meet customer requirements, it may be necessary to sort all products and to scrap or rework any nonconforming items. This must continue until the necessary corrective action on the process has been taken and verified, or until the product specifications have been changed.

It is obvious that inspection followed by action only on the output is a poor substitute for using an effective process performance from the start. Therefore, the discussions that follow focus on gathering process information and analyzing it so that action can be taken to correct the process itself.

Process control plays a very important role in the effort for improvement. When a process is well controlled, analysis and improvement naturally result; and when we try to make an improvement, we naturally come to understand the importance of control. Breakthroughs occur only after achieving control. Without process control, we cannot set appropriate standards or identify needed improvements. Improvement can only be achieved through process analysis.

14.1.1 Control Charts: Recognizing Sources of Variation

A control chart is a type of trend chart (displaying data over time) with statistically determined upper and lower control limits; it is used to determine if the process is under control. A process is said to be under control when the variation within the process is consistently only random and within predictable (control) limits. Random variation results from the interaction of the steps within a process. When the performance falls outside the control limits, assignable variation may be the cause. Assignable variation can be attributed to a number of special causes. A control chart will help determine what type of variation is present within the process. Control charts are also used to assess process variations and their sources and to monitor, control, and improve process performance over time. A control chart focuses attention on detecting and monitoring process variation over time. Using one can allow us to distinguish special causes of variation from common causes of variation. Control charts can serve as an ongoing control tool and help improve a process to perform consistently and predictably. They also provide a common language for discussing process performance.

14.1.2 Sources of Variation

As discussed earlier, the sources of variability in a process are classified into two types: chance or random causes and assignable causes. Chance causes, or common causes, are sources of inherent variability, which cannot be removed easily from the process without fundamental changes in the process itself. Assignable causes, or special causes, arise in somewhat unpredictable fashion, such as operator error, material defects, or machine failure. The variability due to assignable causes is comparably larger than that for chance causes, and can send the process out of control. Table 14.1 compares the two sources of variation, including some examples.

14.1.3 Use of Control Charts for Problem Identification

Control charts by themselves do not correct problems. They indicate that something is wrong and requires corrective action. Assignable causes due to a change in

Table 14.1 Sources of variation

Common or chance causes	Special or assignable causes
Include many individual causes.	Include one or just a few individual causes.
Any one chance cause results in only a minute amount of variation. (However, many chance causes together may result in a substantial amount of variation.)	Any one assignable cause can result in a large amount of variation.
As a practical matter, chance variation cannot be economically eliminated—the process may have to be changed to reduce variability.	The presence of assignable variation can be detected (by control charts), and action to eliminate the causes is usually economically justified.
Examples: ■ Slight variations in raw materials ■ Slight vibrations of a machine ■ Lack of human perfection in reading instruments or setting controls	Examples: ■ Batch of defective raw materials ■ Faulty setup ■ Untrained operator

manpower, materials, machines, or methods, or a combination of these, can cause the process to go out of control.

Assignable causes relating to manpower:

- New or wrong person on the job
- Careless workmanship and attitudes
- Incorrect instructions
- Domestic, personal problems.

Assignable causes relating to materials:

- Improper work handling
- Stock too hard or too soft
- Wrong dimensions
- Contamination, dirt, and so on
- Improper flow of materials.

Assignable causes relating to machines or methods:

- Dull tools
- Poor housekeeping
- Inaccurate machine adjustment
- Improper machine tools, jigs, fixtures
- Improper speeds, feeds, and so on
- Inadequate maintenance
- Worn or improperly placed locators.

When assignable causes are present, as shown in Figure 14.2, the probability of nonconformance may increase, and the process quality deteriorates significantly. The

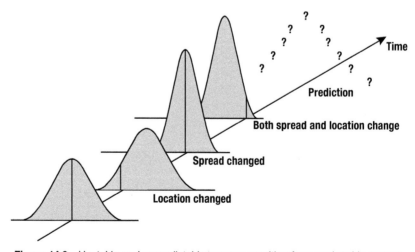

Figure 14.2 Unstable and unpredictable process resulting from assignable causes.

eventual goal of SPC is to improve quality by reducing variability in the process. As one of the primary SPC techniques, the control chart can effectively detect the variation due to assignable causes and reduce process variability if the identified causes can be eliminated from the process.

SPC techniques aim to detect changes over time in the parameters (e.g., mean and standard deviation) of the underlying distribution for the process. In general, the statistical process control problem can be described as follows (Stoumbos et al. 2000). Let X denote a random variable for a quality characteristic with the probability density function $f(x; \theta)$, where θ is a set of parameters. If the process is operating with $\theta = \theta_0$, it is said to be in statistical control; otherwise, it is out of control. The value of θ_0 is not necessarily equal to the target (or ideal) value of the process. Due to experimental design and process adjustment techniques, a process is assumed to start with the in-control state (Box and Luceno 1997; Hicks and Turner 1999; Montgomery 2001). After a random length of time, variability in the process will possibly cause deterioration of or a shift in the process. This shift can be reflected by a change in θ from the value of θ_0; then the process is said to be out of control. Therefore, the basic goal of control charts is to detect changes in θ that can occur over time.

A process is said to be operating in statistical control when the only source of variation is common causes. The status of statistical control is obtained by eliminating special causes of excessive variation one by one.

Process capability is determined by the total variation that comes from common causes. A process must first be brought into statistical control before its capability to meet specifications can be assessed. We will discuss the details of process capability analysis in later sections.

14.2 Control Charts

The basic concept of control charts was proposed by Walter A. Shewhart of the Bell Telephone Laboratories in the 1920s; this was the formal beginning of statistical quality control. The effective use of the control chart involves a series of process improvement activities. For a process variable of interest, someone must observe data from the process over time, monitor the process, and apply a control chart to detect process changes. When the control chart signals the possible presence of an assignable cause, effort should be made to diagnose the assignable cause(s), implement corrective actions to remove them so as to reduce variability, and improve the process quality. The long history of control charting application in many industries has proven the technique's effectiveness in improving productivity, preventing defects, and providing information about diagnostic and process capability.

Control charts must be investigated in order to identify in-control and out-of-control processes and detect common causes and special causes of the out-of-control state. In interpreting control charts, it is important to note that attribute data control charts measure variation among samples. Variations among subgroups over time can be measured by the first variable data control chart, while variations within subgroups over time can be measured by a second chart.

Also, the chart analyst should determine if the process mean (center line) is where it should be relative to production specifications or objectives. If not, then either the process or the objectives have changed. To distinguish between common causes and

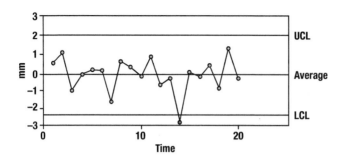

Figure 14.3 A typical control chart (X-bar chart).

special causes, data relative to control limits must be analyzed. Upper and lower control limits (UCL/LCL) are not specification limits and do not imply a value judgment (good, bad, and marginal) about a process. The judgment is derived with other tools, such as benchmarking "stretch" goals. UCL/LCL is only a statistical tool. If a process is consistently performing above the command UCL, the reason must be discovered to enable process improvements. A typical control chart is given in Figure 14.3.

The basic model for Shewhart control charts consist of a center line, an upper control limit (UCL), and a lower control limit (LCL) (ASTM Publication STP-15D 1976).

$$\text{UCL} = \mu_s + L\sigma_s$$
$$\text{Center line} = \mu_s \quad (14.1)$$
$$\text{LCL} = \mu_s - L\sigma_s,$$

where μ_s and σ_s are the mean and standard deviation of the sample statistic, such as the sample mean (X-bar chart), sample range (R chart), and sample proportion defective (p chart). $L\sigma_s$ is the distance of the control limits from the center line, and it is most often set at three times the standard deviation of the sample statistic. Constructing a control chart requires specifying the sample size and sampling frequency. The common wisdom is to take smaller samples at short intervals or larger samples at longer intervals, so that the sampling effort can be allocated economically. An important concept related to sampling scheme is the rational subgroup approach, recommended by Shewhart. In order to maximize the detection of assignable causes between samples, the rational subgroup approach takes samples in a way that the within-sample variability is only due to common causes, while the between-sample variability should indicate assignable causes in the process. Further discussion of the rational subgroup can be found in Montgomery (2005).

An out-of-control signal is given when a sample statistic falls beyond the control limits, or when a nonrandom pattern presents. Western Electric rules are used to identify the nonrandom pattern in the process. According to Western Electric rules (Western Electric 1956), a process is considered out of control if any of the rules given in Table 14.2 are met. More decision rules or sensitizing rules can be found in Montgomery's textbook.

Figure 14.5 gives some examples of an out-of-control condition based on the guidelines given in Table 14.2 and Figure 14.4. The measurements of quality characteristics are typically classified as attributes or variables. Continuous measurements, such as length, thickness, or voltage, are variable data. Discrete measurements, such as the

14.2 Control Charts

Table 14.2 Rules to detect out-of-control processes

1. One or more points fall outside control limits
2. Two out of three consecutive points are in zone A
3. Four out of five consecutive points are in zone A or B
4. Nine consecutive points are on one side of the average
5. Six consecutive points are increasing or decreasing
6. Fourteen consecutive points alternate up and down
7. Fifteen consecutive points within zone C

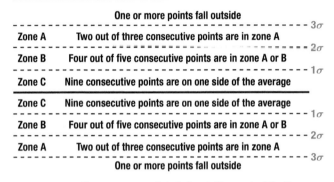

Figure 14.4 Guidelines to distinguish out-of-control process.

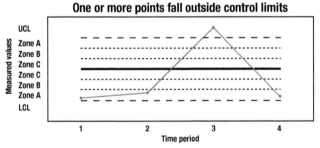

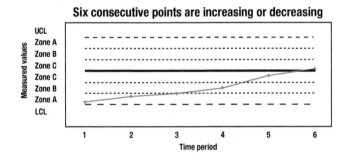

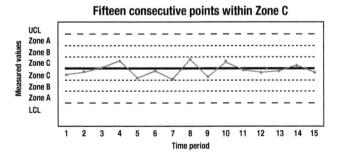

Figure 14.5 Examples of out-of-control situations.

14 Process Control and Process Capability

Table 14.3 The most commonly used Shewhart control charts

Symbol	Description	Sample size
Variable charts		
X-bar and R	The average (mean) and range of measurements in a sample	Must be constant
X-bar and S	The average (mean) and standard deviation of measurements in a sample	May be variable
Attributes charts		
p	The percent of defective (nonconforming) units in a sample	May be variable
np	The number of defective (nonconforming) units in a sample	Must be constant
c	The number of defects in a sample	Must be constant
u	The number of defects per unit	May be variable

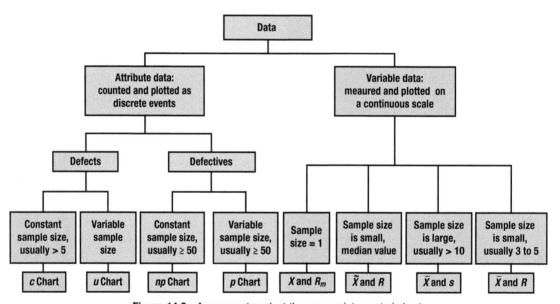

Figure 14.6 A process to select the appropriate control chart.

number of defective units or number of nonconformities per unit, are attributes. The most commonly used Shewhart control charts for both attributes and variables are summarized in Table 14.3.

To draw a control chart a series of guidelines must be considered. Different kinds of control charts can be selected, considering different kinds of data. Figure 14.6 and Figure 14.7 show a guideline to select the control.

To construct a control chart, follow the steps shown in Figure 14.8. To calculate appropriate statistics, it is necessary to know the method being used and the constants for that method. Constants and different formulae used in construction control charts are shown in Table 14.4 and Table 14.5 for variable and attribute data, respectively. Table 14.6 and Table 14.7 give the values of the constants needed for the variable control charts.

14.2 Control Charts

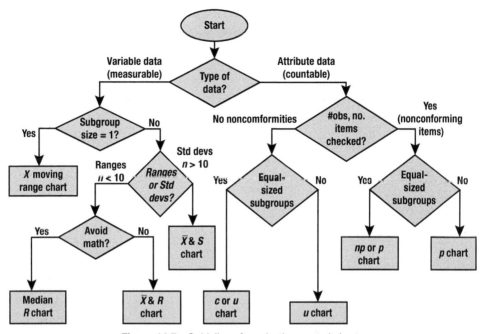

Figure 14.7 Guidelines for selecting control charts.

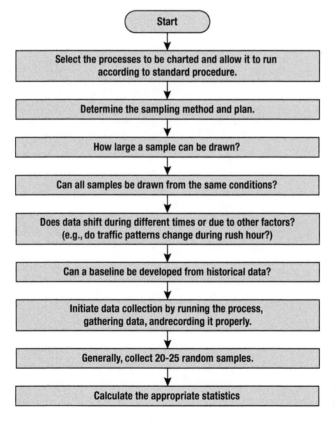

Figure 14.8 Ten steps in control chart construction.

14 Process Control and Process Capability

Table 14.4 Variable data table

Type control chart	Sample size	Central line	Control limits
Average and range X-bar and R	<10, but usually 3–5	$\bar{\bar{X}} = \dfrac{(\bar{X}_1 + \bar{X}_2 + \cdots + \bar{X}_k)}{k}$ $\bar{R} = \dfrac{(R_1 + R_2 + \cdots + R_k)}{k}$	$UCL_{\bar{x}} = \bar{\bar{X}} + A_2\bar{R}$ $LCL_{\bar{x}} = \bar{\bar{X}} - A_2\bar{R}$ $UCL_R = D_4\bar{R}$ $LCL_R = D_3\bar{R}$
Average and standard deviation X-bar and s	Usually > or = 10	$\bar{\bar{X}} = \dfrac{(\bar{X}_1 + \bar{X}_2 + \cdots + \bar{X}_k)}{k}$ $\bar{S} = \dfrac{(S_1 + S_2 + \cdots + S_k)}{k}$	$UCL_{\bar{x}} = \bar{\bar{X}} + A_3\bar{S}$ $LCL_{\bar{x}} = \bar{\bar{X}} - A_3\bar{S}$ $UCL_S = B_4\bar{S}$ $LCL_S = B_3\bar{S}$
Median and range $\tilde{X}$ and R	<10, but usually 3–5	$\bar{\tilde{X}} = \dfrac{(\tilde{X}_1 + \tilde{X}_2 + \cdots + \tilde{X}_k)}{k}$ $\bar{R} = \dfrac{(R_1 + R_2 + \cdots + R_k)}{k}$	$UCL_{\tilde{x}} = \bar{\tilde{X}} + \tilde{A}_2\bar{R}$ $LCL_{\tilde{x}} = \bar{\tilde{X}} - \tilde{A}_2\bar{R}$ $UCL_R = D_4\bar{R}$ $LCL_R = D_3\bar{R}$
Individuals and moving range	1	$\bar{X} = \dfrac{(X_1 + X_2 + \cdots + X_k)}{k}$ $R_m = \|X_{i-1} - X_i\|$ $\bar{R}_m = \dfrac{(R_1 + R_2 + \cdots + R_{k-1})}{k-1}$	$UCL_x = \bar{X} + E_2\bar{R}_m$ $LCL_x = \bar{X} - E_2\bar{R}_m$ $UCL_{R_m} = D_4\bar{R}_m$ $LCL_{R_m} = D_3\bar{R}_m$

Table 14.5 Attribute data table

Type/control chart	Sample size	Central line	Control limits
Fraction defective p Chart	Variable, usually > or = 50	For each subgroup: $p = np/n$ For all subgroups: $\bar{p} = \dfrac{\Sigma np}{\Sigma n}$	$UCL_p = \bar{p} + 3\sqrt{\dfrac{\bar{p}(1-\bar{p})}{n}}$ $LCL_p = \bar{p} - 3\sqrt{\dfrac{\bar{p}(1-\bar{p})}{n}}$
Number defective np Chart	Constant, usually > or = 50	For each subgroup: np = no. of defective units For all subgroups: $n\bar{p} = \dfrac{\Sigma np}{k}$	$UCL_{np} = n\bar{p} + 3\sqrt{n\bar{p}(1-\bar{p})}$ $LCL_{np} = n\bar{p} - 3\sqrt{n\bar{p}(1-\bar{p})}$
Number of defects c Chart	Constant	For each subgroup: c = no. of defects For all subgroups: $\bar{c} = \dfrac{\Sigma c}{k}$	$UCL_c = \bar{c} + 3\sqrt{\bar{c}}$ $LCL_c = \bar{c} - 3\sqrt{\bar{c}}$
Number of defects per unit u Chart	Variable	For each subgroup: $u = c/n$ For all subgroups: $\bar{u} = \dfrac{\Sigma c}{\Sigma n}$	$UCL_u = \bar{u} + 3\sqrt{\dfrac{\bar{u}}{n}}$ $LCL_u = \bar{u} - 3\sqrt{\dfrac{\bar{u}}{n}}$

14.2 Control Charts

Table 14.6 Table of constants for control charts

Sample size n	X-bar and R-bar chart			X-bar and s-bar chart			
	A_2	D_3	D_4	A_3	B_3	B_4	C_4
2	1.880	0	3.267	2.659	0	3.267	0.7979
3	1.023	0	2.574	1.954	0	2.568	0.8862
4	0.729	0	2.282	1.628	0	2.266	0.9213
5	0.577	0	2.114	1.427	0	2.089	0.9400
6	0.483	0	2.004	1.287	0.030	1.970	0.9000
7	0.419	0.076	1.924	1.182	0.118	1.882	0.9594
8	0.373	0.136	1.864	1.099	0.184	1.815	0.9650
9	0.337	0.184	1.816	1.032	0.239	1.761	0.9693
10	0.308	0.223	1.777	0.975	0.284	1.716	0.9727

Table 14.7 Table of constants for charts

Sample size n	X-median and R-bar chart			X and R_m chart			
	$\tilde{A}_2$	D_3	D_4	E_2	D_3	D_4	d_2
2	–	0	3.267	2.659	0	3.267	1.128
3	1.187	0	2.574	1.772	0	2.574	1.693
4	–	0	2.282	1.457	0	2.282	2.059
5	0.691	0	2.114	1.290	0	2.114	2.326
6	–	0	2.004	1.184	0	2.004	2.534
7	0.509	0.076	1.924	1.109	0.076	1.924	2.704
8	–	0.136	1.864	1.054	0.136	1.864	2.847
9	0.412	0.184	1.816	1.010	0.184	1.816	2.970
10	–	0.223	1.777	0.975	0.223	1.777	3.078

After identifying an out-of-control process, a series of actions must be taken in order to bring the process back to under control status. The following are common questions for investigating an out-of-control process. A team should consider any "yes" answer to the question as a potential source of a special cause:

- Are there differences in the measurement accuracy of instruments/methods used?
- Are there differences in the methods used by different personnel?
- Is the process affected by the environment—for example, temperature, and humidity?
- Has there been a significant change in the environment?
- Is the process affected by predictable conditions, such as tool wear?
- Were any untrained personnel involved in the process at the time?
- Has the source of input for the process changed (e.g., raw materials)?
- Is the process affected by employee fatigue?
- Has there been a change in policies or procedures (e.g., maintenance procedures)?
- Is the process adjusted frequently?

- Did the samples come from different parts of the process/shifts/individuals?
- Are employees afraid to report "bad news"?

14.2.1 Control Charts for Variables

When a quality characteristic is measured as a variable, both the process mean and standard deviation must be monitored. For grouped data, use the X-bar chart to detect the process mean shift (between-group variability), and the R or S chart to monitor the process variation (within-group variability). The control limits of each chart are constructed based on the Shewhart model in Equation 14.1. When using X-bar, R, and S charts, assume that the underlying distribution of the quality characteristic is normal, and that the observations exhibit no correlation over time. If the quality characteristic is extremely nonnormal or the observations are autocorrelated, other control charts, such as the exponentially weighted moving average chart (EWMA) or the time series model (ARIMA), may be used instead.

In practice, the parameters of the underlying distribution of a quality characteristic are not known. The process mean and standard deviation are estimated based on the preliminary data. It can be shown that an unbiased estimate of the standard deviation is $\hat{\sigma} = \bar{s}/c_4$, where s-bar is the average sample standard deviation. A more convenient approach in quality control applications is the range method, where the range of the sample, R, is used to estimate the standard deviation, and is obtained as $\hat{\sigma} = \bar{R}/d_2$ where R-bar is the average value of the sample ranges. The resulting control charts using different estimators of standard deviation are the R chart and the S chart, respectively.

14.2.2 X-Bar and R Charts

When the sample size is not very large ($n < 10$), the X-bar and R charts, due to their simplicity of application, are widely used to monitor variable quality characteristics. In order to use the basic Shewhart model for X-bar and R charts, we need to estimate $\mu_{\bar{X}}$ and $\sigma_{\bar{X}}$, μ_R and σ_R first.

It is obvious that we can use the grand average to estimate μ_X and μ_R, that is, $\hat{\mu}_{\bar{X}} = \bar{\bar{X}}$ and $\hat{\mu}_R = \bar{R}$. Using the range method, $\hat{\sigma}_{\bar{X}} = \hat{\sigma}/\sqrt{n} = \bar{R}/(d_2\sqrt{n})$ and $\hat{\sigma}_R = d_3\hat{\sigma} = d_3\bar{R}/d_2$. The control limits for X-bar and R charts are

$$LCL = \bar{\bar{x}} - A_2\bar{R} \qquad LCL = D_3\bar{R}$$
$$CL = \bar{\bar{x}} \quad \text{and} \quad CL = \bar{R}$$
$$ULC = \bar{\bar{x}} + A_2\bar{R} \qquad UCL = D_4\bar{R},$$

respectively, where

$$A_2 = \frac{3}{d_2\sqrt{n}}, \quad D_3 = 1 - \frac{3d_3}{d_2}, \quad \text{and} \quad D_4 = 1 + \frac{3d_3}{d_2}.$$

The values of d_2, d_3, A_2, D_3, and D_4 can be obtained from most books on control charts for n up to 25 (Montgomery 2005). For a sample size up to 10, these values are given in Table 14.6 and Table 14.7. Normally, the preliminary data used to establish the control limits is about 20–25 samples, with a sample size of 3–5. The

14.2 Control Charts

Table 14.8 Organize data in a chart

Group no.	A	B	C	D	E	X-bar	R
1	1.4	1.2	1.3	1.4	1.2		
2	1.3	1.2	1.3	1.5	1.3		
3	1.7	1.3	1.4	1.2	1.2		
4	1.4	1.2	1.3	1.3	1.4		
5	1.5	1.1	1.7	1.3	1.3		
6	1.8	1.2	1.5	1.5	1.4		
7	1.5	1.2	1.3	1.3	1.2		
8	1.7	1.7	1.2	1.2	1.1		
9	1.8	1.8	17	1.8	1.5		
10	1.1	1.2	1.8	1.6	1.3		
11	1.2	1.3	1.4	1.4	1.4		
12	1.3	1.9	1.9	1.5	1.5		
13	1.4	1.8	1.7	1.1	1.3		
14	1.8	1.9	1.5	1.4	1.4		
15	1.1	1.3	1.1	1.8	1.5		
16	1.8	1.9	1.7	1.6	1.3		
17	1.2	1.4	1.3	1.2	1.4		
18	1.1	1.1	1.7	1.2	1.3		
19	1.8	1.6	1.5	1.7	1.8		
20	1.1	1.3	1.3	1.4	1.3		
Total							

established control limits are then used to check if the preliminary samples are in control. The R chart (or S chart) should be checked first to ensure that the process variability is in statistical control, and then the X-bar chart is checked for the process mean shift. Once a set of reliable control limits is constructed, they can be used for process monitoring.

14.2.2.1 X-Bar and R Chart Example In this example, information was needed to analyze the weight of a specific part made in a machine shop. The machine shop sampled the parts at twenty different times (groups), and each group had five measurements (samples), as given in Table 14.8. Since there are variable data with a constant sample size $= 5$, choose the X-bar and R charts.

Compute the Mean and Range for Each Group The mean (X-bar) = the sum of the samples within the group divided by the group size. For example, group 1 has an $=$ $(1.4 + 1.2 + 1.3 + 1.4 + 1.2) / 5 = 1.3$. The range ($R$) = the difference between the largest observation within a group and the smallest observation within that group. The R-value for group 1 is $R_1 = (1.4 - 1.2) = 0.2$. The computed values of X-bar and R are given in Table 14.9.

Compute the Average Mean and Average Range The overall average (X-bar) = the total/total number of groups $= 28.54/20 = 1.427$. This is also called the grand average. This is used as the centerline for the chart. The average of all group ranges (R-bar) = the total R/total number of groups $= 9.0/20 = 0.45$ is used as the centerline (average) for the range chart.

Table 14.9 Add Calculated Data to the Chart

Group no.	A	B	C	D	E	X-bar	R
1	1.4	1.2	1.3	1.4	1.2	1.30	0.2
2	1.3	1.2	1.3	1.5	1.3	1.32	0.3
3	1.7	1.3	1.4	1.2	1.2	1.36	0.5
4	1.4	1.2	1.3	1.3	1.4	1.32	0.2
5	1.5	1.1	1.7	1.3	1.3	1.38	0.6
6	1.8	1.2	1.5	1.5	1.4	1.48	0.6
7	1.5	1.2	1.3	1.3	1.2	1.30	0.3
8	1.7	1.7	1.2	1.2	1.1	1.38	0.3
9	1.8	1.8	1.7	1.8	1.5	1.72	0.3
10	1.1	1.2	1.8	1.6	1.3	1.40	0.7
11	1.2	1.3	1.4	1.4	1.4	1.34	0.2
12	1.3	1.9	1.9	1.5	1.5	1.62	0.6
13	1.4	1.8	1.7	1.1	1.3	1.46	0.7
14	1.8	1.9	1.5	1.4	1.4	1.60	0.5
15	1.1	1.3	1.1	1.8	1.5	1.36	0.7
16	1.8	1.9	1.7	1.6	1.3	1.66	0.6
17	1.2	1.4	1.3	1.2	1.4	1.30	0.2
18	1.1	1.1	1.7	1.2	1.3	1.28	0.6
19	1.8	1.6	1.5	1.7	1.8	1.68	0.3
20	1.1	1.3	1.3	1.4	1.3	1.28	0.3
						28.54	9.0

Determine Control Limits

$$\text{UCL}_{\bar{X}} = \bar{\bar{X}} + A_2\bar{R} = 1.427 + (0.577 \times 0.45) = 1.687$$
$$\text{LCL}_{\bar{X}} = \bar{\bar{X}} - A_2\bar{R} = 1.427 - (0.577 \times 0.45) = 1.168.$$

About 99.73% (3 sigma limits) of the average values should fall between 1.168 and 1.687.

$$\text{UCL}_R = D_4\bar{R} = 2.114 \times 0.45 = 0.951$$
$$\text{LCL}_R = D_3\bar{R} = 0 \times 0.45 = 0.$$

About 99.73% (3 sigma limits) of the sample ranges should fall between 0 and 0.951. The X-bar chart is shown in Figure 14.9, and the R chart is shown in Figure 14.10. This shows that the average based on subgroup 9 is outside the upper control limit, and hence the process is out of control. We have to investigate the reasons for this situation and find the assignable causes and eliminate or remove them from the system.

14.2.3 Moving Range Chart Example

Now, we present an example of the moving range chart. In this example, information was needed to analyze the weights of a specific part made in the machine shop. Only one sample existed per observation. Since there are variable data and only one unit

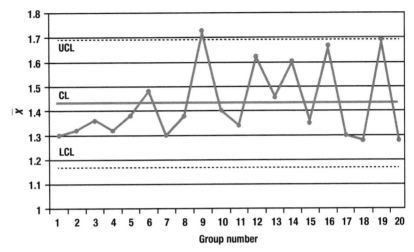

Figure 14.9 X-bar chart.

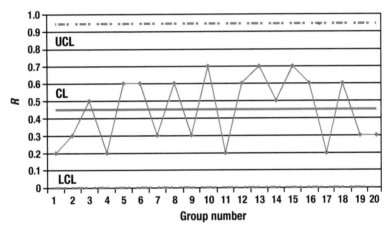

Figure 14.10 Range chart.

in each sample, the moving range chart is most appropriate. The data are given in Table 14.10. We also use the symbol R_m for moving range (MR) and they are used interchangeably in this chapter.

14.2.3.1 Compute the Moving Range (MR) MR = $|R_n - R_{n-1}|$ = Absolute value of the difference between consecutive range values. It is also known as the two-sample moving range (the most common form of moving range.) There is no range for the first observation. The first MR value works out to $MR_1 = |1.4 - 1.3| = 0.1$. The computed values of the MR are given in Table 14.11.

14.2.3.2 Compute the Average Mean and Group Range The overall average (X-bar) = sum of the measurements/number of observations = 28.90/20 = 1.45. Here, X-bar is also called the grand average, and X-bar is used as the centerline for the X chart. The average of all group ranges MR-bar = Total MR/number of ranges = 6.9/19 = 0.36. MR-bar is used as the centerline (average) for the MR chart.

309

Table 14.10 Organize data in a chart

Observation no.	Sample (X)	MR
1	1.4	
2	1.3	
3	1.7	
4	1.4	
5	1.5	
6	1.8	
7	1.5	
8	1.7	
9	1.8	
10	1.1	
11	1.2	
12	1.3	
13	1.4	
14	1.8	
15	1.1	
16	1.8	
17	1.2	
18	1.0	
19	1.8	
20	1.1	
Total	28.0	

Table 14.11 Add calculated data to the chart

Observation no.	Sample (X)	MR
1	1.4	N/A
2	1.3	0.1
3	1.7	0.4
4	1.4	0.3
5	1.5	0.1
6	1.8	0.3
7	1.5	0.3
8	1.7	0.2
9	1.8	0.1
10	1.1	0.7
11	1.2	0.1
12	1.3	0.1
13	1.4	0.1
14	1.8	0.4
15	1.1	0.7
16	1.8	0.7
17	1.2	0.6
18	1.0	0.2
19	1.8	0.8
20	1.1	0.7
Total	28.9	6.9

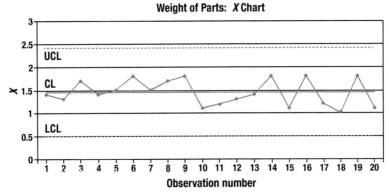

Figure 14.11 X chart.

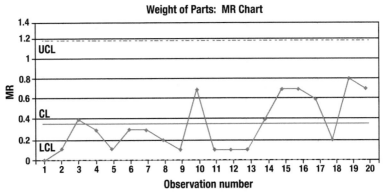

Figure 14.12 MR chart.

14.2.3.3 Determine Control Limits

$$\text{UCL}_X = \bar{X} + (E_2 \times \overline{MR}) = 1.45 + (2.659 \times 0.36) = 2.41$$
$$\text{LCL}_X = \bar{X} - (E_2 \times \overline{MR}) = 1.45 - (2.659 \times 0.36) = 0.49$$

$$\text{UCL}_{MR} = D_4 \times \overline{MR} = 3.267 \times 0.36 = 1.18$$
$$\text{LCL}_{MR} = D_3 \times \overline{MR} = 0 \times 0.36 = 0.$$

The sample size used to obtain the values for E_2, D_3, and D_4 is 2 in this case, since we are using a two-sample moving range. If a three-sample moving range is used, the number of ranges will reduce to 18, and the values of the constants used will change accordingly. The X chart is given in Figure 14.11, and the MR chart is given in Figure 14.12.

14.2.4 X-Bar and S Charts

When the sample size is relatively large ($n > 10$), or the sample size is variable, the X-bar and S charts are preferred to X-bar and R charts. To construct the control limits, first estimate the mean and standard deviation of X-bar and S—that is, $\mu_{\bar{x}}$ and $\sigma_{\bar{x}}$, μ_S and σ_S. We have $\hat{\mu}_{\bar{x}} = \bar{\bar{x}}$ and $\hat{\mu}_S = \bar{S}$. Using $\hat{\sigma} = s/c_4$, we have $\hat{\sigma}_{\bar{x}} = \hat{\sigma}/\sqrt{n} = \bar{s}/(c_4\sqrt{n})$, and $\hat{\sigma}_s = \bar{s}\sqrt{1-c_4^2}/c_4$. Therefore, the control limits for X-bar and S charts are

$$\text{LCL} = \bar{\bar{x}} - A_3 \bar{S} \qquad \text{LCL} = B_3 \bar{S}$$
$$\text{CL} = \bar{\bar{x}} \quad \text{and} \quad \text{CL} = \bar{S}$$
$$\text{ULC} = \bar{\bar{x}} + A_3 \bar{S} \qquad \text{UCL} = B_4 \bar{S},$$

respectively, where

$$A_3 = \frac{3}{c_3 \sqrt{n}}, \quad B_3 = 1 - \frac{3}{c_4}\sqrt{1-c_4^2}, \quad \text{and} \quad B_4 = 1 + \frac{3}{c_4}\sqrt{1-c_4^2}.$$

The values of c_4, A_3, B_3, and B_4 can be obtained from most books on control charts for n up to 25 (see Table 14.6).

14.2.5 Control Charts for Attributes

When quality characteristics are expressed as attribute data, such as defective or conforming items, control charts for attributes are established. Attribute charts can handle multiple quality characteristics jointly because the unit is classified as defective if it fails to meet the specifications on one or more characteristics. The inspection of samples for attribute charts is usually cheaper because it requires less precision. Attribute charts are particularly useful in quality improvement efforts where numerical data are not easily obtained, such as service industrial and health care systems. In the context of quality control, the attribute data include the proportion of defective items and the number of defects per item. A defective unit may have one or more defects due to nonconformance to standards with regard to one or more quality characteristics. Nevertheless, a unit with several defects may not necessarily be classified as a defective unit. This requires two different types of attribute charts: control charts for the proportion defective (p chart and np chart), and control charts for the number of defects (c chart and u chart).

14.2.6 p Chart and np Chart

The proportion defective is defined as the ratio of the number of defective units to the total number of units in a population. We usually assume that the number of defective units in a sample is a binomial variable—that is, each unit in the sample is produced independently, and the probability that a unit is defective is a constant, p. Using preliminary samples, we can estimate the defective rate—that is, $\bar{p} = \sum_{i=1}^{m} D_i / mn$—where D_i is the number of defective units in sample i, n is the sample size, and m is the number of samples taken. The formula used to calculate control limits is then

$$\text{UCL}_p = \bar{p} + 3\sqrt{\frac{\bar{p}(1-\bar{p})}{n}}$$
$$\text{Centerline} = \bar{p}$$
$$\text{LCL}_p = \bar{p} - 3\sqrt{\frac{\bar{p}(1-\bar{p})}{n}}.$$

14.2 Control Charts

Sometimes, it may be easier to interpret the number defective instead of the proportion defective. That is why the *np* chart came into use:

$$\text{UCL} = n\bar{p} + 3\sqrt{n\bar{p}(1-\bar{p})}$$
$$\text{Centerline} = n\bar{p}$$
$$\text{LCL} = n\bar{p} - 3\sqrt{n\bar{p}(1-\bar{p})}.$$

The developed trial control limits are then used to check if the preliminary data are in statistical control, and assignable causes may be identified and removed if a point is out of control. As the process improves, we expect a downward trend in the *p* or *np* control chart.

14.2.7 *np* Chart Example

In this example, 10 weeks of defect data have been collected with a sample size of 50. Since we have attribute data with a constant sample size, we use the *np* chart. The data are given in Table 14.12.

14.2.7.1 Determine the Averages The average percent defective = *p*-bar = total defectives/totaled sampled.

$$\bar{p} = \frac{46}{(n)(\text{weeks})} = \frac{46}{(50)(10)} = 0.092.$$

The grand average = $n \times p$-bar (centerline) also = total defectives/total number of samples.

$$n\bar{p} = (50)(0.092) = 4.6$$
$$n\bar{p} = 46/10 = 4.6.$$

Table 14.12 Organize data in a chart

Week no.	Number of defectives
1	9
2	7
3	4
4	2
5	4
6	5
7	2
8	3
9	5
10	5
Total	46

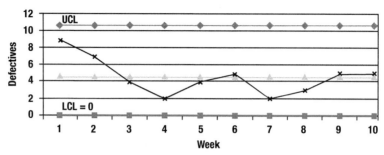

Figure 14.13 *np* chart.

14.2.7.2 Determine Control Limits

$$\mathrm{UCL} = n\bar{p} + 3\sqrt{n\bar{p}(1-\bar{p})} = 4.6 + 3\sqrt{4.6(1-0.092)} = 10.731$$
$$\mathrm{LCL} = n\bar{p} - 3\sqrt{n\bar{p}(1-\bar{p})} = 4.6 - 3\sqrt{4.6(1-0.092)} = 0.$$

Note: Since LCL is less than zero, use zero. The *np* control chart is shown in Figure 14.13.

14.2.8 c Chart and u Chart

Control charts for monitoring the number of defects per sample are constructed based on the Poisson distribution. With this assumption of a reference distribution, the probability of occurrence of a defect at any area is small and constant, the potential area for defects is infinitely large, and defects occur randomly and independently. If the average occurrence rate per sample is a constant, c, both the mean and the variance of the Poisson distribution are the constant c. Therefore, the parameters in the c chart for the number of defects are

$$\mathrm{LCL} = c - 3\sqrt{c}$$
$$\mathrm{CL} = c$$
$$\mathrm{UCL} = c + 3\sqrt{c},$$

where c can be estimated by the average number of defects in a preliminary sample. To satisfy the assumption of a constant rate of occurrence, the sample size must be constant.

For variable sample sizes, a u chart should be used instead of a c chart. Compared with the c chart, which is used to monitor the number of defects per sample, the u chart is designed to check the average number of defects per inspection unit. Usually, a sample may contain one or more inspection units. For example, in a textile finishing plant, dyed cloth is inspected for defects per 50 m^2, which is one inspection unit. A roll of cloth of 500 m^2 is thus one sample with 10 inspection units. Different rolls of cloth may vary in area; hence there is a variable sample size. As a result, it is not appropriate to use a c chart, because the occurrence rate of defects in each sample is not a constant. The alternative is to monitor the average number of defects per

14.2 Control Charts

Table 14.13 Organize data in a chart

Week no.	Number of specification
1	9
2	7
3	4
4	2
5	4
6	15
7	2
8	3
9	5
10	5
Total	56

inspection unit in a sample, $u_i = c_i/n_i$. In this way, the parameters in the u chart are given as

$$\text{LCL} = \bar{u} - 3\sqrt{\frac{\bar{u}}{n}}$$
$$\text{CL} = \bar{u}$$
$$\text{UCL} = \bar{u} + 3\sqrt{\frac{\bar{u}}{n}},$$

where $\bar{u} = \sum_{i=1}^{m} u_i/m$ is an estimation of the average number of defects in an inspection unit. For variable sample sizes, the upper and lower control limits vary for different n values.

14.2.9 c Chart Example

In this example, a company tracks the number of times a specification was changed by either an engineering change proposal (ECP) or by a letter from the contracting officer. The attribute data summarize changes to 50 contracts over a 10-week period (as shown in Table 14.13). Since we have attribute data with a constant sample size, and the number of changes is represented by the number of defects, we use a c chart.

14.2.9.1 Determine Centerline (C bar) and Control Limits
C bar = Total defects found/total number of groups = 56/10 = 5.6 (changes per week). Determine control limits. If LCL is less than zero, set LCL = 0.

$$\text{UCL} = \bar{c} + 3\sqrt{\bar{c}} = 5.6 + 3\sqrt{5.6} = 12.699$$
$$\text{LCL} = \bar{c} - 3\sqrt{\bar{c}} = 5.6 - 3\sqrt{5.6} = 0.$$

14.2.9.2 Draw the c Chart
The c chart is shown in Figure 14.14.

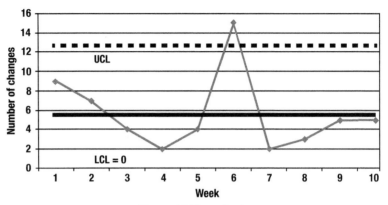

Figure 14.14 c chart.

14.3 Benefits of Control Charts

In this section, we summarize some of the important benefits that can come from using control charts.

- Control charts are simple and effective tools to achieve statistical control. They lend themselves to being maintained at the job station by the operator. They give the people closest to the operation reliable information on when action should and should not be taken.
- When a process is in statistical control, its performance to specification will be predictable. Thus, both producer and customer can rely on consistent quality levels, and both can rely on stable costs for achieving that quality level.
- After a process is in statistical control, its performance can be further improved to reduce variation. The expected effects of proposed improvements in the system can be anticipated, and the actual effects of even relatively subtle changes can be identified through the control chart data. Such process improvements will:
  - Increase the percentage of output that meets customer expectations (improve quality).
  - Decrease the output requiring scrap or rework (improve cost per good unit produced).
  - Increase the total yield of acceptable output through the process (improve effective capacity).
- Control charts provide a common language for communication about the performance of a process between the two or three shifts that operate a process; between line production (operator and supervisor) and support activities (maintenance, material control, process engineering, and quality control); between different stations in the process; between supplier and user; and between the manufacturing/assembly plant and the design engineering activity.
- Control charts, by distinguishing special from common causes of variation, give a good indication of whether any problems are likely to be correctable

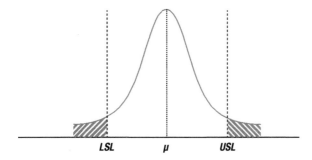

Figure 14.15 Average outgoing quality.

locally or will require management action. This minimizes the confusion, frustration, and excessive cost of misdirected problem-solving efforts.

14.4 Average Outgoing Quality

A measure of part quality is average outgoing quality (AOQ). It is typically defined as the total number of parts per million (ppm) that are outside manufacturer specification limits during the final quality control inspection. A high AOQ indicates a high defective count, and therefore a poor quality level.

For example, manufacturers conduct visual, mechanical, and electrical tests to measure the AOQ of electronic parts. Visual and mechanical tests review marking permanency, dimensions, planarity, solderability, bent leads, and hermeticity (if applicable). Electrical tests include functional and parametric tests at room temperature, high temperature, and low temperature. AOQ is defined in Equation 14.2, referring to Figure 14.15.

$$\text{AOQ} = \frac{\text{Shaded area under the process curve}}{\text{Total area under the process curve}} \times 10^6, \quad (14.2)$$

where USL is the upper specification limit, LSL is the lower specification limit, and μ is the process mean.

The formulae for AOQ calculations may differ among manufacturers. Xilinx provides AOQ based on JEDEC Standard JESD 16–A [2], which is

$$\text{AOQ} = P \times \text{LAR} \times 10^6$$
$$P = \frac{D}{N} \quad (14.3)$$
$$\text{LAR} = \frac{\text{AL}}{\text{TL}}.$$

where D is the total number of defective parts, N is the total number of parts tested, LAR is the lot acceptance rate, AL is the total number of accepted lots, and TL is the total number of lots tested. IDT provided AOQ based on the following formula:

$$\text{AOQ} = P \times 10^6$$
$$P = \frac{D}{N}. \quad (14.4)$$

14 Process Control and Process Capability

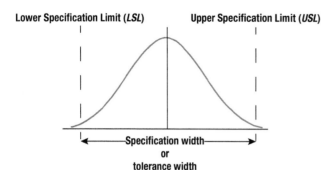

Figure 14.16 Measuring conformance to the customer requirements.

14.4.1 Process Capability Studies

AOQ is a measure of the quality of parts as they leave the production facility. Process capability is a measure of conformance to customer requirements and is typically measured at key process steps. A process capability assessment is conducted to determine whether a process, given its natural variation, is capable of meeting established customer requirements or specifications. It can help to identify changes that have been done in the process, and determine the percent of product or service that does not meet the customer requirements. If the process is capable of making products that conform to the specifications, the specifications can remain the same.

Figure 14.16 shows the specification limits of a product. Specification limits are used to determine if the products will meet the expectations of the customer. Recognize that these specification limits are based solely on the customer requirements and are not meant to reflect on the capability of the process. Figure 14.16 overlays a normal distribution curve on top of the specification limits. In all mathematics related to process capability, an underlying normal distribution of the parameters being examined is assumed.

1. To determine the process capability, the following steps are followed. Determine the process grand average, $\bar{\bar{X}}$, and the average range, R-bar.
2. Determine the USL and the LSL.
3. Calculate the process standard deviation, σ, from the information on the control chart by

$$\hat{\sigma} = \frac{\bar{R}}{d_2} \quad \text{or} \quad \hat{\sigma} = \frac{\bar{s}}{c_4}, \tag{14.5}$$

where R-bar and s-bar are the averages of the subgroup ranges and standard deviation for a period when the process was known to be in control, and d_2 and c_4 are the associated constant values based on the subgroup sample sizes. The process average can be estimated by $\bar{\bar{X}}$ or $\bar{\tilde{X}}$.

A stable process can be represented by a measure of its variation—six standard deviations. Comparing six standard deviations of the process variation to the customer specifications provides a measure of capability. Some measures of capability include C_p, C_r (inverse of C_p), C_{pl}, C_{pu}, and C_{pk}. C_p is calculated using the following formula:

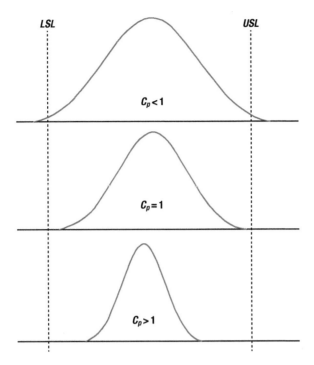

Figure 14.17 Cp, simple process capability.

$$C_P = \frac{USL - LSL}{6\hat{\sigma}}. \tag{14.6}$$

When $C_p < 1$, the process variation exceeds specification and defectives are being made, as shown in Figure 14.17. When $C_p = 1$, the process is just meeting specification. A minimum of 0.27% defectives will be made, more if the process is not centered. When $C_p > 1$, the process variation is less than the specification; however, defectives might be made if the process is not centered on the target value.

The indices C_{pl} and C_{pu} (for single-sided specification limits) and C_{pk} (for two-sided specification limits) measure not only the process variation with respect to the allowable specification, but they also take into account the location of the process average. Capability describes how well centered the curve is in the specification spread and how tight the variation is. C_{pk} is considered a measure of the process capability and is the smaller of either C_{pl} or C_{pu}. If the process is near normal and in statistical control, C_{pk} can be used to estimate the expected percentage of the defective products.

$$C_{pl} = \frac{\bar{\bar{X}} - LSL}{3\hat{\sigma}}, \quad C_{pu} = \frac{USL - \bar{\bar{X}}}{3\hat{\sigma}} \tag{14.7}$$

$$C_{pk} = \min\{C_{pu}, C_{pl}\}. \tag{14.8}$$

Figure 14.18 shows an example of a process not capable of meeting targets. For the process in this figure, $C_p > 1$, but the incapability of the process arises because the process is not centered between *LSL* and *USL*.

If the process is capable of consistently making parts to specification, common causes of the variation in the process must be identified and corrected. Examples of

14 Process Control and Process Capability

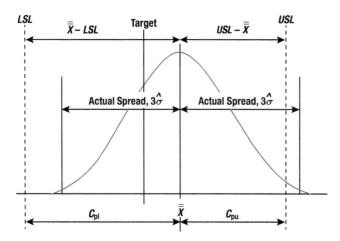

Figure 14.18 Process not capable of meeting specifications.

common remedies include assigning another machine to the process, procuring a new piece of equipment, providing additional training to reduce operator variations, or requiring vendors to implement statistical process controls. In some cases the process may have to be changed, or the specification may have to be relaxed or broadened.

Example 14.1

In a die-cutting process, a control chart was maintained, producing the following statistics: $\bar{\bar{X}} = 212.5$, $\bar{R} = 1.2$, and $n = 5$. The specification limit for this process is 210 ± 3; that means that $USL = 213$, and $LSL = 207$. Calculate C_p and C_{pk} for this process. Also find the number of defects.

Solution:

$$\hat{\sigma} = \frac{\bar{R}}{d_2} = \frac{1.2}{2.326} = 0.516$$

$$C_p = \frac{USL - LSL}{6\hat{\sigma}} = \frac{213 - 207}{6(0.516)} = \frac{6}{3.096} = 1.938$$

$$C_{pl} = \frac{\bar{\bar{X}} - LSL}{3\hat{\sigma}} = \frac{212.5 - 207}{3(0.516)} = \frac{5.5}{1.548} = 3.553$$

$$C_{pu} = \frac{USL - \bar{\bar{X}}}{3\hat{\sigma}} = \frac{213 - 212.5}{3(0.516)} = \frac{0.5}{1.548} = 0.323$$

$$C_{pk} = \min\{C_{pl}, C_{pu}\} = 0.323.$$

Since $C_{pk} < 1$, defective material is being made. Figure 14.19 shows the schematic of the problem.

Defects Calculation:
If the process is near normal and in statistical control, the process of calculating C_{pk} can also be used to estimate the expected percent of defective material. The area under

14.4 Average Outgoing Quality

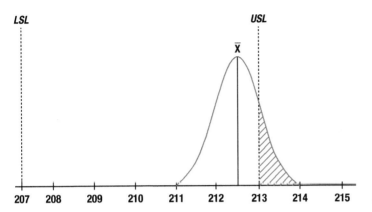

Figure 14.19 Process not capable.

z	0	0.02	0.04	0.06	0.08
-4.00	0.0000	0.0000	0.0000	0.0000	0.0000
-3.80	0.0001	0.0001	0.0001	0.0001	0.0001
:	:	:	:	:	:
-3.00	0.0013	0.0014	0.0015	0.0016	0.0018
-2.90	0.0019	0.0020	0.0021	0.0023	0.0024
-2.80	0.0026	0.0027	0.0029	0.0031	0.0033
-2.70	0.0035	0.0037	0.0039	0.0041	0.0044
:	:	:	:	:	:
0.00	0.5000	0.5080	0.5160	0.5239	0.5319
0.10	0.5398	0.5478	0.5557	0.5636	0.5714
:	:	:	:	:	:
0.70	0.7580	0.7642	0.7704	0.7764	0.7823
0.80	0.7881	0.7939	0.7995	0.8051	0.8106
0.90	0.8159	0.8212	0.8264	0.8315	0.8365
1.00	0.8413	0.8461	0.8508	0.8554	0.8599
:	:	:	:	:	:
2.80	0.9974	0.9976	0.9977	0.9979	0.9980
2.90	0.9981	0.9982	0.9984	0.9985	0.9986
3.00	0.9987	0.9987	0.9988	0.9989	0.9990

$F(z) = P(Z < z)$ is the area under the curve to the left of the z-value (shaded area)

$F(z) = P(Z < z)$
$P(Z < 0.969) \approx 0.832$

Figure 14.20 Sample cumulative normal distribution table.

the curve outside the specification limits is used to determine the number of defects. To determine the area under the curve, the following factors must be calculated:

$$z_1 = \frac{LSL - \bar{\bar{X}}}{\hat{\sigma}} = \frac{207 - 212.5}{0.516} = -10.68$$

$$z_2 = \frac{USL - \bar{\bar{X}}}{\hat{\sigma}} = \frac{213 - 212.5}{0.516} = 0.969.$$

Defects for the value of $z < LSL = \Phi(z_1) = 0$ (approximately). Defects for the value of $z > USL = [1 - \Phi(z_2)]$. Here $[1 - \Phi(z_2)] = [1 - 0.832] = 0.168$. Here, $\Phi(z) = P(Z < z)$

is the cumulative distribution value for any value of z obtained from the standard normal distribution table.

$$\text{Total defects} = \Phi(z_1) + [1 - \Phi(z_2)] = 16.8\%.$$

Example 14.2

We have the following information for a process

$$\hat{\mu} = 0.738, \quad \hat{\sigma} = 0.0725, \quad USL = 0.9, \quad \text{and} \quad LSL = 0.5.$$

Since the process has two-sided specification limits,

$$\begin{aligned} Z_{min} &= \min\left(\frac{USL - \hat{\mu}}{\hat{\sigma}}, \frac{\hat{\mu} - LSL}{\hat{\sigma}}\right) \\ &= \min\left(\frac{0.9 - 0.738}{0.0725}, \frac{0.738 - 0.5}{0.0725}\right) \\ &= \min(2.23, 3.28) = 2.23, \end{aligned}$$

and the proportion of process fallout would be:

$$p = 1 - \Phi(2.23) + \Phi(-3.28) = 0.0129 + 0.0005 = 0.0134.$$

The process capability index would be:

$$C_{pk} = \frac{Z_{min}}{3} = 0.74.$$

If the process could be adjusted toward the center of the specification, the proportion of process fallout might be reduced, even with no change in σ:

$$\begin{aligned} Z_{min} &= \min\left(\frac{USL - \hat{\mu}}{\hat{\sigma}}, \frac{\hat{\mu} - LSL}{\hat{\sigma}}\right) \\ &= \min\left(\frac{0.9 - 0.7}{0.0725}, \frac{0.7 - 0.5}{0.0725}\right) = 2.76, \end{aligned}$$

and the proportion of process fallout would be:

$$p = 2\Phi(-2.76) = 0.0058.$$

The process capability index would be:

$$C_{pk} = \frac{Z_{min}}{3} = 0.92.$$

To improve the actual process performance in the long run, the variation from common causes must be reduced. To consider variability in terms of mean, standard deviation, and the target value, another index is defined as:

$$\hat{C}_{pm} = \frac{USL - LSL}{6\hat{\tau}},$$

where $\hat{\tau}$ is an estimator of the expected square deviation from the target, T, and is given by

$$\tau^2 = E\left[(X-T)^2\right] = \sigma^2 + (\mu - T)^2.$$

Therefore, if we know the estimate of C_p, we can estimate C_{pm} as:

$$\hat{C}_{pm} = \frac{\hat{C}_p}{\sqrt{1 + \left(\frac{\hat{\mu}-T}{\hat{\sigma}}\right)^2}}.$$

At this point, the process has been brought into statistical control and its capability has been described in terms of the process capability index, or $Z_{\min}$. The next step is to evaluate the process capability in terms of meeting customer requirements. The fundamental goal is never-ending improvement in process performance. In the near term, however, priorities must be set as to which processes should receive attention first. This is essentially an economic decision. The circumstances vary from case to case, depending on the nature of the particular process in question. While each such decision could be resolved individually, it is often helpful to use broader guidelines to set priorities and promote consistency of improvement efforts. For instance, certain procedures require $C_{pk} > 1.33$, and further specify $C_{pk} = 1.50$ for new processes. These requirements are intended to assure a minimum performance level that is consistent among characteristics, products, and manufacturing sources.

Whether in response to a capability criterion that has not been met, or in response to the continuing need for improvement in cost and quality performance even beyond the minimum capability requirement, the action required is the same: Improve the process performance by reducing the variation that comes from common causes. This means taking management action to improve the system.

14.5 Advanced Control Charts

In order to effectively detect small process shifts (on the order of 1.5σ or less), a cumulative sum (CUSUM) control chart and the exponentially weighted moving average (EWMA) control chart may be used instead of Shewhart control charts. In addition, there are many situations where we need to simultaneously monitor two or more correlated quality characteristics. The control charts for multivariate quality characteristics will also be discussed in the next section.

The major disadvantage of the Shewhart control chart is that it uses the information in the last plotted point and ignores information given by the sequence of points. This makes it insensitive to small shifts. Thus, either the CUSUM or EWMA charts may be more useful.

14.5.1 Cumulative Sum Control Charts

CUSUM control charts incorporate all the information in the sequence of sample values by plotting the CUSUM of deviations of the sample values from a target value, defined as

$$C_i = \sum_{j=1}^{i}(\bar{x}_j - T). \qquad (14.9)$$

A significant trend developed in C_i is an indication of the process mean shift. Therefore, CUSUM control charts would be more effective than Shewhart charts in detecting small process shifts. Two statistics are used to accumulate deviations from the target, T:

$$\begin{aligned} C_i^+ &= \max[0, \; x_i - (T+K) + C_{i-1}^+] \\ C_i^- &= \max[0, \; (T-K) - x_i + C_{i-1}^-], \end{aligned} \qquad (14.10)$$

where $C_0^+ = C_0^- = 0$, and K is the slack value; it is often chosen about halfway between the target value and the process mean after the shift. If either C^+ or C^- exceeds the decision interval H (a common choice is $H = 5\sigma$), the process is considered to be out of control.

14.5.2 Exponentially Weighted Moving Average Control Charts

As discussed earlier, using Western electric rules increases the sensitivity of Shewhart control charts to detect nonrandom patterns or small shifts in a process. A different approach to highlight small shifts is to use a time average over past and present data values as an indicator of recent performance. The exponentially weighted moving average (EWMA) indicator considers the past data values and remembers them with geometrically decreasing weight. For example, we denote the present and past values of a quality characteristic, x, by $x_t, x_{t-1}, x_{t-2}, \ldots$, then the EWMA y_t with discount factor q is

$$y_t = a(x_t + qx_{t-1} + q^2 x_{t-2} + \cdots), \qquad (14.11)$$

where a is a constant that makes the weights add up to 1 and is equal to $1 - q$. In the practice of process monitoring, the constant $1 - q$ is given the distinguishing symbol λ. Using λ, the EWMA can be expressed as $y_t = \lambda x_t + (1 - \lambda) y_{t-1}$, which is a more convenient formula for updating the value of EWMA at each new observation. It is observed from the formula that a larger value of λ results in weights that die out more quickly and places more emphasis on recent observations. Therefore, a smaller value of λ is recommended to detect small process shifts, usually $\lambda = 0.05, 0.10$, or 0.20. An EWMA control chart with appropriate limits is used to monitor the value of the EWMA. If the process is in statistical control with a process mean of μ and a standard deviation of σ, the mean of the EWMA would be μ, and the standard deviation of the EWMA would be

$$\sigma \left(\frac{\lambda}{2 - \lambda} \right)^{1/2}.$$

Thus, given a value of λ, three-sigma or other appropriate limits can be constructed to monitor the value of EWMA.

14.5.3 Other Advanced Control Charts

The successful use of Shewhart control charts and the CUSUM and EWMA control charts has led to the development of many new techniques over the last 30 years. A brief summary of these techniques and references to more complete descriptions are provided here.

The competitive global market expects lower defect rates and higher quality levels, which requires 100% inspection of output products. The recent advancement of sensing techniques and computer capacity makes one hundred percent inspection more feasible. Due to the reduced intervals between sampling in a 100% inspection, the complete observations will be correlated over time. However, one of the assumptions for Shewhart control charts is the independence between observations over time. When the observations are autocorrelated, Shewhart control charts will give misleading results in the form of many false alarms. Time series models (ARIMA) are used to remove autocorrelation from the data, and then control charts are applied to the residuals. Further discussion of SPC with auto-correlated process data can be found in Box and Luceno (1997) and Montgomery (2005). It is often necessary to simultaneously monitor or control two or more related quality characteristics. Using individual control charts to monitor the independent variables separately can be very misleading. Multivariate SPC control charts were developed based on multivariate normal distribution by Hotelling (1947). The use of control charts requires the selection of sample size, sampling frequency, or interval between samples, and the control limits for the charts. The selection of these parameters has economic consequences in that the cost of sampling, the cost of false alarms, and the cost of removing assignable causes will affect the choice of the parameters. Therefore, the economic design of control charts has also been discussed in the literature.

14.6 Summary

Process control is an effective prevention strategy to manufacture products that will meet the requirements of the customer. There are four elements of process control systems: the process, information about performance, action on the process, and action on the output. The process refers to the combination of people, equipment, input materials, methods, and environment that work together to produce output. The total performance of the process depends on the way the process has been designed and built and on the way it is operated. Information about performance can be learned by studying the process output. Action on the process is future oriented, because it is taken when necessary to prevent the production of nonconforming products. Action on the output is past oriented, because it involves detecting out-of-specification output already produced.

A control chart is a type of trend chart that displays data over time with statistically determined upper and lower control limits; it is used to determine if a process is under control. Control charts by themselves do not correct problems. They indicate that something is wrong and requires corrective action. Assignable causes due to a change in manpower, materials, machines, or methods, or a combination of these, can cause the process to go out of control.

Control charts must be investigated in order to identify in-control and out-of-control processes and detect common causes and special causes of the out-of-control state. A process is said to be operating in statistical control when the only source of variation are common causes. An out-of-control signal is given when a sample statistic falls beyond the control limits, or when a nonrandom pattern is detected.

Process capability is determined by the total variation that comes from common causes. A process must first be brought into statistical control before its capability to meet specifications can be assessed. Process capability is a measure of conformance to customer requirements and is typically measured at key process steps. A process capability assessment is conducted to determine whether a process, given its natural variation, is capable of meeting established customer requirements or specifications. It can help to identify changes that have been done in the process and determine the percent of product or service that does not meet the customer requirements.

The process control and process capability techniques described in this chapter can help to ensure the production of quality products. These techniques can help manufacturers to avoid waste by not producing unacceptable output in the first place, focusing on prevention rather than screening. Statistical process control is an effective prevention strategy to manufacture products that will meet the requirements of customers.

Problems

14.1 For each of the datasets given, identify which of the following control charts should be used to plot the data for process control: c chart, u chart, p chart, np chart, X-bar–R chart, or X–R_m chart. For each case, state why you selected the particular chart type.

a	An equal number of samples of process output have been monitored each week for the last 5 weeks. Ten defective parts were found the first week, eight the second week, six the third week, nine the fourth week, and seven the fifth week.
b	Different numbers of samples (between 40 and 60) of process output have been monitored each week for the last 4 weeks. In the first week, 1.2 defects per sample were observed. In the second week, 1.5 defects per sample were observed. In the third week, 1 defect per sample was observed. In the fourth week, 0.8 defects per sample were observed.
c	The thicknesses of 10 samples were measured each day for a week.
d	An equal number of samples of process output have been monitored each week for the last four weeks. In the first week, 8 defects were observed. In the second week, 12 defects were observed. In the third week, 10 defects were observed. In the fourth week, 9 defects were observed.
e	The thickness of a single sample was measured each day for a week.
f	A process has been observed each week for the last 3 weeks. Ten percent of the parts were found to be defective the first week, 20% were found to be defective the second week, and 15% were found to be defective the third week

14.2 The copper content of a plating bath is measured three times per day and the results are reported in ppm. The X-bar and R-values for 10 days are shown in the following tables.

Day	X-bar	R
1	5.45	1.21
2	5.39	0.95
3	6.85	1.43
4	6.74	1.29
5	5.83	1.35
6	7.22	0.88
7	6.39	0.92
8	6.50	1.13
9	7.15	1.25
10	5.92	1.05

(a) Determine the upper and lower control limits.

(b) Is the process in statistical control?

(c) Estimate the C_p and C_{pk} given that the specification is 6.0 ± 1.0. Is the process capable?

14.3 Printed circuit boards are assembled by a combination of manual assembly and automation. The reflow soldering process is used to make the mechanical and electrical connections of the leaded components to the board. The boards are run through the solder process continuously, and every hour five boards are selected and inspected for process-control purposes. The number of defects in each sample of five boards is noted. The results for 20 samples are shown in the table. What type of control chart is appropriate for this case and why? Construct the control chart limits and draw the chart. Is the process in control? Does it need improvement?

Sample no.	No. of defects	Sample no.	No. of defects
1	6	11	9
2	4	12	15
3	8	13	8
4	10	14	10
5	9	15	8
6	12	16	2
7	16	17	7
8	2	18	1
9	3	19	7
10	10	20	13

14.4 The number of nonconforming switches in samples of size 150 is shown here. Construct a fraction nonconforming control chart for these data. Does the process appear to be in control? If not, assume that assignable causes can be found for all points outside the control limits and calculate the revised control limits.

14 Process Control and Process Capability

Sample no.	No. of noncomformings	Sample no.	No. of noncomformings
1	8	11	6
2	1	12	0
3	3	13	4
4	0	14	0
5	2	15	3
6	4	16	1
7	0	17	15
8	1	18	2
9	10	19	3
10	6	20	0

14.5 The diameter of a shaft with nominal specifications of 60 ± 3 mm is measured six times each hour and the results are recorded. The X-bar and R values for 8 hours are shown in the table below:

Hour	X-bar	R
1	62.54	1.95
2	60.23	2.03
3	58.46	1.43
4	59.95	1.29
5	61.58	0.78
6	57.93	1.48
7	61.56	0.86
8	57.34	1.35

(a) Determine the upper and lower control limits.

(b) Determine if the process is in statistical control.

(c) Estimate the C_p and C_{pk} for the process. Is the process capable?

14.6 The specification for a shaft diameter is 212 ± 2 mm. Provided below are 30 recorded observations for the diameter of a shaft (in mm) taken at 30 different points in time.

First observation: 212.1	214.2	213.7	212.7	212.5	Sixth observation: 212.7
212.8	213.0	212.9	212.3	212.5	212.1
211.8	213.5	212.0	213.0	214.5	212.3
212.2	211.9	213.2	212.7	211.9	212.3
212.0	212.8	213.9	212.6	214.0	Thirtieth observation: 212.4

(a) Develop X and three sample MR charts and determine control limits from the data.

(b) Determine from the control charts whether the process is under control or not.

(c) Determine the capability indices (C_p and C_{pk}) for the process.

(d) Determine the percent defective shafts produced by the process.

14.7 A high-voltage power supply should have a normal output voltage of 350 V. A sample of four units is selected each day and tested for process-control purposes. The data shown give the difference between the observed reading on each unit and the nominal voltage times ten; that is $X_i =$ (observed voltage on unit $i - 350) \times 10$.

Sample no.	X_1	X_2	X_3	X_4
1	6	9	10	15
2	10	4	6	11
3	7	8	10	5
4	8	9	6	13
5	9	10	7	13
6	12	11	10	10
7	16	10	8	9
8	7	5	10	4
9	9	7	8	12
10	15	16	10	13
11	8	12	14	16
12	6	13	9	11
13	16	9	13	15
14	7	13	10	12
15	11	7	10	16
16	15	10	11	14
17	9	8	12	10
18	15	7	10	11
19	8	6	9	12
20	13	14	11	15

(a) Set up X-bar and R charts on this process. Does this process seem to be in statistical control? If necessary, revise the trial control limits.

(b) If specifications are at 350 ± 5 V, what can you say about process capability?

14.8 Vane-opening measurements are as follows. Set up X-bar and s charts on this process. Does this process seem to be in statistical control? If necessary, revise the trial control limits.

Sample no.	X_1	X_2	X_3	X_4	X_5	X-bar	R	s
1	33	29	31	32	33	31.6	4	1.67
2	33	31	35	37	31	33.4	6	2.61
3	35	37	33	34	36	35.0	4	1.58
4	30	31	33	34	33	32.2	4	1.64
5	33	34	35	33	34	33.8	2	0.84
6	38	37	39	40	38	38.4	3	1.14
7	30	31	32	34	31	31.6	4	1.52
8	29	39	38	39	39	36.8	10	4.38
9	28	33	35	36	43	35.0	15	5.43
10	38	33	32	35	32	34.0	6	2.55
11	28	30	28	32	31	29.8	4	1.79
12	31	35	35	35	34	34.0	4	1.73

(*Continued*)

14 Process Control and Process Capability

Sample no.	X_1	X_2	X_3	X_4	X_5	X-bar	R	s
13	27	32	34	35	37	33.0	10	3.81
14	33	33	35	37	36	34.8	4	1.79
15	35	37	32	35	39	35.6	7	2.61
16	33	33	27	31	30	30.8	6	2.49
17	35	34	34	30	32	33.0	5	2.00
18	32	33	30	30	33	31.6	3	1.52
19	25	27	34	27	28	28.2	9	3.42
20	35	35	36	33	30	33.8	6	2.39

14.9 A supply chain engineering group monitors shipments of materials through the company distribution network. Errors on either the delivered material or the accompanying documentation are tracked on a weekly basis. Fifty randomly selected shipments are examined and the errors are recorded. Data for 20 weeks are shown in the table below.

(a) Establish a u chart to monitor this process.

(b) Does this process seem to be in statistical control? If necessary, revise the trial control limits.

(c) Do we need to take any action based on our data? Why? If yes, what action?

Sample no.	Sample size	No. of errors X_i (nonconformities)	Average no. of units per unit, $u_i = x/n$
1	50	2	0.04
2	50	3	0.06
3	50	8	0.16
4	50	1	0.02
5	50	1	0.02
6	50	4	0.08
7	50	1	0.02
8	50	4	0.08
9	50	5	0.10
10	50	1	0.02
11	50	8	0.16
12	50	2	0.04
13	50	4	0.08
14	50	3	0.06
15	50	4	0.08
16	50	1	0.02
17	50	8	0.16
18	50	3	0.06
19	50	7	0.14
20	50	4	0.08
		74	1.48

15 Product Screening and Burn-In Strategies

Burn-in is a screen performed to precipitate defects by exposing the parts to accelerated stress levels. The goal is to prevent failures from occurring in the field (Pecht et al. 1995).

Burn-in as a requirement was instituted during the time of the Minuteman Missile Program where it was shown to be effective in uncovering defects in low-volume immature parts. By 1968, burn-in was incorporated in a military standard, MIL-STD-883 (1968).

Burn-in processes commonly consist of placing parts in a thermal chamber for a specific amount of time under an electrical bias. During and/or after thermal environmental exposure, functional tests are conducted. Parts that fail to meet the device manufacturer's specifications are discarded; parts that pass are used.

The temperature applied during burn-in is higher than the temperature the part will encounter in the field, as a perceived means of reducing the time to precipitate defects. Other accelerated conditions (stresses), which may be part of the burn-in process, include voltage, humidity, electric field, and current density (Lycoudes et al. 1990). To determine which stress condition and stress magnitudes precipitates defect-related failures, the failure modes and mechanisms must be known. The interested reader can find more information on these methods in the book *Quality Conformance and Qualification of Microelectronic Packages and Interconnects* (Pecht et al. 1994).

Over the last decade, there has been scattered evidence that burn-in is not precipitating many defects. For example, plastic parts were failing at a rate of approximately 800 parts per million in 109 hours and 1 part per million in 1975 and 1991, respectively (Slay 1995). In fact, in 1990, Motorola Reliability Group wrote that "The reliability of integrated circuits has improved considerably over the past five years. As a result, burn-in prior to usage, does not remove many failures. On the contrary it may cause failures due to additional handling" (Slay 1995). In 1994, Mark Gorniak of the U.S. Air Force stated that "these end-of-line screens (reference MIL-STD-883) provide a standard series of reliability tests for the industry. Although manufacturers continue to use these screens today, most of the screens are impractical or need

---

Reliability Engineering, First Edition. Kailash C. Kapur and Michael Pecht.
© 2014 John Wiley & Sons, Inc. Published 2014 by John Wiley & Sons, Inc.

Table 15.1 Companies that did not implement burn-in

Company	Product
Hewlett-Packard	PC motherboards
APCD, AHMO, IPO	LAN cards, printer cards, SIMMS
Seagate, Singapore Tech	Disk drive cards
Compaq Asia	Modem cards
TI, NEC Semiconductors	SIMM modules
Exabyte	Tape drive cards
Baxter	Infusion pump PCBAs
Apple	Video tuner cards

modification for new technologies, and add little or no value for mature technologies" (Gorniak 1994). An example of a list of companies that did not implement burn-in is given in Table 15.1.

15.1 Burn-In Data Observations

The first part of our study involved compiling burn-in data from six companies; National Semiconductor, Motorola, Third Party Screening House, Air Transport Systems Division's (ATSD) Third Party Screening House, Honeywell, and Texas Instruments. For confidentiality, the names of the third party screening houses are withheld. The data consist of the total number of parts burned-in along with the number of apparent failures detected via functional tests. Apparent failures are classified as either nonvalid or valid. Valid failures are those that would have occurred in the field if burn-in had not been performed. Nonvalid failures are those that occurred due to handling or other problems that are unique to the burn-in process and thus would not have occurred if burn-in was not performed.

National Semiconductor burn-in data showed that of the 1,119 parts exposed to burn-in conditions, 42 (3.8%) resulted in apparent failures (Plastic Package Availability Program 1995). The apparent failures were due to 35 mechanical rejects and 7 that retested OK. Burn-in did not precipitate any valid failures.

Another study conducted by National Semiconductor showed that the burn-in data, consisting of 169,508 parts, resulted in 6 (0.0035%) apparent failures. Five (83%) of these failures were due to electrical overstress (EOS) and electrostatic discharge (ESD) damage. One (17%) was a valid failure due to AC propagation delay.

Motorola burn-in data showed that parts exposed to burn-in conditions, 186 (0.072%) apparent failures resulted. The apparent failures were due to 182 electrical rejects and four mechanical rejects. Of the apparent failures, none was valid.

A third party burned-in 6105 parts that resulted in 167 (2.7%) apparent failures. The apparent failures were due to 143 mechanical rejects and 24 electrical rejects that were caused by testing errors at the screening facility. Of the apparent failures, none was valid.

Honeywell burned-in a total of 162,940 parts, of which 669 resulted in apparent failures (Scalise 1996). Out of 67 parts that were failure analyzed, five (7%) were valid failures, two (3%) that were process related, and three (4%) that were temperature related. The remaining 62 (93%) were invalid failures, where electrical overstress (EOS) and ESD contributed to 49 (73%) of the failures.

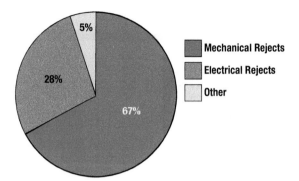

Figure 15.1 Nonvalid failures.

Texas Instruments (TI) burned-in a total of 195,070 different TTL, S, and LS parts (Tang 1996). Of these parts, 25 (0.013%) resulted in apparent failures. Three valid failures (0.0010%) resulted; one due to a die mechanical damage, one from a broken wirebond, and another could not be resolved. The 22 (80%) remaining apparent failures were due to EOS and ESD failures.

Texas Instruments data of HCMOS technology parts showed that out of the 100,165 parts that were burned-in, eight (0.009%) resulted in apparent failures (Scalise 1996). All of the eight (100%) part failures were due to EOS or ESD.

15.2 Discussion of Burn-In Data

Of the total 911,667 parts, the data presented show that burn-in detected 1125 (0.12%) apparent failures, of which 1116 (99.2%) were invalid and nine (0.8%) were valid. The valid failures consisted of: AC propagation delay, a defect in the fabrication process, die mechanical damage, and a broken wirebond.

The breakdown of nonvalid failures is shown in Figure 15.1. Mechanical defects included such things as improper device orientation and bent leads. Electrical rejects include ESD and EOS. ESD is caused by additional handling, whereas EOS occurs due to misapplied power.

The "other" category includes parts that retest OK and parts that are retained by the test lab or lost. Retest OK is defined as parts that do not pass the functional test, initially, but do pass in a subsequent test. These failures are not device related. For example, a contact may have dust and when taken out of its socket and reinserted, the dust particles are removed, creating a better contact that passes the functional test. When these failures are insignificant, the parts are normally discarded. If the failures are believed to be caused by the burn-in process, the burn-in process in reevaluated to aid in preventing such failures from occurring.

Data obtained from Northbrook, in 1991–1992, showed that of 1,017,828 parts burned-in, 70% of the detected defects were due to wafer processing and the remaining were package-related defects. In 1993–1994, 582,480 parts (from the same manufacturers) were burned-in, with 100% of the detected defects due to the wafer process. This suggests that the package quality has improved to the point where burn-in is essentially nonvalue added. In terms of this study, because of the extremely small percentage of valid failures, it is difficult to compare the effectiveness of burn-in to precipitate die-level versus package-level defects. The key point is that 99.2% of the failures could have been avoided if burn-in was not performed.

15.3 Higher Field Reliability without Screening

Honeywell's Air Transport Division had historically screened plastic encapsulated microcircuits (PEMs) for 160 hours at 125°C, followed by a tri-temp screen at −40°C, room temperature and 125°C, believing that this would increase the reliability of the end product due to reduced infant mortality. This has always been a questionable activity because:

- Typically, the integrated circuit (IC) manufacturer does not perform this screen, and doing it at a third-party part screening facility thus becomes suspect and expensive.
- The enormous improvements that semiconductor manufacturers have made in product quality must be addressed with respect to the effect of burn-in. That is, does screening actually decrease field reliability due to part damage occurring during the screening process?

In the previous sections, it was shown that many parts failed during burn-in, the majority being invalid. Two questions to be posed are: were the parts that failed really defective and were the parts that were sent to the field reliable.

It was shown that burn-in caused over 99% of the apparent failures, that is, less than 1% of the parts were really defective. This section attempts to answer the second question.

From previous results, it was believed that the handling required to burn-in parts was causing unacceptable ESD damage. What is particularly unsettling with this conclusion is that it raises the issue of latent ESD damage in fielded equipment that the handling required to burn in parts is causing unacceptable ESD damage. In order to reach a data-driven conclusion as to the necessity of doing part screening, particularly burn-in, data was collected from two sets of data:

- Aircraft field failure data for both military ceramic and commercial plastic parts, where most of the plastic parts were screened, but with a significant and identifiable group that were not screened.
- A ring laser gyro that was built with totally unscreened commercial parts to allow an on-aircraft evaluation of the effects of not screening.

From the first set of data, an examination of the field failures that occurred with the burned-in PEMs is shown in Table 15.2. The results show of the total failures, 6.6% were valid, 31.7% were invalid, 4% could not be determined, and the remaining 57.7% were not failure analyzed. The failures that resulted could be due to the part, sub-system, or system level. For example, the invalid failures do not pertain to those caused by burn-in, as discussed earlier. These invalid failures may be due to a lead that was not soldered to the printed circuit board. A process related failure, which is

15.4 Best Practices

Table 15.2 Field failure results: Ring laser gyro with screening

Part type Failure type	Digital SSM/MSI	Digital LSI/Mem	Linear	Total parts failed	Percent of total failures
Valid					
Fabrication	0	2	2	4	1.8
Temperature	0	0	11	11	4.8
Invalid					
EOS/ESD	8	1	5	14	6.2
Other	7	18	33	58	25.5
Undetermined	2	1	6	9	4.0
No failure analysis	16	70	45	131	57.7
Total parts failed	33	92	102	227	

considered to be a valid failure, could be a solder joint that did not wet properly during the soldering stage.

Of the invalid failures, 6.2% were due to ESD damage. This is far higher than expected based on historical in-service failure data. From this, the conclusion is that burn-in not only adds no value, but it may even increase the field failure rate of the devices built with these burned-in parts. It was also hypothesized that latent ESD effects could be introduced somewhere in the build process of the equipment.

To assess this hypothesis, a decision was made to include in the Honeywell product mix, a ring laser gyro that was built with totally unscreened, commercial parts (second data set). The part types and manufacturers used in this assembly were the same as those used in the devices that had the higher ESD related failure rate, that is, the first data set. This is because the design engineers are required to work from a relatively small list of approved parts and manufacturers. The build facility was also common to all the devices. What this means is that a comparison can be made between devices built with screened parts and devices built with unscreened parts, where the only difference is the part screening. These devices are installed in the same aircraft and in the same equipment bays.

The devices built using the unscreened commercial parts have accrued well over 200 million piece part hours without a failure. The devices that were screened resulted in 669 (0.4%) failures out of the total 162,940 parts. The only difference between these devices is that screening was not done on the devices that had no failures.

What has been shown in this comparison study is that a high percentage of latent ESD failures resulted when the parts were screened, and no failures resulted when the parts were unscreened. Therefore, the parts sent to the field were not as good as originally thought, since burn-in degraded part reliability. That is, burn-in precipitated many invalid failures and degraded part reliability, resulting in field failures. For these reasons, many field failures can be avoided if burn-in is not performed.

15.4 Best Practices

Many companies involved with "critical systems" require burn-in because they believe that the risk to do otherwise is too high. Our findings not only question this so-called safety net viewpoint, but show that this net has become a trap that should be avoided. Burning-in parts from a quality manufacturer will increase the number of field failures. The failures due to the burn-in process can be significant relative to those due to inherent defects and, therefore, an alternate approach to burn-in is necessary.

Since burn-in has shown to be ineffective, many companies have begun implementing a burn-in elimination program. This program is based on burn-in data, which implies that burn-in is being performed. However, our recommendations is that, instead of conducting burn-in and then implementing a burn-in elimination program, manufacturer part family assessment and qualification data should be used to assess the need for burn-in. This approach is based on existing data, where burn-in has not been performed, thereby avoiding part degradation.

Manufacturer part family assessment is dependent on the supplier. Parts must come from a supplier that is periodically certified, implements statistical process control (SPC) or an acceptable documented process, has acceptable qualification testing results, abides by procedures to prevent damage or deterioration (e.g., handling procedures, such as ESD bags), and provides change notifications should not require burn-in. Once a quality part is obtained, results from the qualification tests can be used to determine whether or not burn-in needs to be conducted. If no failures occur in qualification tests and the manufacturing processes are in control, then confidence can be gained to assess the part quality without performing burn-in.

15.5   Summary

Burn-in is a screen performed to precipitate defects by exposing the parts to accelerated stress levels. The goal is to prevent failures from occurring in the field. Burn-in processes commonly consist of placing parts in a thermal chamber for a specific amount of time under an electrical bias. During and/or after thermal environmental exposure, functional tests are conducted. Parts that fail to meet the device manufacturer's specifications are discarded; parts that pass are used.

Many companies involved with "critical systems" require burn-in because they believe that the risk to do otherwise is too high. Burning-in parts from a quality manufacturer will increase the number of field failures. The failures due to the burn-in process can be significant relative to those due to inherent defects and, therefore, an alternate approach to burn-in necessary. Since burn-in has shown to be ineffective, many companies have begun implementing a burn-in elimination program. This program is based on burn-in data, which implies that burn-in is being performed. However, our recommendation is that, instead of conducting burn-in and then implementing a burn-in elimination program, manufacturer part family assessment, and qualification data should be used to assess the need for burn-in. This approach is based on existing data, where burn-in has not been performed, thereby avoiding part degradation.

Manufacturer part family assessment is dependent on the supplier. Once a quality part is obtained, results from the qualification tests can be used to determine whether or not burn-in needs to be conducted. If no failures occur in qualification tests and the manufacturing processes are in control, then confidence can be gained to assess the part quality without performing burn-in.

Problems

15.1 What is screening? What are the steps in conducting screening? What are the benefits of screening?

15.2 Present an example of how to lower the hazard rate during the useful life of a product.

15.3 What is error seeding and why is it used?

15.4 Give examples of defects in electronics that can be detected by screening.

15.5 Explain how screening can be used to validate product reliability during the product development. Provide some examples.

15.6 Define "burn-in" and list some of its pros and cons.

16 Analyzing Product Failures and Root Causes

The root cause is the most basic causal factor or factors that, if corrected or removed, will prevent the recurrence of a problem. It is generally understood that problem identification and correction requires getting to the root cause. This chapter discusses root-cause analysis concepts, presents a methodology for root-cause analysis, and provides guidance for decision making.

Generally, product failures or faults do not "just happen." Products fail due to "failures" in design, manufacture, assembly, screening, storage, transportation, operation, and even in repair and maintenance.

The root cause is the most basic causal factor or factors that, if corrected or removed, will prevent the recurrence of the failure. Identifying root causes is the key to preventing similar occurrences in the future.

Root cause should not be confused with symptoms and apparent causes of failure. A symptom is a sign or indication that a failure exists. For example, a symptom of failure could be a knocking noise made by a washing machine. The apparent cause may be a rocking movement of the machine itself. However, the root cause is the most basic causal factor. In the above example, the root cause could arise because a bearing in the motor is worn, due to a lack of lubricant.

Example 16.1

Consider testing of a printed circuit board (PCB) after fabrication. A symptom of failure upon testing is an open circuit. The apparent cause of the open circuit is that circuit traces on the PCB have discontinuities (scratches). The root cause for the failure could be that during the manufacturing process, the circuit boards are stacked improperly, resulting in scratches to circuit traces on the PCB.

Root-cause analysis is a methodology designed to help describe what happened during a particular occurrence, determine how it happened, and understand why it

---

Reliability Engineering, First Edition. Kailash C. Kapur and Michael Pecht.
© 2014 John Wiley & Sons, Inc. Published 2014 by John Wiley & Sons, Inc.

Table 16.1 Assessment of field removals of aircraft electronic equipment

Removal Assessment	%
From a Boeing study[a]	
Unjustified	30
Assembly errors, handling damage	60
Design, improper installation	9
Parts manufacturing problems	1
Unexplained	<1
From a Rockwell study[b]	
Not verified	36
Apparent electrical	22
Part application	15
Process problem	12
Bad lot	6
Design	6
Miscellaneous	2
Workmanship	1
From a Westinghouse report[c]	
Retest OK	31
Electrically damaged	30
Thermal mechanical stress and switching	21
Manufacturing/quality/design defects	18

[a]Pecht, M., and Ramappan, V., "Review of Electronic System and Device Field Failure Returns," *IEEE Transactions on CHMT*, Vol. 15(6), 1992, pp. 1160–1164.
[b]Brennom, T.R., "Reliability Prediction Methodology Improvements Needed for 1990," presented at Design of Reliable Electronic Packages Workshop, University of Maryland, 1990.
[c]Westinghouse Electric Corp, "Failure Analysis Memos," 1989.

happened (ABS Group, Inc. 1999). The purpose of determining the root cause is to fix the problem at its most basic source so that it does not occur again, even in other products, as opposed to troubleshooting, which is generally employed to merely fix a failure symptom in a given product.

Example 16.2

Consider field removals of aircraft electronic equipment (avionics). Table 16.1 gives identified reasons for the field removals. The reasons are classified in broad categories, according to the life-cycle phase (e.g., design, manufacturing, and application) when the item was removed, or the type of damage (e.g., electrical, thermal-mechanical) incurred by the item. However, the actual root causes are not shown. Certain removals are "unjustified," "unexplained," "apparently" attributed to one type of causal reason, or "not verified," indicating that the root-cause analysis effort has not been successful.

In order to understand failure and the root causes, one needs to have clear definitions of failure-related concepts. Key definitions are listed in Table 16.2.

Table 16.2 Definitions of failure-related concepts

Failure	A product no longer performs the function for which it was intended.
Failure mode	The effect by which a failure is observed.
Failure site	The location of the failure.
Failure mechanism	The physical, chemical, thermodynamic, or other process or combination of processes that result in failure.
Fault	Event or weak process (e.g., design), which may or may not cause failure.
Load	Application and environmental conditions (electrical, thermal, mechanical, and chemical) which can precipitate a failure mechanism.
Stress	Intensity of the load at the failure site.

16.1 Root-Cause Analysis Processes

Only when investigators truly understand the question, "why a failure occurred," will they be able to specify proper corrective measures. A well-structured root-cause analysis will provide added benefits over time by focusing resources on preventing failures.

Figure 16.1 is a flowchart of the root-cause analysis methodology. The process begins by establishing a root-cause culture within the organization, which must be prepared to effectively and efficiently investigate and correct failures. This preplanning phase involves preparing root-cause analysis methodologies and procedures that are specific to the organization and its products. Once a failure incident occurs, the root-cause investigation begins with data collection and assessment of immediate cause(s). Analysis techniques to hypothesize root causes include formal evaluation methods, such as Ishikawa diagram, failure modes and effects analysis, and fault tree analysis. The hypotheses formulated are then assessed based on the evidence gathered, design reviews and physical evaluation of the failed system. Root-cause identification and the development of corrective actions are then conducted. Finally, the implemented corrective actions are assessed with emphasis on cost and benefit analysis.

16.1.1 Preplanning

The preplanning establishes the investigation procedures and teams that can be activated as soon as an incident occurs. The goal is to introduce foresight and execution activities that will make failure analysis effectively prevent equipment failures and lessen the consequence of failure.

Preplanning begins by establishing a root-cause culture with management support and responsibilities, through awareness and education. It is essential that management commits to support root-cause analysis. The organizational or corporate culture must provide an environment where employees are encouraged to report faults, and where suitable mechanisms have been implemented to evaluate existing and potential problems.

Procedures to report product failures should be defined, such as verbal and written notification reports. The use of an event-reporting format can help bound the potential problem and determine the effort required for problem resolution (Mobley 1999). The incident report form should specify the person reporting the incident, the incident

16 Analyzing Product Failures and Root Causes

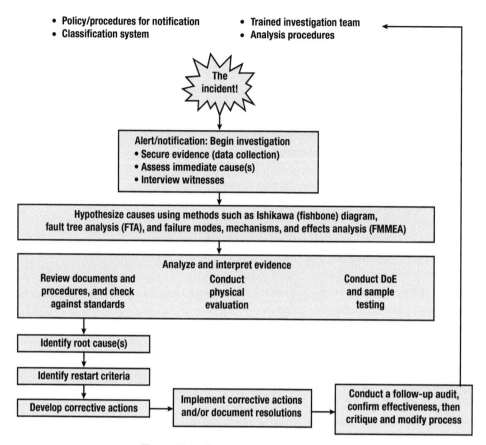

Figure 16.1 Root-cause analysis flowchart.

location and date, product or system affected, and any photographs, probable cause(s) perceived and any corrective actions taken. The recorded information on the failed product should be confirmed with the customers, as appropriate. These steps are to ensure that the appropriate product and failure mechanism is being investigated and that basic information on the circumstances of the event is not lost.

A classification system of failures, failure symptoms, and apparent causes should be developed to aid in the documentation of failures and their root causes, and help identify suitable preventive methods against future incidents. By having a common classification system, it may be easier for engineers, designers, and managers to identify and share information on vulnerable areas in the design, manufacture, assembly, storage, transportation, and operation of the product. Broad failure classifications include product damage or failure, loss in operating performance (e.g., deviation in product quality, plant production, or capacity rate) or economic performance (e.g., high production or maintenance cost), safety (e.g., events that have the potential to cause personal injury), and regulatory compliance (e.g., safety, health, or environmental compliance). Failures categorized as product damage can be further categorized according to the failure mode and mechanism. Different categories of failures may require different root-cause analysis approaches and tools.

Simple failures do not necessarily mean simple problems. For instance, even a simple problem may require investigating shipping, handling, assembly, and usage processes and conditions. Thus, before starting root-cause analysis, a team of investigators needs to be formed. An ideal root-cause analysis team consists of a principal analyst, experts, and the vendors of the failed product. Diversity among the team members is essential, and the respective team experts' backgrounds should be in relation with the issue being analyzed. Training will also benefit those who identify and report faults, by helping them to understand their role in the root-cause analysis process. Managers, who will use the results of the investigation, should understand the vocabulary, develop meaningful expectations from root-cause analysis training, and an understanding of the long-term benefits of fault identification and correction (Wilson et al. 1993).

Analysis strategies and procedures should also be defined, including when root-cause analysis should be performed (i.e., For every failure? Repeated failures? For which type of detected faults?). Answers to these questions may depend on the impact of the failure, in terms of cost, reliability requirements for the product category (e.g., defense and medical electronics), customer, and scheduling. Based on these considerations, failures or faults can be prioritized.

Other procedures that can be identified in the preplanning phase are the root-cause hypothesization techniques (e.g., failure modes, mechanisms, and effects analysis [FMMEA], fault tree analysis [FTA]) that are most suited to investigate specific failures. Performing an FMMEA or FTA analysis for each product proactively (i.e., before any failure occurs) will help to more rapidly identify the possible root causes of a failure.

16.1.2 Collecting Data for Analysis and Assessing Immediate Causes

Once an incident has occurred, the product design, manufacturing and marketing teams must be notified based on the procedures defined in the pre-planning phase.

The first step of root-cause analysis is collecting data to understand the events that lead to the failure. The evidence gathered is used to identify the critical aspects of failure, including the failure mode, site, and mechanism, time in the life-cycle where failure occurred, length of time for the failure to initiate, and periodicity of the failure. The investigators must be confident that they have been able to identify the major contributors and potential causal factors associated with the incident. It is important to note that events are rarely caused by one causal factor.

Complete data collection may not be feasible depending on the product or the phase of the life cycle where failure occurred. For example, failures or faults detected during manufacturing may be easier to trace than failures which occur in the field. In such instances, experimental approaches may be needed to replicate the failure and to identify the root cause(s).

Data gathering must be performed as soon as possible after the event occurs in order to prevent loss or alteration of data. An interview process should be followed to gather data from people. The people to be interviewed may include personnel directly involved in the incident, supervisors and managers of those involved in the incident, personnel who have similar background and experience, and applicable technical experts (Mobley 1999). The following key questions should be answered: what happened, where (e.g., machine, system, or area) and when (time frame and sequence of events that bound the event) did it happen, what changed (e.g., equipment failure, product performance, practices, and environment), who was involved (personnel directly or indirectly involved, including supervisory and management personnel), what is the impact (e.g., in terms of injury, reliability, and finance), what is the probability of a recurrence of the event or of similar events, and whether the recurrence can be prevented (Mobley 1999). Follow-up interviews may be needed to answer additional questions that will arise during the course of the analysis.

When investigating an event involving equipment damage or failure, the highest priority is to preserve physical evidence. If possible, the scene of the failure (e.g., the failed item and the system) should be isolated from service and stored in a secure area until a full investigation can be conducted. If this approach is not practical, the scene of the failure should be fully documented before the item is removed from its installations. Photographs, sketches, instrumentation, and control settings should be documented to ensure that all data are preserved for the investigating team. Physical data includes information of parts, residues, and chemical samples. For different industries, physical data has different meanings. For example, in the automobile industry, parts can be motors, pumps, and processing equipment. In health care, parts can be medicines, syringes, and surgical tools. All documentation records, logs, data-recording results, procedures, memos, and program manuals should also be available for analysis.

16.1.3 Root-Cause Hypothesization

There are many analysis techniques available to hypothesize root causes. These techniques range from relatively simple, unstructured approaches to the more elaborate, formal evaluation methods, such as Ishikawa diagram (fishbone analysis), FMMEA, and FTA.

Which analysis techniques should be utilized is determined by the specific problem. Some techniques seem to work better than others in a particular situation. Even less structured approaches, such as intuition, experience, communication, and brainstorming, can be extremely valuable. These kinds of techniques can be faster in solving problems than structured techniques, but they have a greater degree of subjectivity, and there is a chance of identifying an incorrect root cause.

Structured approaches usually generate some logic tables or flow diagrams that help in the analysis process. Since structured methods are organized processes, they are repeatable, and provide better documentation. Some of the structured techniques are discussed in the following sections.

Once root causes have been hypothesized, additional evidence is often required to validate or invalidate the hypotheses formulated. New evidence can be gathered by conducting additional interviews, reviewing documents and procedures against standards, conducting experiments, and further evaluating physical evidence.

16.1.3.1 Ishikawa Diagram (Fishbone Analysis) The Ishikawa diagram, developed in the 1950s, is a graphical tool to discover root cause(s). The Ishikawa diagram is also known as a "fishbone diagram" due to its shape, and also as a "cause and effect diagram" according to its function. It allows an individual or teams to identify, explore, and display the relationship between an effect and all of its potential causes. The diagram also identifies factors that can contribute to design or process improvements.

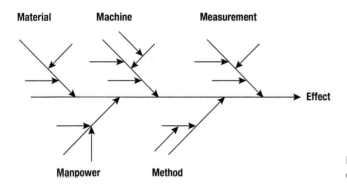

Figure 16.2 Typical structure of a fishbone diagram.

Figure 16.2 illustrates the general structure of a fishbone diagram. The diagram has two sides: "effect" and "cause." The effect is typically on the right side at the end of the main "bone," while the causes appear on the left side as subbones. Effects are listed as particular quality characteristics or problems resulting from work, such as problems involving product quality, cost, delivery, workplace safety, and quality control circle activities. Causes are the factors that influence the stated effect or characteristics, and are arranged according to their level of importance or detail, resulting in a depiction of relationships and hierarchies. This can help in identifying potential causes.

There are typically various subbones joining the main bone in the diagram. These are the major categories of causes to the effect. In a production process, the traditional categories are: machines (equipment), methods (how work is done), materials (components or raw materials), and people (the human element). In a service process, the traditional categories are: policies (high level decision rules), procedures (steps in a task), plant (equipment and space), and people. In both types of processes, environment (buildings, logistics, and space), and measurement (calibration and data collection) are also frequently listed.

Once the fishbone diagram is complete, one has a rather complete picture of all the possible causes. Figure 16.3 is an example of a fishbone diagram of electrostatic discharge in electronic devices. Figure 16.4 is another example of a fishbone diagram used to analyze the thermosonic ball bonding process to attach a wire inside a semiconductor component.

The advantage of the fishbone diagram is that it forces the investigating team to logically group and consider each of the factors identified during the investigation. This process helps uncover issues that must be addressed. Once all the factors have been identified, the team can systematically evaluate each one.

16.1.3.2 Failure Modes, Mechanisms and Effects Analysis FMMEA is a systematic approach to identify failure mechanisms and models for the potential failures modes. FMMEA process begins by defining the product to be analyzed. The product is divided into various subsystems or levels and it continues to the lowest possible level, which is a "component" or "element."

At the initial level of analysis, every possible failure mode and site for each component of each subsystem of the product is identified. Each identified failure mode

16 Analyzing Product Failures and Root Causes

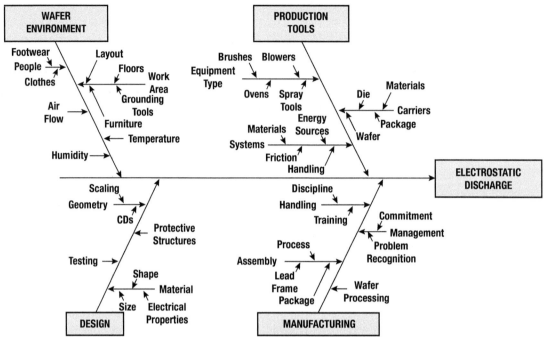

Figure 16.3 Cause and effect diagram example: electrostatic discharge in electronic devices.

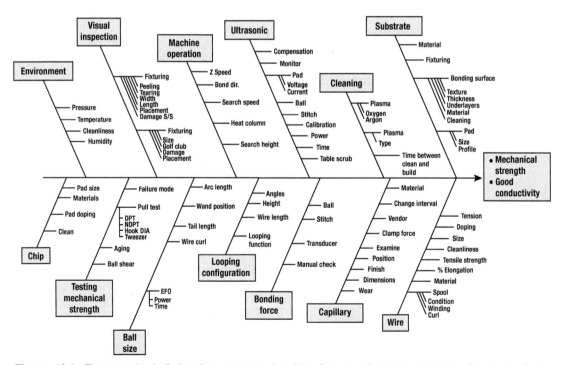

Figure 16.4 Thermosonic ball bonding: process baseline for an electronic package die to leadframe interconnection.

is analyzed to determine its effect on a given subsystem and the whole product. For some components, modes of failure are easy to determine. For example, a capacitor may have three failure modes: shorts, opens, and capacitance change.

Once the failure modes are identified, the next step is to identify the relevant failure mechanisms and models that can explain why the failure manifested as a particular failure mode. Information about the environmental and operating conditions on a product helps in identifying the stresses that can propagate a failure mechanism. Information from the failure models can help prioritize the most significant failure causes.

Example 16.3

Consider an electronic product, when an electronic component mounted on a printed circuit board inside the product is observed to exhibit an electrical open failure mode. The electrical open could be due to the component, the electrical (solder joint) connection of the component to the circuit board or the circuit board itself. If the component and the board have no faults, then the focus can be on the solder joints.

The cause of solder joint failure could be fatigue could be due to loads including vibration and temperature cycling. Data from the life-cycle conditions of the product may indicate that the product sees high random vibrations but no temperature cycling effects. Thus the root cause of the failure of the electronic product can be attributed to the high random vibrations experience by the product.

16.1.3.3 Fault Tree Analysis Fault tree analysis (FTA) is a deductive methodology to determine the potential causes of failures and estimate failure probabilities. It was first introduced at Bell Telephone Laboratories in connection with the safety analysis of the Minuteman missile launch control system in 1962. The method was further developed at the Boeing Company in the mid-1960s. Since then, fault tree techniques have been widely used to investigate system reliability and safety.

In contrast with the "bottom-up" assessment of FMMEA, FTA is a "top-down" approach that starts qualitatively to determine the failure modes that contribute to an undesirable top-level event. FTA provides structure from which simple logical relationships can be used to express the probabilistic relationships among the various events that lead to the failure of the system. FTA is also different from FMMEA in that it focuses on identifying subsystem elements and events that lead to a particular undesired event.

The fault tree analysis begins with the identification of the top events, which are usually the set of system failure modes that could occur. Data are then obtained and analyzed to determine which faults (events) will result in the top level failure events. The faults (events) are connected to the top failure modes (events) by Boolean logical links (AND/OR gates). The process is continued for each intermediate event until all basic events (called primary events) have been developed. If a quantitative analysis is desired, failure rates for the basic events need to be determined. Then working from basic events upward, failure probability data can be determined.

A fault tree shows only the failure modes that could individually or collectively cause the undesired primary event. In some cases, probabilities can be allocated, and they may help identify more probable root causes.

16.1.4 Analysis and Interpretation of Evidence

Reviewing in-house procedures (e.g., design, manufacturing process, procurement, storage, handling, quality control, maintenance, environmental policy, safety, communication, or training procedures) against corresponding standards, regulations, or part- and equipment vendor documentation (e.g., part datasheet and application notes and equipment operating and maintenance manuals) can help identify factors that could precipitate the failure. For example, misapplication of a component could arise from its use outside the vendor specified operating conditions (e.g., current, voltage, or temperature). Equipment (e.g., assembly, rework, or inspection equipment) misapplication can result from, for example, uncontrolled modifications or improper changes in the operating requirements of the machine.

A design review should be conducted to determine the design limitations, acceptable operating envelope, and specific indices that quantify the actual operating condition of the part, equipment, machine, or process associated with an event or failure. Unless the investigators understand what the part, equipment, or process was designed to do and its inherent limitations, it may be impossible to isolate the root cause of a problem or event (Mobley 1999).

Physical evaluation of a failed part should begin with the least destructive tests to preserve evidence that might be otherwise destroyed by destructive tests. For example, for microelectronic equipment, a failed part might be inspected visually or using magnification equipment, such as an optical or scanning electron microscopy (SEM). All anomalies, regardless of their à priori relevance to the failure, are documented. The part is then characterized and parametric behavior is recorded. To ensure a nondestructive evaluation, any applied loads should be within corresponding vendor-specified maximum allowable values (e.g., voltage and current). The investigation then proceeds to minimally invasive procedures. These can include internal examination of the materials and interfaces using acoustic-, thermal-, or magnetic-based imaging techniques. The use of conditioning environments, such as imposing a specific ambient temperature or relative humidity, may be required to replicate field conditions. Examples of other physical characterization techniques include energy dispersive spectroscopy, X-ray microscopy, scanning acoustic microscopy, scanning magnetic microscopy, Fourier transform infrared spectroscopy, and atomic force microscopy.

Selection of the appropriate test sequence is crucial to minimize alteration or destruction of evidence, depending upon the failure mechanism investigated. However, at some time, destructive physical evaluation techniques will be needed, such as microsectioning and decapsulation, followed by imaging of the exposed analysis plane or conducting material characterization.

A design of experiment (DoE) approach is recommended to minimize the number of tests without losing critical evidence that could lead to root cause. It is important to consider the test sequence, sample design attributes and manufacturing process parameters, sample size for each test, loads for parameter characterization, and the location of imaging analysis planes for physical evaluation.

16.1.5 Root-Cause Identification and Corrective Actions

After the completion of data analysis and identification of all the causal factors, root-cause identification begins. At this point, each individual and groups of causal factors are investigated. By analyzing the causal factors separately and in combinations, the probability of important details being overlooked decreases.

Once the root causes have been identified, they should be categorized. Failures with severe consequences (e.g., safety) may require processes such as manufacturing and distribution to be interrupted after discovery of the failure. Depending upon the identified root cause, processes may be restarted if corrective action(s) can be implemented that will prevent the recurrence of the failure, or sufficiently minimize its impact.

In the root-cause identification process, one or more potential corrective actions that would resolve the incident are defined. These actions include a wide range of possibilities from doing nothing to replacing or redesigning the product or system. Then, the next step is to determine which of these corrective actions should be implemented. There may be constraints to a proposed solution, in terms of, for example, cost, schedule, or difficulty of implementation if the solution is too elaborate. In some instances, breaking the chain of events at any point may be sufficient. For example, enforcing the correct application of safety procedures may be sufficient to prevent accidents associated with the use of hazardous equipment, although a corrective action that eliminates the root cause could involve the redesign of the equipment or process during which the accident occurred.

Many of the failures or events having a direct impact on production require immediate corrective actions that will minimize the system's downtime. As a result, temporary corrective actions, or immediate corrective actions, are often required to permit resumption of production. However, temporary solutions may not be financially justifiable over the "long haul." The rationale for any decision must describe the limitations or restrictions that the partial correction will have on equipment reliability, plant performance, life-cycle costs, schedule, and other factors of plant operation and maintenance.

Not all of the actions are financially justifiable. In some cases, the impact of the incident or event is lower than the cost of the corrective action.

Although the temporary solution is often unavoidable, every effort should be made to implement permanent corrective actions. The goal is to eliminate all negative factors associated with the event or incident. While there generally is a corrective action that meets this goal, the challenge is to find the best acceptable action that also is cost effective.

A cost–benefit analysis serves to compare the cost with the benefits (including risk) derived from the corrective actions being considered. A cost-benefit analysis is simply a direct comparison of the actual total costs associated with the activity with the benefits to be derived from the change. The cost–benefit analysis contains three parts: cost analysis (quantify costs), benefit analysis (quantify benefits), and cost-benefit comparison. The analysis should include the effect of the problem on downstream production facilities. For example, a problem in one area of the plant generally has direct impact on all downstream plant areas. The loss of revenue, potential increase in conversion costs, and reluctant benefits of the corrective action on the downstream areas must be considered in the cost–benefit analysis. A full cost-benefit analysis should be conducted before recommending a course of action.

The cost analysis consists of two parts. The first part quantifies the impact of the problem, incident, or event on the process. The impact must be defined in financial terms rather than in terms of delays, downtime, and other traditional tracking mechanisms. The second part of the analysis defines all direct and indirect costs associated with actually implementing the recommended corrective actions and the benefits (in terms of dollars) from the correction actions.

The benefit analysis defines the benefits derived from implementing specific corrective actions. In root-cause analysis, the goal is to quantify the actual improvement to ensure that the potential benefits are real and significant. Benefits generally can be quantified as improvements in process-related costs, increased revenue generation, and cost avoidance of maintenances and liability.

The format of the benefits analysis should mirror the cost categories (e.g., material and labor costs) so that a comparison can be made. After a valid corrective action is implemented, there will be a measurable improvement process-related cost.

One possible benefit is a reduction in the total production and maintenance costs per unit. In some cases, this improvement may occur simply because the capacity of the replacement machine is greater than the one placed. This increase in capacity should reduce the total cost per unit because greater volume is produced. In addition to the capacity gain, the increase in availability has the additional benefit of reducing the cost per unit of both production and maintenance. Increased capacity, as discussed in the preceding paragraphs, is a major benefit that may result from implementing corrective actions. If there is a market for the additional product, this increased capacity will also provide additional revenue for the plant. A second type of benefit that should be considered is cost avoidance (e.g., risk of failure) or the eliminating of unnecessary or excessive costs, such as high maintenance costs created by a machine with a history of chronic problems. To establish this type of benefit, the cost history of the machine needs to be gathered. These data will provide the reference for existing costs. Then the projected costs will be calculated. Using the vendor's recommendation for routine maintenance and upkeep, along with the internal labor and material costs, the annual and lifetime costs of the upgrade or modification can be calculated.

Once the costs and the benefits have been quantified, an assessment should be made to determine the value of the potential corrective actions. Although different companies have different payback periods, the cost–benefit analysis must clearly show that the recommended corrective action will offset incurred costs and generate a measurable improvement over this life cycle. In general, the cost portion should include 3–5 years of historical costs, and the benefits should be projected over an equal stated period. This method will provide a more accurate picture of the real life-cycle improvements. A valid action will result in greater benefits than the costs.

16.1.6 Assessment of Corrective Actions

Once the corrective action has been approved and implemented, the final task in a root-cause analysis is to verify that this action actually corrected the problem.

Feasibility, in terms of technical difficulty and time of implementation, and manufacturability are factors that should be considered in evaluating the effectiveness of a solution. In addition, the completeness of the solution is critical. For example, revising procedures may not be a complete solution if personnel do not follow the procedures, and a more complete solution may require staff training on the revised procedures and verification that they are observed. The intended scope of the solution should be clearly understood, in terms of its specificity or generic impact. For example, training may focus on the proper use of manufacturing equipment, or on the overall manufacturing process (Wilson et al. 1993). Furthermore, both the long- and short-terms effects of the solution should be considered. For example, if training is part of the solution, its implementation, on either an isolated or continued basis, and personnel turnover, would impact on the long-term effectiveness of training.

Although compliance with standards or regulations is a well-defined indicator of the solution's acceptability, it does not guarantee its suitability.

16.2 No-Fault-Found

No-fault-found (NFF) implies that a failure (fault) occurred or was reported to have occurred during a product's use. The product was analyzed or tested to confirm the failure, but "a failure or fault" could be not found.

A failure in a product occurs when it no longer performs its intended function (Thomas et al. 2002). An intermittent failure can be defined as failure for a limited period of time, and then recovers its ability to perform its required function (IEEE 100 2000). The "failure" of the product may not be easily predicted, nor is it necessarily repeatable. However, an intermittent failure can be, and often is, recurrent.

Intermittent failures can be a cause of NFF occurrences in electronic products and systems. NFF implies that a failure (fault) occurred or was reported to have occurred during a product's use. The product was analyzed or tested to confirm the failure, but "a failure or fault" could be not found. A common example of the NFF phenomenon occurs when your computer "hangs up." Clearly, a "failure" has occurred. However, if the computer is rebooted, it often works again.

Terms related to NFF include trouble-not-identified (TNI), cannot duplicate (CND), no- trouble-found (NTF), or retest OK (RTOK) (IEEE 100 2000; Johnson and Rabe 1995; Kimseng et al. 1999; Maher 2000; Pecht and Ramappan 1992b; Sorensen et al. 1994; Thomas et al. 2002). These terms can be defined as follows:

- *Trouble Not Identified (TNI)*. A failure occurred or was reported to have occurred in service or in manufacturing of a product. But diagnostic testing could not identify the exact problem.
- *Cannot Duplicate (CND)*. Failures that occur during manufacture or field operation of a product that cannot be verified or assigned
- *No Problem Found (NPF)*. A problem occurred or was reported to have occurred in field, but the problem was not found during verification testing.
- *Retest OK*. A failure occurred or was reported to have occurred in a product. On retesting the product at the factory, test results indicated that it was OK.

The commonality of these terms is that a failure may have occurred but cannot be verified, replicated at will, or attributed to a specific root cause, failure site, or failure mode. In this chapter, we will use the generic term NFF.

The impact of NFF and intermittent failures can be profound. Due to their characteristics, manufacturers may assume a cause(s) rather than spend the time and cost to determine a root cause. For example, a hard drive supplier claimed NFFs were not failures and allowed all NFF products to be return back into the field. Later, it was determined that these products had a significantly higher return rate, suggesting that the NFF condition was actually a result of intermittent failures in the product. The result was increased maintenance costs, decreased equipment availability, increased customer inconvenience, reduced customer confidence, damaged company reputation, and in some cases, potential safety hazards.

NFF and intermittent failures have been reported in the automotive, avionics and telecommunications, computer, and consumer industries where they represent a significant percentage of reported warranty returns and field returns (Chan and Englert 2001), resulting in significant costs. The percentage of NFF and intermittent failures varies with the industry and products. For example, Sorensen (2003) stated the occurrence of intermittent and NFF failures on military aircraft can be as high as 50% percentage based on information from *Defense Electronics Magazine*.

An avionics field failure study, conducted in 1990, showed that NFF observations of electronics in avionics systems represented 21–70% of the total failures depending on the application (Sorensen 2003). NFF observations reported by commercial airlines and military repair depots have been found to be as high as 50–60% (Pecht and Ramappan 1992b).

Ford's thick film ignition (TFI) module may well be the most widespread intermittent fault problem ever reported. The failed TFI module could cause the vehicle to stall and "die" on the highway at any time (Castelli et al. 2003). However, the failed ignition modules often passed the required engineering tests established to reflect the design intent. In October 2001, Ford agreed to the largest automotive class-action settlement in history, promising to reimburse drivers for the TFI that could cause their cars to stall.

Kimseng et al. (1999) studied an intermittent failure in digital electronic cruise control modules (CCM) used in automobiles. This intermittent failure was corroborated by the fact that 96% of the modules returned to the vehicle manufacturer due to customer complaints operated properly and passed the bench tests. Kimseng concluded that the tests conducted by the vehicle manufacturer on returned products were not representative of the actual automotive environments, nor were they conducted in a manner to assess actual failures. This inappropriate testing may also lead to the high percentage of NFF.

An NFF implies that a failure was reported to have occurred during a product's use, but upon subsequent use, analysis, and/or testing, the failure was no longer observable. An intermittent failure is a failure of a product function or performance characteristic over some period of time, followed by the subsequent recovery, generally without any known external corrective action. Intermittent failures are thus common causes of NFF. A typical NFF situation occurs when the user of a product under warranty reports an intermittent failure, but the manufacturer's tests of the returned product cannot detect failure or faults in the product.

In many companies, the NFF category does not constitute a product failure statistic, because it is not considered that a failure has occurred. In some cases, the manufacturer may not understand the need or have little incentive to uncover the root cause of the problem encountered by the user of the product. The impact of this lost opportunity can be profound and generally leads to increased product cost due to extra shipping costs, warranty, and diagnostic and labor time. In addition, there can be unknown reliability and potential safety hazards of this product if the NFF product is put back in use. In addition, a high NFF rate in a product can cause customer inconvenience, loss of customer confidence, and can damage a company's reputation.

The cause-and-effect diagram is an efficient approach to analyze NFF observations in removed or returned electronic product from the field. Factors to be considered include people-related causes, such as communication and knowledge of products and equipments, capability of test equipments, test environment and test conditions, test methods, and materials related causes. Using such diagrams, the engineering team can identify the potential causes for the occurrence of intermittent failures or NFF, and isolate the most probable cause by conducting specific tests based on the possible failure mechanisms. The test results can be fed back to manufacturing, design, and logistics support teams to improve product reliability.

The cause-and-effect diagram is also an efficient approach to analyze intermittent failure observations in electronic products. The major causes of intermittent failures in electronic assemblies can be placed into four categories: printed circuit board, connectors, components and component–PCB interconnects, and specific failure mechanisms in each category that are prone to cause intermittent failures in electronic assemblies can be identified. The characteristics of intermittent failures can be summarized as unpredictable, unnecessarily repeatable, and often recurrent. As with NFF, root-cause analysis of intermittent failures can help in manufacturing and design. The analysis can also be used to help tailor tests to locate the intermittent failures in the product qualification process.

It is a good business and engineering practice to start with the premise that field returns are field failures, unless some alternative reason can be verified. It must not be assumed that a returned product that passes tests is necessarily free from faults. Companies should start with the premise that field returns are field failures. NFF statistics should not be used to ignore, mitigate, or transfer the field return problem to a nonfailure status. Once this premise is accepted, then the comprehensive cause-and-effect diagram can help identify all the possible causes for a field failure.

16.2.1 An Approach to Assess NFF

To evaluate all the possible causes of intermittent failures and NFF, a systematic method is required. One such systematic method is the development of a cause-and-effect diagram or Ishikawa diagram or fishbone diagram (Ishikawa 1985). This diagram is a graphical tool which allows an individual or team(s) to identify, explore, and display the relationship between an effect and all of its potential causes.

The general causes of NFF can be categorized into people (human), machine, methods, and intermittent failures. In each category, the causes are further broken down into subcategories. For example, people causes are subdivided into communication, skills, and behavior. Machine causes are subdivided into measurement tools and test equipments. Methods causes are subdivided into test methods, handling methods and failure analysis methods. Figure 16.5 is a fishbone diagram for NFF in electronic products. Intermittent failure causes are grouped under five categories: PCBs, components, interconnects, connectors, and software. There may be other categories and subcategories based on the specific product and its life-cycle profile.

16.2.1.1 People Category Lack of skills or proper behavior of engineers, maintenance personnel, and technicians can contribute to NFF. For example, "sneak circuits" are well-known electrical design causes of latent failures. A sneak circuit is an unexpected path or logic flow within a system which, under certain unique conditions, can initiate an undesired function or inhibit a desired function. The path may arise from hardware, software, operator actions, or combinations of these elements. Sneak circuits are not the result of hardware failure but are latent conditions, inadvertently

16 Analyzing Product Failures and Root Causes

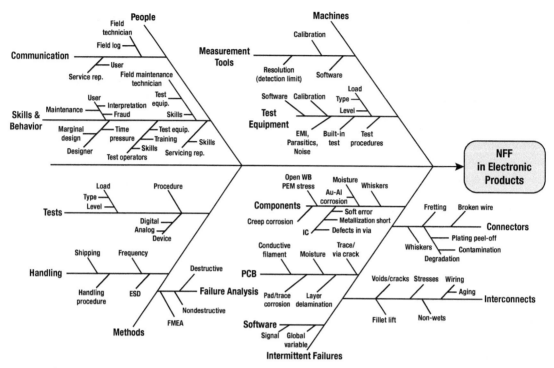

Figure 16.5 An example cause and effect diagram of no-fault-found (NFF) in electronic products.

designed into the system, coded into the software program, or triggered by human error. These conditions generally require rigorous system-level tests to uncover (OPS ALACARTE Company 2006).

Communications can also precipitate NFF observations. For example, consider the customer of a car that experiences an intermittent failure. The customer reports this problem to the service depot, where the service clerk documents the problem in a service request. At this step alone, people's ability to communicate, understand, and summarize the problem, as well as the condition leading to the problem, play a key role in its correct documentation of the problem. This includes the ability of the customer to clearly explain the problem, the service clerk's ability to understand the customer's problem, and the service clerk's ability to document the problem correctly. Without correct problem identification, the service technician could come to the erroneous conclusion of NFF, even if the problem actually exists.

NFF observations can also occur by people driven by warranty or insurance claims, who replace multiple items to repair one fault or resort to fraudulent reporting of a problem that never occurred. In some cases, car dealers may replace multiple subsystems under warranty to find a problem. Those subsystems that were removed but had not failed will be identified as NFF when tested.

16.2.1.2 Machine Category The limitations of test equipment and measurement tools can lead to the conclusion of NFF. For example, limits on measurement resolutions and measurement intervals can result in the inability to detect intermittent failures. In addition, the incapability of test equipment to simulate the actual loading conditions (e.g., vibrations, temperatures, and human operations) exhibited in the field can also contribute the NFF conclusion.

The regular and proper calibration on equipments is also the key to avoid NFF conclusions. Without accurate, capable, and reliable test equipment and measurement tools, the failed products from the field may pass bench tests.

16.2.1.3 Method Category An inadequate or improper test procedure may lead to NFF. For example, complex products such as microprocessors may not be 100% testable if the allowable test time is too short. In addition, if the combinations of diagnostic test conditions (e.g., temperature, humidity, and vibration) are different from the field conditions where failure occurred, the failure may not be duplicated. A failure can occur due to a unique combination of loading conditions. Thus diagnostic tests may require the combined loading conditions that can occur in the field.

Properly handling the returned electronic hardware is also a key to detect NFF conditions. Handling methods should follow the established procedures in preventing uncontrolled effects. For example, anti-electrostatic discharge (ESD) packages are required for ESD sensitive products. To prevent deformation due to handling, the vibration environment should be well controlled. Clean room storage may be also necessary for debris sensitive products. Improper handling can both "heal" some failure mechanisms and induce problems that can mask the real problem reported in the field.

The failure analysis methods also play a key role in the verification of failure. Failure analysis should include nondestructive methods followed by destructive analysis as needed. Before embarking on failure analysis, it is good practice to develop a FMMEA document (Ganesan et al. 2005a; Mishra et al. 2004). Incomplete or insufficient failure analysis can lead to erroneous conclusions.

16.2.1.4 Intermittent Failure Category Intermittent failure is one of the main causes of NFF. Intermittent failures in products can be caused by software or hardware. For example, if a global variable is read and rewritten over another global variable, a miscalculation can arise and lead to product failure (Boehm 2004). However, when the global variables are reset, perhaps upon rebooting the computer, the product can return to normal function. Intermittent failures of hardware, for example, electronic assemblies, can be divided into four categories: printed circuit board, components, connectors, and interconnects. In each category, the causes can be furthered subdivided as shown in Figure 16.6.

16.2.2 Common Mode Failure

Reliability is a key issue in critical applications such as aircraft control systems, and nuclear power plants. It is imperative that systems used in these applications perform reliably for a specified time. To increase the reliability of such systems, redundancy is commonly employed. Redundancy can be at component level or at system level. In spite of redundancy employed at various levels, systems still fail. The failure is due to common mode failures (CMFs).

CMFs occur when a single root cause, with the help of some dependencies between the components in a system or systems themselves, causes failure in multiple components or systems. Thus, an apparent "fail-safe" redundant system can easily fail within

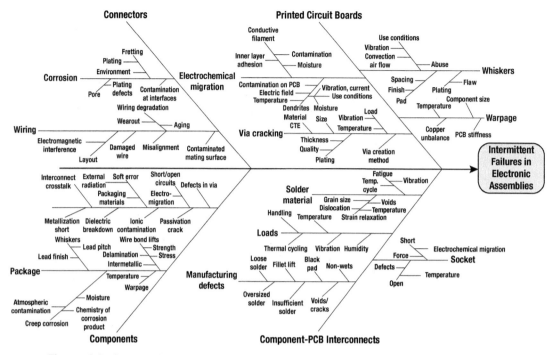

Figure 16.6 An example cause-and-effect diagram of intermittent failures in electronic assemblies.

a short period of time. Such failures can be catastrophic especially in case of critical applications as mentioned earlier. It is therefore necessary to design these systems taking CMF causes and interdependencies into account.

A very well-known example of a CMF is the massive fire of 1975 at the Browns Ferry nuclear power plant in Alabama, USA. The source of the CMF observed in this case was due to human errors. Two power plant operators used a candle to check for air leak between the cable room and one of the reactor buildings. The reactor building had a negative air pressure. Due to this negative pressure, the candle's flame was sucked along the conduct, and the methane seal for these conducts at the walls, caught fire. More than 2000 cables were damaged, including those of automatic emergency shutdown systems and of all manually operated valves except four. Thus, it can be seen in this case that one root cause, that is, using fire inside the reactor, along with coupling factors such as location of all the cables along with the negative pressure resulted in simultaneous failure of different devices in the reactor.

This section discusses the various sources of CMFs in redundant systems with emphasis on electronics systems. Various analysis tools and models to predict and analyze CMFs are presented along with several strategies that can be employed to reduce the occurrence and impact of CMFs in redundant systems. Finally, a case study of CMFs in avionics industry is discussed.

16.2.3 Concept of Common Mode Failure

CMF, in its simplest form, is defined as a failure that occurs when a single fault results in the failure of multiple components. The failure of these components may be simultaneous or over a short period of time.

CMF is a special class of dependent failures. In most of the current complex systems, different parts or components have interdependent functionality. This means that functional performance of one part/component determines the functional performance of other part/component. Thus, there is dependence between different parts/components of a system. This is not restricted to a single system. In some critical applications, such as aviation and nuclear power plants, to ensure reliable function of a system, more than one system are connected in parallel to form an overall redundant system. There is also interdependence between these systems in parallel, such as shared load and shared power.

In complex systems, reliability and related risk are impacted by the interfaces between components, subsystems, systems, and the environment. Thus some parts of the system depends on or interacts with another part of the system. Failures arising from dependencies are often difficult to quantify and model. Consider A as failure of one of the parts/components in a system and B as failure of another part/component. Dependent failure occurs when both the failure events A and B occurs simultaneously or within a short period of time (Mauri 2000). The probability of this happening is given by conditional probability as follows:

$$P(A \text{ and } B) = P(A)P(B|A) = P(B)P(A|B). \qquad (16.1)$$

Here, $P(A)$ and $P(B)$ are the probabilities of A and B, respectively and $P(A|B)$ is the conditional probability that simply states that probability of failure event A occurring given that the failure event B has occurred. If one assumes that there is no interdependence between different parts/components of the systems or different modules of a redundant system, that is, the failures in those parts/components are independent then the likelihood of failure events A and B occurring simultaneously is the product of individual probabilities of failure events as shown as given below

$$P(A \text{ and } B) = P(A)P(B). \qquad (16.2)$$

If A and B are dependent, then

$$P(A \text{ and } B) \neq P(A)P(B). \qquad (16.3)$$

In many situations, based on dependencies, $P(A \text{ and } B) > P(A)P(B)$, and we end up overestimating the reliability of the system. There are many different ways to classify dependencies (Guenzi 2010). They may be due to intended functional and physical characteristics of the system or due to external factors and unintended characteristics. Thus, we could say that dependencies are either intrinsic or extrinsic.

> *Intrinsic Dependency.* When the functional state of one component is affected by the functional state of another component, it is called intrinsic dependency, and this stems from the way the system is designed to perform its intended function. Following are some of the subclasses of intrinsic dependency:
> 1. *Functional Requirement Dependency.* This refers to the case where the functional state of component A effects or related to the functional state or requirements of component B. Some examples are the following:

- B is not needed when A works.
- B is not needed when A does not work.
- B is needed when A works.
- B is needed when A fails.
- Load on B increases when A does not work or vice versa.

2. *Functional Input Dependency (or Functional Unavailability).* Here, the functional status of B depends on the functional status of A. In this case, A must work for B to work. An example is the dependence of a pump on electric power.

3. *Cascade Failure.* In this case, the failure of A leads to the failure of B. One example is that an over-current failure of a power supply may cause failure of all the components it feeds. After the power supply is made operable, the components that have been damaged by the overcurrent will still remain inoperable.

Other types of dependencies are based on the combination of the earlier three dependencies. When several components are dependent on the same component, it is called shared equipment dependency. For example, if both B and C are functionally dependent on A, then B and C have shared equipment dependency.

Extrinsic Dependency. These dependencies are not inherent to the design of the system but are physically external to the system. The following are some examples of extrinsic dependencies:

1. *Physical and Environmental Dependency.* This category includes dependencies due to common environmental factors, including harsh or abnormal environment created by a component. For example, high vibrations of A cause failure of B.

2. *Human Interactions.* An example is the failure of multiple components due to the same maintenance error.

Mauri (2000) has presented a table of definitions of dependent failures and its classification into common cause failures and cascade failures (see Table 16.3).

It can be seen from the Table 16.3 that CMF is a subset of common cause failure, which in turn is a subset of dependent failure. This means that all CMFs are common cause failures but not vice versa. In other words, if two or more parts/components in a system or modules of a redundant system fail via same mode then it implies that there was a common cause that brought about the CMFs. On the other hand, if two or more components in a system or modules of a redundant system face a common cause, then this does not mean that the components or modules will definitely fail in common mode. The structural integrity, design, type of components, and layout of the different components will decide whether the components or modules will fail in common mode.

Figure 16.7 provides a graphical presentation of CMF as a part of dependent failures. Thus, now it can be inferred that a CMFs should have common cause that initiates this failure. This common cause is also known as the root cause, which through coupling factors results in failure of more than one component or module, simultaneously resulting in CMF (Figure 16.8). The root cause determines the

16.2 No-Fault-Found

Table 16.3 Definitions and hierarchy of dependent failures, common cause failures, and common mode failures

Dependent failure (DF)		The likelihood of a set of events, the probability of which cannot be expressed as a simple product of the unconditional failure probabilities of the individual events.
	Common cause failure (CCF)	This is a specific type of dependent failure that arises in redundant components where simultaneous (or near simultaneous) multiple failures result in different channels from a single shared source.
	Common mode failure (CMF)	This term is reserved for common cause failures in which multiple items fail in the same mode.
	Cascade failure (CMF)	These are all those dependent failures that are not common cause, that is, they do not affect redundant components.
Note: The term "dependent failure" as defined earlier is designed to cover all definitions of failures that are not independent. From this definition of dependent failure, it is clear that an independent failure is one where the failure of a set of events is expressible as simple product of individual-event unconditional failure probabilities.		

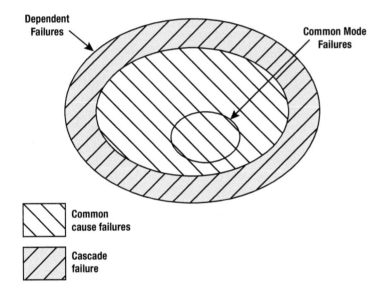

Figure 16.7 Schematic representation of relationship between dependent failures, common cause and common mode failures.

mechanism with which CMF occurs. For example, consider a highly redundant system in a space exploration aircraft sent on a planet exploration mission. If the temperature on the planet of interest is not considered properly, then all the components used in the system will be designed for maximum temperature rating less than that is present on the planet. Thus, no matter how much redundancy is present, all the components will fail, simultaneously resulting in a CMF. In this case, temperature is the root cause of failure, and designing of all the components at the same temperature rating is a coupling factor. Root causes can be external or internal to the system. They can also arise from failures of operating equipment or human errors. For example, environmental loading is an external root cause, whereas the ill-functioning of a component

359

in a system resulting in failures of other components is an internal root cause. Generally, four different types of failure mechanisms are recognized. They are as follows:

1. Coincidence of failures of two or more identical components in separate channels of a redundant system, due to common cause (may have common failure mode too).
2. Coincidence of failures of two or more different components in separate channels of a redundant system, due to common cause (may have common failure mode too)
3. Failures of one or more components that result in coincidence of failures of one or more other components not necessarily of the same type, as the consequence of single initial cause.
4. Failure of some single component or service common to all channels in a redundant system.

Thus, a CMF as defined by Edwards and Watson (1979) is "the result of an event, which because of dependencies, causes a coincidence of failure states of components in two or more separate channels of a redundancy system, leading to the defined system failing to perform its intended function."

16.2.4 Modeling and Analysis for Dependencies for Reliability Analysis

Intrinsic dependencies are considered when we develop system logic models and fault tree analysis for risk and reliability analysis. Functional dependencies arising from the dependence of systems on electric power are included in the logic model by including basic events, which represent component failure modes associated with the failure of the electric power supply system. Failures resulting with the failure of another component (cascading or propagating failures) are also explicitly modeled. Human error is included as branches on event trees or as basic events on fault trees. Maintenance errors are considered in fault trees or included in overall component failure probability.

Extrinsic dependencies based on physical or environmental factors such as temperature, vibrations, and radiation, are modeled as part of the physical processes for degradation or failure. Dependent failures whose root cause is not explicitly considered are also known as common cause failures and this category can be accounted in probabilistic risk analysis by considering common basic cause event. A common basic cause event is defined as the failure of more than one component due to shared cause during the system mission. Components that fail due to a shared cause may fail in the same functional mode. CMF is the failure of multiple components in the same mode and thus is a subset of common cause failures as shown in Figure 16.7. The following are some examples of common cause failures:

- All redundant auxiliary feedwater pumps failed at the Three Mile Island nuclear power plant.
- Hydrazine leaks leading to two APU explosions on STS-9.
- Failure of two O-rings causing hot gas blow-by in the space shuttle solid rocket booster of Shuttle Flight 51L.

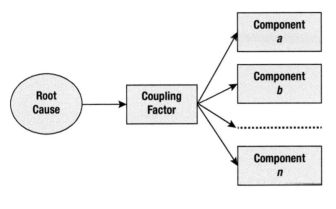

Figure 16.8 Graphical representation of common mode failure occurrence.

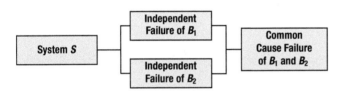

Figure 16.9 β factor system block diagram.

- Two redundant circuit boards failed due to electrostatic shock by technician during replacement of an adjustment unit.
- Worker accidently tripped two redundant pumps by placing a ladder near pump motors to paint the ceiling.
- Check valves installed backward, blocked flow in two redundant lines.

Common cause failures may also be viewed as being caused by the presence of two factors: (1) a root cause which is reason for failure of the components which failed in the common cause failure event and (2) a coupling factor which is responsible for the event to involve multiple components (Figure 16.8).

Failure of two identical redundant electronic devices due to exposure to excessively high temperatures shows that heat is the root cause and the coupling factor is that both of them are identical and are being subjected to the same harsh environment.

One simple way to model dependencies and the impact of common cause failures is the beta-factor model. Consider a simple redundancy of two identical components B_1 and B_2. Then we can further evaluate the failure of components as failing independently of each other and also failing due to common cause failures. Let us assume that the total component failure frequency [λ_T] is the sum of independent failure frequency [λ_I] and the common cause failure frequency [λ_C] over some given period. Then the β factor is defined as

$$\beta = \frac{\lambda_C}{\lambda_T}. \qquad (16.4)$$

Thus for this system with two identical components, the β factor system block diagram is shown in Figure 16.9.

16.2.5 Common Mode Failure Root Causes

As mentioned above, CMFs have a common cause (root cause) that initiates failures in more than one components of a redundant system. The following is a discussion of CMF root causes. The root causes can be classified into three broad categories, namely, design and manufacturing errors, catastrophic events, and human interaction errors.

16.2.5.1 Design and Manufacturing Errors As the name suggests, these errors are induced in a redundant system due to design inadequacies and manufacturing defects. These errors are qualitative in nature, meaning that their severity can be measured using some unit. The design errors are difficult to predict as the prediction power depends on the extent of knowledge about the possible loads a system will face in its life cycle. Thus, it is hard to completely remove design errors from a system. Design errors include the following:

- Inability of a component or system to perform its intended function. Limitations on carrying out exhaustive tests or simulate actual operating conditions can prohibit detection of all the design errors.
- Inadequate, poorly designed or harmful periodic tests can precipitate errors in the design that would not have been observed in case of regular operating conditions.
- Poor optimization regarding CMFs. Protection against one CMF can increase susceptibility to other CMFs.
- Inadequate study during design procedure.

Let us look at some of the examples of design errors that result in CMFs. Consider a variation of the example provided earlier about the planetary mission. Consider a redundant system installed in a room where the relative humidity is controlled with the help of an air conditioner. Now, suppose some classes of components in parallel paths of the redundant system are designed to function properly until some relative humidity level in the room. If the air conditioning system fails, then the relative humidity level in the room will increase and if it crosses the designed limit of the components, then they will fail inadvertently. Here, the root cause is the increase in relative humidity and the coupling factor is location of all the parallel paths in the same room.

In case of high safety requirements in systems, periodic tests are carried out to ensure that all the components in the parallel paths are functioning properly. If the tests are poorly designed, in a way that some loads applied during were not considered during designing of the component, then there is a possibility of weakening of the components. Thus, the components might not be able to withstand loads or function properly when put in operation. For same components, it is possible that this will occur in more than one component, leading to CMF in a redundant system.

Similarly, if a redundant system is designed to be "fail-safe" in high temperature applications, then there is possibility of weakening of the system in a moderate to high radiation applications. For instance, if an electronic component on a board is designed to better dissipate and radiate heat to the surroundings, then the size of the component has to be increased. This will lead to reduction in physical separation of components on board. Thus, there is a possibility of more interference between the components on same or nearby boards, leading to malfunctioning of more than one component.

Manufacturing errors arise in components due to inferior quality of manufacturing processes or materials. Most of the time, these errors are due to improper technical specifications or technological errors. Manufacturing errors can affect few or all components produced. Now, if the same components are used in a system, then the errors induced during their manufacturing will precipitate and eventually lead to failure. Since the same manufacturing error is present in similar components, it is quite likely that these components will fail simultaneously or within a short period of time.

These causes can be detected and rectified to get a better redundant system. With increase in technical know-how, designers will be better equipped to model and simulate comprehensive life-cycle operating conditions for improved design of components or parts.

16.2.5.2 Human Interaction Errors Even if a system is well designed to eliminate all design errors and perfectly manufactured to remove all the manufacturing errors, it is still prone to CMFs due to human interaction induced failures. No system is completely automatic and requires some kind of human control and interaction. For example, a person might be required to operate some control, and human activities are required for maintenance and repair of a system. Human interaction with the system can occur at various phases, such as installation, operation, and maintenance. If errors induced during these phases affect functioning of more than one component or module, then those components and modules can fail simultaneously during the system's operation period.

Systems are designed to operate at some predefined operating conditions. Designed systems are tested for reliability at those operating conditions. If an operator takes an incorrect action, then the system can be subjected to conditions different from the initially considered operating conditions during the design process. Depending upon the compliance of system design, the system will be able to withstand these conditions. If the system is not compliant enough, then it will fail.

Human errors during maintenance activities, such as improper implementation of maintenance activities and improper handling of the system can create conditions suitable for more than one component. For instance, if during handling of an electronic board with components mounted on it, a crucial component is damaged. The output for this component is used by more than one components, then those components in the system will fail, that is, will not function. Especially in case of the current high density electronic circuitry, where components with delicate structures are used, induction of failures due to human interactions is a key concern.

In case of software, human errors are the most common and important ones. Software is completely developed by humans. Any mistakes made by the programmer will lead to software malfunctioning. This is analogous to manufacturing defects. These errors are hard to detect due to the complex nature of today's software. Moreover, these errors can spread from one version of software to another.

16.2.5.3 Catastrophic Events Catastrophic events, such as lightning, flooding, and earthquakes, can act as root cause for CMFs in electronic systems. The probability of these root causes occurring is fairly less. Moreover, catastrophic events are also random in nature. Thus, it is difficult to predict the occurrence of these events and therefore is hard to model.

Consider a room consisting of numerous servers storing critical information. If the room is at ground level, then it is prone to flooding. Now if the room gets flooded, then all the servers will go under water. Shorting of all the circuits in all the computers will occur simultaneously, and they will fail at the same time, losing all the important information. Here, flooding of the room is the root cause, and the location of all the computers in a room at lower level is the coupling factor.

Another example could be a control module on an aircraft. Consider a redundant wing control module placed at back of the plane. Thus, all the paths in the redundant system are at the same location. Suppose the components in the module are rated for a specific voltage. While passing through clouds, lightning strikes the back part of plane. If the voltage induced by that lightning is greater than that of the voltage ratings, then the components in the redundant system will fail simultaneously leading to CMF. Here lightning is the root cause, while the location of all the parallel paths of the redundant system at the same location is the coupling factor.

In case of electronic components, common examples of catastrophic events are electrostatic discharge, electrical overstress, mechanical overstress, thermal overstress. These root causes can be external or internal to the system.

16.2.6 Common Mode Failure Analysis

CMFs as discussed previously are the type of failures in which one cause results in failures of more than one system either simultaneously or within a short period of time. CMFs become especially critical in redundant systems where safety is the key issue. An example of this could be various control systems on an aircraft. Thus, it is imperative to analyze the possible CMFs in the earlier stages so that effective preventive actions can be taken well in advance. This section discusses the different models employed to analyze CMFs.

16.2.6.1 Failure Mode and Effect Analysis (FMEA) Thus, FMEA is technique used to identify potential failure modes and evaluate the effects of those failures. This approach (see Figure 16.10) can be applied to a component, assembly, or a system. This analysis starts with an intense review of component failure modes and its effect on the assemblies and various further dependent systems. The main intention in this analysis is to determine the root cause of failures in a component that can initiate CMFs. Thus, once the root cause of failure is identified, then corrective actions can be taken so as to avoid component failures that can lead to CMFs. For example, consider a case of a redundant electronics system. In this system, electronic components are held on the same type of printed wiring board that has a high CTE mismatch with most of the electronic components. Now, if this redundant system is placed in an environment with high operating temperatures with little thermal isolation, then similar components will fail due to cracks in solders resulting in open circuit. Here, the root cause, CTE mismatch-induced stresses, in the presence of the coupling factor (location), brings about a CMF of a redundant system. Hence, it is essential to take into account coupling factors while performing FMEA. The following is the list of coupling factors that can be used in FMEA analysis (Childs and Mosleh 1999).

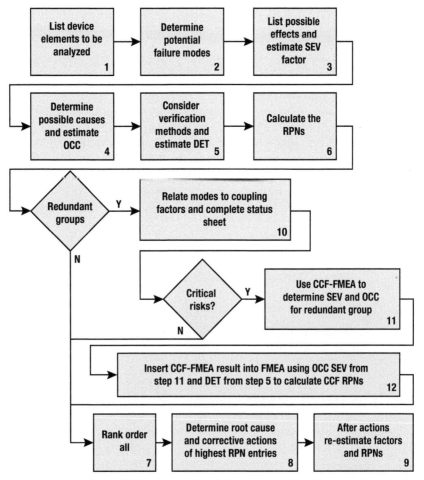

Figure 16.10 FMEA process flow for common mode failure analysis.

Common Operation Usage

Age. The age of the components in a redundant system can act as a coupling factor in CMFs. Suppose components in a redundant system are used for a considerable amount of time and they are in the wearout phase, then there is higher probability of their failure. If similar components are used in different parallel blocks of a redundant system, then the probability of their simultaneous failure goes up, which can lead to CMF in the redundant system.

Maturity. This is related to the compatibility issues. Suppose if worn-out components in a redundant system are replaced with new components with higher maturity level, then there is a possibility that the new components may not be compatible with the old systems. This can lead to simultaneous failure of different subsystems in a redundant system.

Shared Work Environment

Proximity. If the different elements of a redundant system are placed near each other, then it is possible that operating conditions of one element can affect operation of another closely placed element. For instance, electromagnetic field produced by one electronic component in a system can adversely affect functioning of another electronic component. Heat produced by a high powered device can adversely affect a heat sensitive device if it is placed close to the high power producing device.

Medium Similarity. If the surrounding medium for all the redundant element is same, then there is a possibility of simultaneous failure of different redundant elements. For example consider, a redundant control system in an aircraft is placed at the nose of the aircraft. If this system consists of components that are highly sensitive to pressure, then there is a high probability of simultaneous failure of those components. This can lead to CMFs in the redundant system. Another example could be failure of humidity sensitive components in a redundant system place in a highly humid environment.

Functional Coupling

Same Energy Source. Suppose if the different elements of a redundant system are operating on the same energy source then in case of source failure, energy to all the elements will cease simultaneously.

Same Input/Output. If different parallel paths of a redundant system have a common input, then in case of a false input all the element of the system will perform improperly leading to CMF.

Same Load. In case a redundant system shares a common load, then failure in the load can lead to failure in all the paths of the redundant system.

Common Personnel If common personnel are not used for design, installation, construction, operations, and maintenance, then there is a possibility of dissimilarities occurring, which can lead to failure precipitation in different components of a redundant system leading to CMF.

Documentation Incomplete or incorrect procedures, displays, and drawings can lead to design defects, and defect produced during installation, maintenance in various elements of a redundant system leading to a possible CMF.

Similarity in Components If all the components used in a redundant system are similar, then the probability of them failing simultaneously goes up as their response to the operating conditions will be similar and will also fail under same conditions.

FMEA thus can provide a tool for component selection in a redundant system, as well as provide a guideline for proper design and maintenance of the system so as to prevent or reduce CMFs. However, FMEA is not a replacement for a more detailed analysis method.

16.2.7 Common Mode Failure Occurrence and Impact Reduction

CMFs can produce failures in redundant systems that can lead to catastrophes. This is especially critical in case of applications such as airplanes and nuclear plants. Hence, it is necessary to reduce the occurrence of CMFs in such applications or reduce their severity of impact. This can be done during the initial stages of design and development. The following are some strategies that can be employed to reduce occurrence and impact of CMFs.

16.2.7.1 Fault Detection Systems It is possible to establish defense against CMFs in systems by using programmable electronic systems. Fault detection systems are indented to detect faults in components or systems before they can cause failures, so that it increases the robustness of the system. Fault detection systems are further divided into two classes:

Periodic Tests. Periodic tests test the correct operation of a system resource at a given time. These tests are run on power-on mode and at periodic intervals during operation. It is thus unable to detect transient faults or an error occurring between two tests. Periodic tests are basically used to detect accumulation of faults that can result in CMFs.

Online Monitoring. This technique continuously monitors the functioning of all or part of a system. Thus, faults are detected instantaneously within a very short period of time. Thus, with a system that can effectively compare the performance of all or part of the system with some preset standard, it is able to detect any faults that might occur during operation of a system.

16.2.7.2 Separation Separation is an effective technique that can be employed in redundant systems to reduce the propagation of failures from one function to another or one component to another. Following are some methods of separation.

Physical Separation In this type of separation, different paths of the redundant system or different components in the same path are physically separated to avoid interaction between them that can cause malfunctioning. Consider a component that produces large electromagnetic field around it. If another component that is more susceptible to electromagnetic field is in vicinity of this component, then there can be some malfunctioning of the component that is sensitive to electromagnetic field. Thus, in order to avoid this interference effect, it is necessary to separate these components with proper distance.

Electrical Isolation It is necessary to have electrical isolation between various paths of redundant systems to avoid failure due to events such as sudden increase in current through the paths. If all the paths are electrically connected, the sudden rise in current will affect similar components that are sensitive to high currents in the different parallel paths. This can result in CMF in the redundant system.

Power Supply Separation If all the channels of a redundant system are connect to a same power supply, then in case of power supply failure all the channels in the system will fail leading to CMF of the system. To avoid this, different channels of the redundant system can be put on different power supply thus ensuring that at least one of the channel will work in case of a power supply failure.

16.2.7.3 Diversity Diversity implies variation in different components, subsystems of multiple path redundant system so as to reduce the occurrence of CMF in such systems. Diversity reduces the probability of simultaneous failure of similar components due to common root cause. The following are different types of diversities that can be employed so as to reduce CMFs.

Design Diversity Different designs for different paths in a redundant system can reduce CMFs that occur due to design flaws. If a similar design in terms of layout, placement of components, and so on is employed for all the paths in a redundant system, and if there is an inherent flaw in that design, then there is a high chance that all the paths will fail simultaneously, causing CMFs in the redundant system. Thus, different designs for the paths can reduce the susceptibility of the redundant system to CMFs. For example, digital instrumentation and control systems in nuclear power plants employ independent protection systems.

Functional Diversity Functionality of components in a redundant system depends upon the operating conditions and environmental conditions. Thus if similar kinds of components are used in a system, then under certain operating conditions or environmental conditions that cause malfunctioning of the components, it will result in simultaneous or nearly simultaneous failure of all the similar components leading to CMF in the system. Functional diversity consists of different parameters, different technologies, and so on.

Manufacturing Diversity One of the major sources of CMFs is manufacturing defects induced in the components or assemblies. Thus, if all the components in all the channels are from the same manufacturers that have manufacturing defect, then there is a high probability that all the components in different channels will fail, simultaneously leading to CMF in the system. Thus, using components in different paths from different manufacturers reduces the probability that the components have similar manufacturing defects that can lead to CMFs. For example, most airplane manufacturing companies like Airbus and Boeing use components designed and manufactured by different equipment manufacturers.

Equipment Diversity If similar equipment is employed in steps such as maintenance, testing that can produce defects in components, then it is possible that components in all the paths of a redundant system will incept defects from the equipment. These defects can nucleate during the operation of the system leading to simultaneous failure of the components in different paths of the redundant systems. This can lead to CMFs in the redundant system. For example Airbus (EADS) uses one channel for control and another channel for monitoring in their flight control system.

Signal Diversity If all the channels of a redundant system use same input signal, then in case of an incorrect signal, all the channels will fail or operate improperly. This is nothing but CMF that occurs due to wrong input signals.

Human Diversity It was explained earlier that one of the sources of CMFs is human interaction-induced defects. By employing human diversity, it is possible to reduce these defects. Human interaction effects vary during different phases in the life cycle of a system. Thus, one can have different groups working on the same module or subsystem.

16.2.7.4 Fail-Safe Design
The notion of fail-safe is defined here as a theoretical condition that is attained if a safety function were to remain unchanged in case of a failure in the power supply or in a component contributing to this function. A fail-safe design will ensure that short-circuits, open-circuits, and all such sources of failures are minimized to ensure that the safety function of the system is not altered. Moreover, dynamic signals are preferred over static signals. The earlier-mentioned methods, such as FMEA and fault tree and event tree analysis, can also be used to develop a fail-safe design.

16.2.7.5 Protection against Environmental Aggressions
The system must be designed in such a way that CMFs due to environmental aggression, such as lightning, voltage surges, or flooding, can be reduced. For this purpose, methods such as electrical isolation or mechanical shielding can be employed.

16.2.7.6 Reducing Uncertainty about Common Mode Failures
The ultimate goal for any system is for it to perform exactly to its specifications, producing the correct output reliably for its intended life. If the purpose of the system is to process information and output a solution based on the inputs, given an ideal system, one input would result in the exact same output an infinite number of times or until the useful life of the system had expired. Similarly, the system would be expected to produce the correct output given an infinite number of varied inputs during its life. In other words, the goal is for the system to be perfect.

While perfection such as this is almost always desired, it is very difficult to attain. This presents the following design problem: how do you design a system to be as close to perfect, or error-free, as possible? This is the situation with many of the systems used in the fields of avionics, automotive, and nuclear power control. In these systems, even minor failures cannot be tolerated, because the consequences of the failure could be very severe.

Error reduction or elimination is a complex engineering problem on its own, but alone it is not enough to satisfy the safety and performance standards of many systems used in avionics, automotive, and nuclear control applications. In order for these systems to be effective, they have to not only be error free, but also resilient and fault-tolerant. Being fault-tolerant means that the system is able to withstand some unanticipated event, and remain operating without error. In these industries, many systems are only useful if there is no uncertainty involving their operation, and if they are error free and fault tolerant.

Although these systems can be resilient and can possess as few errors as possible, it is still very difficult to eliminate CMF. CMF is defined as: "Failure of two or more structures, systems or components in the same manner or mode due to a single event or cause." The susceptibility of these systems to CMF arises from the interdependence of one system on many others. This interdependence causes problems when one component of the system fails, because it in turn may cause all the components dependent on it to fail. Consider the guidance and control system in an aircraft: a failure of the component powering all of the guidance and control systems would result in a failure of all of the components depending on power. If this system is not designed to withstand this example of CMF, then the result would be disastrous.

16.2.7.7 Designing to Reduce Uncertainty
In order to design the optimal system, it is necessary to understand the source of the problem that causes an error. In many

16 Analyzing Product Failures and Root Causes

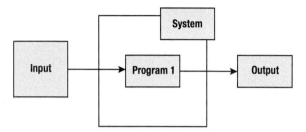

Figure 16.11 A basic system with only one version of the program.

systems, the source of the problem arises from uncertainties within the system's operation, which eventually result in aberrant or incorrect outputs, also known as errors. These uncertainties, which must be minimized in order to achieve the optimal system, show up in the form of anomalies.

An anomaly can be described as an event that occurs in the execution of software that causes the output of the software to differ from what the result would have been without the anomaly being present. More simply, an anomaly is an unexpected error in a part of the system that causes an incorrect output. Anomalies are usually present in three forms: (1) internal defects present in the code, (2) ambiguities and errors in specification, and (3) external events (not involving an error in the code) that modify either the input or the output of the code. In order to reduce the uncertainty of either category of anomaly occurring, the system must be designed to deal with each type individually. To design the systems with the least amount of uncertainty involving the occurrence of CMFs, it is important to understand how the design of the systems reduces the effects of the three types of anomalies.

The following four systems illustrate how proper design can limit the uncertainty about CMF. They also show what design techniques are used to counter the effect of each of the three types of anomalies.

In this system, the input signal is received by program 1, which performs the necessary function producing the output. This system is an example of a system that has a very high uncertainty about the possibility for CMF. It is obvious that any anomalous events that occur during the operation of this system, either an error in program 1's code, an error in specification, or an external event could all singularly cause the system to output an incorrect result, and thus fail. The use of this system in avionics would not be tolerable because there is obviously no design techniques used to reduce the effect of any possible anomalies.

One technique that is used to reduce the uncertainty involved in the system and to reduce the possibilities of CMF is replication. A system designed using replication is shown in Figure 16.12.

This system is much improved over the system shown in Figure 16.11. The goal of designing using replication is to achieve fault tolerance using many identical copies of the necessary program. This design lessens the chance that an external event anomaly, such as a hardware failure, will cause failure of the entire system. Even should one copy of the program become corrupted, the remaining replications of the program should still output the correct result.

Since this system produces four separate outputs, a voter is used to determine which output is the correct one, should there be a discrepancy. In most cases, the voter will compare the output from each copy or program 1 and accept the majority answer as the correct one. This is reasonable because it is unlikely that all of the

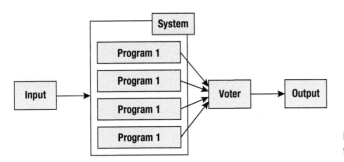

Figure 16.12 A system exhibiting replication through multiple copies of program 1.

replicas will fail at the same time in the same way. The voter could also identify errors in specification by comparing the outputs to an expected range of outputs. If the outputs all lie outside of the expected range, there may be an anomaly in the form of a specification error.

While replication increases the resiliency and fault tolerance of the system by a great deal, it fails to counter all possible anomalies, and thus leaves some amount of uncertainty. Although this system is resilient against some forms of external events, the resilience could be greatly improved if the replicas of the program were stored in spatially separate and isolated locations. This would lessen the likelihood of damage to one area of an aircraft, rendering all of the replicas of the program as failed.

The other uncertainty and the most severe is the susceptibility to CMF. Protection against two out of the three anomalous events affords the system nothing if the third kind arises and causes total failure. This event is an error in the code of program 1. The replications of program 1, even in spatially separate locations, do nothing to counter the effects of an error in the code. Since an error in the code will be present in all the replicas of program 1, an input that causes an error in one replica will similarly cause all of the programs to produce an identical, incorrect output otherwise known as a CMF. In his case, the voter receives four identical outputs, and either assumes that they are all in agreement so they are all correct, or decides that none lies within the range so none of them is correct.

Either way, the system fails to achieve its function without the correct output. This failure is very severe because one error in the code of program 1 can disable the entire system, and with the programs used in these systems, it is nearly impossible to remove all of the errors present. In order to reduce the uncertainty about CMF and the system, the system is designed with diversity and replication.

In this final example, diversity is used to create a more fault tolerant system than the system with only replicated programs. Diversity in the programs provides protection against the CMFs that the previous system was susceptible to. In a diverse system, critical components are created in different ways, with the least number of similarities between the two.

The goal of a diverse system is to eliminate the possibility of CMF between the diverse components of the system. Consider the system shown in Figure 16.13. The diversity is made possible by "implementing the individual functions in a diverse manner." This system exhibits both replication and diversity, therefore countering all three types of anomalies that may arise during the life of the system.

The first type of anomaly, internal defects in the code, will not be present in all of the varied versions of the program (the likelihood that unrelated errors in each version of the program would produce the same output are very slim). This is not to say that

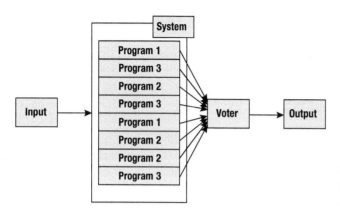

Figure 16.13 System exhibiting diversity and replication through repeated differing versions of the program.

each varied version will be error-free. This would once again be impossible to achieve, rather the hope is that the errors made in each version do not lead to the same incorrect output for any one input, thus creating a CMF.

The second type of anomaly is ambiguities and errors in specification. Once again, the voter algorithm should be able to detect outputs in the correct range, and make a guess as to in the specifications might be incorrect. Of course, the only true way to prevent this kind of anomaly from occurring is to make sure all of the specifications are correct.

The third anomaly is external events, which change the output of the program. A system employing both diversity and replication is able to handle failures of individual components, providing that there is still a majority of votes left for the voter to decide on. The resiliency of this system could also be increased by separating the differing and replicated versions spatially within the environment. In this kind of system, there are only two ways that failure can occur: CMFs between multiple, diverse programs, or errors in the voter that cause it to select the correct outputs reliably.

16.2.7.8 Simulating Errors or Faults in the System   In order to predict and reduce the uncertainty about CMF occurrence, it is helpful to set up a system and simulate multiple faults, and then analyze if the system is resilient or susceptible to CMFs. This simulation can be accomplished by a computer program using fault injection. The process begins by simulating the system without the presence of defects. Then the program randomly inserts errors in the systems' various programs and compares the output to the previous output. If the two are different, then failure has occurred.

Voas et al. (1997) present an algorithm to reduce uncertainty about CMF. The algorithm they propose calls for the execution of a set of inputs in the original run without state perturbations and the perturbed run with fault injection and the outputs for each input are temporarily stored. Then perturbed run is made with same input but with fault injection and again output is stored. The outputs from perturbed run and the original run are compared. The comparison of all perturbed outputs for a given input can help us determine if a single input common-mode failure resulted in these simulations. For a single input, if the perturbed outputs are identical and different from the original output, a single input CMF is counted.

Using this process, it is possible to predict the performance of a system in the event that anomalies are encountered, and to "fix a problem before it occurs." The more modeling and testing that is done on a system, the greater the reduction in uncertainty. The ultimate goal of this testing is to prevent even a single failure from occurring during the operational life of the system, because it is much easier to fix a problem during simulation than when the system is in operation.

16.3 Summary

Generally, product failures or faults do not "just happen." Products fail due to "failures" in design, manufacture, assembly, screening, storage, transportation, or operation. The root cause of a failure is the most basic causal factor or factors that, if corrected or removed, will prevent the recurrence of a problem. Getting at the root cause involves problem identification and correction requires getting to the root cause. Root-cause analysis is a methodology designed to help describe what happened during a particular occurrence, determine how it happened, and understand why it happened. Only when investigators truly understand why a failure occurred will they be able to specify proper corrective measures. A well-structured root-cause analysis will provide added benefits over time by focusing resources on preventing failures.

The process begins by establishing a root-cause culture within the organization, which must be prepared to effectively and efficiently investigate and correct failures. This preplanning phase involves preparing root-cause analysis methodologies and procedures that are specific to the organization and its products. Once a failure incident occurs, the root-cause investigation begins with data collection and assessment of immediate cause(s). Analysis techniques to hypothesize root causes include formal evaluation methods, such as Ishikawa diagram, failure modes and effects analysis, and FTA. The hypotheses formulated are then assessed based on the evidence gathered, design reviews, and physical evaluation of the failed system. Root-cause identification and the development of corrective actions are then conducted. Finally, the implemented corrective actions are assessed with emphasis on cost and benefit analysis.

NFF implies that a failure (fault) occurred or was reported to have occurred during a product's use. The product was analyzed or tested to confirm the failure, but "a failure or fault" could be not found. Intermittent failures can be a cause of NFF occurrences in electronic products and systems. NFF implies that a failure (fault) occurred or was reported to have occurred during a product's use. The product was analyzed or tested to confirm the failure, but "a failure or fault" could be not found. The general causes of NFF can be categorized into people (human), machine, methods, and intermittent failures. In each category, the causes are further broken down into subcategories. Particular categories and subcategories will differ based on the specific product and its life-cycle profile.

This chapter provides guidelines for determining failure mechanisms and root causes, including difficult NFF failures. Analysis of failure mechanisms and root causes is essential for both proper repair and maintenance, as well as product development.

Problems

16.1 What causes a product to fail? List as many causes as possible.

16.2 What is a root cause? How can physics of failure help root-cause analysis?

16.3 Describe three methodologies for root-cause analysis.

16.4 What is a cause-and-effect diagram? Draw a cause-and-effect diagram for the failure of an electronic device.

16.5 Can nondestructive testing cause permanent changes to a product? Why or why not?

17 System Reliability Modeling

To design, analyze, and evaluate the reliability and maintainability characteristics of a system, there must be an understanding of the system's relationships to all the subsystems, assemblies, and components. Many times, this can be accomplished through logical and mathematical models of the system that show the functional relationships among all the components, the subsystems, and the overall system. The reliability of the system is a function of the reliabilities of its components and building blocks.

17.1 Reliability Block Diagram

Engineering analysis of the system has to be conducted in order to develop a reliability model. The engineering analysis consists of the following steps:

1. Develop a functional block diagram of the system based on physical principles governing the operations of the system.
2. Develop the logical and topological relationships between functional elements of the system.
3. Determine the extent to which a system can operate in a degraded state, based on performance evaluation studies.
4. Define the spare and repair strategies (for maintenance systems).

Based on the preceding analysis, a reliability block diagram is developed, which can be used to calculate various measures of reliability and maintainability. The reliability block diagram (RBD) is a pictorial way of showing the success or failure combinations for a system. A system reliability block diagram presents a logical relationship of the system, subsystems, and components. Some of the guidelines for drawing these diagrams are as follows:

Reliability Engineering, First Edition. Kailash C. Kapur and Michael Pecht.
© 2014 John Wiley & Sons, Inc. Published 2014 by John Wiley & Sons, Inc.

1. A group of components that are essential for the performance of the system and/or its mission are drawn in series (Figure 17.1).
2. Components that can substitute for other components are drawn in parallel (Figure 17.3).
3. Each block in the diagram is like a switch: it is closed when the component it represents is working and is opened when the component has failed. Any closed path through the diagram is a success path.

The failure behavior of all the redundant components must be specified. Some of the common types of redundancies are:

1. *Active Redundancy or Hot Standby.* The component has the same failure rate as if it was operating in the system.
2. *Passive Redundancy, Spare, or Cold Standby.* The standby component cannot fail. This is generally assumed for spare or shelf items.
3. *Warm Standby.* The standby component has a lower hazard rate than the operating component. This is usually a realistic assumption.

This chapter describes how to design, analyze, and evaluate the reliability of a system based on the parts, assemblies, and subsystems that compose a system. Most of the concepts in this chapter are explained using one level of the system hierarchical process. For example, we will illustrate how to compute system reliability if we know the reliabilities of the subsystems. Then the same methods and logic can be used to combine assemblies of the subsystem, and so on.

17.2 Series System

In a series system, all subsystems must operate successfully if the system is to function or operate successfully. This implies that the failure of any subsystem will cause the entire system to fail.

The reliability block diagram of a series system is shown in Figure 17.1. The reliability of each block is represented by $R_i(t)$ and the times to failure are represented by $TTF(i)$. The units need not be physically connected in series for the system to be called a series system.

System reliability can be determined using the basic principles of probability theory. We make the assumption that all the subsystems are probabilistically independent. This means that whether or not one subsystem works does not depend on the success or failure of other subsystems.

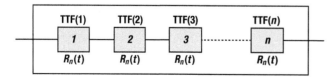

Figure 17.1 Series system representation.

17.2 Series System

Let us first consider the static case. Let R_i be the reliability of the ith subsystem, $i = 1, 2, \ldots, n$. Let E_s be the event that the system functions successfully and E_i be the event that each subsystem i functions successfully ($i = 1, 2, \ldots, n$). Then

$$R_S = P[E_S] = P[E_1 \cap E_2 \cap \cdots \cap E_n] \tag{17.1}$$

because the system will function if and only if all the subsystems function. If all the events E_i, $i = 1, 2, \ldots, n$, are probabilistically independent, then

$$R_S = P[E_1]P[E_2]\cdots P[E_n] = \prod_{i=1}^{n} P[E_i] = \prod_{i=1}^{n} R_i. \tag{17.2}$$

Equation 17.2 can be generalized for time-dependent or dynamic reliability models. If we denote the time to failure random variable for the ith subsystem by T_i, $i = 1, 2, \ldots, n$. Then for the series system, the system reliability is given by

$$R_S(t) = P[(T_1 > t) \cap (T_2 > t) \cap \cdots \cap (T_n > t)]. \tag{17.3}$$

If we assume that all the random variables, T_i, $i = 1, 2, \ldots, n$, are independent, then

$$R_S(t) = P(T_1 > t) P(T_2 > t) \cdots P(T_n > t). \tag{17.4}$$

Hence, we can state the following equation:

$$R_S(t) = R_1(t) \cdot R_2(t) \cdots R_n(t) = \prod_{i=1}^{n} R_i(t). \tag{17.5}$$

From Equation 17.2, it is clear that the reliability of the system reduces with an increase in the number of subsystems or components in the series system (see Figure 17.2).

Assume that the time-to-failure distribution for each subsystem/component of a system is exponential and has a constant failure rate, λ_i. For the exponential distribution, the component reliability is

$$R_i(t) = e^{\lambda_i t}. \tag{17.6}$$

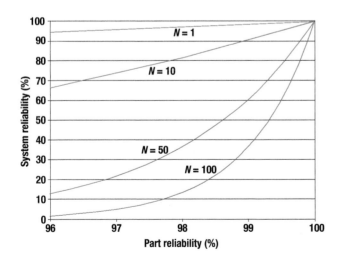

Figure 17.2 Effects of part reliability and number of parts on system reliability in series configuration.

Hence, the system reliability is given by:

$$R_S(t) = \prod_{i=1}^{n} R_i(t) = \prod_{i=1}^{n} e^{-\lambda_i t} = e^{-\left(\sum_{i=1}^{n} \lambda_i\right)t}. \qquad (17.7)$$

The system also has an exponential time-to-failure distribution, and the constant system failure rate is given by:

$$\lambda_S = \sum_{i=1}^{n} \lambda_i, \qquad (17.8)$$

and the mean time between failures for the system is

$$\text{MTBF} = \frac{1}{\lambda_S} = \frac{1}{\sum_{i=1}^{n} \lambda_i}. \qquad (17.9)$$

The system hazard rate is constant if all the components of the system are in series and have constant hazard rates. The assumptions of a constant hazard rate and a series system make the mathematics simple, but this is rarely the case in practice.

For the general case, taking the log of both sides of Equation 17.5, we have

$$\ln R_S(t) = \sum_{i=1}^{n} \ln R_i(t). \qquad (17.10)$$

Also recall that

$$R(t) = \exp\left[-\int_0^t h(\tau)d\tau\right], \qquad (17.11)$$

which means that

$$\int_0^t h(\tau)d\tau = -\ln R(t) \qquad (17.12)$$

or

$$h(t) = -\frac{d}{dt} \ln R(t). \qquad (17.13)$$

Applying this to Equation 17.10, we have

$$h_S(t) = \sum_{i=1}^{n} h_i(t). \qquad (17.14)$$

Thus, the hazard rate for the system is the sum of the hazard rates of the subsystems under the assumption that the time-to-failure random variables for all the subsystems

17.2 Series System

are independent, regardless of the form of the pdf's for the time-to-failure random variables for all the subsystems.

Example 17.1

An electronic system consists of two parts that operate in series. Assuming that failures are governed by a constant failure rate λ_i for the ith part, determine (1) the system failure rate, (2) the system reliability for a 1000-hour mission, and (3) the system mean time to failure (MTTF).

The failure rates of the parts for this problem are given by:

$$\lambda_1 = 6.5 \text{ failures}/10^6 \text{ hours}$$

$$\lambda_2 = 26.0 \text{ failures}/10^6 \text{ hours}.$$

Solution:
For a constant failure rate, the reliability R_i for the ith part has the form:

$$R_i(t) = e^{-\int_0^t \lambda_i d\tau} = e^{-\lambda_i t}.$$

The reliability, R_S, of the series system is

$$R_s = e^{-\sum_{i=1}^{n} \lambda_i t} = e^{-\lambda_s t}$$

$$\lambda_s = \sum_{i=1}^{n} \lambda_i,$$

for a series system with parts assumed to have a constant failure rate. Substituting the given values:

$$\lambda_S = 32.5/10^6 \text{ hours}.$$

The reliability for a 1000-hour mission is thus:

$$R_S(1000) = e^{-(32.5 \times 10^6) \times 1000} = 0.968.$$

The MTTF for the system is:

$$\text{MTTF} = \int_0^\infty R_s(t)\,dt = \int_0^\infty e^{-\lambda_s t}\,dt = 1/\lambda_s$$

$$= 30{,}770 \text{ hours}.$$

Example 17.2

Two subsystems of a system functionally operate in series and have the time to failure random variable with the pdfs given by

17 System Reliability Modeling

$$f_i(t) = \left(\frac{t}{\eta_i}\right)\exp\left[-\frac{t^2}{2\eta_i}\right] \quad t \geq 0, \quad i = 1, 2.$$

where η_i is the parameter for the pdf for the ith subsystem. Time is measured in hours. We want to answer the following five parts.

(a) Find the system hazard function, $h_S(t)$.
(b) Find the system reliability function, $R_S(t)$.
(c) Find the pdf, $f_S(t)$, for the time to failure for the system.
(d) If $\eta_1 = 300$ hours and $\eta_2 = 400$ hours, find $R_S(20$ hours$)$.
(e) For the values in (d) find t^* such that $R_S(t^*) = 0.90$.

Solution:
We can easily notice that $f_i(t)$ is a Weibull distribution with

$$\beta = 2,\ \eta = (2\eta_i)^{1/\beta}.$$

So, the reliability function and the hazard function for each subsystem are

$$R_i(t) = \exp\left[-\frac{t^2}{2\eta_i}\right]$$

$$h_i(t) = \frac{f_i(t)}{R_i(t)} = \frac{t}{\eta_i}.$$

(a) Find the system hazard function, $h_S(t)$. Using Equation 17.14, we have

$$h_S(t) = \sum_{i=1}^{2} h_i(t) = \frac{t}{\eta_1} + \frac{t}{\eta_2} = \frac{t}{\eta},$$

where $1/\eta = 1/\eta_1 + 1/\eta_2$.

(b) Find the system reliability function, $R_S(t)$.
From part (a), and using Equation 17.5, we have

$$R_S(t) = \prod_{i=1}^{2} R_i(t)$$

$$= \exp\left[-\frac{t^2}{2\eta_1}\right] \times \exp\left[-\frac{t^2}{2\eta_2}\right]$$

$$= \exp\left[-\frac{t^2}{2}\left(\frac{1}{\eta_1} + \frac{1}{\eta_2}\right)\right] = \exp\left[-\frac{t^2}{2\eta}\right].$$

(c) Find the system pdf, $f_S(t)$.

$$f_S(t) = h_S(t) \cdot R_S(t) = \frac{t}{\eta}\exp\left[-\frac{t^2}{2\eta}\right].$$

(d) If $\eta_1 = 300$ hours and $\eta_2 = 400$ hours, find R_S (20 hours). First, $1/\eta = 1/\eta_1 + 1/\eta_2 = 1/300 + 1/400$, so $\eta = 171.4$ hours.

Now,

$$R_S(20) = \exp\left[-\frac{t^2}{2\eta}\right]$$

$$= \exp\left[-\frac{20}{2 \times 171.4}\right]$$

$$= 0.3113.$$

(e) For the values in (d), find t^* such that $R_S(t^*) = 0.90$:

$$R_S(t^*) = \exp\left[-\frac{(t^*)^2}{2\eta}\right] = 0.9$$

$$\frac{(t^*)^2}{2\eta} = -\ln(0.9)$$

$$t^* = \sqrt{2\eta(-\ln(0.9))} = \sqrt{2*1714*(-\ln(0.9))}$$

$$t^* = 6.01 \text{ hours.}$$

17.3 Products with Redundancy

Redundancy exists when one or more of the parts of a system can fail and the system will still be able to function with the parts that remain operational. Two common types of redundancy are active and standby. In active redundancy, all the parts are energized and operational during the operation of a system. In active redundancy, the parts will consume life at the same rate as the individual components.

In standby redundancy, some parts do not contribute to the operation of the system, and they get switched on only when there are failures in the active parts. In standby redundancy, the parts in standby ideally should last longer than the parts in active redundancy.

There are three conceptual types of standby redundancy: cold, warm, and hot. In cold standby, the secondary parts are shut down until needed. This lowers the number of hours that the part is active and typically assumes negligible consumption of useful life, but the transient stresses on the parts during switching may be high. This transient stress can cause faster consumption of life during switching. In warm standby, the secondary parts are usually active, but are idling or unloaded. In hot standby, the secondary parts form an active parallel system. The life of the hot standby parts are assumed to be consumed at the same rate as active parts.

17.3.1 Active Redundancy

An active redundant system is a standard "parallel" system. That fails only when all components have failed. Sometimes, the parallel system is called a 1-out-of-n or $(1, n)$ system, which implies that only one (or more) out of n subsystems has to operate

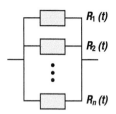

Figure 17.3 Active redundant system.

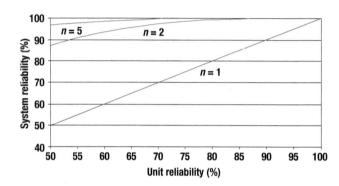

Figure 17.4 Effect of part reliability and number of parts on system reliability in an active redundant system.

for the system to be operational or functional. Thus, a series system is an *n*-out-of-*n* system. The reliability block diagram of a parallel system is given in Figure 17.3.

The units need not be physically connected in parallel for the system to be called a parallel system. The system will fail if all of the subsystems or all of the components fail by time *t*, or the system will survive the mission time, *t*, if at least one of the units survives by time *t*. Then, the system reliability can be expressed as

$$R_s(t) = 1 - Q_s(t), \tag{17.15}$$

where $Q_s(t)$ is the probability of system failure, or

$$Q_S(t) = [1 - R_1(t)] \times [1 - R_2(t)] \times \cdots \times [1 - R_n(t)]$$
$$= \prod_{i=1}^{n} [1 - R_i(t)], \tag{17.16}$$

under the assumption that the time to failure random variables for all the subsystems are probabilistically independent.

The system reliability for a mission time, *t*, is

$$R_S(t) = 1 - \prod_{i=1}^{n} [1 - R_i(t)]. \tag{17.17}$$

For the static situation or for an implied fixed value of *t*, we have an equation similar to Equation 17.2, which is given by

$$R_S = 1 - \prod_{i=1}^{n} [1 - R_i]. \tag{17.18}$$

Figure 17.4 shows the effect of component reliability on system reliability for an active parallel system for a static situation.

17.3 Products with Redundancy

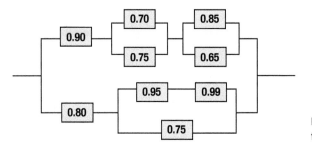

Figure 17.5 Reliability block diagram for series-parallel system.

We can use Equation 17.2 and Equation 17.18 to calculate the reliability of systems that have subsystems in series and in parallel. This is illustrated in Example 17.3.

Example 17.3

The reliability block diagram of a system is given in Figure 17.5. The number in each box is the reliability of the component. This system has nine components. Find the reliability of the system.

Solution:
We can consider the top, consisting of five components, as subsystem A and the bottom with four components as subsystem B. Then, using step by step Equation 17.2 and Equation 17.18, we have

$$R_A = [0.90][1-(1-0.70)(1-0.75)][1-(1-0.85)(1-0.65)] = 0.788794$$

$$R_B = [1-[1-0.95 \times 0.99][1-0.75]] \times 0.80 = 0.7881,$$

and because A and B are in parallel, we have

$$R_S = 1-(1-R_A)(1-R_B) = 1-(1-0.788794)(1-0.7881) = 0.955245.$$

After we know the system reliability function from Equation 17.17, the system hazard rate is given by:

$$h_S(t) = \frac{f_S(t)}{R_S(t)}, \quad (17.19)$$

where $f_S(t)$ is the system time-to-failure probability density function (pdf). The mean life, or the expected life, of the system is determined by:

$$E[T_S] = \int_0^\infty R_S(t) = \int_0^\infty \left(1 - \prod_{i=1}^n (1-R_i(t))\right) dt, \quad (17.20)$$

where T_S is the time to failure for the system.

For example, if the system consists of two units ($n = 2$) with an exponential failure distribution with constant failure rates λ_1 and λ_2, then the system mean life is given by Equation 17.21. Note that the system mean life is not equal to the reciprocal of the sum of the component's constant failure rates, and we can prove that the hazard rate is not constant over time, although the individual unit failure rates are constant.

$$E[T_S] = \frac{1}{\lambda_1} + \frac{1}{\lambda_2} - \frac{1}{\lambda_1 + \lambda_2}. \tag{17.21}$$

Example 17.4

Consider an electronics system consisting of two parts with constant failure rates as given below:

$$\lambda_1 = 6.5 \text{ failures}/10^6 \text{ hours}$$

$$\lambda_2 = 26.0 \text{ failures}/10^6 \text{ hours}.$$

Assume that failures are governed by a constant failure rate λ_i for the ith part. Determine:

(a) The system reliability for a 1000-hour mission
(b) The system MTTF
(c) The failure probability density function
(d) The system "failure rate."

Solution:
For a constant failure rate, the reliability R_i of the ith part has the form:

$$R_i(t) = e^{-\lambda_i t}.$$

For a parallel system:

$$R_S(t) = 1 - \prod_{i=1}^{2}\left(1 - e^{-\lambda_i(t)}\right) = e^{-\lambda_1 t} + e^{-\lambda_2 t} - e^{-(\lambda_1+\lambda_2)t}.$$

The failure probability density function is:

$$f_S(t) = -\frac{d[R_S(t)]}{dt} = \lambda_1 e^{-\lambda_1 t} + \lambda_2 e^{-\lambda_2 t} - (\lambda_1 + \lambda_2)e^{-(\lambda_1+\lambda_2)t}.$$

Substituting numbers in the equation for system reliability, we get the answer for part (a):

$$R_S(1000) = 0.99352 + 0.97434 - 0.96802 = 0.99983.$$

The MTTF (part b) for the parallel system is

$$\text{MTTF}_S = \int_0^\infty R_S(t)\,dt = \frac{1}{\lambda_1} + \frac{1}{\lambda_2} - \frac{1}{(\lambda_1 + \lambda_2)} = 161,538 \text{ hours}.$$

17.3 Products with Redundancy

The failure probability density function (part c) for the parallel system is

$$f_S(t) = -\frac{d[R_S(t)]}{dt}$$
$$= 6.5 \times 10^{-6} e^{-6.5 \times 10^{-6} t} + 26.0 \times 10^{-6} e^{-26.0 \times 10^{-6} t} - 32.5 \times 10^{-6} e^{-32.5 \times 10^{-6} t}.$$

The system hazard rate for the parallel system is given by:

$$h_S(t) = \frac{f_S(t)}{R_S(t)}.$$

The system failure rate for the parallel system (part (d)) can be obtained by substituting the results in the equation stated above. We will find that $h_S(t)$ is a function of time and is not constant over time.

If the time to failure for all n components is exponentially distributed with MTBF θ, then the MTBF for the system is given by

$$E[T_S] = \sum_{i=1}^{n} \frac{\theta}{i}. \tag{17.22}$$

Here, θ = MTBF for every component or subsystem. Thus, each additional component increases the expected life of the system but at a slower and slower rate. This motivates us to consider standby redundant systems in the next section.

17.3.2 Standby Systems

A standby system consists of an active unit or subsystem and one or more inactive (standby) units that become active in the event of the failure of the functioning unit. The failures of active units are signaled by a sensing subsystem, and the standby unit is brought to action by a switching subsystem. The simplest standby configuration is a two-unit system, as shown in Figure 17.6. In general, there will be n number of units with $(n - 1)$ of them in standby.

Let us now develop the system reliability models for the standby situation with two subsystems. Let $f_i(t)$ be the pdf for the time to failure random variable, T_i, for the ith unit, $i = 1, 2$, and $f_S(t)$ be the pdf for the time to failure random variable, T_S, for the system. Let us first consider a situation with only two units under the assumption that the sensing and the switching mechanisms are perfect. Thus, the second unit is switched on when the first component fails. Thus, $T_S = T_1 + T_2$, and T_S is nothing but a convolution of two random variables. Hence,

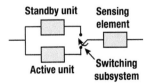

Figure 17.6 Stand-by system.

$$f_S(t) = \int_0^t f_1(x) f_2(t-x) dx. \tag{17.23}$$

Similarly, if we have a primary active component and two standby components, we have

$$f_S(t) = \int_0^t \int_0^x f_1(y) f_2(x-y) f_3(t-x) dy dx. \tag{17.24}$$

We can evaluate Equation 17.23 when both T_1 and T_2 have the exponential distribution as below:

$$f_S(t) = \int_0^t \lambda_1 e^{-\lambda_1 x} \lambda_2 e^{-\lambda_2 (t-x)} dx = \frac{\lambda_1 \lambda_2}{\lambda_1 - \lambda_2} e^{-\lambda_2 t} + \frac{\lambda_1 \lambda_2}{\lambda_2 - \lambda_1} e^{-\lambda_1 t}. \tag{17.25}$$

From Equation 17.25, we have

$$R_S(t) = \int_t^\infty f_S(x) dx = \frac{\lambda_1}{\lambda_1 - \lambda_2} e^{-\lambda_2 t} + \frac{\lambda_2}{\lambda_2 - \lambda_1} e^{-\lambda_1 t}. \tag{17.26}$$

The MTBF$_S$, θ_S, for the system is given by

$$\theta_S = \frac{1}{\lambda_1} + \frac{1}{\lambda_2}, \tag{17.27}$$

as is expected since $T_S = T_1 + T_2$ and $E[T_S] = E[T_1] + E[T_2]$.

When the active and the standby units have equal constant failure rates, λ, and the switching and sensing units are perfect, the reliability function for such a system is given by

$$R_S(t) = e^{-\lambda t}(1 + \lambda t). \tag{17.28}$$

We can rewrite Equation 17.26 in the form

$$R_S(t) = e^{-\lambda_1 t} + \frac{\lambda_1}{\lambda_1 - \lambda_2} \left(e^{-\lambda_2 t} - e^{-\lambda_1 t} \right), \tag{17.29}$$

or as shown in Equation 17.30, where $AR_{(2)}$ is the contribution to the reliability value of the system by the second component

$$R_S(t) = e^{-\lambda_1 t} + AR_{(2)}. \tag{17.30}$$

This can easily be generalized to a situation where we have one primary component and two or more standby components. For example, if we have one primary component and $(n - 1)$ standby components, and all have exponential time to failure with a constant failure rate of λ, then the system reliability function is given by

$$R_S(t) = e^{-\lambda t} \sum_{i=0}^{n-1} (\lambda t)^i / i!. \tag{17.31}$$

17.3.3 Standby Systems with Imperfect Switching

Switching and sensing systems are not perfect. There are many ways these systems can fail. Let us look at a situation where the switching and sensing unit simply fails to operate when called upon to do its job. Let the probability that the switch works when required be p_{SW}. Then, the system reliability for one primary component and one standby is given by

$$R_S(t) = R_1(t) + p_{SW} \int_0^t f_1(x) R_2(t-x) dx. \tag{17.32}$$

When the main and the standby units have exponential time-to-failure distributions, we can use Equation 17.30 to develop the following equation:

$$R_S(t) = e^{-\lambda t} + p_{SW} A R_{(2)}. \tag{17.33}$$

Now, let us generalize Equation 17.32, where the switching and sensing unit is dynamic and the switching and sensing unit starts its life at the same time the active or the main unit starts its life. Let T_{SW} denote the time to failure for the switching and sensing unit, where its pdf and reliability functions are denoted by $f_{SW}(t)$ and $R_{SW}(t)$, respectively. Then the reliability of the system is given by

$$R_S(t) = R_1(t) + \int_0^t f_1(x) R_{SW}(x) R_2(t-x) dx. \tag{17.34}$$

If the time to failure of the switching and sensing unit follows an exponential distribution with a failure rate of λ_{SW}, then Equation 17.34 reduces to

$$R_S(t) = R_1(t) + \int_0^t f_1(x) e^{-\lambda_{SW} x} R_2(t-x) dx. \tag{17.35}$$

If we consider a special case where both the main unit and the standby units have exponential time-to-failure distributions with parameter λ, then Equation 17.35 reduces to

$$R_S(t) = e^{-\lambda t} \left[1 + \frac{\lambda}{\lambda_{SW}} \left(1 - e^{-\lambda_{SW} t} \right) \right], \quad t \geq 0. \tag{17.36}$$

Example 17.5

Suppose the constant failure rates for both the main and the standby units are constant and are given by $\lambda = 0.02$ per hour, and the constant failure rate of the switching and sensing unit is $\lambda_{SW} = 0.01$ per hour. Find the reliability of this system for an operating time of 50 hours.

Solution:
Using Equation 17.36, we have

$$R_S(50) = e^{-0.02 \times 50}\left[1 + \frac{0.02}{0.01}\left(1 - e^{-0.01 \times 50}\right)\right]$$
$$= 0.6574.$$

Example 17.6

A B7XX plane has two similar computers onboard for flight control functions: one that is operational and the second as an active standby. The time to failure for each computer follows an exponential distribution with an MTBF of 4000 hours.

(a) Find the reliability of the computer system (consisting of both computers) for 800 hours when the switching is perfect and the second computer is instantaneously switched on when the first computer fails. Also find the MTBF of the computer system.

Solution:
We have, using Equation 17.28,

$$\theta = 4000$$
$$\lambda = \frac{1}{4000} = 0.00025$$
$$R_S(t) = e^{-\lambda t}(1 + \lambda t) = e^{-0.00025 \times 800}(1 + 0.00025 \times 800) = 0.982477$$
$$\text{MTBF} = \int_0^\infty e^{-\lambda t}(1 + \lambda t)dt = \int_0^\infty e^{-\lambda t}dt + \int_0^\infty \lambda t e^{-\lambda t}dt$$
$$= \frac{1}{\lambda} + \frac{1}{\lambda} = \frac{2}{\lambda} = 8000 \text{ hours.}$$

(b) Find the MTBF of the computer system when the switching and sensing unit is not perfect and the switching mechanism has a reliability of 0.98 when it is required to function.

Solution:
We have

$$R_S(t) = e^{-\lambda t} + p_{SW}\left(\lambda t e^{-\lambda t}\right)$$
$$\text{MTBF} = \int_0^\infty R_S(t)dt = \frac{1}{\lambda} + p_{SW}\frac{1}{\lambda} = 4000 + 0.98 * 4000 = 7920 \text{ hours.}$$

(c) Find the reliability of the computer system for 800 hours when the switching mechanism is not perfect and is dynamic. The time to failure for the switching mechanism also has exponential distribution with MTBF of 12,000 hours.

Solution:
We have

$$\lambda = \frac{1}{4000} \quad \lambda_{SW} = \frac{1}{12,000}$$

$$R_S(t) = e^{-\lambda t}\left[1 + \frac{\lambda}{\lambda_{SW}}(1 - e^{-\lambda_{SW} t})\right]$$

$$R_S(800) = e^{-0.00025 \times 800}\left[1 + 3(1 - e^{-800/12000})\right] = 0.977138.$$

Example 17.7

Consider a DC power supply consisting of a generator with a constant failure rate of $\lambda_1 = 0.0003$ and a standby battery with a failure rate $\lambda_2 = 0.0005$.

Assume that both the generator and the stand-by battery have exponential time to failure distributions. Assume that the switching circuit has a known reliability of 0.98 for one switching operation. When the generator fails, then the switch turns on the standby battery. The reliability block diagram of the circuit is shown in Figure 17.7.

(a) Find the reliability of the above system for 15 hours of operation.

Solution:
Using Equation 17.32, we have

$$R_S(t) = R_1(t) + p_{SW}\int_0^t f_1(x)R_2(t-x)dx$$

$$= e^{-\lambda_1 t} + p_{SW}\int_0^t \lambda_1 e^{-\lambda_1 x} e^{-\lambda_2(t-x)}dx$$

$$= e^{-\lambda_1 t} + p_{SW}\lambda_1 e^{-\lambda_2 t}\int_0^t e^{-(\lambda_1 - \lambda_2)x}dx$$

$$= e^{-\lambda_1 t} + p_{SW}\frac{\lambda_1}{\lambda_2 - \lambda_1}\left[e^{-\lambda_1 t} - e^{-\lambda_2 t}\right].$$

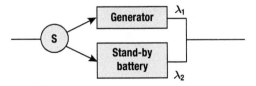

Figure 17.7 Generator and standby battery system.

Hence,

$$R_S(15) = e^{-0.0003 \times 15} + 0.98 \frac{0.0003}{0.0005 - 0.0003}\left[e^{-0.0003 \times 15} - e^{-0.0005 \times 15}\right]$$
$$= 0.99551 + 0.98 \times 1.5 \times (0.99551 - 0.992528)$$
$$= 0.99989.$$

(b) Find the MTBF for the system.

Solution:

$$\text{MTBF} = \int_0^\infty R_S(t)dt = \int_0^\infty \left[e^{-\lambda_1 t} + p_{SW}\frac{\lambda_1}{\lambda_2 - \lambda_1}\left[e^{-\lambda_1 t} - e^{-\lambda_2 t}\right]\right]dt$$
$$= \frac{1}{\lambda_1} + p_{SW}\frac{\lambda_1}{\lambda_2 - \lambda_1}\left[\frac{1}{\lambda_1} - \frac{1}{\lambda_2}\right]$$
$$= \frac{1}{\lambda_1} + p_{SW}\frac{1}{\lambda_2} = \frac{1}{0.0003} + 0.98\frac{1}{0.0005} = 5293.333 \text{ hours.}$$

17.3.4 Shared Load Parallel Models

A situation that is common in engineering systems and their design is called a shared load parallel model. In this case, the two parallel components/units share a load together. Thus, the load on each unit is half of the total load. When one of the units fails, the other unit must take the full load. An example of a shared load parallel configuration is one in which two bolts are used to hold a machine element, and if one of the bolts fails, the other bolt must take the full load. The stresses on the bolt now will be doubled, and this will result in an increased hazard rate for the surviving bolt.

Let $f_{1h}(t)$ and $f_{2h}(t)$ be pdfs for the time to failure for the two units under half or shared load, and $f_{1F}(t)$ and $f_{2F}(t)$ be the pdfs under the full load for each unit, respectively. In this case, we can prove that the pdf for the time to failure of the system is

$$f_S(t) = \int_0^t f_{1h}(x)R_{2h}(x)f_{2F}(t-x)dx + \int_0^t f_{2h}(x)R_{1h}(x)f_{1F}(t-x)dx. \quad (17.37)$$

The reliability function for the system if both units are identical (such as identical bolts), where we have $f_{1h}(t) = f_{2h}(t) = f_h(t)$ and $f_{1F}(t) = f_{2F}(t) = f_F(t)$, can be shown as

$$R_S(t) = [R_h(t)]^2 + 2\int_0^t f_h(x)R_h(x)R_F(t-x)dx. \quad (17.38)$$

If both $f_h(t)$ and $f_F(t)$ follow exponential distributions with parameters λ_h and λ_F, respectively, then it can be shown that the reliability function for the system is

$$R_S(t) = e^{-2\lambda_h t} + \frac{2\lambda_h}{2\lambda_h - \lambda_F}\left[e^{-\lambda_F t} - e^{-2\lambda_h t}\right]. \quad (17.39)$$

Example 17.8

Consider a two-unit shared load parallel system where

$$f_h(t) = \lambda e^{-\lambda t} \quad t \geq 0 \quad \text{pdf for time to failure under half load}$$

$$f_F(t) = 5\lambda e^{-5\lambda t} \quad t \geq 0 \quad \text{pdf for time to failure under full load}$$

(a) Find the system reliability function
(b) Find the MTTF.

Solution:

(a) With $\lambda_h = \lambda$ and $\lambda_F = 5\lambda$, using Equation (17.39), we have:

$$R_S(t) = e^{-2\lambda t} + \frac{2\lambda}{2\lambda - 5\lambda}\left[e^{-5\lambda t} - e^{-2\lambda t}\right] = \frac{5}{3}e^{-2\lambda t} - \frac{2}{3}e^{-5\lambda t}.$$

(b) MTTF is given by:

$$\begin{aligned}
\text{MTTF} &= \int_0^\infty R_S(t)\,dt \\
&= \int_0^\infty \left(\frac{5}{3}e^{-2\lambda t} - \frac{2}{3}e^{-5\lambda t}\right)dt \\
&= \frac{5}{3}\int_0^\infty e^{-2\lambda t}\,dt - \frac{2}{3}\int_0^\infty e^{-5\lambda t}\,dt \\
&= \frac{5}{3}\frac{1}{2\lambda} - \frac{2}{3}\frac{1}{5\lambda} = \frac{7}{10\lambda}.
\end{aligned}$$

17.3.5 (k, n) Systems

A system consisting of n components is called a *k-out-of-n* or *(k, n)* system if the system only operates when at least k or more components are in an operating state. The reliability block diagram (Figure 17.8) for the *k*-out-of-*n* system is drawn similar to the parallel system, but in this case at least k items need to be operating for the system to be functional.

In this configuration, the system works if and only if at least k components out of the n components work, $1 \leq k \leq n$. When $R_i = R(t)$ for all i, with the assumption that the time to failure random variables are independent, we have

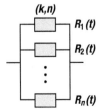

Figure 17.8 *k*-out-of-*n* system.

17 System Reliability Modeling

$$R_S(t) = \sum_{i=k}^{n} \binom{n}{i} [R(t)]^i [1-R(t)]^{n-i}, \tag{17.40}$$

and the probability of system failure, where $Q(t) = 1 - R(t)$, is

$$Q_S(t) = 1 - R_S(t) = 1 - \sum_{i=k}^{n} \binom{n}{i} \cdot [1-Q(t)]^i \cdot [Q(t)]^{n-i}$$

$$= \sum_{i=0}^{k-1} \binom{n}{i} \cdot [1-Q(t)]^i \cdot [Q(t)]^{n-i}. \tag{17.41}$$

The probability density function can be determined by

$$f_S(t) = \frac{dQ_S(t)}{dt}, \tag{17.42}$$

and the system hazard rate is given by

$$h_S(t) = \frac{f_S(t)}{R_S(t)}. \tag{17.43}$$

If $R(t) = e^{-t/\theta}$, for an exponential case, the MTBF for the system is given by

$$\sum_{i=k}^{n} \frac{\theta}{i}. \tag{17.44}$$

The reliability function for the system is mathematically complex to compute in a closed form when the components have different failure distributions. We will present the methodology later on in this chapter to solve this problem.

Example 17.9

Compute the reliability of an active redundant configuration system with two out of three units (all with identical reliability R) required for success.

Solution:
In this case, $n = 3$ and $k = 2$. The reliability for a k-out-of-n active redundancy reliability is obtained from Equation 17.40:

$$R_{2 \text{ out of } 3} = \frac{3!}{(1!)(2!)} R^2 Q^1 + \frac{3!}{(0!)(3!)} R^3 Q^0$$

$$R_{2 \text{ out of } 3} = 3R^2(1-R) + R^3.$$

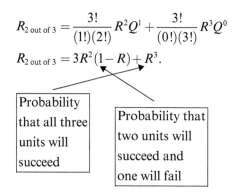

Example 17.10

Consider a system that has eight components and the system will work if at least any five of the eight components work (5-out-of-8 system). Each component has a reliability of 0.87 for a given period. Find the reliability of the system.

Solution:

$$R_S = \sum_{i=k}^{n} \binom{n}{i} R^i (1-R)^{n-i} = \sum_{i=5}^{n} \binom{8}{i} R^i (1-R)^{8-i}$$

$$= \binom{8}{5} 0.87^5 (0.13)^3 + \binom{8}{6} 0.87^6 (0.13)^2 + \binom{8}{7} 0.87^7 (0.13)^1 + \binom{8}{8} 0.87^8 (0.13)^0$$

$$= 56(0.001095) + 28(0.007328) + 8(0.049043) + 0.328212$$

$$= 0.06132 + 0.205192 + 0.392345 + 0.328212 = 0.98707.$$

17.3.6 Limits of Redundancy

It is often difficult to realize the benefits of redundancy if there are common mode failures, load sharing, and switching and standby failures. Common mode failures are caused by phenomena that create dependencies between two or more redundant parts and which then cause them to fail "simultaneously." Common mode failures can be caused by many things, such as common electric connections, shared environmental stresses, and common maintenance problems.

Load sharing failures occur when the failure of one part increases the stress level of other parts. This increased stress level can affect the life of the active parts. For redundant engines, motors, pumps, structures, and many other systems and devices in active parallel setup, the failure of one part may increase the load on the other parts and decrease their times to failure (or increase their hazard rates).

Several common assumptions are made regarding the switching and sensing of a standby system. Regarding switching, it is often assumed that switching is in one direction only, that switching devices respond only when directed to switch by the monitor, and that switching devices do not fail if not energized. Regarding standby, the general assumption is that standby nonoperating units cannot fail if not energized. When any of these idealizations are not met, switching and standby failures occur. Monitor or sensing failures include both dynamic (failure to switch when active path fails) and static (switching when not required) failures.

17.4 Complex System Reliability

If the system architecture cannot be decomposed into some combination of series-parallel structures, it is deemed a complex system. There are three methods for reliability analysis of a complex system using Figure 17.9 as an example.

17.4.1 Complete Enumeration Method

The complete enumeration method is based on a list of all possible combinations of states of the subsystems. Table 17.1 lists $2^5 = 32$ system states, which are all the

17 System Reliability Modeling

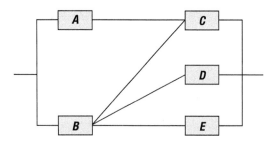

Figure 17.9 A complex system.

Table 17.1 Complete enumeration example

System description	System condition	System status
All components operable	ABCDE	O
One unit in failed state	aBCDE	O
	AbCDE	O
	ABcDE	O
	ABCdE	O
	ABCDe	O
Two units in failed state	abCDE	F
	aBcDE	O
	aBCdE	O
	aBCDe	O
	AbcDE	F
	AbCdE	O
	AbCDe	O
	ABcdE	O
	ABcDe	O
	ABCde	O
Three units in failed state	ABcde	F
	AbCde	O
	AbcDe	F
	AbcdE	F
	aBCde	O
	aBcDe	O
	aBcdE	O
	abCDe	F
	abCdE	F
	abcDE	F
Four units in failed state	Abcde	F
	aBcde	F
	abCde	F
	abcDe	F
	abcdE	F
All five units in failed state	abcde	F

possible states of the system given in Figure 17.9 based on the states of the subsystems. The symbol O stands for "system in operating state," and F stands for "system in failed state." Letters in uppercase denote a unit in an operating state, and lowercase letters denote a unit in a failed state.

Each combination representing the system status can be written as a product of the probabilities of units being in a given state; for example, the second combination in

Table 17.1 can be written as $(1 - R_A)R_B R_C R_D R_E$, where $(1 - R_A)$ denotes the probability of failure of unit A by time t. The system reliability can be written as the sum of all the combinations for which the system is in operating state, O, that is,

$$\begin{aligned} R_S = {} & R_A R_B R_C R_D R_D R_E + (1 - R_A) R_B R_C R_D R_E + R_A (1 - R_B) R_C R_D R_E \\ & + R_A R_B (1 - R_C) R_D R_E + R_A R_B R_C (1 - R_D) R_E + R_A R_B R_C R_D (1 - R_E) \\ & + (1 - R_A) R_B (1 - R_C) R_D R_E + (1 - R_A) R_B R_C (1 - R_D) R_E \\ & + (1 - R_A) R_B R_C R_D (1 - R_E) \\ & + \cdots \\ & \vdots \\ & + (1 - R_A) R_B (1 - R_C)(1 - R_D) R_E. \end{aligned} \qquad (17.45)$$

After simplification, the system reliability can be represented as

$$\begin{aligned} R_S = {} & R_B R_C R_D R_E - R_A R_B R_C - R_B R_C R_D - R_B R_C R_E \\ & - R_B R_D R_E + R_A R_C + R_B R_C + R_B R_D + R_B R_E. \end{aligned} \qquad (17.46)$$

17.4.2 Conditional Probability Method

The conditional probability method is based on the law of total probability, which allows system decomposition by a selected unit and its state at time t. For example, system reliability is equal to the reliability of the system given that unit A is in its operating state at time t, denoted by $R_S|A_S$, times the reliability of unit A, plus the reliability of the system, given that unit A is in a failed state at time t, $R_S|A_F$, times the unreliability of unit A, or

$$R_S = (R_S | A_S) \cdot R_A + (R_S | A_F) \cdot Q_A. \qquad (17.47)$$

This decomposition process continues until each term is written in terms of the reliability and unreliability of each of the units.

As an example, consider the system given in Figure 17.9 and decompose the system using unit C. Then, the system reliability can be written as

$$R_S = (R_S | C_S) \cdot R_C + (R_S | C_F) \cdot Q_C. \qquad (17.48)$$

If the unit C is in the operating state at time t, the system reduces to the configuration shown in Figure 17.10. Therefore, the system reliability, given that unit C is in its operating state at time t, is equal to the series-parallel combination as shown above, or

$$R_S | C_S = [1 - (1 - R_A) \cdot (1 - R_B)]. \qquad (17.49)$$

If unit C is in a failed state at time t, the system reduces to the configuration given in Figure 17.11. Then the system reliability, given that unit C is in a failed state, is given by

$$R_s | C_F = R_B \cdot [1 - (1 - R_D) \cdot (1 - R_E)]. \qquad (17.50)$$

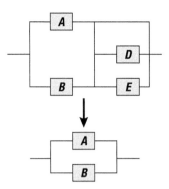

Figure 17.10 System reduction when unit C is operating.

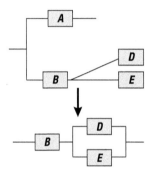

Figure 17.11 System reduction when unit C fails.

The system reliability is obtained by substituting Equation 17.49 and Equation 17.50 into Equation 17.48:

$$R_S = (R_S/C_S) \cdot R_C + (R_S/C_F) \cdot Q_C \\ = [1-(1-R_A) \cdot (1-R_B)] \cdot R_C + R_B \cdot [1-(1-R_D) \cdot (1-R_E)] \cdot (1-R_C). \quad (17.51)$$

The system reliability is expressed in terms of the reliabilities of its components. Simplification of Equation 17.51 gives the same expression as Equation 17.46.

17.4.3 Concept of Coherent Structures

In general, the concept of coherent systems can be used to determine the reliability of any system (Barlow and Proschan 1975; Leemis 1995; Rausand and Hoyland 2003). The performance of each of the n components in the system is represented by a binary indicator variable, x_i, which takes the value 1 if the ith component functions and 0 if the ith component fails. Similarly, the binary variable ϕ indicates the state of the system, and ϕ is a function of $x = (x_1, \ldots, x_n)$.

The function $\phi(x)$ is called the structure function of the system. The structure function is represented by using the concept of minimal paths and minimal cuts. A minimal path is the minimal set of components whose functioning ensures the functioning of the system. A minimal cut is the minimal set of components whose failures would cause the system to fail. Let $\alpha_j(x)$ be the jth minimal path series structure for path $A_j, j = 1, \ldots, p$, and $\beta_k(x)$ be the kth minimal parallel cut structure for cut B_k, $k = 1, \ldots, s$. Then we have

17.4 Complex System Reliability

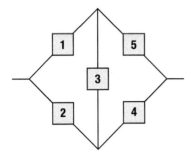

Figure 17.12 Reliability block diagram of a bridge structure.

$$\alpha_j(x) = \prod_{i \in A_j} x_i, \qquad (17.52)$$

and

$$\beta_k(x) = 1 - \prod_{i \in B_k}(1 - x_i). \qquad (17.53)$$

The structure function of the system using minimum cuts is given by Equation 17.54, and the structure function using minimum cuts is given by Equation 17.55, as follows:

$$\phi(x) = 1 - \prod_{j=1}^{p}[1 - \alpha_j(x)] \qquad (17.54)$$

$$\phi(x) = \prod_{k=1}^{s} \beta_k(x). \qquad (17.55)$$

Let us consider the following bridge structure given in Figure 17.12. For the bridge structure (Figure 17.12), we have four minimal paths and four minimal cuts, and their structure functions are given below:

$$\alpha_1 = x_1 x_5 \qquad \beta_1 = 1 - (1 - x_1)(1 - x_2)$$
$$\alpha_2 = x_2 x_4 \qquad \beta_2 = 1 - (1 - x_4)(1 - x_5)$$
$$\alpha_3 = x_1 x_3 x_4 \qquad \beta_3 = 1 - (1 - x_1)(1 - x_3)(1 - x_4)$$
$$\alpha_4 = x_2 x_3 x_5 \qquad \beta_4 = 1 - (1 - x_2)(1 - x_3)(1 - x_5).$$

Then the reliability of the system is given by

$$R_S = P[\phi(X) = 1] = E[\phi(X)], \qquad (17.56)$$

where X is the random vector of the states of the components $(X_1, \ldots, X_n)$.

We can develop the structure function by putting structure functions of minimum paths and minimum cuts in Equation 17.54 and Equation 17.55, respectively. When we do the expansion, we should remember that each x_i is a binary variable that takes values of 0 or 1, and hence, x_i^n for any positive integer n is also a binary variable and takes the value of 0 or 1. If we do the expansion using Equation 17.54 or Equation 17.55, we can prove that the structure function for the system in Figure 17.12 is

$$\phi(x) = x_1 x_5 + x_1 x_3 x_4 + x_2 x_3 x_5 + x_2 x_4 - x_1 x_3 x_4 x_5 - x_1 x_2 x_3 x_5 \\ - x_1 x_2 x_4 x_5 - x_1 x_2 x_3 x_4 - x_2 x_3 x_4 x_5 + 2 x_1 x_2 x_3 x_4 x_5. \quad (17.57)$$

If R_i is the reliability of the ith component, then we know that

$$R_i = E[X_i] = P[X_i = 1], \quad (17.58)$$

and the system reliability for the bridge structure is given by

$$R_S = R_1 R_5 + R_1 R_3 R_4 + R_2 R_3 R_5 + R_2 R_4 - R_1 R_3 R_4 R_5 - R_1 R_2 R_3 R_5 \\ - R_1 R_2 R_4 R_5 - R_1 R_2 R_3 R_4 - R_2 R_3 R_4 R_5 + 2 R_1 R_2 R_3 R_4 R_5. \quad (17.59)$$

If all $R_i = R = 0.9$, we have

$$\begin{aligned} R_S &= 2R^2 + 2R^3 - 5R^4 + 2R^5 \\ &= 0.9785. \end{aligned} \quad (17.60)$$

The exact calculations for R_S are generally very tedious because the paths and the cuts are dependent, since they may contain the same component. Bounds on system reliability are given by

$$\prod_{k=1}^{s} P[\beta_k(X) = 1] \leq P[\phi(X) = 1] \leq 1 - \prod_{j=1}^{p} \{1 - P[\alpha_j(X)] = 1\}. \quad (17.61)$$

Using these bounds for the bridge structure, we have, when $R_i = R = 0.9$, the upper bound, R_U, on system reliability, R_S, is

$$\begin{aligned} R_U &= 1 - (1 - R^2)^2 (1 - R^3)^2 \\ &= 0.9973, \end{aligned} \quad (17.62)$$

and the lower bound, R_L, is

$$\begin{aligned} R_L &= \left[1 - (1 - R)^2\right]^2 \left[1 - (1 - R)^3\right]^2 \\ &= 0.9781. \end{aligned} \quad (17.63)$$

The bounds on system reliability using the concepts of minimum paths and cuts can be improved.

17.4 Complex System Reliability

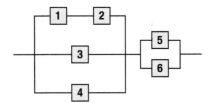

Figure 17.13 Six component series-parallel model.

Example 17.11

Consider a system, shown in Figure 17.13, with six components, which has the following reliability block diagram.

The reliabilities of the components are as follows:

$$R_1 = 0.95$$
$$R_2 = 0.90$$
$$R_3 = 0.80$$
$$R_4 = 0.85$$
$$R_5 = 0.75$$
$$R_6 = 0.90.$$

(a) Find the exact reliability of the system using the series-parallel model.

Solution:

$$\begin{aligned} R_S &= R_{1234} \times R_{56} \\ &= [1-(1-R_1 R_2)(1-R_3)(1-R_4)] \times [1-(1-R_5)(1-R_6)] \\ &= [1-(1-0.95 \times 0.90)(1-0.80)(1-0.85)] \times [1-(1-0.75)(1-0.90)] \\ &= 0.970759. \end{aligned}$$

(b) Find all the minimum paths and minimum cuts for the above system.

Solution:

Components for minimal paths	Components for minimal cuts
1, 2, 5	5, 6
1, 2, 6	1, 3, 4
3, 5	2, 3, 4
3, 6	
4, 5	
4, 6	

(c) Find the lower bound and the upper bound on the system reliability using the equations for the bounds on system reliability, which uses the minimum paths and minimum cuts.

17 System Reliability Modeling

Solution:
Using Equation 17.61, we have

$$R_U = 1 - (1 - R_1 R_2 R_5)(1 - R_1 R_2 R_6)(1 - R_3 R_5)(1 - R_3 R_6)(1 - R_4 R_5)(1 - R_4 R_6)$$
$$= 1 - (1 - 0.95 * 0.90 * 0.75)(1 - 0.95 * 0.90 * 0.90)(1 - 0.80 * 0.75)$$
$$* (1 - 0.80 * 0.90)(1 - 0.85 * 0.75)(1 - 0.85 * 0.90)$$
$$= 0.999211.$$

and

$$R_L = [1 - (1 - R_5)(1 - R_6)][1 - (1 - R_1)(1 - R_3)(1 - R_4)][1 - (1 - R_2)(1 - R_3)(1 - R_4)]$$
$$= [1 - 0.25 * 0.10][1 - 0.05 * 0.2 * 0.15][1 - 0.1 * 0.2 * 0.15]$$
$$= 0.970617.$$

Thus, the reliability bounds are $0.970617 \leq R_S \leq 0.999211$.

The lower bound is much better because there is less dependency between the minimum cuts (fewer components share different minimum cuts) than for minimum paths (where some components are part of several minimum paths).

Example 17.12

Consider a 4-out-of-5 system with the reliabilities of the five components as given below:

$$R_1 = 0.90$$

$$R_2 = 0.95$$

$$R_3 = 0.85$$

$$R_4 = 0.80$$

$$R_5 = 0.75.$$

(a) Develop the structure function $\phi(x)$ for this 4-out-of-5 system.

Solution:
There are five minimum paths for this system. Hence, we can develop the structure function using minimum paths (Eq. 17.54):

$$\phi(x) = 1 - (1 - x_1 x_2 x_3 x_4)(1 - x_1 x_2 x_3 x_5)(1 - x_1 x_2 x_4 x_5) \times (1 - x_1 x_3 x_4 x_5)(1 - x_2 x_3 x_4 x_5),$$

which can be simplified as

$$= 1-(1-x_1x_2x_3x_4 - x_1x_2x_3x_5 + x_1x_2x_3x_4x_5)$$
$$(1-x_1x_2x_4x_5 - x_1x_3x_4x_5 + x_1x_2x_3x_4x_5)(1-x_2x_3x_4x_5)$$
$$= 1-(1-x_1x_2x_3x_4 - x_1x_2x_3x_5 + x_1x_2x_3x_4x_5 - x_1x_2x_4x_5 + x_1x_2x_3x_4x_5 + x_1x_2x_3x_4x_5$$
$$- x_1x_2x_3x_4x_5 - x_1x_3x_4x_5 + x_1x_2x_3x_4x_5 + x_1x_2x_3x_4x_5 - x_1x_2x_3x_4x_5$$
$$+ x_1x_2x_3x_4x_5 - x_1x_2x_3x_4x_5 - x_1x_2x_3x_4x_5 + x_1x_2x_3x_4x_5)(1-x_2x_3x_4x_5)$$

$$= 1-(1-x_1x_2x_3x_4 - x_1x_2x_3x_5 - x_1x_2x_4x_5 - x_1x_3x_4x_5 + 3x_1x_2x_3x_4x_5) \times (1-x_2x_3x_4x_5)$$

$$= 1-(1-x_1x_2x_3x_4 - x_1x_2x_3x_5 - x_1x_2x_4x_5 - x_1x_3x_4x_5$$
$$+ 3x_1x_2x_3x_4x_5 - x_2x_3x_4x_5 + 4x_1x_2x_3x_4x_5 - 3x_1x_2x_3x_4x_5)$$

$$= x_1x_2x_3x_4 + x_1x_2x_3x_5 + x_1x_3x_4x_5 + x_2x_3x_4x_5 + x_1x_2x_4x_5 - 4x_1x_2x_3x_4x_5.$$

(b) Find the system reliability with the reliability values of the components given above.

Solution:
Taking the expected value of the structure function, we can calculate the system reliability as

$$R_S = R_1R_2R_3R_4 + R_1R_2R_3R_5 + R_1R_3R_4R_5 + R_2R_3R_4R_5 + R_1R_2R_4R_5 - 4R_1R_2R_3R_4R_5$$

$$= (0.90)(0.95)(0.85)(0.80) + (0.90)(0.95)(0.85)(0.75) + (0.90)(0.85)(0.80)(0.75)$$
$$+ (0.90)(0.85)(0.80)(0.75) + (0.90)(0.95)(0.85)(0.75)$$
$$- 4(0.90)(0.95)(0.85)(0.85)(0.75)$$
$$= 0.5814 + 0.5451 + 0.4590 + 0.4845 + 0.5130 - 1.744$$
$$= 0.8388.$$

17.5 Summary

The reliability of the system is a function of the reliabilities of its components and building blocks. To design, analyze, and evaluate the reliability and maintainability characteristics of a system, there must be an understanding of the system's relationships to all the subsystems, assemblies, and components. Many times this can be accomplished through logical and mathematical models. Engineering analysis of a system has to be conducted in order to develop a reliability model. Based on this analysis, a reliability block diagram is developed, which can be used to calculate various measures of reliability and maintainability. A reliability block diagram is a pictorial way of showing the success or failure combinations for a system. A system reliability block diagram presents a logical relationship of the system, subsystems, and components.

In a series system, all subsystems must operate successfully if the system is to function or operate successfully. This implies that the failure of any subsystem will cause the entire system to fail. Redundancy is a strategy to resolve this problem.

17 System Reliability Modeling

Redundancy exists when one or more of the parts of a system can fail and the system will still be able to function with the parts that remain operational. Two common types of redundancy are active and standby. In active redundancy, all the parts are energized and operational during the operation of a system. In standby redundancy, some parts do not contribute to the operation of the system, and they get switched on only when there are failures in the active parts. In standby redundancy, the parts in standby ideally should last longer than the parts in active redundancy. It is often difficult to realize the benefits of redundancy if there are common mode failures, load sharing, and switching and standby failures. In addition to series systems, there are complex systems. If the system architecture cannot be decomposed into some combination of series-parallel structures, it is deemed a complex system. These two types of systems, series-parallel and complex, require different strategies for monitoring and evaluating system reliability.

Problems

17.1 The reliability block diagram of a system is given below. The number in each box is the reliability of the component. Find the reliability of the system.

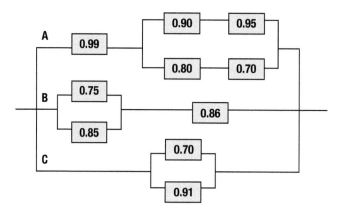

Thus, A, B, and C are three subsystems that are in parallel.

17.2 The reliability block diagram of a system is given below. The number in each box is the reliability of the component. Find the reliability of the system.

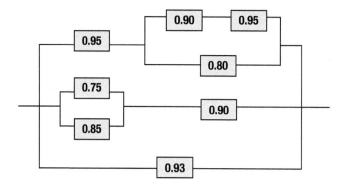

17.3 There are three components, A, B, and C, and they are represented by different blocks in the following two reliability block diagrams. Both reliability block diagrams use the same component twice. Let the reliabilities of the components be denoted by R_A, R_B, and R_C.

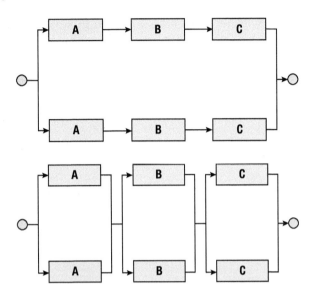

(a) Is there a difference in reliability between the two configurations when the failures or success of all the components are independent of each other? Which system configuration or reliability block diagram has higher reliability? Explain your answer.

(b) Which configuration is more susceptible to common mode failure and why? Assume that each component (A, B, and C) can fail primarily by different mechanisms and those mechanisms are affected by different loads.

17.4 The reliability block diagram shown below is a complex system that cannot be decomposed into a "series-parallel" configuration. We want to determine the reliability equation for the system using the conditional probability method. We have decided to use the component B for the decomposition. Draw the two reliability block diagrams that result from "B operating" and "B failed" conditions.

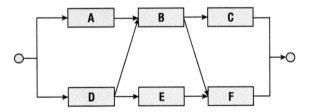

17.5 Consider the system shown in the block diagram and derive an equation for the reliability of the system. R_X denotes the reliability of each component in the system, where X is the name of the component. For stage 3 (four C components in parallel), and it is a two-out-of-four system, that is, two components need to operate for the system to operate.

17 System Reliability Modeling

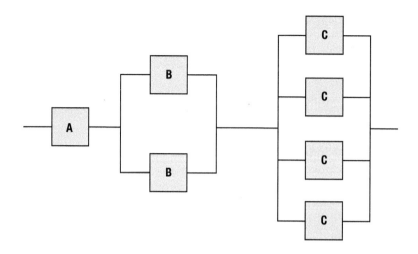

17.6 Derive (manually) the reliability equation of the system shown below. This is a complex dynamic system and the failure distribution for each component is shown in the table.

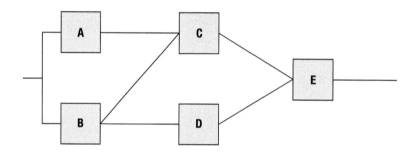

Component	Failure Distribution	Parameter (in Hour or Equivalent)
A	Weibull 3 parameter	$\beta = 3$, $\eta = 1000$, $\gamma = 100$
B	Exponential	MTBF $= 1000$
C	Lognormal	Mean $= 6$, standard deviation $= 0.5$
D	Weibull 3 parameter	$\beta = 0.7$, $\eta = 150$, $\gamma = -100$
E	Normal	Mean $= 250$, standard deviation $= 15$

Find the following for this complex system:

(a) System reliability at 100 hours
(b) System reliability at 0 hours
(c) Failure rate at 1000 hours
(d) Time when wearout region begins (use the graph)
(e) How long does it take for 75% of the system to fail?

What happens to the results if you switch the properties of component C and D?

Problems

17.7 Consider a series system composed of two subsystems where the first subsystem has a Weibull time to failure distribution with parameters $\eta = 2$ and $\theta = 200$ hours. The second subsystem has an exponential time to failure distribution with $\theta = 300$ hours. Develop the following functions for the system:

(a) Find the hazard rate function.
(b) Find the reliability function.

17.8 Consider a parallel system composed of two identical subsystems where the subsystem failure rate is λ, a constant.

(a) Assume a pure parallel arrangement and plot the reliability function using a normalized time scale for the abscissa as

$$t' = t\lambda.$$

(b) Assume a standby system with perfect switching and plot this reliability function on the same graph.
(c) Assume that the standby system has a switch with a probability of failure of 0.2, and plot this reliability function on the same graph.
(d) Compare the three systems.

17.9 A system consists of a basic unit and two standby units. All units (basic and the two standby) have an exponential distribution for time to failure with a failure rate of $\lambda = 0.02$ failures per hour. The probability that the switch will perform when required is 0.98.

(a) What is the reliability of the system at 50 hours?
(b) What is the expected life or MTTF for the system?

17.10 Consider a two-unit pure parallel arrangement where each subsystem has a constant failure rate of λ, and compare this to a standby redundant arrangement that has a constant switch failure rate of λ_{SW}. Specifically, what is the maximum permissible value of λ_{SW} such that the pure parallel arrangement is superior to the standby arrangement?

17.11 Consider a system that has seven components and the system will work if any five of the seven components work (5-out-of-7 system). Each component has a reliability of 0.92 for a given period. Find the reliability of the system.

17.12 Consider the following system, which consists of five components. The reliabilities of the components are as follows:

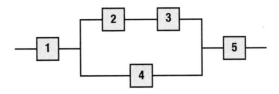

17 System Reliability Modeling

$$R_1 = 0.95$$

$$R_2 = 0.95$$

$$R_3 = 0.96$$

$$R_4 = 0.85$$

$$R_5 = 0.87.$$

(a) Find the exact reliability of the system using the concepts of series and parallel models.
(b) Find all the minimum paths and minimum cuts for the system.
(c) Find the structure function $\phi(x)$ using the minimum paths.
(d) Fine the structure function $\phi(x)$ using minimum cuts and show that you get the same answer as in part (c).
(e) Find an expression for the reliability of the system based on the structure function developed in part (c). Find the reliability using this equation and show that you get the same answer as you get in part (a).
(f) Find the lower bound, R_L, and the upper bound, R_U, on the system reliability using minimum paths and minimum cuts.

17.13 A system has four components with the following reliability block diagram:

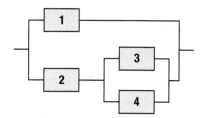

The reliability of the four components is as follows:

$$R_1 = 0.85$$

$$R_2 = 0.95$$

$$R_3 = 0.80$$

$$R_4 = 0.90.$$

(a) Find the exact reliability of the above system using the concepts of series and parallel models.
(b) Find all the minimum paths and minimum cuts for the above system.

(c) Find the structure function, $\phi(x)$, of the system using (1) minimum paths and (2) minimum cuts. Show that you get the same answer in both cases. Use the structure function to find the exact value of the system reliability.

(d) Find the lower bound and upper bound on system reliability with the above reliability numbers of the components using all the minimum paths and minimum cuts.

18 Health Monitoring and Prognostics

As a result of intense global competition, companies are considering novel approaches to enhance the operational efficiency of their products. For many products and systems, high in-service reliability can be a means to ensure customer satisfaction. In addition, global competitive demands for increased warranties, and the severe liability of product failures, is encouraging manufacturers to improve field reliability and operational availability,[1] provide knowledge of in-service use, and life-cycle operational and environmental conditions.

The American Heritage Dictionary defines prognostic as an adjective that relates to prediction or foretelling and as a noun for a sign or symptom indicating the future course of a disease or sign or forecast of some future occurrence. Hippocrates founded the 21 axioms of prognostics some 2400 years ago.[2] The goal of prognostics is to foretell (predict) the future health (or state) of a system. Health for human beings is defined as a state of complete physical, mental, and social well-being. These ideas can also be applied for the overall health or quality of products and systems. Interest has been growing in monitoring the ongoing health of products and systems in order to provide advance warning failure, and assist in administration and logistics. Here, health is defined as the extent of degradation or deviation from an expected normal condition. Prognostics is the prediction of the future state of health based on current and historical health conditions (Vichare and Pecht 2006). Prognostics deals with prediction of quality in systems. Quality is defined in dictionaries as the essential character or attribute of an entity. It's the inherent characteristic or attribute of something. Thus prognostics deals with prediction of some desired quality or characteristic of a system. Prognostics is based on understanding the science of degradation of the underlying system. This is also called as physics or chemistry or biology or psychology of failure from the viewpoint of the customer. As such, development of sensors and monitoring devices are key for Prognostics and System Health Management (PSHM).

---

[1] Operational availability is the degree (expressed as a decimal between 0 and 1, or the percentage equivalent) to which a piece of equipment or system can be expected to work properly when required. Operational availability is often calculated by dividing uptime by the sum of uptime and downtime.

[2] See http://classics.mit.edu/Hippocrates/ prognost.html (MIT 2010) (accessed February 2010).

Reliability Engineering, First Edition. Kailash C. Kapur and Michael Pecht.
© 2014 John Wiley & Sons, Inc. Published 2014 by John Wiley & Sons, Inc.

18 Health Monitoring and Prognostics

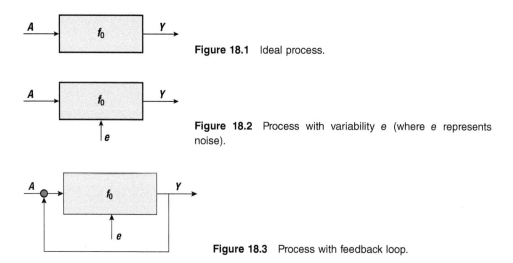

Figure 18.1 Ideal process.

Figure 18.2 Process with variability e (where e represents noise).

Figure 18.3 Process with feedback loop.

Electronics are integral to the functionality of most systems today, and their reliability is often critical for system reliability (Vichare et al. 2007). This chapter provides a basic understanding of prognostics and health monitoring of products and systems and the techniques being developed to enable prognostics for electronic systems.

18.1 Conceptual Model for Prognostics

Figure 18.1 shows a typical system with inputs A [single or vector] and response variable (or output) represented by Y. If we have perfect knowledge about this system, and we know the transfer function $f_0(A) = Y$, then inputs A can be determined as $A = f_0^{-1}(Y)$. If we know the system (the transfer function), then we can predict the response variable Y and adjust the inputs A to maintain the output within the desired range. This is the ideal deterministic process and is shown in Figure 18.1.

A perfect prognostics is the situation where we know the transfer function and we have perfect knowledge of the system. In that case we can foretell many measures of Y when A is the system input. If we know what output Y is desired, we can determine how input A should be adjusted. There are challenges to achieve this goal:

(a) The inverse problem is not unique and not easy to determine.
(b) We often lack knowledge (or there is uncertainty) about our model.
(c) The real-world systems might be very complex and cause output Y to appear as a random variable.

A cause of variation in Y is due to error or noise factors represented by e in Figure 18.2. Thus, e makes the output Y a random variable. In terms of reliability, Y may be the time to failure random variable.

Thus, if the output is not closer to the ideal value or the target due to the presence of noise factors, we can measure it and one way to overcome these deviations is to use feedback as shown in Figure 18.3. It is clear that feedback is reactive and can be too late in order to adjust the input properly. Instead, we want to be proactive and

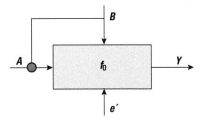

Figure 18.4 Prognostics P with feed forward.

prevent deviations from ideal or target value occurring in the response Y in the first place.

The traditional approaches have been based on feedback to correct the system behavior. Many of the reliability methodologies are based on feedback, like reliability growth through testing.

In prognostics, we need to understand the underlying causes of error. We decompose error e further to identify the disturbance or another factor B that is part of e and that affect the output. Thus, now we have

$$e = (e', B). \tag{18.1}$$

We can measure this disturbance B (though we cannot change or control disturbance) so we can determine how to change the other input variables to create feed forward and maintain the system response variable Y closer to the target.

We can track trends and make forecasts to identify future system behavior when disturbances are measured. The trend is to use artificial neural networks to model prognostics. Based on the system knowledge, we can use feed forward to provide prognostics. We account for error (uncertainties, perturbations, and disturbances) using feedback, and we use the known disturbance and its measurement for feed forward and prognostics. Figure 18.4 illustrates the role of prognostic system P and its feed forward loop B.

The design of prognostics P, relies on feed forward model to properly regulate inputs A. The role of prognostic system P is to develop the relationship $B = g(X_1, X_2, X_3, \ldots, X_n)$ such that the response variable $Y = f_0(A, B)$ remains in the expected or desired range.

The design of prognostic systems is challenging. Finding the inverse function is not always easy and not necessarily unique. The actual system and its environment in the future can be quite different from what we perceive today. In addition, there is a lack of knowledge about the inner workings of the system in operation. This lack of knowledge stems from uncertainty and could be caused by the initial model error or not considering all factors. As a result the response variable Y appears as a random variable to the observer.

In Figure 18.5, another illustration for a feed forward design is shown. The feed forward prognostics model incorporates the concept of disturbance B and error e'. We monitor and measure disturbance B and want to know its effect on the output. Likewise, we are interested in understanding the reasons for error e'; for example, whether it is due to variation or uncertainty that stems from incomplete or fuzzy understanding of the system's operating and environmental factors. Our goal is to decompose the causes of error into informative variables with the aid of the subject matter experts. Finally, we design the prognostic system such that by monitoring B,

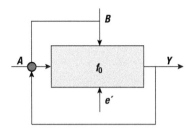

Figure 18.5 Process with feedback and feed forward.

when values of B change, we adjust the input variables A accordingly in order to prevent a problem or to prolong system lifetime. This is illustrated in Figure 18.5, where both feedback loop and feed forward are used to maintain the response Y in the expected range.

Identifying prognostics information is based on considering a system's use information in several areas: in environmental and operational conditions, measured process values, monitored residuals, and fault alarms to identify fault modes. We have to keep in mind that different fault modes can progress to manifest into different failures and hence require different prognostic models.

Among many methods of prognostics is estimating the remaining useful life (RUL) of a system. RUL can be estimated from historical and operational data collected from a system. Various methods are used to determine system degradation and predict RUL.

Another method is estimating the probability of failure (POF). POF is the failure probability distribution of the system or a component. Additionally, we can study time to failure (TTF), the time a component is expected to fail. TTF defines the time when a system no longer meets its design specifications.

Prognostic methods combine RUL, TTF, and POF with other techniques to extend system life, ensure mission completion, and improve corporate profitability.

18.2 Reliability and Prognostics

Reliability is the ability of a product or system to perform as intended (i.e., without failure and within specified performance limits) for a specified time, in its life-cycle environment. Traditional reliability prediction methods for electronic products include Mil-HDBK-217 (U.S. Department of Defense 1965), 217-PLUS, Telcordia (Telcordia Technologies 2001), PRISM (Denson 1999), and FIDES (FIDES Group 2004). These methods rely on the collection of failure data and generally assume the components of the system have failure rates (most often assumed to be constant) that can be modified by independent "modifiers" to account for various quality, operating, and environmental conditions. There are numerous well-documented concerns with this type of modeling approach (Cushing et al. 1993; Leonard 1991b; Talmor and Arueti 1997; Wong 1990). The general consensus is that these handbooks should never be used, because they are inaccurate for predicting actual field failures and provide highly misleading predictions, which can result in poor designs and logistics decisions (Morris 1990; Wong 1990).

The traditional handbook method for the reliability prediction of electronics started with Mil-Hdbk-217A, published in 1965. In this handbook, there was only a single-point failure rate for all monolithic integrated circuits, regardless of the stresses, the materials, or the architecture. Mil-Hdbk-217B was published in 1973, with the RCA/Boeing models simplified by the U.S. Air Force to follow a statistical exponential (constant failure rate) distribution. Since then, all the updates were mostly "band-aids" for a modeling approach that was proven to be flawed (Pecht and Nash 1994). In 1987–1990, the Center for Advanced Life Cycle Engineering (CALCE) at the University of Maryland was awarded a contract to update Mil-Hdbk-217. It was concluded that this handbook should be cancelled and the use of this type of modeling approach discouraged.

In 1998, IEEE 1413 standard, "IEEE Standard Methodology for Reliability Prediction and Assessment for Electronic Systems and Equipment," was approved to provide guidance on the appropriate elements of a reliability prediction (IEEE Standard 1413–1998 1998). A companion guidebook, IEEE 1413.1, "IEEE Guide for Selecting and Using Reliability Predictions Based on IEEE 1413," provides information and an assessment of the common methods of reliability prediction for a given application (IEEE Standard 1413.1-2002 2003). It is shown that the Mil-Hdbk-217 is flawed. There is also discussion of the advantage of reliability prediction methods that use stress and damage physics-of-failure (PoF) technique.

The PoF approach and design-for-reliability (DfR) methods have been developed by CALCE (Pecht and Dasgupta 1995) with the support of industry, government, and other universities. PoF is an approach that utilizes knowledge of a product's life-cycle loading and failure mechanisms to perform reliability modeling, design, and assessment. The approach is based on the identification of potential failure modes, failure mechanisms, and failure sites for the product as a function of its life-cycle loading conditions. The stress at each failure site is obtained as a function of both the loading conditions and the product geometry and material properties. Damage models are then used to determine fault generation and propagation.

Prognostics and health management (PHM) is a method that permits the assessment of the reliability of a product (or system) under its actual application conditions. When combined with physics-of-failure models, it is thus possible to make continuously updated predictions based on the actual environmental and operational conditions. PHM techniques combine sensing, recording, interpretation of environmental, operational, and performance-related parameters to indicate a system's health. PHM can be implemented through the use of various techniques to sense and interpret the parameters indicative of:

- Performance degradation, such as deviation of operating parameters from their expected values
- Physical or electrical degradation, such as material cracking, corrosion, interfacial delamination, or increases in electrical resistance or threshold voltage
- Changes in a life-cycle profile, such as usage duration and frequency, ambient temperature and humidity, vibration, and shock.

The framework for prognostics is shown in Figure 18.6. Performance data from various levels of an electronic product or system can be monitored in situ and analyzed using prognostic algorithms. Different implementation approaches can be adopted individually or in combination. These approaches will be discussed in the subsequent sections. Ultimately, the objective is to predict the advent of failure in

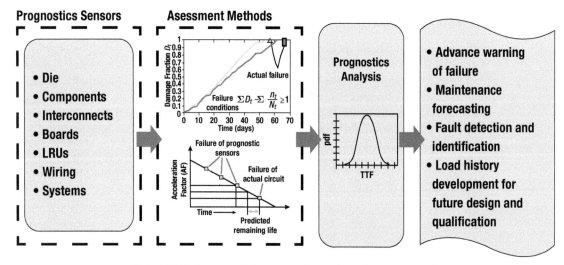

Figure 18.6 Framework for prognostics and health management.

terms of a distribution of remaining life, level of degradation, or probability of mission survival.

18.3 PHM for Electronics

Most products and systems contain significant electronics content to provide needed functionality and performance. If one can assess the extent of deviation or degradation from an expected normal operating condition for electronics, this information can be used to meet several powerful goals, which include:

1. Providing advanced warning of failures;
2. Minimizing unscheduled maintenance, extending maintenance cycles, and maintaining effectiveness through timely repair actions;
3. Reducing the life-cycle cost of equipment by decreasing inspection costs, downtime, and inventory; and
4. Improving qualification and assisting in the design and logistical support of fielded and future systems (Vichare and Pecht 2006).

In other words, since electronics are playing an increasingly large role in providing operational capabilities for today's products and systems, prognostic techniques have become highly desirable.

Some of first efforts in diagnostic health monitoring of electronics involved the use of built-in test (BIT), defined as an onboard hardware–software diagnostic means to identify and locate faults. A BIT can consist of error detection and correction circuits, totally self-checking circuits, and self-verification circuits (Vichare and Pecht 2006). Two types of BIT concepts are employed in electronic systems: interruptive BIT

(I-BIT) and continuous BIT (C-BIT). The concept behind I-BIT is that normal equipment operation is suspended during BIT operation. The concept behind C-BIT is that equipment is monitored continuously and automatically without affecting normal operation.

Several studies (Johnson 1996; Pecht et al. 2001) conducted on the use of BIT for fault identification and diagnostics showed that BIT can be prone to false alarms and can result in unnecessary costly replacement, requalification, delayed shipping, and loss of system availability. BIT concepts are still being developed to reduce the occurrence of spurious failure indications. However, there is also reason to believe that many of the failures actually occurred, but were intermittent in nature (DoD 5000.2 Policy Document 2004). The persistence of such issues over the years is perhaps because the use of BIT has been restricted to low volume systems. Thus, BIT has generally not been designed to provide prognostics or remaining useful life due to accumulated damage or progression of faults. Rather, it has served primarily as a diagnostic tool.

PHM has also emerged as one of the key enablers for achieving efficient system-level maintenance and lowering life-cycle costs in military systems. In November 2002, the U.S. Deputy Under Secretary of Defense for Logistics and Materiel Readiness released a policy called condition-based maintenance plus (CBM+). CBM+ represents an effort to shift unscheduled corrective equipment maintenance of new and legacy systems to preventive and predictive approaches that schedule maintenance based upon the evidence of need. A 2005 survey of 11 CBM programs highlighted "electronics prognostics" as one of the most needed maintenance-related features or applications, without regard for cost (Cutter and Thompson 2005), a view also shared by the avionics industry (Kirkland et al. 2004). Department of Defense 5000.2 policy document on defense acquisition, which states that "program managers shall optimize operational readiness through affordable, integrated, embedded diagnostics and prognostics, embedded training and testing, serialized item management, automatic identification technology, and iterative technology refreshment" (DoD 5000.2 Policy Document 2004). Thus, a prognostics capability has become a requirement for any system sold to the U.S. Department of Defense.

Prognostics and health management is also emerging as a high priority issue in space applications. NASA's Ames Research Center (ARC) in California is focused on conducting fundamental research in the field of Integrated Systems Health Management (ISHM). ARC is involved in design of health management systems, selection and optimization of sensors, in situ monitoring, data analysis, prognostics, and diagnostics. The prognostics center for excellence at ARC develops algorithms to predict the remaining life of NASA's systems and subsystems. ARC's current prognostics projects involve power semiconductor devices (investigation of the effects of ageing on power semiconductor components, identification of failure precursors to build a PoF model and development of algorithms for end-of-life prediction), batteries(algorithms for batteries prognosis), flight actuators (PoF modeling and development of algorithms for estimation of remaining life), solid rocket motor failure prediction, and aircraft wiring health management (Korsmeyer 2013).

In addition to in-service reliability assessment and maintenance, health monitoring can also be effectively used to support product take-back and end-of-life decisions. Product take-back indicates the responsibility of manufacturers for their products over the entire life cycle, including disposal. The motivation driving product take-back is the concept of Extended Producer Responsibility (EPR) for postconsumer electronic waste (Rose et al. 1999). The objective of EPR is to make manufacturers and distributors financially responsible for their products when they are no longer needed.

End-of-life product recovery strategies include repair, refurbishing, remanufacturing, reuse of components, material recycling, and disposal. One of the challenges in end-of-life decision-making is to determine whether product lines can be extruded, whether any components could be reused and what subset should be disposed of in order to minimize system costs (Sandborn and Murphy 1999). Several interdependent issues must be considered concurrently to properly determine the optimum component reuse ratio, including assembly/disassembly costs and any defects introduced by either process, product degradation incurred in the original life cycle, and the waste stream associated with the life cycle. Among these factors, the estimate of the degradation of the product in its original life cycle could be the most uncertain input to end-of-life decisions. This could be effectively carried out using health monitoring, with knowledge of the entire history of the product's life cycle.

Scheidt and Zong (1994) proposed the development of special electrical ports, referred to as green ports, to retrieve product usage data that could assist in the recycling and reuse of electronic products. Klausner et al. (1998a, 1998b) proposed the use of an integrated electronic data log (EDL) for recording parameters indicative of product degradation. The EDL was implemented on electric motors to increase the reuse of motors. In another study (Simon et al. 2000), domestic appliances were monitored for collecting usage data by means of electronic units fitted on the appliances. This work introduced the Life Cycle Data Acquisition Unit, which can be used for data collection and also for diagnostics and servicing. Middendorf et al. (2002) suggested developing life-information modules to record the cycle conditions of products for reliability assessment, product refurbishing, and reuse.

Designers often establish the usable life of products and warranties based on extrapolating accelerated test results to assumed usage rates and life-cycle conditions. These assumptions may be based on worst-case scenarios of various parameters composing the end user environment. Thus, if the assumed conditions and actual use conditions are the same, the product would last for the designed time, as shown in Figure 18.7(a). However, this is rarely true, and usage and environmental conditions could vary significantly from those assumed. For example, consider products equipped with life-consumption monitoring systems for providing in situ assessment of remaining life. In this situation, even if the product is used at a higher usage rate and in harsh conditions, it can still avoid unscheduled maintenance and catastrophic failure, maintain safety, and ultimately save cost. These are typically the motivational factors for use of health monitoring or life consumption monitoring (LCM), as shown in Figure 18.7(b).

One of the vital inputs in making end-of-life decisions is the estimate of degradation and the remaining life of the product. Figure 18.7c illustrates a scenario in which a working product is returned at the end of its designed life. Using the health monitors installed within the product, the reusable life can be assessed. Unlike testing conducted after the product is returned, this estimate can be made without having to disassemble the product. Ultimately, depending on other factors, such as cost of the product, demand for spares, cost, and yield in assembly and disassembly, the manufacturer can choose to reuse or dispose.

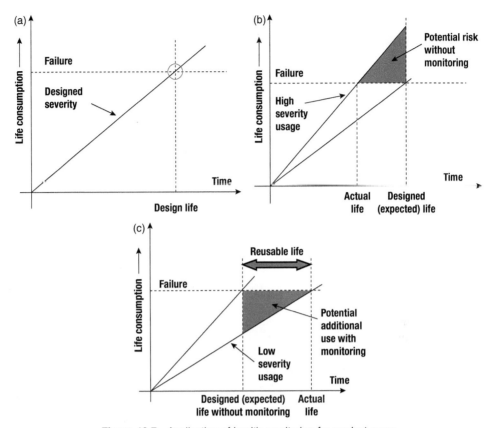

Figure 18.7 Application of health monitoring for product reuse.

18.4 PHM Concepts and Methods

The general PHM methodology is shown in Figure 18.8 (Gu and Pecht 2007). The first step involves a virtual life assessment, where design data, expected life-cycle conditions, failure modes, mechanisms, and effects analysis (FMMEA), and PoF models are the inputs to obtain a reliability (virtual life) assessment. Based on the virtual life assessment, it is possible to prioritize the critical failure modes and failure mechanisms. The existing sensor data, bus monitor data, maintenance, and inspection record can also be used to identify the abnormal conditions and parameters. Based on this information, the monitoring parameters and sensor locations for PHM can be determined.

Based on the collected operational and environmental data, the health status of the products can be assessed. Damage can also be calculated from the PoF models to obtain the remaining life. Then PHM information can be used for maintenance forecasting and decisions that minimize life-cycle costs, maximize availability, or some other utility function.

The different approaches to prognostics and the state of research in electronics PHM are presented here. Three current approaches include:

18 Health Monitoring and Prognostics

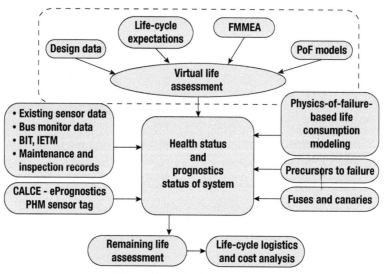

Figure 18.8 PHM methodology.

1. The use of fuses and canary devices
2. Monitoring and reasoning of failure precursors
3. Monitoring environmental and usage condition for stress and damage PoF modeling.

18.4.1 Fuses and Canaries

Expendable devices, such as fuses and canaries, have been a traditional method of protection for structures and electrical power systems. Fuses and circuit breakers are examples of elements used in electronic products to sense excessive current drain and to disconnect power. Fuses within circuits safeguard parts against voltage transients or excessive power dissipation, and protect power supplies from shorted parts. For example, thermostats can be used to sense critical temperature limiting conditions, and to shut down the product, or a part of the system, until the temperature returns to normal. In some products, self-checking circuitry can also be incorporated to sense abnormal conditions and to make adjustments to restore normal conditions, or to activate switching means to compensate for a malfunction (Ramakrishnan et al. 2000).

The word "canary" is derived from one of coal mining's earliest systems for warning of the presence of hazardous gas using the canary bird. Because the canary is more sensitive to hazardous gases than humans, the death or sickening of the canary was an indication to the miners to get out of the shaft. The canary thus provided an effective early warning of catastrophic failure that was easy to interpret. The same approach has been employed in prognostic health monitoring. Canary devices mounted on the actual product can also be used to provide advance warning of failure due to specific wearout failure mechanisms.

Mishra and Pecht (2002) studied the applicability of semiconductor-level health monitors by using precalibrated cells (circuits) located on the same chip with the actual circuitry. The prognostics cell approach, known as Sentinel Semiconductor™ technology, has been commercialized to provide an early warning sentinel for

18.4 PHM Concepts and Methods

upcoming device failures (Ridgetop Semiconductor-Sentinel Silicon™ Library 2004). The prognostic cells are available for 0.35-, 0.25-, and 0.18-μm CMOS processes; the power consumption is approximately 600 microwatts. The cell size is typically 800 μm² at the 0.25-μm process size. Currently, prognostic cells are available for semiconductor failure mechanisms, such as electrostatic discharge (ESD), hot carrier, metal migration, dielectric breakdown, and radiation effects.

The time to failure of prognostic canaries can be precalibrated with respect to the time to failure of the actual product. Because of their location, these canaries contain and experience substantially similar dependencies, as does the actual product. The stresses that contribute to degradation of the circuit include voltage, current, temperature, humidity, and radiation. Since the operational stresses are the same, the damage rate is expected to be the same for both the circuits. However, the prognostic canary is designed to fail faster through increased stress on the canary structure by means of scaling.

Scaling can be achieved by controlled increase of the stress (e.g., current density) inside the canaries. With the same amount of current passing through both circuits, if the cross-sectional area of the current-carrying paths in the canary is decreased, a higher current density is achieved. Further control in current density can be achieved by increasing the voltage level applied to the canaries. A combination of both of these techniques can also be used. Higher current density leads to higher internal (joule) heating, causing greater stress on the canaries. When a current of higher density passes through the canaries, they are expected to fail faster than the actual circuit (Mishra and Pecht 2002).

Figure 18.9 shows the failure distribution of the actual product and the canary health monitors. Under the same environmental and operational loading conditions, the canary health monitors wear out faster to indicate the impending failure of the actual product. Canaries can be calibrated to provide sufficient advance warning of failure (prognostic distance) to enable appropriate maintenance and replacement activities. This point can be adjusted to some other early indication level. Multiple trigger points can also be provided, using multiple canaries spaced over the bathtub curve.

Goodman et al. (2006) used a prognostic canary to monitor time-dependent dielectric breakdown (TDDB) of the metal-oxide semiconductor field-effect transistor (MOSFET) on the integrated circuits. The prognostic canary was accelerated to

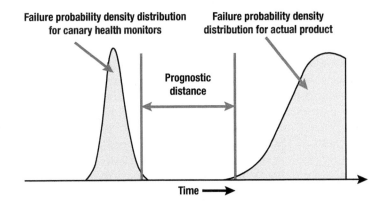

Figure 18.9 Advanced warning of failure using canary structures.

failure under certain environmental conditions. Acceleration of the breakdown of an oxide could be achieved by applying a voltage higher than the supply voltage, to increase the electric field across the oxide. When the prognostics canary failed, a certain fraction of the circuit lifetime was used up. The fraction of consumed circuit life was dependent on the amount of overvoltage applied and could be estimated from the known distribution of failure times.

The extension of this approach to board-level failures was proposed by Anderson and Wilcoxon (2004), who created canary components (located on the same printed circuit board) that include the same mechanisms that lead to failure in actual components. Anderson et al. identified two prospective failure mechanisms: (1) low cycle fatigue of solder joints, assessed by monitoring solder joints on and within the canary package; and (2) corrosion monitoring, using circuits that are susceptible to corrosion. The environmental degradation of these canaries was assessed using accelerated testing, and degradation levels were calibrated and correlated to actual failure levels of the main system. The corrosion test device included an electrical circuitry susceptible to various corrosion-induced mechanisms. Impedance spectroscopy was proposed for identifying changes in the circuits by measuring the magnitude and phase angle of impedance as a function of frequency. The change in impedance characteristics can be correlated to indicate specific degradation mechanisms.

There remain unanswered questions with the use of fuses and canaries for PHM. For example, if a canary monitoring a circuit is replaced, what is the impact when the product is reenergized? What protective architectures are appropriate for postrepair operations? What maintenance guidance must be documented and followed when fail-safe protective architectures have or have not been included? The canary approach is also difficult to implement in legacy systems, because it may require requalification of the entire system with the canary module. Also, the integration of fuses and canaries with the host electronic system could be an issue with respect to real estate on semiconductors and boards. Finally, the company must ensure that the additional cost of implementing PHM can be recovered through increased operational and maintenance efficiencies.

18.5 Monitoring and Reasoning of Failure Precursors

A failure precursor is a data event or trend that signifies impending failure. A precursor indication is usually a change in a measurable variable that can be associated with subsequent failure. For example, a shift in the output voltage of a power supply might suggest impending failure due to a damaged feedback regulator and opto-isolator circuitry. Failures can then be predicted by using causal relationships between measured variables that can be correlated with subsequent failure, and for PoF.

A first step in failure precursor PHM is to select the life-cycle parameters to be monitored. Parameters can be identified based on factors that are crucial for safety, that are likely to cause catastrophic failures, that are essential for mission completeness, or that can result in long downtimes. Selection can also be based on knowledge of the critical parameters established by past experience, field failure data on similar products, and on qualification testing. More systematic methods, such as FMMEA (Ganesan et al. 2005b), can also be used to determine parameters that need to be monitored.

18.5 Monitoring and Reasoning of Failure Precursors

Table 18.1 Potential failure precursors for electronics

Electronic subsystem	Failure precursor
Switching power supply	■ DC output (voltage and current levels) ■ Ripple ■ Pulse width duty cycle ■ Efficiency ■ Feedback (voltage and current levels) ■ Leakage current ■ RF noise
Cables and connectors	■ Impedance changes ■ Physical damage ■ High-energy dielectric breakdown
CMOS IC	■ Supply leakage current ■ Supply current variation ■ Operating signature ■ Current noise ■ Logic level variations
Voltage-controlled oscillator	■ Output frequency ■ Power loss ■ Efficiency ■ Phase distortion ■ Noise
Field effect transistor	■ Gate leakage current/resistance ■ Drain-source leakage current/resistance
Ceramic chip capacitor	■ Leakage current/resistance ■ Dissipation factor ■ RF noise
General purpose diode	■ Reverse leakage current ■ Forward voltage drop ■ Thermal resistance ■ Power dissipation ■ RF noise
Electrolytic capacitor	■ Leakage current/resistance ■ Dissipation factor ■ RF noise
RF power amplifier	■ Voltage standing wave ratio (VSWR) ■ Power dissipation ■ Leakage current

Pecht et al. (1999) proposed several measurable parameters that can be used as failure precursors for electronic products, including switching power supplies, cables and connectors, CMOS integrated circuits, and voltage-controlled high-frequency oscillators (see Table 18.1).

In general, to implement a precursor reasoning-based PHM system, it is necessary to identify the precursor variables for monitoring, and then develop a reasoning algorithm to correlate the change in the precursor variable with the impending failure. This characterization is typically performed by measuring the precursor variable under an expected or accelerated usage profile. Based on the characterization, a model is developed—typically a parametric curve-fit, neural network, Bayesian network, or a time-series trending of a precursor signal. This approach assumes that there is one or more expected usage profiles that are predictable and can be simulated, often in a laboratory setup. In some products, the usage profiles are predictable, but this is not always true.

For a fielded product with highly varying usage profiles, an unexpected change in the usage profile could result in a different (noncharacterized) change in the precursor signal. If the precursor reasoning model is not characterized to factor in the uncertainty in life-cycle usage and environmental profiles, it may provide false alarms. Additionally, it may not always be possible to characterize the precursor signals under all possible usage scenarios (assuming they are known and can be simulated). Thus, the characterization and model development process can often be time-consuming and costly and may not always work.

There are many examples of the monitoring and trending of failure precursor to assess health and product reliability. Some key studies are presented in the next section.

Smith and Campbell (2000) developed a quiescent current monitor (QCM) that can detect elevated Iddq current in real time during operation.[3] The QCM performed leakage current measurements on every transition of the system clock to get maximum coverage of the IC in real time. Pecuh et al. (1999) and Xue and Walker (2004) proposed a low-power built-in current monitor for CMOS devices. In the Pecuh et al. study, the current monitor was developed and tested on a series of inverters for simulating open and short faults. Both fault types were successfully detected and operational speeds of up to 100 MHz were achieved with negligible effect on the performance of the circuit under test. The current sensor developed by Xue and Walker enabled Iddq monitoring at a resolution level of 10 pA. The system translated the current level into a digital signal with scan chain readout. This concept was verified by fabrication on a test chip.

GMA Industries (Wright and Kirkland 2003; Wright et al. 2001, 2003) proposed embedding molecular test equipment (MTE) within ICs to enable them to continuously test themselves during normal operation and to provide a visual indication that they have failed. The molecular test equipment could be fabricated and embedded within the individual integrated circuit in the chip substrate. The molecular-sized sensor "sea of needles" could be used to measure voltage, current, and other electrical parameters, as well as sense changes in the chemical structure of integrated circuits that are indicative of pending or actual circuit failure. This research focuses on the development of specialized doping techniques for carbon nanotubes to form the basic structure comprising the sensors. The integration of these sensors within conventional IC circuit devices, as well as the use of molecular wires for the interconnection of sensor networks, is an important factor in this research. However, no product or prototype has been developed to date.

---

[3] The power supply current (Idd) can be defined by two elements: the Iddq-quiescent current and the Iddt-transient or dynamic current. Iddq is the leakage current drawn by the CMOS circuit when it is in a stable (quiescent) state. Iddt is the supply current produced by circuits under test during a transition period after the input has been applied. Iddq has been reported to have the potential for detecting defects such as bridging, opens, and parasitic transistor defects. Operational and environmental stresses, such as temperature, voltage, and radiation, can quickly degrade previously undetected faults and increase the leakage current (Iddq). There is extensive literature on Iddq testing, but little has been done on using Iddq for in situ PHM. Monitoring Iddq has been more popular than monitoring Iddt.

18.5 Monitoring and Reasoning of Failure Precursors

Kanniche and Mamat-Ibrahim (2004) developed an algorithm for health monitoring of voltage source inverters with pulse width modulation. The algorithm was designed to detect and identify transistor open circuit faults and intermittent misfiring faults occurring in electronic drives. The mathematical foundations of the algorithm were based on discrete wavelet transform (DWT) and fuzzy logic (FL). Current waveforms were monitored and continuously analyzed using DWT to identify faults that may occur due to constant stress, voltage swings, rapid speed variations, frequent stop/start-ups, and constant overloads. After fault detection, "if-then" fuzzy rules were used for VLSI fault diagnosis to pinpoint the fault device. The algorithm was demonstrated to detect certain intermittent faults under laboratory experimental conditions.

Self-monitoring analysis and reporting technology (SMART), currently employed in select computing equipment for hard disk drives (HDD), is another example of precursor monitoring (Hughes et al. 2002; Self-Monitoring Analysis and Reporting Technology (SMART) 2001). HDD operating parameters, including the flying height of the head, error counts, variations in spin time, temperature, and data transfer rates, are monitored to provide advance warning of failures (see Table 18.2). This is achieved through an interface between the computer's start-up program (BIOS) and the hard disk drive.

Systems for early fault detection and failure prediction are being developed using variables such as current, voltage, and temperature, continuously monitored at various locations inside the system. Sun Microsystems refers to this approach as continuous system telemetry harnesses. Along with sensor information, soft performance parameters, such as loads, throughputs, queue lengths, and bit error rates, are tracked. Prior to PHM implementation, characterization is conducted by monitoring the signals of different variables to establish a multivariate state estimation technique (MSET) model of the "healthy" systems. Once the "healthy" model is established using this

Table 18.2 Monitoring parameters based on reliability concerns in hard drives

Reliability issues	Parameters monitored
■ Head assembly 　■ Crack on head 　■ Head contamination or resonance 　■ Bad connection to electronics module ■ Motors/bearings 　■ Motor failure 　■ Worn bearing 　■ Excessive run-out 　■ No spin ■ Electronic module 　■ Circuit/chip failure 　■ Interconnection/solder joint failure 　■ Bad connection to drive or bus ■ Media 　■ Scratch/defects 　■ Retries 　■ Bad servo ■ ECC corrections	■ Head flying height: A downward trend in flying height will often precede a head crash. ■ Error checking and correction (ECC) use and error counts: The number of errors encountered by the drive, even if corrected internally, often signals problems developing with the drive. ■ Spin-up time: Changes in spin-up time can reflect problems with the spindle motor. ■ Temperature: Increases in drive temperature often signal spindle motor problems. ■ Data throughput: Reduction in the transfer rate of data can signal various internal problems.

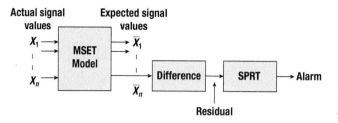

Figure 18.10 Sun Microsystems' approach to PHM.

data, it is used to predict the signal of a particular variable based on learned correlations among all variables (Whisnant et al. 2005). Based on the expected variability in the value of a particular variable during application, a sequential probability ratio test (SPRT) is constructed. During actual monitoring, SPRT is used to detect deviations of the actual signal from the expected signal based on distributions (and not on a single threshold value) (Cassidy et al. 2002; Mishra and Gross 2003). This signal is generated in real time based on learned correlations during characterization (Figure 18.10). A new signal of residuals is generated, which is the arithmetic difference of the actual and expected time-series signal values. These differences are used as input to the SPRT model, which continuously analyzes the deviations and provides an alarm if the deviations are of concern (Whisnant et al. 2005). The monitored data is analyzed to provide alarms based on leading indicators of failure, and enable use of monitored signals for fault diagnosis, root-cause analysis, and analysis of faults due to software aging (Vaidyanathan and Gross 2003).

Brown et al. (2005) demonstrated that the remaining useful life of a commercial global positioning system (GPS) can be predicted by using a precursor-to-failure approach. The failure modes for GPS included precision failure due to an increase in position error, and solution failure due to increased outage probability. These failure progressions were monitored in situ by recording system-level features reported using the National Marine Electronics Association (NMEA) Protocol 0183. The GPS was characterized to collect the principal feature value for a range of operating conditions. Based on experimental results, parametric models were developed to correlate the offset in the principal feature value with solution failure. During the experiment, the BIT provided no indication of an impending solution failure (Brown et al. 2005).

18.5.1 Monitoring Environmental and Usage Profiles for Damage Modeling

The life-cycle profile of a product consists of manufacturing, storage, handling, operating, and nonoperating conditions. The life-cycle loads (Table 18.3), either individually or in various combinations, may lead to performance or physical degradation of the product and reduce its service life (Ramakrishnan and Pecht 2003). The extent and rate of product degradation depends upon the magnitude and duration of exposure (usage rate, frequency, and severity) to such loads. If one can measure these loads in situ, the load profiles can be used in conjunction with damage models to assess the degradation due to cumulative load exposures.

The assessment of the impact of life-cycle usage and environmental loads on electronic structures and components was studied by Ramakrishnan and Pecht (2003). This study introduced the LCM methodology (Figure 18.11), which combined in situ

18.5 Monitoring and Reasoning of Failure Precursors

Table 18.3 Examples of life-cycle loads

Load	Load conditions
Thermal	Steady-state temperature, temperature ranges, temperature cycles, temperature gradients, ramp rates, heat dissipation
Mechanical	Pressure magnitude, pressure gradient, vibration, shock load, acoustic level, strain, stress
Chemical	Aggressive versus inert environment, humidity level, contamination, ozone, pollution, fuel spills
Physical	Radiation, electromagnetic interference, altitude
Electrical	Current, voltage, power, resistance

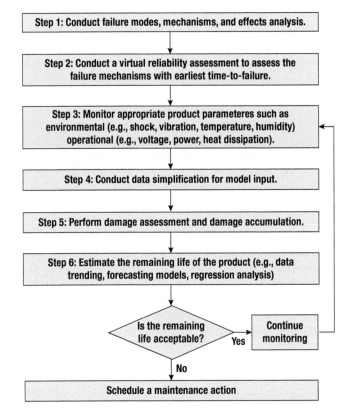

Figure 18.11 CALCE life consumption monitoring methodology.

measured loads with physics-based stress and damage models to assess remaining product life.

Mathew et al. (2006) applied the LCM methodology to conduct a prognostic remaining-life assessment of circuit cards inside a space shuttle solid rocket booster (SRB). Vibration-time history, recorded on the SRB from the prelaunch stage to splashdown, was used in conjunction with physics-based models to assess damage. Using the entire life-cycle loading profile of the SRBs, the remaining life of the components and structures on the circuit cards were predicted. It was determined that an electrical failure was not expected within another 40 missions. However, vibration and shock analysis exposed an unexpected failure of the circuit card due to a broken aluminum bracket mounted on the circuit card. Damage accumulation analysis determined that the aluminum brackets had lost significant life due to shock loading.

Shetty et al. (2002) applied the LCM methodology to conduct a prognostic remaining-life assessment of the end effector electronics unit (EEEU) inside the robotic arm of the space shuttle remote manipulator system (SMRS). A life-cycle loading profile of thermal and vibrational loads was developed for the EEEU boards. Damage assessment was conducted using physics-based mechanical and thermomechanical damage models. A prognostic estimate using a combination of damage models, inspection, and accelerated testing showed that there was little degradation in the electronics and they could be expected to last another 20 years.

Gu et al. (2007) developed a methodology for monitoring, recording, and analyzing the life-cycle vibration loads for remaining-life prognostics of electronics. The responses of printed circuit boards to vibration loading in terms of bending curvature were monitored using strain gauges. The interconnect strain values were then calculated from the measured PCB response and used in a vibration failure fatigue model for damage assessment. Damage estimates were accumulated using Miner's rule after every mission and then used to predict the life consumed and remaining life. The methodology was demonstrated for remaining-life prognostics of a printed circuit board assembly. The results were also verified by checking the resistance data.

In case studies (Mishra et al. 2002; Ramakrishnan and Pecht 2003), an electronic component-board assembly was placed under the hood of an automobile and subjected to normal driving conditions. Temperature and vibrations were measured in situ in the application environment. Using the monitored environmental data, stress and damage models were developed and used to estimate consumed life. Figure 18.12 shows estimates obtained using similarity analysis, and the actual measured life. Only LCM accounted for this unforeseen event because the operating environment was being monitored in situ.

Vichare and Pecht (2006) outlined generic strategies for in situ load monitoring, including selecting appropriate parameters to monitor and designing an effective monitoring plan. Methods for processing the raw sensor data during in situ monitoring to reduce the memory requirements and power consumption of the monitoring device were presented. Approaches were also presented for embedding intelligent front-end data processing capabilities in monitoring systems to enable data reduction

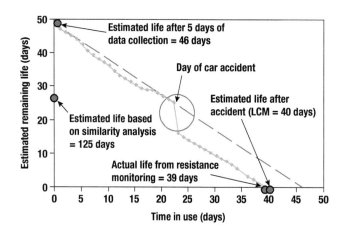

Figure 18.12 Remaining-life estimation of test board.

18.5 Monitoring and Reasoning of Failure Precursors

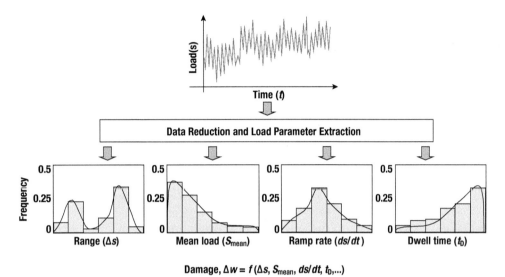

Figure 18.13 Load feature extraction.

and simplification (without sacrificing relevant load information) prior to input in damage models for health assessment and prognostics.

To reduce on-board storage space, power consumption, and uninterrupted data collection over longer durations, Vichare et al. (2006) suggested embedding data reduction and load parameter extraction algorithms into sensor modules. As shown in Figure 18.13, a time-load signal can be monitored in situ using sensors, and further processed to extract cyclic range(s), cyclic mean load (S_{mean}), and rate of change of load (ds/dt), using embedded load extraction algorithms. The extracted load parameters can be stored in appropriately binned histograms to achieve further data reduction. After the binned data are downloaded, it can be used to estimate the distributions of the load parameters. The usage history is used for damage accumulation and remaining life prediction.

Efforts to monitor life-cycle load data on avionics modules can be found in time-stress measurement device (TSMD) studies. Over the years, TSMD designs have been upgraded using advanced sensors, and miniaturized TSMDs are being developed with advances in microprocessor and nonvolatile memory technologies (Rouet and Foucher 2004).

Searls et al. (2001) undertook in situ temperature measurements in both notebook and desktop computers used in different parts of the world. In terms of the commercial applications of this approach, IBM has installed temperature sensors on hard drives (Drive-TIP) to mitigate risks due to severe temperature conditions, such as thermal tilt of the disk stack and actuator arm, off-track writing, data corruptions on adjacent cylinders, and outgassing of lubricants on the spindle motor. The sensor is controlled using a dedicated algorithm to generate errors and control fan speeds.

Strategies for efficient in situ health monitoring of notebook computers were provided by Vichare et al. (2004). In this study, the authors monitored and statistically analyzed the temperatures inside a notebook computer, including those experienced during usage, storage, and transportation, and discussed the need to collect such data

427

both to improve the thermal design of the product and to monitor prognostic health. The temperature data was processed using ordered overall range (OOR) to convert an irregular time–temperature history into peaks and valleys and also to remove noise due to small cycles and sensor variations. A three-parameter Rainflow algorithm was then used to process the OOR results to extract full and half cycles with cyclic range, mean, and ramp rates. The effects of power cycles, usage history, CPU computing resources usage, and external thermal environment on peak transient thermal loads were characterized.

In 2001, the European Union funded a 4-year project, "Environmental Life-Cycle Information Management and Acquisition" (ELIMA), which aimed to develop ways to manage the life cycles of products (Bodenhoefer 2004). The objective of this work was to predict the remaining life time of parts removed from products, based on dynamic data, such as operation time, temperature, and power consumption. As a case study, the member companies monitored the application conditions of a game console and a household refrigerator. The work concluded that in general, it was essential to consider the environments associated with all life intervals of the equipment. These included not only the operational and maintenance environments, but also the preoperational environments, when stresses maybe imposed on the parts during manufacturing, assembly, inspection, testing, shipping, and installation. Such stresses are often overlooked, but can have a significant impact on the eventual reliability of equipment.

Skormin et al. (2002) developed a data-mining model for failure prognostics of avionics units. The model provided a means of clustering data on parameters measured during operation, such as vibration, temperature, power supply, functional overload, and air pressure. These parameters are monitored in situ on the flight using time–stress measurement devices. Unlike the physics-based assessments made by Ramakrishnan (Ramakrishnan and Pecht 2003), the data-mining model relies on statistical data of exposures to environmental factors and operational conditions.

Tuchband and Pecht (2007) presented the use of prognostics for a military line replaceable units (LRU) based on their life-cycle loads. The study was part of an effort funded by the Office of Secretary of Defense to develop an interactive supply chain system for the U.S. military. The objective was to integrate prognostics, wireless communication, and databases through a web portal to enable cost-effective maintenance and replacement of electronics. The study showed that prognostics-based maintenance scheduling could be implemented into military electronic systems. The approach involves an integration of embedded sensors on the LRU, wireless communication for data transmission, a PoF-based algorithm for data simplification and damage estimation, and a method for uploading this information to the Internet. Finally, the use of prognostics for electronic military systems enabled failure avoidance, high availability, and reduction of life-cycle costs.

The PoF models can be used to calculate the remaining useful life, but it is necessary to identify the uncertainties in the prognostic approach and assess the impact of these uncertainties on the remaining life distribution in order to make risk-informed decisions. With uncertainty analysis, a prediction can be expressed as a failure probability.

Gu et al. (2007) implemented the uncertainty analysis of prognostics for electronics under vibration loading. Gu identified the uncertainty sources and categorized them into four different types: measurement uncertainty, parameter uncertainty, failure criteria uncertainty, and future usage uncertainty (see Figure 18.14). Gu et al. (2007)

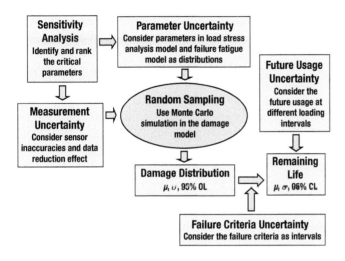

Figure 18.14 Uncertainty implementation for prognostics.

utilized a sensitivity analysis to identify the dominant input variables that influence the model output. With information of input parameter variable distributions, a Monte Carlo simulation was used to provide a distribution of accumulated damage. From the accumulated damage distributions, the remaining life was then predicted with confidence intervals. A case study was also presented for an electronic board under vibration loading and a step-by-step demonstration of the uncertainty analysis implementation. The results showed that the experimentally measured failure time was within the bounds of the uncertainty analysis prediction.

18.6 Implementation of PHM in a System of Systems

System of systems is the term used to describe a complex system comprising of many different subsystems that may be structurally or functionally connected. These different subsystems might themselves be made up of different subsystems. In a system of systems, many independent subsystems are integrated such that the individual functions of the subsystems are combined to achieve a capability/function beyond the capability of the individual subsystems. For example, a military aircraft is made up of subsystems, including airframe, body, engines, landing gear, wheels, weapons, radar, and avionics. Avionic subsystems could include the communication navigation and identification (CNI) system, global positioning system (GPS), inertial navigation system (INS), identification friend or foe (IFF) system, landing aids, and voice and data communication systems.

Implementing an effective PHM strategy for a complete system of systems requires integrating different prognostic and health monitoring approaches. Because the systems are so complex, the first step in implementation of prognostics is to determine the weak link(s) in the system. One of the ways to achieve this is by conducting a FMMEA for the product. Once the FMMEA have been identified, a combination of canaries, precursor reasoning, and life-cycle damage modeling may be implemented for different subsystems of the product, depending on their failure attributes. Once the monitoring techniques have been decided, then the next step is to analyze the data.

Different data analysis approaches, such as data-driven models, PoF-based models, or hybrid data analysis models, can be used to analyze the same recorded data. For example, operational loads of computer system electronics, such as temperature, voltage, current, and acceleration, can be used with PoF damage models to calculate the susceptibility to electromigration between metallization, and thermal fatigue of interconnects, plated-through holes, and die attach. Also, data about the CPU usage, current, and CPU temperature, and so on, can be used to build a statistical model that is based on the correlations between these parameters. This data-driven model can be appropriately trained to detect thermal anomalies and identify signs for certain transistor degradation.

Implementation of prognostics for system of systems is complicated and in the very initial stages of research and development. But there has been tremendous development in the certain areas related to prognostics and health management. Advances in sensors, microprocessors, compact nonvolatile memory, battery technologies, and wireless telemetry have already enabled the implementation of sensor modules and autonomous data loggers. Integrated, miniaturized, low power, reliable sensor systems operated using portable power supplies (such as batteries) are being developed. These sensor systems have a self-contained architecture requiring minimum or no intrusion into the host product, in addition to specialized sensors for monitoring localized parameters. Sensors with embedded algorithms will enable fault detection, diagnostics, and remaining life prognostics, which will ultimately drive the supply chain. The prognostic information will be linked via wireless communications to relay needs to maintenance officers. Automatic identification techniques such as radio frequency identification (RFID) will be used to locate parts in the supply chain, all integrated through a secure web portal to acquire and deliver replacement parts quickly on an as-needed basis.

Research is being conducted in the field of algorithm development to analyze, trend and isolate large-scale multivariate data. Methods, such as projection pursuit using principal component analysis and support vector machines, mahanalobis distance analysis, symbolic time series analysis, neural networks analysis, and Bayesian networks analysis, can be used to process multivariate data.

Even though there are advances in certain areas related to prognostics, many challenges still remain. The key issues with regard to implementing PHM for a system of systems include decisions of which systems within the system of systems to monitor, which system parameters to monitor, selection of sensors to monitor parameters, power supply for sensors, on board memory for storage of sensed data, in situ data acquisition, and feature extraction from the collected data. It is also a challenge to understand how failures in one system affect another system within the system of systems and how it affects the functioning of the overall system of systems. Getting information from one system to the other could be hard especially when the systems are made by different vendors. Other issues that should be considered before implementation of PHM for system of systems are the economic impact due to such a program, contribution of PHM implementation to a condition-based maintenance, and logistics.

The elements necessary for a PHM application are available, but the integration of these components to achieve the prognostics for a system of systems is still in the works. In the future, electronic system designs will integrate sensing and processing modules that will enable in situ PHM. A combination of different PHM implementations for different subsystems of a system of systems will be the norm for the industry.

18.7   Summary

Due to the increasing amount of electronics in the world and the competitive drive toward more reliable products, prognostics and health management is being looked upon as a cost-effective solution for the reliability prediction of electronic products and systems. Approaches for implementing prognostics and health management in products and systems include (1) installing built-in structures (fuses and canaries) that will fail faster than the actual product when subjected to application conditions, (2) monitoring and reasoning of parameters (e.g., system characteristics, defects, and performance) that are indicative of an impending failure, and (3) monitoring and modeling environmental and usage data that influence the system's health and converting the measured data into life consumed. A combination of these approaches may be necessary to successfully assess the degradation of a product or system in real time and subsequently provide estimates of remaining useful life.

Problems

18.1 One of the potential investment returns (cost avoidances) listed for PHM was associated with warranties. Explain the ways in which the cost of warranties could be decreased by using PHM.

18.2 Why does unscheduled maintenance cost more than scheduled maintenance?

18.3 What is "remaining useful life"? How can remaining useful life prognosis improve system reliability?

18.4 What is a failure precursor and how can a failure precursor be identified? Explain with examples.

18.5 Explain the methods for PHM.

18.6 Discuss the pros and cons for data-driven prognostic methods and PoF prognostic methods.

18.7 Suppose you are designing a PHM system for batteries. Discuss the steps and factors for the implementation of the PHM system.

19 Warranty Analysis

> *Money back guaranties or the Good Housekeeping seal of approval is of damn little comfort to the rifleman on the battlefield when his rifle jams. It is of little comfort to the commander who has a five-year or fifty-thousand-mile warranty on a truck if it breaks down on the way to the front with ammunition that may be critical to winning the battle.*
>
> —Former Commandant General Paul X. Kelly, USMC,
> June 1994 (Brennan 1994)

A warranty is a guarantee by the manufacturer or seller, usually in the form of a contract with a buyer, that defines a responsibility with respect to the product or service provided. Manufacturers provide warranties to attract customers and assure them of the high quality of their products. The primary role of warranties is protective, for both buyer and seller. Protection for the buyer is provided by warranty terms that require compensation in some form for defective items, namely repair of the item or replacement by a new item at reduced cost to the consumer or at no cost at all. The manufacturer is protected by specifications in the warranty terms of conditions under which the warranty is invalidated (e.g., use of the product for a purpose other than that intended, failure to perform proper maintenance, and use of a consumer product in a commercial application), and by limiting the amount of compensation, specifying the length of coverage, excluding specified components, limiting or precluding claims for ancillary damages, and so forth.

Customers value a good warranty as an economic protection, but a product is generally not considered good if it fails during the customer's perception of the product's useful life, regardless of the warranty. In other words, customers want the product that they purchase to function properly for some specified time in its application environment. Consumers would prefer that the product they purchase never needs to be returned for repair or replacement during its useful life. This is especially true if the malfunction involves safety, such as the loss of drivability of a vehicle.

Reliability Engineering, First Edition. Kailash C. Kapur and Michael Pecht.
© 2014 John Wiley & Sons, Inc. Published 2014 by John Wiley & Sons, Inc.

19 Warranty Analysis

19.1 Product Warranties

To protect consumers from defective products and misleading product specifications, Congress empowered the Federal Trade Commission (FTC), along with state and local governments, to enforce consumer protection mandates. The Magnusson–Moss Warranty Act of July 1975 defined the terms for a "full warranty" (Federal Consumer Product Warranties Law 1975). The act states that a defective product manufactured with a warranty labeled "full" must be fixed within a reasonable amount of time and without charge to the customer for parts and labor. In addition, the warrantor cannot impose any unreasonable repair conditions on a customer, unless approved by the FTC. All warranties that offer anything less than the "full warranty" requirements must be designated as "limited." In such instances, should the product ever need servicing, the consumer may be responsible for labor costs.

A warranty can also be categorized based on whether it is express or implied. An express warranty is any specific statement in the manufacturer's written warranty, including those in brochures, advertisements, and other documents provided by a sales representative about a specific product. Claiming that a car "purrs like a kitten" is not considered an express warranty due to the general, as opposed to specific, nature of the specification, and is not recoverable under warranty. Specifying that a car has a six-cylinder engine constitutes an express warranty because it is a measurable specification subject to evaluation for accuracy. Thus, express warranties are based upon specific factual aspects of the product, such as "four vs. six-cylinder" engines, which are subject to evaluation for compliance to the specifications of the product.

An implied warranty is an unwritten warranty that is implied by state law. Accepted by all 50 states, an implied warranty can be based either on merchantability or fitness for a particular purpose. Under the Uniform Commercial Code (UCC), merchantability is defined as being fit for the regular purposes for which the products are intended. An example of merchantability is the ability of an automobile to stop when the brakes are applied. On the other hand, the implied warranty of fitness for a particular purpose is based on the seller's skill or judgment. An example of an implied warranty for fitness for a particular purpose might be selling a truck and verbally stating it is capable of towing a large boat of a particular weight. The consumer is relying upon the professional judgment of the salesperson to evaluate the capability of the product to perform a specific task.

Three remedy options are typically offered in warranties: a free replacement, a pro rata refund, or a combination of free replacement and pro rata refund. Free replacement warranties require the seller to pay the entire cost of the remedy if the product fails before the end of the warranty period. Under a pro rata warranty, if a product fails before the end of the warranty period, the seller is responsible for repair or replacement, and the cost extent of the seller's obligation is determined based upon the age or wear of the product at the time of failure. For example, it is common to purchase tires with the condition that the replacement cost will be determined by measurement of tread wear and determination of the proportional amount of use of the tire prior to the appearance of the defect or malfunction. Warranty policies can be a combination of a free replacement and pro rata refund when an initial free replacement period is followed by a pro rata period. Full warranties, under the Magnusson–Moss Act, place an obligation on the manufacturer to provide remedies for claims within a reasonable period of time at no charge, or refund the purchase price less a reasonable depreciation if the defect cannot be remedied within a reasonable number of attempts.

A warranty pertains to the obligations of the manufacturer in relation to materials, workmanship, and fitness of a product for ordinary use or reasonably intended purposes throughout the duration of the warranty period defined in the express warranty. "Nonconformity" means a defect or other condition that substantially impairs the use, value, or safety of the product, but does not include a defect or condition that is the result of abuse, neglect, or unauthorized modification or alteration of the product. Thus, if a vehicle had been driven off-road, when it was expressly intended by the manufacturer for on-road use, there may be no liability under a consumer or commercial warranty.

The purchase price of a product generally incorporates the projected warranty cost, which includes repair cost, administration cost, and field service cost. The repair cost consists of all labor and material costs to accomplish repair of the warranted product to restore it to a serviceable condition, including the costs associated with the return of the product to the manufacturer, failure verification, repair, testing, and possible shipping to return the product to the consumer. Repair cost is typically the most significant component of the total warranty cost. Administration cost includes all cost for repair documentation, processing of warranty data, and the preparation of required status reports. Field service cost includes the cost incurred by the manufacturer during the repair, such as electricity, transportation, loaner items, diagnosis by an independent contractor of the problem, and other general administrative and overhead expenses.

19.2 Warranty Return Information

Projections from early warranty data may not capture pending problems if there is an increasing failure rate with time. Wearout failure mechanisms may initially exhibit an extremely low rate of occurrence, but have an increasing failure rate. The root cause of failure must therefore be assessed to determine if failures are caused by wearout failure mechanisms. Constant failure rate statistics or prediction (e.g., Mil-Hdbk-217, Telcordia, and PRISM) should not be used. Furthermore, even if early returns are not a result of wearout mechanisms, it should not be assumed that wearout will not occur.

Warranty data will underestimate problems if owners go to sources that do no document repairs for warranty claims. Thus, even small warranty return rates must be assessed, especially if the product is driveability, emissions, or safety related.

A manufacturer should assume that all field returns are field failures and treat them as such. For example, a company that produces a drivability-, safety-. or emission-regulated product should assume that every return of that product is a failure, and take on full responsibility of ascertaining the root cause. A thorough root-cause analysis must be conducted so that potential and nonobvious failure mechanisms are not overlooked.

Root-cause analysis should include diagnostic monitoring, troubleshooting ,and "ABA" testing.[1] The manufacturer has the responsibility to conduct both tests and

---

[1] In ABA testing, a good module is used to replace the "apparently failed" module, and then the failed module is reinstalled to ensure that a system integration problem (i.e., connector) was not the cause of the problem.

teardowns of every product (regardless of warranty stages), or determine why an apparently good product has been returned.

Testing of a product will often require combining multiple and time operating and environmental loading (stress) conditions for extended periods of time.[2] (Mitra and Patankar 1997) It must be recognized that a unique sequence of environmental and operating conditions can trigger a fault; for example, computer users are familiar with such situations where a unique and often complex sequence of keystrokes can initiate an unpredictable fault, which appears intermittent, and cannot be duplicated.

Both nondestructive and destructive evaluations must be conducted after testing and must involve investigating every potential failure mode and failure site to assess degradation. If no defect can be determined and there is no other verifiable and auditable explanation, then the module should still be considered a field failure, because electrical failures can often be quite elusive and intermittent in nature. Terms such as trouble not identified (TNI) or no-fault-found (NFF) should not be used to ignore problems. It is misleading to state that an unidentified problem means there was no problem, just because limited investigations, tests, and failure analyses have not yet enabled observation of the defect.

Incentives should be provided to the supply-chain participants to track all failure statistics of products that are out of warranty. Just because a product is out of warranty does not mean that the hazard rates and the root causes of any failures should not be determined. Out of warranty wear-out mechanisms that cause problems cannot be ignored. The additional data that these products provide can also help engineers to design more reliable products in the future.

The warranty and maintenance plan of a product should directly correspond to the "customer's expectations" of useful life. The customer must be told that the product must be replaced or maintained in some manner at the end of the warranty, especially if a failure of the product can affect safety.

19.3 Warranty Policies

Two types of warranty policies have been widely applied in practice: replacement warranty policy and pro rata warranty policy. For a replacement warranty policy, the seller is required to either repair the product or provide a new product at no cost to the buyer from the time of initial purchase. Such a policy is usually offered with repairable products. Pro rata warranty policy requires the seller to provide replacement at some cost, which is called pro rata cost, to the buyer. Pro rata cost can be either a linear or nonlinear function of the remaining time in the warranty length. Pro rata warranty policy is usually offered with nonrepairable products.

A combination of the replacement and pro rata warranty policies is also common because it has a significant promotional value to the seller while at the same time providing adequate control over costs for both buyer and seller in most applications

---

[2]The engineering specification of a module should specify the worst-case operating conditions, and backed by tests conducted to assess all combinations of worst-case applications and environments. If any module fails to meet the engineering specifications, no further modules should be placed into a vehicle. It is strongly recommended that IEC Standard 60134 be followed and applied to all electronic products that are drivability, emissions, or safety related.

(Chukova et al. 2004). One type of the combinations is the fully renewing combination free replacement and pro rata warranty (FRW/PRW) policy. Denote W as the warranty length of the product. Under the FRW/PRW policy, the seller agrees to replace the product that fails prior to the time point W', where $W' < W$, from the time of purchase with a new product at no cost to the buyer; meanwhile, any failure in the time interval from W' to W results in a pro rata replacement, that is, any product is replaced with a new item at pro rata cost to the buyer.

19.4 Warranty and Reliability

Topics that will be dealt with in the sections that follow are warranty cost analysis, the relationship between warranty and reliability, and management of warranty.

Consumers are seldom in a position to evaluate all products adequately and have little information concerning product performance and reliability before making a purchase decision. Warranty terms provide signals of these characteristics, generous warranty terms conveying the message that the risk is low. As a result, warranty is used as an advertising tool in competition with other manufacturers. Note that the manufacturer must use caution in setting warranty terms. Terms that are too generous may lead to excessive future warranty costs.

There are several technical business and legal decisions that need to be made by the seller and buyer in determining warranty policies and warranty contracts. The seller has to consider issues related to warranty policies (terms, length, and so forth), costs, a function of the warranty parameters (e.g., length, amount of rebate, and repair vs. replace options), servicing products and warranty, data (historic, test, field, claims, etc.) needed for warranty management and how to obtain them, and impact on product and process design.

The buyer has to take into account the cost of warranty, including the additional cost options (e.g., length of warranty), comparison with other sellers, needed lifetime, and options of extended warranty. Addressing these issues is a difficult problem for both manufacturer and buyer. In addition, there are many other aspects of warranty, and many disciplines are involved in the analysis of these issues. Product reliability has an impact, directly or indirectly, in nearly all of these areas (Blischke and Murthy 1996).

Some other variations and definitions regarding warranty should also be mentioned. A warranty is renewing if on failure of an item, a replacement is made and this replacement item carries a warranty that is identical with that of the original item. In effect, the warranty period begins anew.

There are three main categories of warranties, including consumer, commercial, and defense acquisition warranties. For consumer goods, the most common warranties are various versions of the free replacement and pro rata warranties (which involve repair or replacement at no cost or prorated cost); cash rebates on failure of the item; and a combination free replacement/pro rata warranty.

Commercial and industrial warranties are those offered in sales by a manufacturer to another company. Sales of aircraft engines, seats, windshields, radar systems, and so forth, to an aircraft manufacturer are examples. These warranties often are of the same basic type as those offered on consumer products, but additional features may be involved. For example, groups or lots of items may be warranted rather than individual items. Warranties of this type are called cumulative or fleet warranties (Berke and Zaino 1991; Guin 1984; Zaino and Berke 1994).

Warranties on items procured by the government include all of the above plus some special warranties, particularly in acquisition of defense products. The best known of these special warranties is the reliability improvement warranty, which includes provisions for product development and improvement.

There are of course many different combinations of warranties when we take into account all the possibilities.

- *Nonrenewing Free Replacement Warranty (FRW)*. Under this warranty, the manufacturer will repair or replace a failed item free of charge up to time W from the time of initial purchase. The repaired or replaced item is warrantied only for the time remaining in the original warranty period. The nonrenewing FRW is most often offered on repairable items (with repair almost always involving replacement of a faulty part or component). Examples of consumer products are household appliances, electronic items such as PCs and television sets, and automobile parts. Commercial products include tools, motors, and heavy equipment. This warranty is one of the more costly to the manufacturer.

- *Renewing Pro Rata Warranty (PRW)*. Under the renewing PRW, on failure at time X of an item, the manufacturer will provide a replacement item at cost to the buyer of $[1 - X/W]C$. The replacement item is covered under warranty identical to that of the item originally purchased. This warranty features linear proration. Although not use in practice, the proration function could be nonlinear as well, for example, quadratic, pro rata warranties are most commonly used for nonrepairable items. Nearly all automobile tires and batteries are covered by warranties of this type. This warranty favors the seller, that is, is less costly than the FRW.

- *Renewing Combination FRW/PRW.* The manufacturer will provide a free replacement for a failed item up to time $W_1 < W$ from the time of initial purchase, and at pro-rated cost from W_1 to W. The replacement item is warrantied under the same terms as the original purchased item.

  This is the most common combination warranty. It is typically offered on items ranging from automobile tires to appliances. It is a compromise between the FRW and PRW in that the cost of this warranty is less than that of the FRW and more than the PRW.

Other combination warranties are those that cover different components for different periods of time (common in TV sets where the picture tube carries a separate warranty, refrigerators, where the compressor is warrantied for a longer period of time that other components, and automobiles, where there are many separate warranties.)

A classification of warranties in accordance with the various features offered is given in Blischke and Murthy (1993). This taxonomy of warranties is given in Figure 19.1 and is based on the following characteristics:

Whether or not the warranty includes product development (redesign, design or process improvements, and so forth). This is typically only found in reliability improvement warranties (RIW).

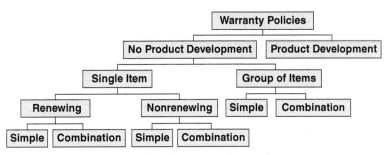

Figure 19.1 Taxonomy of warranties.

- Whether the warranty covers a single item purchased or a lot of items.
- Whether or not the warranty is renewing.
- Whether or not the warranty is a combination warranty.
- Whether the warranty is one-dimensional or more.
- Whether the warranty is single-attribute or covers multiple attributes.
- Whether or not the warranty has the option of extended coverage.

Burn-in can be considered as a part of the production process in which the manufactured products are operated under accelerated stresses for a short time period, which is called burn-in time, before their release. The principal motivation of burn-in is to detect the situation that latent defects exist in the early stage of introducing products. According to Nguyen and Murthy (1982), for the products with an initially high failure rate, burn-in can be used to reduce the warranty cost. Kar and Nachlas (1997) presented a model to study the warranty policy and burn-in together in order to examine the benefits for product management. Determination of optimal burn-in time to minimize the associated cost is always an interesting topic in the literature. Nguyen and Murthy (1982) derived the optimal burn-in time for repairable and non-repairable products sold under the failure-free and rebate policy. Yun et al. (2002) determined the optimal burn-in time to minimize the total mean cost, which is the sum of the manufacturing cost, the burn-in cost, and the cumulative warranty cost, under free replacement warranty policy. Sheu and Chien (2005) developed a generalized burn-in and field-operation model for the repairable products sold under warranty.

19.5 Warranty Cost Analysis

There are many aspects to the analysis of the cost of a warranty. First, it is necessary to develop adequate cost models. These depend on the perspective (buyer or seller), the basis on which the costs are to be assessed, and the probabilistic structure of the random elements involved. In this section, we discuss these factors and present cost models for the FRW, PRW, and a few additional warranties. The various types of information needed to estimate the models will also be discussed.

19.5.1 Elements of Warranty Cost Models

There are many types of models used in analysis of warranties. These include sales and demand models is marketing, cost and other models in economics, engineering models for analysis of various aspects of design, product reliability, production and quality control, operational models for servicing of warranties, and so forth. As is apparent, cost models must be developed separately for seller and buyer.

The cost of warranty depends on a number of factors, including at least the following:

- *Type of Warranty.* All other aspects being equal, generally manufacturer's costs for the FRW are higher than for a combination FRW/PRW, and these are higher than those for the PRW. Length of the warranty, renewability, and other features of the warranty can also have a significant impact on cost. Buyer's costs, at least in the long run, are affected in an inverse fashion to those of the seller.
- *Failure Pattern of New Items.* The mean time to failure (MTTF) is an important measure of product performance. Different statistical distributions of time to failure, however, can significantly affect costs, even when the MTTF is the same.
- *Repairability of Failed Items.* If an item is repairable and can be repaired at less cost than providing a replacement item, warranty costs may be reduced.
- *Failure Pattern of Repaired or Replaced Items.* The distribution of failure times of replacement items (which may be different from that of the original item) or repaired items (which may depend on the type of repair and the number of times an item has been repaired) may also significantly impact costs.
- *Incidental Costs.* These include warranty administration, shipping, the cost of service centers, cost of spare's storage, and many related items.

19.5.2 Failure Distributions

Item failure is a random process. As such, it is modeled by a probability distribution. Which distribution is appropriate depends on a number of factors, including product characteristics (determined by engineering design, raw materials, process design, decisions regarding outsourcing and selection of suppliers, and perhaps other factors, such as time of production and production rate), type and rate of usage, determined primarily by the buyer/user, age of the item, maintenance, again determined primarily by the buyer, but influenced by the warranty requirements, environmental factors, some under control or partial control of the user (e.g., protection from weather, extremes of heat and cold, and moisture), and some not (the weather itself).

19.5.3 Cost Modeling Calculation

There are a number of approaches to the costing of warranty. Costs clearly are different for buyer and seller. In, addition, the following are some of the methods for calculating costs that might be considered:

- Cost, to the seller, per item sold. This per unit cost may be calculated as the total cost of warranty, as determined by general principles of accounting, divided by number of items sold.
- Cost per item to the buyer, averaging over all items purchased plus those obtained free or at reduced price under warranty.
- Life-cycle cost of ownership of an item with or without warranty, including purchase price and operating and maintenance cost, and finally including cost of disposal.
- Life-cycle cost of an item and its replacements, whether purchased at full price or replaced under warranty, over a fixed time horizon.
- Cost per unit of time.

The selection of an appropriate cost basis depends on the product, the context, and perspective. The type of customer—individual, corporation, government—is important, as are many other factors.

The cost of offering a warranty clearly depends on the reliability of the item. The precise role of product reliability will become more apparent in this section and in the remainder of the presentation. Warranty expenses also depend on a number of other factors. These include at least the following:

- *The Proportion of Legitimate Claims That Are Made.* This is called warranty execution. For various reasons (too much trouble, not worth the effort near the end of a pro rata warranty, desire to switch brands), claims are not made for all items that fail within the warranty period.
- *The Proportion of Claims That Are Not Legitimate.* Some typical bogus claims are those made after expiration of the warranty and not verified as such, failures due to misuse, and intentionally failed items.
- *Servicing Policy.* Factors include whether buyer or seller pays shipping costs, number and location of service centers, coverage of parts and labor or parts only, repair versus replace decisions, company versus contracted warranty service, and so forth. The list is a long one and many management decisions must be made.
- *Administrative Costs.* Processing of claims, cost of setting up and maintaining a warranty department, and so forth are included.
- *Incidental Costs.* Setting up a warranty information system; general overhead items.

19.5.4 Modeling Assumptions and Notation

As is always the case in constructing mathematical models, many assumptions are made. Models that are useful are those for which the assumptions provide a reasonable approximation of the true nature of the phenomenon being studied and are not unduly sensitive to minor violations of these assumptions. In modeling seller's warranty costs, we make the following assumptions, some of which can be relaxed but at the expense of, for our purposes, unnecessarily increased complexity:

- Warranty claims are made on all items that fail within the warranty period (warranty execution is 100%).
- No illegitimate claims are made.
- Claims are made immediately on failure of the item. In practice, this means that the time from item failure to warranty claim is small relative to the warranty period and the lifetime of the product.
- Compensation under warranty (repair or replacement) is instantaneous. Again, this is interpreted as in Item 3.
- Constant, fixed cost per claim. In practice, this cost is a random variable. The analysis in effect uses an average repair cost.
- Identical items. All repaired or replaced items have the same life distribution as that of the original item. Thus no design changes that would affect the lifetime of the item have been made and any repairs bring the item back to "good as new." (Some of the models can be modified in a relatively straightforward manner to accommodate other repair or replace regimes.)
- Statistical independence. This in effect says that there is no relationship between lifetimes of items.
- No brand switching. The buyer is assumed to purchase the same make and model item on failure of a product, whether or not it is covered by warranty.
- All parameters are known. Costs of supplying an item to the buyer, repair or replacement costs, life distribution of the items and their parameters, and so forth, are all known to the analyst or can be reliably estimated.

19.5.5 Cost Models Examples

In expressing the cost models, we require the following additional notation:

$$\mu_T = \int_0^T x \, dF(x), \tag{19.1}$$

μ_T is called the partial expectation of X; it is the average time to failure of all items that fail with lifetimes less than some given time value T, for the time to failure random variable with cumulative distribution function $F(x)$. In addition, we have

$$M(T) = \text{Expected number of failures in the interval } (0, T),$$

where $M(T)$ is the renewal function. It is defined as the solution to an integral equation, and can be obtained in closed form only for a few distributions. For the exponential distribution, for example, $M(t) = \lambda t$. In other cases, it can be evaluated by means of computer algorithms and has been extensively tabulated (Baxter et al. 1982; Blischke and Murthy 1994).

Finally, we use C_s to denote the seller's average cost per item (including development, production, distribution, marketing, etc.), C_b to denote the buyer's cost, and C_r to denote the average cost of repair.

19.5.5.1 Nonrenewing FRW We look first at the seller's cost per unit sold for nonrepairable items sold under nonrenewing FRW with warranty period W. The analysis of this warranty by Menke (1969) and Lowerre (1968) was one of the first theoretical analyses of warranty costs. The analyses presented were first-failure models, ignoring the possibility of multiple failures during the warranty period. Nonetheless, some useful first approximations were obtained. These results were extended by Blischke and Scheuer (1975, 1981), providing the model for seller's expected cost, say $E[C_m(W)]$, in this case as

$$E[C_m(W)] = C_s[1 + M(W)]. \qquad (19.2)$$

Example 19.1

Suppose that a TV picture tube has a lifetime that is exponentially distributed with MTTF = 6.5 years. Then the failure rate is $\lambda = 1/6.5 = 0.1538$ per year. Suppose that the cost of supplying a new tube (original or under warranty) is $67.20. Compare the costs of 6-month and 1-year nonrenewing free replacement warranties.

For this distribution, the renewal function is $M(t) = \lambda t$. For a 6-month warranty, the average cost to the seller per unit sold is

$$\$67.20[1 + 0.1538(0.5)] = \$72.37.$$

For 1-year warranty:

$$\text{Avg. cost} = \$67.20[1 + 0.1538(1)] = \$77.54.$$

For a 1-year warranty, this cost would be $77.54. Note that the warranty cost to the seller has doubled, from about 7.7% of the production price to about 15%.

For repairable items, the situation is somewhat more complicated. The average cost to the seller depends on the repair policy, the average cost of repairing an item, and the life distribution of repaired items. Some useful results for repairable items are given by Nguyen and Murthy (1984).

If repairable items are repaired good-as-new, that is, repaired items have the same failure distribution as new items (which may occur if failure of the item is due to failure of a component that has a much higher failure rate than any other component), then the expected cost is given by:

$$E[C_m(W)] = C_s + C_r M(W). \qquad (19.3)$$

Many additional models for FRW, from buyer's and seller's points of view, and under various assumptions regarding costs and repairability and a number of other factors, are given in Newman and Nesbitt (1978).

19.5.5.2 Nonrenewing PRW A nonrenewing warranty is equivalent to a warranty that provides a rebate on failure of an item. The buyer mayor may not use the rebate to buy an identical replacement item of the same brand. Whether or not this occurs, from the seller's point of view the cost is based on the cost of supplying an item plus the selling price, because that is what the rebate is based on. The expected cost to the seller, per item, for this type of pro rata warranty is

$$E[C_m(W)] = C_s + C_b[F(W) - \mu_W/W]. \qquad (19.4)$$

Formulas for evaluation of the partial expectation are available for a number of distributions, but again lead to computational difficulties in most cases. Computation of the CDF is also sometimes difficult. Computer analysis of these cost equations, however, in reasonably straightforward.

For the exponential distribution, the partial expectation is given by

$$\mu_W = \lambda^{-1}\left[1-(1+\lambda W)e^{-\lambda W}\right]. \tag{19.5}$$

Example 19.2

There is a TV picture tube with an exponential lifetime and MTTF of 6.5 years, and cost to seller of $67.20. Suppose that the selling price is $C_b = \$105$ and that the tube has a one-year pro rata warranty, with a pro-rated rebate rather than a replacement at pro-rated price. From Equation (19.4) and Equation (19.5), we find that the expected cost to the seller, including warranty, is

$$E(C_m) = \$67.20 + 105\left[1 - e^{-0.1538} - 0.0674\right] = \$74.88.$$

Note that this is less than the cost to the seller of the nonrenewing FRW. (The warranty cost is 11% rather than 15%.) The cost for the renewing PRW would be even less, because it involves a new item (with cost based on C_s, rather than C_b) rather than a cash rebate.

19.5.6 Information Needs

Information regarding both the technical and commercial aspects is necessary for effective warranty management. Technical aspects include product design, manufacturing, test results, quality control, and many related issues. Commercial aspects include marketing strategy, pricing, warranty, service policy, and so forth. Here we discuss a few key issues.

19.5.6.1 Requirements for Successful Application of the Models In order to apply the models discussed earlier and the many other cost models in Newman and Nesbitt (1978), and elsewhere, it is necessary to know:

- *The Form of the Failure Distribution.* This may be obtained from theoretical considerations based on an understanding of the physical mechanism of failure, or empirically, through fitting of various distributions graphically or by use of other statistical methods.
- *Type of Warranty.* The examples given in the previous section are two of the more simple warranties. There are many other possibilities (Blischke and Murthy 1993, 1996; Guin 1984) and many forms of each, involving renewability, length of the warranty period, and so forth. The decision as to warranty policy is an important managerial responsibility, and there are many potential cost ramifications.
- *Parameter Values.* As is apparent, the cost models require many inputs, some of which may be poorly known or unknown. These include cost parameters as well as the parameters of the life distribution.

- *Rectification Policy.* Here we are concerned with servicing policy, for example, repair versus replacement, and under what conditions. The type of warranty influences this and parameter values as listed in Item 3 must be determined for each scenario under consideration.
- *Other Cost Information.* Many direct and indirect costs are included in the models. A number of these were discussed previously. The impact of these and how they are assessed depend on the company and its policies and methods of accounting, the product itself, and many other factors.
- *The Broader Picture.* For a proper interpretation of the model output (and input), the analyst, engineer, and manager must be aware of the overall company perspective. (Here we have looked at the process from the seller's point of view. Similar considerations are relevant in the case of the buyer.) Important factors include the overall organizational strategy, management objectives for this product in that context, the marketplace, including demand and competition, marketing strategy, and many more.

19.5.6.2 Information Sources In any organization, a great deal of information is available. The problem is to determine exactly what is, indeed, accessible and the relevance of available data to the problem at hand. For assessment of product reliability and warranty costs, at least the following should be sought:

- *Test Data.* This may include data on testing of prototypes, data on various designs under consideration, and so forth. Data on tests of this specific product are, of course, of most relevance. It is important that the data result from a well-designed, comprehensive experiment.
- *Part and Component Data.* Much useful information can be obtained from tests at the part and component level. Some tests of this type will be done on parts and components obtained from suppliers, as well, for example, acceptance testing data. In reliability studies, the use of this information requires detailed models of the relationship between the reliability of these items and that of the system (product). See any text on reliability, for example, Blischke and Murthy (1998), Kapur and Lamberson (1977), Barlow and Proschan (1965). Data of this type are invaluable in Bayesian reliability analysis (Martz and Waller 1982, 1990), as they form the basis of the prior distribution that is used in this analysis.
- *Data on Similar Products.* In many cases, much information will have been collected on prior versions of the product being analyzed. In fact, it is often the case that some of the same parts and components used previously will be included in the current product design. Data of this type may be useful in predicting the reliability of the new product. Again, Bayesian methods may be an effective approach.
- *Vendor Data.* When parts or components are produced by a supplier, extensive testing may not be done by the manufacturer of the product but may have been done by the producer of the part. These data may be very useful for the purposes described earlier, and should be requested of the supplier or made a part of the purchase contract.

- *Subjective Information.* "Engineering judgment" often plays an important role in reliability analysis, particularly when little information of other types is available. It is also frequently used in Bayesian analysis.
- *Claims Data.* After introduction of a product into the marketplace, other types of data become available. The most important of these is data on warranty claims. Although there are often problems with this type of data (e.g., validity of claims, claims execution, actual time of failure, identification of failure cause, and censoring at time W), it can provide valuable information concerning the validity of models and assessment of model parameters.
- *Operational Data.* Field data on operation of an item, when available, is another important source of information that can be obtained after product introduction. Operational data provide information on failure causes and rates in real operational conditions rather than simulated laboratory environments.

19.5.7 Other Cost Models

A schematic representation of warranty cost is shown in Figure 19.2. In the previous sections, we have looked at only a few specific warranties and even fewer cost models. Among the many extension of these warranties, models and related results are the following (Blischke and Murthy 1993, 1994, 1996; Murthy and Blischke 1991a, 1991b):

Other Types of Warranties. There are many versions of the basic FRW and PRW, many combinations of these, and a nearly unlimited number of exclusions and limitations on coverage. Cumulative and extended warranties offer still additional possibilities. In competitive markets, careful management and analysis of the possibilities is essential.

More General Failure Distributions. The exponential distribution, which features a constant failure rate, is realistic in many situations. (It has the added advantage of being mathematically tractable where most other distributions are not.) When the assumption of constant failure rate is not realistic, however, other distributions should be used in modeling failures. Distributions that may be considered include the Weibull, gamma, lognormal, inverse Gaussian, truncated normal, and extreme value, to name a few.

Higher Dimensional Warranties. Two-dimensional warranties (e.g., based on calendar time and usage) as well as some three-dimensional warranties have been discussed previously. In theory, there are an endless number of possible versions of these, and the modeling and analysis problems become even more difficult.

Life-Cycle Cost Models. Models that represent life-cycle costs may be defined in various ways and have been discussed briefly previously. These typically present additional analytical difficulties, for example, involving even more complex renewal-type equations for solution, but are essential in any analysis of long term costs.

Various Cost Bases. The two models given in the previous section look at costs from the seller's point of view. Cost models from the buyer's point of view, including life-cycle costs, have also been developed. Other cost bases—unit,

19.5 Warranty Cost Analysis

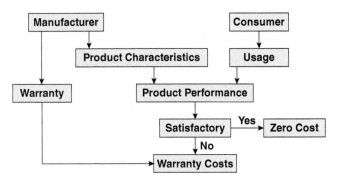

Figure 19.2 Schematic representation of warranty cost.

time, and so forth—have been discussed previously and models of these types have been developed as well.

Discounted Costs. Discounting of future costs to the present value of money is an important aspect of the analysis of long-term warranties. A number of models include this feature. The selection of an appropriate discount rate is always an additional uncertainty, but comparing the results of analyses assuming various possibilities can be enlightening.

Indifference Price Structures. A different approach to determining to cost or value of a warranty to either buyer or seller is to look at the indifference price. The idea is as follows: Suppose that a product is sold without warranty at price C_b; determine a price c^* ($>C_b$) for selling an item with warranty such that the buyer (or seller, as the case may be) would incur the same cost if the item is sold with or without warranty. Again, this may be calculated in various ways, depending on whether it is the indifference price for the buyer or seller, the cost basis, whether or not discounting to present value is involved, and so forth.

In analysis of the process, important considerations are:

- *Cause and Effect Relationships.* Here, as noted, the most important is the relationship between product reliability and warranty cost.
- *Sequence of Model Elements.* The sequence in the warranty context is apparent in the chart describing the system characterization of the warranty process. Both buyer and seller have an impact on ultimate warranty costs.
- *Static versus Dynamic Elements.* Some engineering design changes that may affect reliability can be made after release of the product. Some cannot, at least not without excessive cost.
- *Level of Complexity.* Realistic models are nearly always complex. The result is that realistic models are difficult to analyze and compromises must be made.
- *Level of Realism.* The tradeoff between analytical tractability and realism is well known. The analyst must be aware of the reality of the models used. In warranty analysis, the problems are quite difficult, but the models may be a fair representation of the process. As has been said by many analysts, "No model is correct, but some are useful!"
- *Deterministic versus Stochastic Models.* In warranty analysis, many elements that are stochastic are modeled as though they are deterministic (expected

values are used). If the distribution has small variance, this approach is reasonable, but the analyst must be aware of the possible implications.

- *Generalizability.* Many models developed for a specific application can, in fact, be applied much more generally. To determine whether or not this is the case, studies of the sensitivity of the results to the assumptions made must be undertaken. One such study in the context of warranty is given in Blischke and Vij (1997). It was found that, indeed, distributional assumptions can have a significant impact on predicted costs.

19.6 Warranty and Reliability Management

There is an inverse relationship between the reliability of a product and the future cost of warranty to the manufacturer of the product. Analysis of this cost trade-off is essential for effective management of both issues.

A systems analysis provides valuable insight in that, as usual, the quantification of the problem that it imposes on the analyst, engineer, and manager tends to focus on the longer term and the overall objectives as well as on the methodology. We briefly summarize the systems approach, emphasizing some of the details that are particularly important in the warranty context. The important aspects are:

- *System Characterization.* The first step in an analysis is determining a basis for description of the system, that is, a system characterization. This involves definition of the important variables that can be used to describe the system and attempting to list the possible relationships between them. How much detail is included will depend on the level of understanding of the system at this point and the point of view—buyer, seller, and so on.

- *Mathematical Modeling.* There is inherent uncertainty in the warranty process, as is evident in our previous discussion of warranty analysis. This must be taken into consideration in model formulation, and introduces probabilistic (stochastic) elements into the analysis. The life distribution of the items models an essential part of this uncertainty and plays an important role in the analysis.

Many of these were discussed in Chapter 3 (e.g., exponential and Weibull).

- *Analysis.* The exponential distribution can be dealt with analytically in many cases. Most other important life distributions cannot, particularly with regard to evaluation of renewal functions and other complex functions encountered in warranty analysis, and numerical methods are required, but this is not a major problem—either numerical evaluation of some of the complex integrals or simulation can be used.

- *Model Validation.* Having formulated a mathematical model representing the warranty process, including its stochastic aspects, validation by use of data is required. Claims data are important, as are the other data sets mentioned previously.

- *Interpretation.* Interpretation of the results of all of the above in the context of the real problem in the real world is essential. Even more important is:
- *Implementation.* In order to achieve actual gains in reducing warranty costs, it is necessary to understand the tradeoffs, in the long term, between the cost of increasing reliability and the savings realized from reducing the cost of warranty. This is a managerial decision that must be made early on, and implemented beginning at the product concept and next at the product design stage.

A simplified system characterization of the important elements involved in the analysis of warranties is given in the following chart. The roles of both manufacturer and buyer are indicated and both are clearly important. More detailed characterizations of the warranty process, with extensive flow charts, are given by Murthy and Blischke (1991a, 1991b) and Blischke and Murthy (1993, 1994).

19.7 Summary

In this chapter, we have looked at warranty from several points of view, buyer/seller, engineer/manager (or both), warranty versus reliability and the cost trade-offs of the two, and so forth. Some of the important points we have made with regard to the management of warranty (for both buyer and seller) are the following:

- Determine precisely what the warranty terms are, or are to be.
- Are these negotiable?
- What is the failure distribution of the item?
- How are costs related to this?
- What is (or should be) the warranty servicing policy?
- How shall warranty be managed (for both seller and buyer)?
- There are many other important considerations, for example, what data are available.

Attention must be paid to warranty costs and their relation to product reliability and to methods for predicting and managing both. This has always been important but is crucial in today's marketplace, where products are often nearly indistinguishable and warranty in used as a competitive tool to increase market share.

Problems

19.1 Discuss the role of a warranty from different points of view. Consider both the seller's and the buyer's viewpoints.

19.2 Draw a tree diagram of warranty policies for consumer products.

19.3 How can a burn-in test reduce warranty cost?

19.4 What are the factors that should be considered when developing a warranty cost model?

19.5 Suppose that the cost of producing an electronic instrument is $2000. The instrument is repaired good-as-new after failures under an ordinary 1-year free-replacement warranty. The average cost of servicing a warranty claim is $100, and the MTTF of the instruments is 5 years. Assuming that the failures are exponentially distributed, what is the average cost to the manufacturer per unit sold?

19.6 A manufacturer plans to provide a 12-month ordinary free-replacement warranty for a new laptop. Assume that all failures result in replacements instead of repair. The lifetimes of the laptops are exponentially distributed with MTBF of 7.5 years. The manufacturer's average cost of servicing a warranty claim is $120. The fixed cost of providing warranty coverage for this laptop is also considered. (The fixed cost can include administrative costs to run the warranty department for this laptop.) Assume that the fixed cost is $10,000 for 500,000 laptops sold. What warranty reserve should be put in place (discount rate is ignored)? That is, how much money should the manufacturer of the laptop budget to satisfy the promised warranty?

19.7 Suppose that the electronic instrument in Problem 19.5 has a 1-year pro rata warranty with a pro-rated rebate rather than replacement at a pro-rated price. What sales price will the manufacturer set if the same cost is expected per unit sold?

19.8 Suppose that the electronic instrument in Problem 19.5 has a 1-year pro rata warranty with a replacement at a prorated price and that multiple failures are considered within the warranty period. What sales price will the manufacturer set if the same cost is expected per unit sold?

19.9 List as many sources as possible where information can be obtained for warranty cost modeling and effective warranty management.

19.10 Examples of prognostics-based warranty models have been provided in this chapter. Think of one more warranty model that could be enabled by PHM, and describe how this warranty model could be implemented.

Appendix A: Some Useful Integrals

1. $\int u\,dv = u\int dv - \int v\,du = uv - \int v\,du$, useful for integration by parts

2. $\int x^n dx = \dfrac{x^{n+1}}{n+1}$, except when $n = -1$

3. $\int \dfrac{f'(x)dx}{f(x)} = \log f(x), (df(x) = f'(x)dx)$

4. $\int \dfrac{dx}{x} = \log x$

5. $\int \dfrac{f'(x)dx}{2\sqrt{f(x)}} = \sqrt{f(x)}$

6. $\int e^x dx = e^x$

7. $\int e^{ax} dx = \dfrac{e^{ax}}{a}$

8. $\int \log x\,dx = x\log x - x$

9. $\Gamma(n) = \int_0^\infty x^{n-1} e^x dx$, Gamma function

10. $B(m,n) = \int_0^1 x^{m-1}(1-x)^{n-1} dx$, Beta function

Appendix A: Some Useful Integrals

11. $\int \sin x\, dx = -\cos x$

12. $\int \cos x\, dx = \sin x$

13. $\int_0^\infty \sqrt{x} e^{-x} dx = \frac{1}{2}\sqrt{\pi}\left[\Gamma\left(\frac{1}{2}\right) = \sqrt{\pi}\right]$

14. $\int_0^\infty e^{-ax^2} dx = \frac{1}{2}\sqrt{\frac{\pi}{a}}$, Gaussian integral

Appendix B: Table for Gamma Function

$$\Gamma(n) = \int_0^\infty e^{-x} x^{n-1} dx, \quad 1 \leq n \leq 2.$$

n	$\Gamma(n)$	n	$\Gamma(n)$	n	$\Gamma(n)$	n	$\Gamma(n)$
1.00	1.00000	1.25	0.90640	1.50	0.88623	1.75	0.91906
1.01	0.99433	1.26	0.90440	1.51	0.88659	1.76	0.92137
1.02	0.98884	1.27	0.90250	1.52	0.88704	1.77	0.92376
1.03	0.98355	1.28	0.90072	1.53	0.88757	1.78	0.92623
1.04	0.97844	1.29	0.89904	1.54	0.88818	1.79	0.92877
1.05	0.97350	1.30	0.89747	1.55	0.88887	1.80	0.93138
1.06	0.96874	1.31	0.89600	1.56	0.88964	1.81	0.93408
1.07	0.96415	1.32	0.89464	1.57	0.89049	1.82	0.93685
1.08	0.95973	1.33	0.89338	1.58	0.89142	1.83	0.93969
1.09	0.95546	1.34	0.89222	1.59	0.89243	1.84	0.94261
1.10	0.95135	1.35	0.89115	1.60	0.89352	1.85	0.94561
1.11	0.94739	1.36	0.89018	1.61	0.89468	1.86	0.94869
1.12	0.94359	1.37	0.88931	1.62	0.89592	1.87	0.95184
1.13	0.93993	1.38	0.88854	1.63	0.89724	1.88	0.95507
1.14	0.93642	1.39	0.88785	1.64	0.89864	1.89	0.95838
1.15	0.93304	1.40	0.88726	1.65	0.90012	1.90	0.96177
1.16	0.92980	1.41	0.88676	1.66	0.90167	1.91	0.96523
1.17	0.92670	1.42	0.88636	1.67	0.90330	1.92	0.96878
1.18	0.92373	1.43	0.88604	1.68	0.90500	1.93	0.97240
1.19	0.92088	1.44	0.88580	1.69	0.90678	1.94	0.97610
1.20	0.91817	1.45	0.88565	1.70	0.90864	1.95	0.97988
1.21	0.91558	1.46	0.88560	1.71	0.91057	1.96	0.98374
1.22	0.91311	1.47	0.88563	1.72	0.91258	1.97	0.98768
1.23	0.91075	1.48	0.88575	1.73	0.91466	1.98	0.99171
1.24	0.90852	1.49	0.88595	1.74	0.91683	1.99	0.99581
						2.00	1.00000

Reliability Engineering, First Edition. Kailash C. Kapur.
© 2014 John Wiley & Sons, Inc. Published 2014 by John Wiley & Sons, Inc.

Appendix B: Table for Gamma Function

Some properties of the gamma function:

$$\Gamma(n+1) = n\Gamma(n), n > 0,$$

and when $n =$ integer > 0, we have $\Gamma(n) = (n-1)!$

The gamma function is related to the beta function, $B(m,n)$, as follows:

$$B(m,n) = \int_0^1 x^{m-1}(1-x)^{n-1}\,dx$$

$$B(m,n) = B(n,m) = \frac{\Gamma(m)\Gamma(n)}{\Gamma(m+n)}.$$

Appendix C: Table for Cumulative Standard Normal Distribution

The following table gives values for the cumulative standard normal distribution function. The probability density function for the standard normal random variable, z, is:

$$\phi(z) = \frac{1}{\sqrt{2\pi}} e^{z^2/2} dx, -\infty < z < \infty.$$

The cumulative distribution function is given by:

$$\Phi(z) = \frac{1}{\sqrt{2\pi}} \int_{-\infty}^{z} e^{x^2/2} dx, -\infty < z < \infty.$$

The table has values for $\Phi(z)$ for nonnegative values for z (for the range $0 \leq z \leq 4.99$). The values for negative values for z can be found by using the following equation because standard normal distribution is symmetrical:

$$\Phi(-z) = 1 - \Phi(z), 0 \leq z \leq \infty.$$

We read values such as $\Phi(3.39) = 0.9^3 6505 = 0.9996505$:

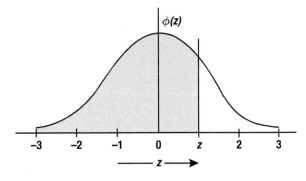

Reliability Engineering, First Edition. Kailash C. Kapur.
© 2014 John Wiley & Sons, Inc. Published 2014 by John Wiley & Sons, Inc.

Appendix C: Table for Cumulative Standard Normal Distribution

$$\Phi(z) = \frac{1}{\sqrt{2\pi}} \int_{-\infty}^{z} e^{-x^2/2} dx.$$

z	0.00	0.01	0.02	0.03	0.04	0.05	0.06	0.07	0.08	0.09
0.0	0.5000	0.5040	0.5080	0.5120	0.5160	0.5199	0.5239	0.5279	0.5319	0.5359
0.1	0.5398	0.5438	0.5478	0.5517	0.5557	0.5596	0.5636	0.5675	0.5714	0.5753
0.2	0.5793	0.5832	0.5871	0.5910	0.5948	0.5987	0.6026	0.6064	0.6103	0.6141
0.3	0.6179	0.6217	0.6255	0.6293	0.6331	0.6368	0.6406	0.6443	0.6480	0.6517
0.4	0.6554	0.6591	0.6628	0.6664	0.6700	0.6736	0.6772	0.6808	0.6844	0.6879
0.5	0.6915	0.6950	0.6985	0.7019	0.7054	0.7088	0.7123	0.7157	0.7190	0.7224
0.6	0.7257	0.7291	0.7324	0.7357	0.7389	0.7422	0.7454	0.7486	0.7517	0.7549
0.7	0.7580	0.7611	0.7642	0.7673	0.7703	0.7734	0.7764	0.7794	0.7823	0.7852
0.8	0.7881	0.7910	0.7939	0.7967	0.7995	0.8023	0.8051	0.8078	0.8106	0.8133
0.9	0.8159	0.8186	0.8212	0.8238	0.8264	0.8289	0.8315	0.8340	0.8365	0.8389
1.0	0.8413	0.8438	0.8461	0.8485	0.8508	0.8531	0.8554	0.8577	0.8599	0.8621
1.1	0.8643	0.8665	0.8686	0.8708	0.8729	0.8749	0.8770	0.8790	0.8810	0.8830
1.2	0.8849	0.8869	0.8888	0.8907	0.8925	0.8944	0.8962	0.8980	0.8997	0.9^{0}147
1.3	0.9^{0}320	0.9^{0}490	0.9^{0}658	0.9^{0}824	0.9^{0}988	0.91149	0.91309	0.91466	0.91621	0.91774
1.4	0.91924	0.92073	0.92220	0.92364	0.92507	0.92647	0.92785	0.92922	0.93056	0.93189
1.5	0.93319	0.93448	0.93574	0.93699	0.93822	0.93943	0.94062	0.94179	0.94295	0.94408
1.6	0.94520	0.94630	0.94738	0.94845	0.94950	0.95053	0.95154	0.95254	0.95352	0.95449
1.7	0.95543	0.95637	0.95728	0.95818	0.95907	0.95994	0.96080	0.96164	0.96246	0.96327
1.8	0.96407	0.96485	0.96562	0.96638	0.96712	0.96784	0.96856	0.96926	0.96995	0.97062
1.9	0.97128	0.97193	0.97257	0.97320	0.97381	0.97441	0.97500	0.97558	0.97615	0.97670
2.0	0.97725	0.97778	0.97831	0.97882	0.97932	0.97982	0.98030	0.98077	0.98124	0.98169
2.1	0.98214	0.98257	0.98300	0.98341	0.98382	0.98422	0.98461	0.98500	0.98537	0.98574
2.2	0.98610	0.98645	0.98679	0.98713	0.98745	0.98778	0.98809	0.98840	0.98870	0.98899
2.3	0.98928	0.98956	0.98983	0.9^{2}0097	0.9^{2}0358	0.9^{2}0613	0.9^{2}0863	0.9^{2}1106	0.9^{2}1344	0.9^{2}1576
2.4	0.9^{2}1802	0.9^{2}2024	0.9^{2}2240	0.9^{2}2451	0.9^{2}2656	0.9^{2}2857	0.9^{2}3053	0.9^{2}3244	0.9^{2}3431	0.9^{2}3613
2.5	0.9^{2}3790	0.9^{2}3963	0.9^{2}4132	0.9^{2}4297	0.9^{2}4457	0.9^{2}4614	0.9^{2}4766	0.9^{2}4915	0.9^{2}5060	0.9^{2}5201
2.6	0.9^{2}5339	0.9^{2}5473	0.9^{2}5604	0.9^{2}5731	0.9^{2}5855	0.9^{2}5975	0.9^{2}6093	0.9^{2}6207	0.9^{2}6319	0.9^{2}6427
2.7	0.9^{2}6533	0.9^{2}6636	0.9^{2}6736	0.9^{2}6833	0.9^{2}6928	0.9^{2}7020	0.9^{2}7110	0.9^{2}7197	0.9^{2}7282	0.9^{2}7365
2.8	0.9^{2}7445	0.9^{2}7523	0.9^{2}7599	0.9^{2}7673	0.9^{2}7744	0.9^{2}7814	0.9^{2}7882	0.9^{2}7948	0.9^{2}8012	0.9^{2}8074
2.9	0.9^{2}8134	0.9^{2}8193	0.9^{2}8250	0.9^{2}8305	0.9^{2}8359	0.9^{2}8411	0.9^{2}8462	0.9^{2}8511	0.9^{2}8559	0.9^{2}8605
3.0	0.9^{2}8650	0.9^{2}8694	0.9^{2}8736	0.9^{2}8777	0.9^{2}8817	0.9^{2}8856	0.9^{2}8893	0.9^{2}8930	0.9^{2}8965	0.9^{2}8999
3.1	0.9^{3}0324	0.9^{3}0646	0.9^{3}0957	0.9^{3}1260	0.9^{3}1553	0.9^{3}1836	0.9^{3}2112	0.9^{3}2378	0.9^{3}2636	0.9^{3}2886
3.2	0.9^{3}3129	0.9^{3}3363	0.9^{3}3590	0.9^{3}3810	0.9^{3}4024	0.9^{3}4230	0.9^{3}4429	0.9^{3}4623	0.9^{3}4810	0.9^{3}4991
3.3	0.9^{3}5166	0.9^{3}5335	0.9^{3}5499	0.9^{3}5658	0.9^{3}5811	0.9^{3}5959	0.9^{3}6103	0.9^{3}6242	0.9^{3}6376	0.9^{3}6505
3.4	0.9^{3}6631	0.9^{3}6752	0.9^{3}6869	0.9^{3}6982	0.9^{3}7091	0.9^{3}7197	0.9^{3}7299	0.9^{3}7398	0.9^{3}7493	0.9^{3}7585
3.5	0.9^{3}7674	0.9^{3}7759	0.9^{3}7842	0.9^{3}7922	0.9^{3}7999	0.9^{3}8074	0.9^{3}8146	0.9^{3}8215	0.9^{3}8282	0.9^{3}8347
3.6	0.9^{3}8409	0.9^{3}8469	0.9^{3}8527	0.9^{3}8583	0.9^{3}8637	0.9^{3}8689	0.9^{3}8739	0.9^{3}8787	0.9^{3}8834	0.9^{3}8879
3.7	0.9^{3}8922	0.9^{3}8964	0.9^{4}0039	0.9^{4}0426	0.9^{4}0799	0.9^{4}1158	0.9^{4}1504	0.9^{4}1838	0.9^{4}2159	0.9^{4}2568
3.8	0.9^{4}2765	0.9^{4}3052	0.9^{4}3327	0.9^{4}3593	0.9^{4}3848	0.9^{4}4094	0.9^{4}4331	0.9^{4}4558	0.9^{4}4777	0.9^{4}4988
3.9	0.9^{4}5190	0.9^{4}5385	0.9^{4}5573	0.9^{4}5753	0.9^{4}5926	0.9^{4}6092	0.9^{4}6253	0.9^{4}6406	0.9^{4}6554	0.9^{4}6696
4.0	0.9^{4}6833	0.9^{4}6964	0.9^{4}7090	0.9^{4}7211	0.9^{4}7327	0.9^{4}7439	0.9^{4}7546	0.9^{4}7649	0.9^{4}7748	0.9^{4}7843
4.1	0.9^{4}7934	0.9^{4}8022	0.9^{4}8106	0.9^{4}8186	0.9^{4}8263	0.9^{4}8338	0.9^{4}8409	0.9^{4}8477	0.9^{4}8542	0.9^{4}8605
4.2	0.9^{4}8665	0.9^{4}8723	0.9^{4}8778	0.9^{4}8832	0.9^{4}8882	0.9^{4}8931	0.9^{4}8978	0.9^{5}0226	0.9^{5}0655	0.9^{5}1066
4.3	0.9^{5}1460	0.9^{5}1837	0.9^{5}2199	0.9^{5}2545	0.9^{5}2876	0.9^{5}3193	0.9^{5}3497	0.9^{5}3788	0.9^{5}4066	0.9^{5}4332
4.4	0.9^{5}4587	0.9^{5}4831	0.9^{5}5065	0.9^{5}5288	0.9^{5}5502	0.9^{5}5706	0.9^{5}5902	0.9^{5}6089	0.9^{5}6268	0.9^{5}6439
4.5	0.9^{5}6602	0.9^{5}6759	0.9^{5}6908	0.9^{5}7051	0.9^{5}7187	0.9^{5}7318	0.9^{5}7442	0.9^{5}7561	0.9^{5}7675	0.9^{5}7784
4.6	0.9^{5}7888	0.9^{5}7987	0.9^{5}8081	0.9^{5}8172	0.9^{5}8258	0.9^{5}8340	0.9^{5}8419	0.9^{5}8494	0.9^{5}8566	0.9^{5}8634
4.7	0.9^{5}8699	0.9^{5}8761	0.9^{5}8821	0.9^{5}8877	0.9^{5}8931	0.9^{5}8983	0.9^{6}0320	0.9^{6}0789	0.9^{6}1235	0.9^{6}1661
4.8	0.9^{6}2067	0.9^{6}2453	0.9^{6}2822	0.9^{6}3173	0.9^{6}3508	0.9^{6}3827	0.9^{6}4131	0.9^{6}4420	0.9^{6}4696	0.9^{6}4958
4.9	0.9^{6}5208	0.9^{6}5446	0.9^{6}5673	0.9^{6}5889	0.9^{6}6094	0.9^{6}6289	0.9^{6}6475	0.9^{6}6652	0.9^{6}6821	0.9^{6}6981

Example: $\Phi(3.39) = 0.9996505$, $\Phi(0.98) = 0.8365$.

Appendix D: Values for the Percentage Points $t_{\alpha,\nu}$ of the t-Distribution

Let $X_1, X_2, \ldots, X_n$ be a random sample from a normal distribution with unknown mean μ and also unknown variance σ^2. Then the following random variable

$$T = \frac{\bar{X} - \mu}{\frac{s}{\sqrt{n}}},$$

where

$$s^2 = \frac{\sum_{i=1}^{n}(X_i - \bar{X})^2}{n-1}$$

and

$$\bar{X} = \frac{1}{n}\sum_{i=1}^{n} X_i,$$

has a t-distribution with $n - 1$ degrees of freedom. The t probability density function is given by

$$f(x) = \frac{\Gamma[(k+1)/2]}{\sqrt{\pi k}\,\Gamma(k/2)} \cdot \frac{1}{[(x^2/k)+1]^{(k+1)/2}}, \quad -\infty < x < \infty,$$

where k is the number of degrees of freedom for the underlying random variable, written as T_k, for the t-distribution. The mean and variance of the t-distribution are 0 and $k/(k-2)$ (for $k > 2$), respectively. The t-distribution is symmetrical like the standard normal distribution, is unimodal, and the mode is at 0.

Reliability Engineering, First Edition. Kailash C. Kapur.
© 2014 John Wiley & Sons, Inc. Published 2014 by John Wiley & Sons, Inc.

Appendix D: Values for the Percentage Points $t_{\alpha,\nu}$ of the t-Distribution

The percentage points $t_{\alpha,\nu}$ of the t-distribution are defined as follows:

$$P(T_\nu > t_{\alpha,\nu}) = \alpha.$$

Because the t-distribution is symmetrical, we have $t_{1-\alpha,\nu} = -t_{\alpha,\nu}$.

$t_{\alpha,\nu}$

ν \ α	0.100	0.050	0.025	0.010	0.005	0.001
1	3.078	6.314	12.706	31.821	63.657	318.313
2	1.886	2.920	4.303	6.965	9.925	22.327
3	1.638	2.353	3.182	4.541	5.841	10.215
4	1.533	2.132	2.776	3.747	4.604	7.173
5	1.476	2.015	2.571	3.365	4.032	5.893
6	1.440	1.943	2.447	3.143	3.707	5.208
7	1.415	1.895	2.365	2.998	3.499	4.782
8	1.397	1.860	2.306	2.896	3.355	4.499
9	1.383	1.833	2.262	2.821	3.250	4.296
10	1.372	1.812	2.228	2.764	3.169	4.143
11	1.363	1.796	2.201	2.718	3.106	4.024
12	1.356	1.782	2.179	2.681	3.055	3.929
13	1.350	1.771	2.160	2.650	3.012	3.852
14	1.345	1.761	2.145	2.624	2.977	3.787
15	1.341	1.753	2.131	2.602	2.947	3.733
16	1.337	1.746	2.120	2.583	2.921	3.686
17	1.333	1.740	2.110	2.567	2.898	3.646
18	1.330	1.734	2.101	2.552	2.878	3.610
19	1.328	1.729	2.093	2.539	2.861	3.579
20	1.325	1.725	2.086	2.528	2.845	3.552
21	1.323	1.721	2.080	2.518	2.831	3.527
22	1.321	1.717	2.074	2.508	2.819	3.505
23	1.319	1.714	2.069	2.500	2.807	3.485
24	1.318	1.711	2.064	2.492	2.797	3.467
25	1.316	1.708	2.060	2.485	2.787	3.450
26	1.315	1.706	2.056	2.479	2.779	3.435
27	1.314	1.703	2.052	2.473	2.771	3.421
28	1.313	1.701	2.048	2.467	2.763	3.408
29	1.311	1.699	2.045	2.462	2.756	3.396
30	1.310	1.697	2.042	2.457	2.750	3.385
31	1.309	1.696	2.040	2.453	2.744	3.375
32	1.309	1.694	2.037	2.449	2.738	3.365
33	1.308	1.692	2.035	2.445	2.733	3.356
34	1.307	1.691	2.032	2.441	2.728	3.348
35	1.306	1.690	2.030	2.438	2.724	3.340
36	1.306	1.688	2.028	2.434	2.719	3.333
37	1.305	1.687	2.026	2.431	2.715	3.326
38	1.304	1.686	2.024	2.429	2.712	3.319
39	1.304	1.685	2.023	2.426	2.708	3.313
40	1.303	1.684	2.021	2.423	2.704	3.307
41	1.303	1.683	2.020	2.421	2.701	3.301
42	1.302	1.682	2.018	2.418	2.698	3.296

Appendix D: Values for the Percentage Points $t_{\alpha,\nu}$ of the t-Distribution

ν \ α	0.100	0.050	0.025	0.010	0.005	0.001
43	1.302	1.681	2.017	2.416	2.695	3.291
44	1.301	1.680	2.015	2.414	2.692	3.286
45	1.301	1.679	2.014	2.412	2.690	3.281
46	1.300	1.679	2.013	2.410	2.687	3.277
47	1.300	1.678	2.012	2.408	2.685	3.273
48	1.299	1.677	2.011	2.407	2.682	3.269
49	1.299	1.677	2.010	2.405	2.680	3.265
50	1.299	1.676	2.009	2.403	2.678	3.261
51	1.298	1.675	2.008	2.402	2.676	3.258
52	1.298	1.675	2.007	2.400	2.674	3.255
53	1.298	1.674	2.006	2.399	2.672	3.251
54	1.297	1.674	2.005	2.397	2.670	3.248
55	1.297	1.673	2.004	2.396	2.668	3.245
56	1.297	1.673	2.003	2.395	2.667	3.242
57	1.297	1.672	2.002	2.394	2.665	3.239
58	1.296	1.672	2.002	2.392	2.663	3.237
59	1.296	1.671	2.001	2.391	2.662	3.234
60	1.296	1.671	2.000	2.390	2.660	3.232
61	1.296	1.670	2.000	2.389	2.659	3.229
62	1.295	1.670	1.999	2.388	2.657	3.227
63	1.295	1.669	1.998	2.387	2.656	3.225
64	1.295	1.669	1.998	2.386	2.655	3.223
65	1.295	1.669	1.997	2.385	2.654	3.220
66	1.295	1.668	1.997	2.384	2.652	3.218
67	1.294	1.668	1.996	2.383	2.651	3.216
68	1.294	1.668	1.995	2.382	2.650	3.214
69	1.294	1.667	1.995	2.382	2.649	3.213
70	1.294	1.667	1.994	2.381	2.648	3.211
71	1.294	1.667	1.994	2.380	2.647	3.209
72	1.293	1.666	1.993	2.379	2.646	3.207
73	1.293	1.666	1.993	2.379	2.645	3.206
74	1.293	1.666	1.993	2.378	2.644	3.204
75	1.293	1.665	1.992	2.377	2.643	3.202
76	1.293	1.665	1.992	2.376	2.642	3.201
77	1.293	1.665	1.991	2.376	2.641	3.199
78	1.292	1.665	1.991	2.375	2.640	3.198
79	1.292	1.664	1.990	2.374	2.640	3.197
80	1.292	1.664	1.990	2.374	2.639	3.195
81	1.292	1.664	1.990	2.373	2.638	3.194
82	1.292	1.664	1.989	2.373	2.637	3.193
83	1.292	1.663	1.989	2.372	2.636	3.191
84	1.292	1.663	1.989	2.372	2.636	3.190
85	1.292	1.663	1.988	2.371	2.635	3.189
86	1.291	1.663	1.988	2.370	2.634	3.188
87	1.291	1.663	1.988	2.370	2.634	3.187
88	1.291	1.662	1.987	2.369	2.633	3.185
89	1.291	1.662	1.987	2.369	2.632	3.184
90	1.291	1.662	1.987	2.368	2.632	3.183
∞	1.282	1.645	1.960	2.326	2.576	3.090

Appendix E: Percentage Points $\chi^2_{\alpha,\nu}$ of the Chi-Square Distribution

Let $X_1, \ldots, X_n$ be a random sample from a normal distribution with mean μ and variance σ^2, and let S^2 be the sample variance. Then the random variable

$$\chi^2 = \frac{(n-1)S^2}{\sigma^2}$$

has a chi-square (χ^2) distribution with $n-1$ degrees of freedom.

The probability density function of a χ^2 random variable is

$$f(x) = \frac{1}{2^{k/2}\Gamma(k/2)} x^{(k/2)-1} e^{-x/2}, \quad x > 0,$$

where k is the number of degrees of freedom. The mean and variance of the χ^2 distribution are k and $2k$, respectively. The limiting form of χ^2 distribution as $k \to \infty$ is the normal distribution.

The percentage points of the χ^2 distribution are given in the following table. We define $\chi^2_{\alpha,\nu}$ as the percent point or value of the chi-square random variable with ν degrees of freedom such that the probability that χ^2 exceeds this value is α. We can write it as

$$P(\chi^2 > \chi^2_{\alpha,\nu}) = \alpha.$$

The χ^2 distribution is skewed, and hence we need to find separate value for $\chi^2_{1-\alpha,\nu}$ from the table.

Reliability Engineering, First Edition. Kailash C. Kapur.
© 2014 John Wiley & Sons, Inc. Published 2014 by John Wiley & Sons, Inc.

Appendix E: The Percentage Points $\chi^2_{\alpha,\nu}$ of the Chi-Square Distribution

$$\chi^2_{\alpha,\nu}$$

ν \ α	0.100	0.050	0.025	0.010	0.001
1	2.706	3.841	5.024	6.635	10.828
2	4.605	5.991	7.378	9.210	13.816
3	6.251	7.815	9.348	11.345	16.266
4	7.779	9.488	11.143	13.277	18.467
5	9.236	11.070	12.833	15.086	20.515
6	10.645	12.592	14.449	16.812	22.458
7	12.017	14.067	16.013	18.475	24.322
8	13.362	15.507	17.535	20.090	26.125
9	14.684	16.919	19.023	21.666	27.877
10	15.987	18.307	20.483	23.209	29.588
11	17.275	19.675	21.920	24.725	31.264
12	18.549	21.026	23.337	26.217	32.910
13	19.812	22.362	24.736	27.688	34.528
14	21.064	23.685	26.119	29.141	36.123
15	22.307	24.996	27.488	30.578	37.697
16	23.542	26.296	28.845	32.000	39.252
17	24.769	27.587	30.191	33.409	40.790
18	25.989	28.869	31.526	34.805	42.312
19	27.204	30.144	32.852	36.191	43.820
20	28.412	31.410	34.170	37.566	45.315
21	29.615	32.671	35.479	38.932	46.797
22	30.813	33.924	36.781	40.289	48.268
23	32.007	35.172	38.076	41.638	49.728
24	33.196	36.415	39.364	42.980	51.179
25	34.382	37.652	40.646	44.314	52.620
26	35.563	38.885	41.923	45.642	54.052
27	36.741	40.113	43.195	46.963	55.476
28	37.916	41.337	44.461	48.278	56.892
29	39.087	42.557	45.722	49.588	58.301
30	40.256	43.773	46.979	50.892	59.703
31	41.422	44.985	48.232	52.191	61.098
32	42.585	46.194	49.480	53.486	62.487
33	43.745	47.400	50.725	54.776	63.870
34	44.903	48.602	51.966	56.061	65.247
35	46.059	49.802	53.203	57.342	66.619
36	47.212	50.998	54.437	58.619	67.985
37	48.363	52.192	55.668	59.893	69.347
38	49.513	53.384	56.896	61.162	70.703
39	50.660	54.572	58.120	62.428	72.055
40	51.805	55.758	59.342	63.691	73.402
41	52.949	56.942	60.561	64.950	74.745
42	54.090	58.124	61.777	66.206	76.084
43	55.230	59.304	62.990	67.459	77.419
44	56.369	60.481	64.201	68.710	78.750
45	57.505	61.656	65.410	69.957	80.077
46	58.641	62.830	66.617	71.201	81.400

Appendix E: The Percentage Points $\chi^2_{\alpha,\nu}$ of the Chi-Square Distribution

$\chi^2_{\alpha,\nu}$

ν \ α	0.100	0.050	0.025	0.010	0.001
47	59.774	64.001	67.821	72.443	82.720
48	60.907	65.171	69.023	73.683	84.037
49	62.038	66.339	70.222	74.919	85.351
50	63.167	67.505	71.420	76.154	86.661
51	64.295	68.669	72.616	77.386	87.968
52	65.422	69.832	73.810	78.616	89.272
53	66.548	70.993	75.002	79.843	90.573
54	67.673	72.153	76.192	81.069	91.872
55	68.796	73.311	77.380	82.292	93.168
56	69.919	74.468	78.567	83.513	94.461
57	71.040	75.624	79.752	84.733	95.751
58	72.160	76.778	80.936	85.950	97.039
59	73.279	77.931	82.117	87.166	98.324
60	74.397	79.082	83.298	88.379	99.607
61	75.514	80.232	84.476	89.591	100.888
62	76.630	81.381	85.654	90.802	102.166
63	77.745	82.529	86.830	92.010	103.442
64	78.860	83.675	88.004	93.217	104.716
65	79.973	84.821	89.177	94.422	105.988
66	81.085	85.965	90.349	95.626	107.258
67	82.197	87.108	91.519	96.828	108.526
68	83.308	88.250	92.689	98.028	109.791
69	84.418	89.391	93.856	99.228	111.055
70	85.527	90.531	95.023	100.425	112.317
71	86.635	91.670	96.189	101.621	113.577
72	87.743	92.808	97.353	102.816	114.835
73	88.850	93.945	98.516	104.010	116.092
74	89.956	95.081	99.678	105.202	117.346
75	91.061	96.217	100.839	106.393	118.599
76	92.166	97.351	101.999	107.583	119.850
77	93.270	98.484	103.158	108.771	121.100
78	94.374	99.617	104.316	109.958	122.348
79	95.476	100.749	105.473	111.144	123.594
80	96.578	101.879	106.629	112.329	124.839
81	97.680	103.010	107.783	113.512	126.083
82	98.780	104.139	108.937	114.695	127.324
83	99.880	105.267	110.090	115.876	128.565
84	100.980	106.395	111.242	117.057	129.804
85	102.079	107.522	112.393	118.236	131.041
86	103.177	108.648	113.544	119.414	132.277
87	104.275	109.773	114.693	120.591	133.512
88	105.372	110.898	115.841	121.767	134.746
89	106.469	112.022	116.989	122.942	135.978
90	107.565	113.145	118.136	124.116	137.208
100	118.498	124.342	129.561	135.807	149.449

Appendix E: The Percentage Points $\chi^2_{\alpha,\nu}$ of the Chi-Square Distribution

$$\chi^2_{\alpha,\nu}$$

ν \ α	0.900	0.950	0.975	0.990	0.999
1	0.016	0.004	0.001	0.000	0.000
2	0.211	0.103	0.051	0.020	0.002
3	0.584	0.352	0.216	0.115	0.024
4	1.064	0.711	0.484	0.297	0.091
5	1.610	1.145	0.831	0.554	0.210
6	2.204	1.635	1.237	0.872	0.381
7	2.833	2.167	1.690	1.239	0.598
8	3.490	2.733	2.180	1.646	0.857
9	4.168	3.325	2.700	2.088	1.152
10	4.865	3.940	3.247	2.558	1.479
11	5.578	4.575	3.816	3.053	1.834
12	6.304	5.226	4.404	3.571	2.214
13	7.042	5.892	5.009	4.107	2.617
14	7.790	6.571	5.629	4.660	3.041
15	8.547	7.261	6.262	5.229	3.483
16	9.312	7.962	6.908	5.812	3.942
17	10.085	8.672	7.564	6.408	4.416
18	10.865	9.390	8.231	7.015	4.905
19	11.651	10.117	8.907	7.633	5.407
20	12.443	10.851	9.591	8.260	5.921
21	13.240	11.591	10.283	8.897	6.447
22	14.041	12.338	10.982	9.542	6.983
23	14.848	13.091	11.689	10.196	7.529
24	15.659	13.848	12.401	10.856	8.085
25	16.473	14.611	13.120	11.524	8.649
26	17.292	15.379	13.844	12.198	9.222
27	18.114	16.151	14.573	12.879	9.803
28	18.939	16.928	15.308	13.565	10.391
29	19.768	17.708	16.047	14.256	10.986
30	20.599	18.493	16.791	14.953	11.588
31	21.434	19.281	17.539	15.655	12.196
32	22.271	20.072	18.291	16.362	12.811
33	23.110	20.867	19.047	17.074	13.431
34	23.952	21.664	19.806	17.789	14.057
35	24.797	22.465	20.569	18.509	14.688
36	25.643	23.269	21.336	19.233	15.324
37	26.492	24.075	22.106	19.960	15.965
38	27.343	24.884	22.878	20.691	16.611
39	28.196	25.695	23.654	21.426	17.262
40	29.051	26.509	24.433	22.164	17.916
41	29.907	27.326	25.215	22.906	18.575
42	30.765	28.144	25.999	23.650	19.239
43	31.625	28.965	26.785	24.398	19.906
44	32.487	29.787	27.575	25.148	20.576
45	33.350	30.612	28.366	25.901	21.251
46	34.215	31.439	29.160	26.657	21.929

Appendix E: The Percentage Points $\chi^2_{\alpha,\nu}$ of the Chi-Square Distribution

$$\chi^2_{\alpha,\nu}$$

ν \ α	0.900	0.950	0.975	0.990	0.999
47	35.081	32.268	29.956	27.416	22.610
48	35.949	33.098	30.755	28.177	23.295
49	36.818	33.930	31.555	28.941	23.983
50	37.689	34.764	32.357	29.707	24.674
51	38.560	35.600	33.162	30.475	25.368
52	39.433	36.437	33.968	31.246	26.065
53	40.308	37.276	34.776	32.018	26.765
54	41.183	38.116	35.586	32.793	27.468
55	42.060	38.958	36.398	33.570	28.173
56	42.937	39.801	37.212	34.350	28.881
57	43.816	40.646	38.027	35.131	29.592
58	44.696	41.492	38.844	35.913	30.305
59	45.577	42.339	39.662	36.698	31.020
60	46.459	43.188	40.482	37.485	31.738
61	47.342	44.038	41.303	38.273	32.459
62	48.226	44.889	42.126	39.063	33.181
63	49.111	45.741	42.950	39.855	33.906
64	49.996	46.595	43.776	40.649	34.633
65	50.883	47.450	44.603	41.444	35.362
66	51.770	48.305	45.431	42.240	36.093
67	52.659	49.162	46.261	43.038	36.826
68	53.548	50.020	47.092	43.838	37.561
69	54.438	50.879	47.924	44.639	38.298
70	55.329	51.739	48.758	45.442	39.036
71	56.221	52.600	49.592	46.246	39.777
72	57.113	53.462	50.428	47.051	40.519
73	58.006	54.325	51.265	47.858	41.264
74	58.900	55.189	52.103	48.666	42.010
75	59.795	56.054	52.942	49.475	42.757
76	60.690	56.920	53.782	50.286	43.507
77	61.586	57.786	54.623	51.097	44.258
78	62.483	58.654	55.466	51.910	45.010
79	63.380	59.522	56.309	52.725	45.764
80	64.278	60.391	57.153	53.540	46.520
81	65.176	61.261	57.998	54.357	47.277
82	66.076	62.132	58.845	55.174	48.036
83	66.976	63.004	59.692	55.993	48.796
84	67.876	63.876	60.540	56.813	49.557
85	68.777	64.749	61.389	57.634	50.320
86	69.679	65.623	62.239	58.456	51.085
87	70.581	66.498	63.089	59.279	51.850
88	71.484	67.373	36.941	60.103	52.617
89	72.387	68.249	64.793	60.928	53.386
90	73.291	69.126	65.647	61.754	54.155
100	82.358	77.929	74.222	70.065	61.918

Appendix F: Percentage Points for the F-Distribution

Let W and Y be independent chi-square random variables with u and v degrees of freedom, respectively. Then the ratio

$$F = \frac{W/u}{Y/v},$$

has the probability density function

$$f(x) = \frac{\Gamma\left(\frac{u+v}{2}\right)\left(\frac{u}{v}\right)^{u/2} x^{(u/2)-1}}{\Gamma\left(\frac{u}{2}\right)\Gamma\left(\frac{v}{2}\right)\left[\left(\frac{u}{v}\right)x+1\right]^{(u+v)/2}}, \quad 0 < x < \infty.$$

This is the probability density for the F distribution with u degrees of freedom for the numerator and v degrees of freedom for the denominator and is denoted by $F_{u,v}$.

The percentage points of the F distribution are given in the following table. If f_{α,ν_1,ν_2} be the percentage point of the F distribution with the numerator degrees of freedom of ν_1 and the denominator degrees of freedom of ν_2, then

$$P(F_{\nu_1,\nu_2} > f_{\alpha,\nu_1,\nu_2}) = \alpha.$$

The lower tail percentage points $f_{1-\alpha, u, v}$ can be found using the following equation

$$f_{1-\alpha,\nu_1,\nu_2} = \frac{1}{f_{\alpha,\nu_2,\nu_1}}.$$

Reliability Engineering, First Edition. Kailash C. Kapur.
© 2014 John Wiley & Sons, Inc. Published 2014 by John Wiley & Sons, Inc.

$\alpha = 0.25$ degrees of freedom for the numerator (ν_1)

ν_2 \ ν_1	1	2	3	4	5	6	7	8	9	10	12	15	20	24	30	40	60	120	∞
1	5.83	7.50	8.20	8.58	8.82	8.98	9.10	9.19	9.26	9.32	9.41	9.49	9.58	9.63	9.67	9.71	9.76	9.80	9.85
2	2.57	3.00	3.15	3.23	3.28	3.31	3.34	3.35	3.37	3.38	3.39	3.41	3.43	3.44	3.45	3.45	3.46	3.47	3.48
3	2.02	2.28	2.36	2.39	2.41	2.42	2.43	2.44	2.44	2.44	2.45	2.46	2.46	2.47	2.47	2.47	2.47	2.47	2.47
4	1.81	2.00	2.05	2.06	2.07	2.08	2.08	2.08	2.08	2.08	2.08	2.08	2.08	2.08	2.08	2.08	2.08	2.08	2.08
5	1.69	1.85	1.88	1.89	1.89	1.89	1.89	1.89	1.89	1.89	1.89	1.89	1.88	1.88	1.88	1.88	1.87	1.87	1.87
6	1.62	1.76	1.78	1.79	1.79	1.78	1.78	1.78	1.77	1.77	1.77	1.76	1.76	1.75	1.75	1.75	1.74	1.74	1.74
7	1.57	1.70	1.72	1.72	1.71	1.71	1.70	1.70	1.70	1.69	1.68	1.68	1.67	1.67	1.66	1.66	1.65	1.65	1.65
8	1.54	1.66	1.67	1.66	1.66	1.65	1.64	1.64	1.63	1.63	1.62	1.62	1.61	1.60	1.60	1.59	1.59	1.58	1.58
9	1.51	1.62	1.63	1.63	1.62	1.61	1.60	1.60	1.59	1.59	1.58	1.57	1.56	1.56	1.55	1.54	1.54	1.53	1.53
10	1.49	1.60	1.60	1.59	1.59	1.58	1.57	1.56	1.56	1.55	1.54	1.53	1.52	1.52	1.51	1.51	1.50	1.49	1.48
11	1.47	1.58	1.58	1.57	1.56	1.55	1.54	1.53	1.53	1.52	1.51	1.50	1.49	1.49	1.48	1.47	1.47	1.46	1.45
12	1.46	1.56	1.56	1.55	1.54	1.53	1.52	1.51	1.51	1.50	1.49	1.48	1.47	1.46	1.45	1.45	1.44	1.43	1.42
13	1.45	1.55	1.55	1.53	1.52	1.51	1.50	1.49	1.49	1.48	1.47	1.46	1.45	1.44	1.43	1.42	1.42	1.41	1.40
14	1.44	1.53	1.53	1.52	1.51	1.50	1.49	1.48	1.47	1.46	1.45	1.44	1.43	1.42	1.41	1.41	1.40	1.39	1.38
15	1.43	1.52	1.52	1.51	1.49	1.48	1.47	1.46	1.46	1.45	1.44	1.43	1.41	1.41	1.40	1.39	1.38	1.37	1.36
16	1.42	1.51	1.51	1.50	1.48	1.47	1.46	1.45	1.44	1.44	1.43	1.41	1.40	1.39	1.38	1.37	1.36	1.35	1.34
17	1.42	1.51	1.50	1.49	1.47	1.46	1.45	1.44	1.43	1.43	1.41	1.40	1.39	1.38	1.37	1.36	1.35	1.34	1.33
18	1.41	1.50	1.49	1.48	1.46	1.45	1.44	1.43	1.42	1.42	1.40	1.39	1.38	1.37	1.36	1.35	1.34	1.33	1.32
19	1.41	1.49	1.49	1.47	1.46	1.44	1.43	1.42	1.41	1.41	1.40	1.38	1.37	1.36	1.35	1.34	1.33	1.32	1.30
20	1.40	1.49	1.48	1.47	1.45	1.44	1.43	1.42	1.41	1.40	1.39	1.37	1.36	1.35	1.34	1.33	1.32	1.31	1.29
21	1.40	1.48	1.48	1.46	1.44	1.43	1.42	1.41	1.40	1.39	1.38	1.37	1.35	1.34	1.33	1.32	1.31	1.30	1.28
22	1.40	1.48	1.47	1.45	1.44	1.42	1.41	1.40	1.39	1.39	1.37	1.36	1.34	1.33	1.32	1.31	1.30	1.29	1.28
23	1.39	1.47	1.47	1.45	1.43	1.42	1.41	1.40	1.39	1.38	1.37	1.35	1.34	1.33	1.32	1.31	1.30	1.28	1.27
24	1.39	1.47	1.46	1.44	1.43	1.41	1.40	1.39	1.38	1.38	1.36	1.35	1.33	1.32	1.31	1.30	1.29	1.28	1.26
25	1.39	1.47	1.46	1.44	1.42	1.41	1.40	1.39	1.38	1.37	1.36	1.34	1.33	1.32	1.31	1.29	1.28	1.27	1.25
26	1.38	1.46	1.45	1.44	1.42	1.41	1.39	1.38	1.37	1.37	1.35	1.34	1.32	1.31	1.30	1.29	1.28	1.26	1.25
27	1.38	1.46	1.45	1.43	1.42	1.40	1.39	1.38	1.37	1.36	1.35	1.33	1.32	1.31	1.30	1.28	1.27	1.26	1.24
28	1.38	1.46	1.45	1.43	1.41	1.40	1.39	1.38	1.37	1.36	1.34	1.33	1.31	1.30	1.29	1.28	1.27	1.25	1.24
29	1.38	1.45	1.45	1.43	1.41	1.40	1.38	1.37	1.36	1.35	1.34	1.32	1.31	1.30	1.29	1.27	1.26	1.25	1.23
30	1.38	1.45	1.44	1.42	1.41	1.39	1.38	1.37	1.36	1.35	1.34	1.32	1.30	1.29	1.28	1.27	1.26	1.24	1.23
40	1.36	1.44	1.42	1.40	1.39	1.37	1.36	1.35	1.34	1.33	1.31	1.30	1.28	1.26	1.25	1.24	1.22	1.21	1.19
60	1.35	1.42	1.41	1.38	1.37	1.35	1.33	1.32	1.31	1.30	1.29	1.27	1.25	1.24	1.22	1.21	1.19	1.17	1.15
120	1.34	1.40	1.39	1.37	1.35	1.33	1.31	1.30	1.29	1.28	1.26	1.24	1.22	1.21	1.19	1.18	1.16	1.13	1.10
∞	1.32	1.39	1.37	1.35	1.33	1.31	1.29	1.28	1.27	1.25	1.24	1.22	1.19	1.18	1.16	1.14	1.12	1.08	1.00

$\alpha = 0.10$ degrees of freedom for the numerator (ν_1)

ν_2 \ ν_1	1	2	3	4	5	6	7	8	9	10	12	15	20	24	30	40	60	120	∞
1	39.86	49.50	53.59	55.83	57.24	58.20	58.91	59.44	59.86	60.19	60.71	61.22	61.74	62.00	62.26	62.53	62.79	63.06	63.33
2	8.53	9.00	9.16	9.24	9.29	9.33	9.35	9.37	9.38	9.39	9.41	9.42	9.44	9.45	9.46	9.47	9.47	9.48	9.49
3	5.54	5.46	5.39	5.34	5.31	5.28	5.27	5.25	5.24	5.23	5.22	5.20	5.18	5.18	5.17	5.16	5.15	5.14	5.13
4	4.54	4.32	4.19	4.11	4.05	4.01	3.98	3.95	3.94	3.92	3.90	3.87	3.84	3.83	3.82	3.80	3.79	3.78	3.76
5	4.06	3.78	3.62	3.52	3.45	3.40	3.37	3.34	3.32	3.30	3.27	3.24	3.21	3.19	3.17	3.16	3.14	3.12	3.10
6	3.78	3.46	3.29	3.18	3.11	3.05	3.01	2.98	2.96	2.94	2.90	2.87	2.84	2.82	2.80	2.78	2.76	2.74	2.72
7	3.59	3.26	3.07	2.96	2.88	2.83	2.78	2.75	2.72	2.70	2.67	2.63	2.59	2.58	2.56	2.54	2.51	2.49	2.47
8	3.46	3.11	2.92	2.81	2.73	2.67	2.62	2.59	2.56	2.54	2.50	2.46	2.42	2.40	2.38	2.36	2.34	2.32	2.29
9	3.36	3.01	2.81	2.69	2.61	2.55	2.51	2.47	2.44	2.42	2.38	2.34	2.30	2.28	2.25	2.23	2.21	2.18	2.16
10	3.29	2.92	2.73	2.61	2.52	2.46	2.41	2.38	2.35	2.32	2.28	2.24	2.20	2.18	2.16	2.13	2.11	2.08	2.06
11	3.23	2.86	2.66	2.54	2.45	2.39	2.34	2.30	2.27	2.25	2.21	2.17	2.12	2.10	2.08	2.05	2.03	2.00	1.97
12	3.18	2.81	2.61	2.48	2.39	2.33	2.28	2.24	2.21	2.19	2.15	2.10	2.06	2.04	2.01	1.99	1.96	1.93	1.90
13	3.14	2.76	2.56	2.43	2.35	2.28	2.23	2.20	2.16	2.14	2.10	2.05	2.01	1.98	1.96	1.93	1.90	1.88	1.85
14	3.10	2.73	2.52	2.39	2.31	2.24	2.19	2.15	2.12	2.10	2.05	2.01	1.96	1.94	1.91	1.89	1.86	1.83	1.80
15	3.07	2.70	2.49	2.36	2.27	2.21	2.16	2.12	2.09	2.06	2.02	1.97	1.92	1.90	1.87	1.85	1.82	1.79	1.76
16	3.05	2.67	2.46	2.33	2.24	2.18	2.13	2.09	2.06	2.03	1.99	1.94	1.89	1.87	1.84	1.81	1.78	1.75	1.72
17	3.03	2.64	2.44	2.31	2.22	2.15	2.10	2.06	2.03	2.00	1.96	1.91	1.86	1.84	1.81	1.78	1.75	1.72	1.69
18	3.01	2.62	2.42	2.29	2.20	2.13	2.08	2.04	2.00	1.98	1.93	1.89	1.84	1.81	1.78	1.75	1.72	1.69	1.66
19	2.99	2.61	2.40	2.27	2.18	2.11	2.06	2.02	1.98	1.96	1.91	1.86	1.81	1.79	1.76	1.73	1.70	1.67	1.63
20	2.97	2.59	2.38	2.25	2.16	2.09	2.04	2.00	1.96	1.94	1.89	1.84	1.79	1.77	1.74	1.71	1.68	1.64	1.61
21	2.96	2.57	2.36	2.23	2.14	2.08	2.02	1.98	1.95	1.92	1.87	1.83	1.78	1.75	1.72	1.69	1.66	1.62	1.59
22	2.95	2.56	2.35	2.22	2.13	2.06	2.01	1.97	1.93	1.90	1.86	1.81	1.76	1.73	1.70	1.67	1.64	1.60	1.57
23	2.94	2.55	2.34	2.21	2.11	2.05	1.99	1.95	1.92	1.89	1.84	1.80	1.74	1.72	1.69	1.66	1.62	1.59	1.55
24	2.93	2.54	2.33	2.19	2.10	2.04	1.98	1.94	1.91	1.88	1.83	1.78	1.73	1.70	1.67	1.64	1.61	1.57	1.53
25	2.92	2.53	2.32	2.18	2.09	2.02	1.97	1.93	1.89	1.87	1.82	1.77	1.72	1.69	1.66	1.63	1.59	1.56	1.52
26	2.91	2.52	2.31	2.17	2.08	2.01	1.96	1.92	1.88	1.86	1.81	1.76	1.71	1.68	1.65	1.61	1.58	1.54	1.50
27	2.90	2.51	2.30	2.17	2.07	2.00	1.95	1.91	1.87	1.85	1.80	1.75	1.70	1.67	1.64	1.60	1.57	1.53	1.49
28	2.89	2.50	2.29	2.16	2.06	2.00	1.94	1.90	1.87	1.84	1.79	1.74	1.69	1.66	1.63	1.59	1.56	1.52	1.48
29	2.89	2.50	2.28	2.15	2.06	1.99	1.93	1.89	1.86	1.83	1.78	1.73	1.68	1.65	1.62	1.58	1.55	1.51	1.47
30	2.88	2.49	2.28	2.14	2.03	1.98	1.93	1.88	1.85	1.82	1.77	1.72	1.67	1.64	1.61	1.57	1.54	1.50	1.46
40	2.84	2.44	2.23	2.09	2.00	1.93	1.87	1.83	1.79	1.76	1.71	1.66	1.61	1.57	1.54	1.51	1.47	1.42	1.38
60	2.79	2.39	2.18	2.04	1.95	1.87	1.82	1.77	1.74	1.71	1.66	1.60	1.54	1.51	1.48	1.44	1.40	1.35	1.29
120	2.75	2.35	2.13	1.99	1.90	1.82	1.77	1.72	1.68	1.65	1.60	1.55	1.48	1.45	1.41	1.37	1.32	1.26	1.19
∞	2.71	2.30	2.08	1.94	1.85	1.77	1.72	1.67	1.63	1.60	1.55	1.49	1.42	1.38	1.34	1.30	1.24	1.17	1.00

$\alpha = 0.05$ degrees of freedom for the numerator (v_1)

v_2 \ v_1	1	2	3	4	5	6	7	8	9	10	12	15	20	24	30	40	60	120	∞
1	161.4	199.5	215.7	224.6	230.2	234.0	236.8	238.9	240.5	241.9	243.9	245.9	248.0	249.1	250.1	251.1	252.2	253.3	254.3
2	18.51	19.00	19.16	19.25	19.30	19.33	19.35	19.37	19.38	19.40	19.41	19.43	19.45	19.45	19.46	19.47	19.48	19.49	19.50
3	10.13	9.55	9.28	9.12	9.01	8.94	8.89	8.85	8.81	8.79	8.74	8.70	8.66	8.64	8.62	8.59	8.57	8.55	8.53
4	7.71	6.94	6.59	6.39	6.26	6.16	6.09	6.04	6.00	5.96	5.91	5.86	5.80	5.77	5.75	5.72	5.69	5.66	5.63
5	6.61	5.79	5.41	5.19	5.05	4.95	4.88	4.82	4.77	4.74	4.68	4.62	4.56	4.53	4.50	4.46	4.43	4.40	4.36
6	5.99	5.14	4.76	4.53	4.39	4.28	4.21	4.15	4.10	4.06	4.00	3.94	3.87	3.84	3.81	3.77	3.74	3.70	3.67
7	5.59	4.74	4.35	4.12	3.97	3.87	3.79	3.73	3.68	3.64	3.57	3.51	3.44	3.41	3.38	3.34	3.30	3.27	3.23
8	5.32	4.46	4.07	3.84	3.69	3.58	3.50	3.44	3.39	3.35	3.28	3.22	3.15	3.12	3.08	3.04	3.01	2.97	2.93
9	5.12	4.26	3.86	3.63	3.48	3.37	3.29	3.23	3.18	3.14	3.07	3.01	2.94	2.90	2.86	2.83	2.79	2.75	2.71
10	4.96	4.10	3.71	3.48	3.33	3.22	3.14	3.07	3.02	2.98	2.91	2.85	2.77	2.74	2.70	2.66	2.62	2.58	2.54
11	4.84	3.98	3.59	3.36	3.20	3.09	3.01	2.95	2.90	2.85	2.79	2.72	2.65	2.61	2.57	2.53	2.49	2.45	2.40
12	4.75	3.89	3.49	3.26	3.11	3.00	2.91	2.85	2.80	2.75	2.69	2.62	2.54	2.51	2.47	2.43	2.38	2.34	2.30
13	4.67	3.81	3.41	3.18	3.03	2.92	2.83	2.77	2.71	2.67	2.60	2.53	2.46	2.42	2.38	2.34	2.30	2.25	2.21
14	4.60	3.74	3.34	3.11	2.96	2.85	2.76	2.70	2.65	2.60	2.53	2.46	2.39	2.35	2.31	2.27	2.22	2.18	2.13
15	4.54	3.68	3.29	3.06	2.90	2.79	2.71	2.64	2.59	2.54	2.48	2.40	2.33	2.29	2.25	2.20	2.16	2.11	2.07
16	4.49	3.63	3.24	3.01	2.85	2.74	2.66	2.59	2.54	2.49	2.42	2.35	2.28	2.24	2.19	2.15	2.11	2.06	2.01
17	4.45	3.59	3.20	2.96	2.81	2.70	2.61	2.55	2.49	2.45	2.38	2.31	2.23	2.19	2.15	2.10	2.06	2.01	1.96
18	4.41	3.55	3.16	2.93	2.77	2.66	2.58	2.51	2.46	2.41	2.34	2.27	2.19	2.15	2.11	2.06	2.02	1.97	1.92
19	4.38	3.52	3.13	2.90	2.74	2.63	2.54	2.48	2.42	2.38	2.31	2.23	2.16	2.11	2.07	2.03	1.98	1.93	1.88
20	4.35	3.49	3.10	2.87	2.71	2.60	2.51	2.45	2.39	2.35	2.28	2.20	2.12	2.08	2.04	1.99	1.95	1.90	1.84
21	4.32	3.47	3.07	2.84	2.68	2.57	2.49	2.42	2.37	2.32	2.25	2.18	2.10	2.05	2.01	1.96	1.92	1.87	1.81
22	4.30	3.44	3.05	2.82	2.66	2.55	2.46	2.40	2.34	2.30	2.23	2.15	2.07	2.03	1.98	1.94	1.89	1.84	1.78
23	4.28	3.42	3.03	2.80	2.64	2.53	2.44	2.37	2.32	2.27	2.20	2.13	2.05	2.01	1.96	1.91	1.86	1.81	1.76
24	4.26	3.40	3.01	2.78	2.62	2.51	2.42	2.36	2.30	2.25	2.18	2.11	2.03	1.98	1.94	1.89	1.84	1.79	1.73
25	4.24	3.39	2.99	2.76	2.60	2.49	2.40	2.34	2.28	2.24	2.16	2.09	2.01	1.96	1.92	1.87	1.82	1.77	1.71
26	4.23	3.37	2.98	2.74	2.59	2.47	2.39	2.32	2.27	2.22	2.15	2.07	1.99	1.95	1.90	1.85	1.80	1.75	1.69
27	4.21	3.35	2.96	2.73	2.57	2.46	2.37	2.31	2.25	2.20	2.13	2.06	1.97	1.93	1.88	1.84	1.79	1.73	1.67
28	4.20	3.34	2.95	2.71	2.56	2.45	2.36	2.29	2.24	2.19	2.12	2.04	1.96	1.91	1.87	1.82	1.77	1.71	1.65
29	4.18	3.33	2.93	2.70	2.55	2.43	2.35	2.28	2.22	2.18	2.10	2.03	1.94	1.90	1.85	1.81	1.75	1.70	1.64
30	4.17	3.32	2.92	2.69	2.53	2.42	2.33	2.27	2.21	2.16	2.09	2.01	1.93	1.89	1.84	1.79	1.74	1.68	1.62
40	4.08	3.23	2.84	2.61	2.45	2.34	2.25	2.18	2.12	2.08	2.00	1.92	1.84	1.79	1.74	1.69	1.64	1.58	1.51
60	4.00	3.15	2.76	2.53	2.37	2.25	2.17	2.10	2.04	1.99	1.92	1.84	1.75	1.70	1.65	1.59	1.53	1.47	1.39
120	3.92	3.07	2.68	2.45	2.29	2.17	2.09	2.02	1.96	1.91	1.83	1.75	1.66	1.61	1.55	1.55	1.43	1.35	1.25
∞	3.84	3.00	2.60	2.37	2.21	2.10	2.01	1.94	1.88	1.83	1.75	1.67	1.57	1.52	1.46	1.39	1.32	1.22	1.00

$\alpha = 0.025$ degrees of freedom for the numerator (ν_1)

ν_2 \ ν_1	1	2	3	4	5	6	7	8	9	10	12	15	20	24	30	40	60	120	∞
1	647.8	799.5	864.2	899.6	921.8	937.1	948.2	956.7	963.3	968.6	976.7	984.9	993.1	997.2	1001	1006	1010	1014	1018
2	38.51	39.00	39.17	39.25	39.30	39.33	39.36	39.37	39.39	39.40	39.41	39.43	39.45	39.46	39.46	39.47	39.48	39.49	39.50
3	17.44	16.04	15.44	15.10	14.88	14.73	14.62	14.54	14.47	14.42	14.34	14.25	14.17	14.12	14.08	14.04	13.99	13.95	13.90
4	12.22	10.63	9.98	9.60	9.36	9.20	9.07	8.98	8.90	8.84	8.75	8.66	8.56	8.51	8.46	8.41	8.36	8.31	8.26
5	10.01	8.43	7.76	7.39	7.15	6.98	6.85	6.76	6.68	6.62	6.52	6.43	6.33	6.28	6.23	6.18	6.12	6.07	6.02
6	8.81	7.26	6.60	6.23	5.99	5.82	5.70	5.60	5.52	5.46	5.37	5.27	5.17	5.12	5.07	5.01	4.96	4.90	4.85
7	8.07	6.54	5.89	5.52	5.29	5.12	4.99	4.90	4.82	4.76	4.67	4.57	4.47	4.42	4.36	4.31	4.25	4.20	4.14
8	7.57	6.06	5.42	5.05	4.82	4.65	4.53	4.43	4.36	4.30	4.20	4.10	4.00	3.95	3.89	3.84	3.78	3.73	3.67
9	7.21	5.71	5.08	4.72	4.48	4.32	4.20	4.10	4.03	3.96	3.87	3.77	3.67	3.61	3.56	3.51	3.45	3.39	3.33
10	6.94	5.46	4.83	4.47	4.24	4.07	3.95	3.85	3.78	3.72	3.62	3.52	3.42	3.37	3.31	3.26	3.20	3.14	3.08
11	6.72	5.26	4.63	4.28	4.04	3.88	3.76	3.66	3.59	3.53	3.43	3.33	3.23	3.17	3.12	3.06	3.00	2.94	2.88
12	6.55	5.10	4.47	4.12	3.89	3.73	3.61	3.51	3.44	3.37	3.28	3.18	3.07	3.02	2.96	2.91	2.85	2.79	2.72
13	6.41	4.97	4.35	4.00	3.77	3.60	3.48	3.39	3.31	3.25	3.15	3.05	2.95	2.89	2.84	2.78	2.72	2.66	2.60
14	6.30	4.86	4.24	3.89	3.66	3.50	3.38	3.29	3.21	3.15	3.05	2.95	2.84	2.79	2.73	2.67	2.61	2.55	2.49
15	6.20	4.77	4.15	3.80	3.58	3.41	3.29	3.20	3.12	3.06	2.96	2.86	2.76	2.70	2.64	2.59	2.52	2.46	2.40
16	6.12	4.69	4.08	3.73	3.50	3.34	3.22	3.12	3.05	2.99	2.89	2.79	2.68	2.63	2.57	2.51	2.45	2.38	2.32
17	6.04	4.62	4.01	3.66	3.44	3.28	3.16	3.06	2.98	2.92	2.82	2.72	2.62	2.56	2.50	2.44	2.38	2.32	2.25
18	5.98	4.56	3.95	3.61	3.38	3.22	3.10	3.01	2.93	2.87	2.77	2.67	2.56	2.50	2.44	2.38	2.32	2.26	2.19
19	5.92	4.51	3.90	3.56	3.33	3.17	3.05	2.96	2.88	2.82	2.72	2.62	2.51	2.45	2.39	2.33	2.27	2.20	2.13
20	5.87	4.46	3.86	3.51	3.29	3.13	3.01	2.91	2.84	2.77	2.68	2.57	2.46	2.41	2.35	2.29	2.22	2.16	2.09
21	5.83	4.42	3.82	3.48	3.25	3.09	2.97	2.87	2.80	2.73	2.64	2.53	2.42	2.37	2.31	2.25	2.18	2.11	2.04
22	5.79	4.38	3.78	3.44	3.22	3.05	2.93	2.84	2.76	2.70	2.60	2.50	2.39	2.33	2.27	2.21	2.14	2.08	2.00
23	5.75	4.35	3.75	3.41	3.18	3.02	2.90	2.81	2.73	2.67	2.57	2.47	2.36	2.30	2.24	2.18	2.11	2.04	1.97
24	5.72	4.32	3.72	3.38	3.15	2.99	2.87	2.78	2.70	2.64	2.54	2.44	2.33	2.27	2.21	2.15	2.08	2.01	1.94
25	5.69	4.29	3.69	3.35	3.13	2.97	2.85	2.75	2.68	2.61	2.51	2.41	2.30	2.24	2.18	2.12	2.05	1.98	1.91
26	5.66	4.27	3.67	3.33	3.10	2.94	2.82	2.73	2.65	2.59	2.49	2.39	2.28	2.22	2.16	2.09	2.03	1.95	1.88
27	5.63	4.24	3.65	3.31	3.08	2.92	2.80	2.71	2.63	2.57	2.47	2.36	2.25	2.19	2.13	2.07	2.00	1.93	1.85
28	5.61	4.22	3.63	3.29	3.06	2.90	2.78	2.69	2.61	2.55	2.45	2.34	2.23	2.17	2.11	2.05	1.98	1.91	1.83
29	5.59	4.20	3.61	3.27	3.04	2.88	2.76	2.67	2.59	2.53	2.43	2.32	2.21	2.15	2.09	2.03	1.96	1.89	1.81
30	5.57	4.18	3.59	3.25	3.03	2.87	2.75	2.65	2.57	2.51	2.41	2.31	2.20	2.14	2.07	2.01	1.94	1.87	1.79
40	5.42	4.05	3.46	3.13	2.90	2.74	2.62	2.53	2.45	2.39	2.29	2.18	2.07	2.01	1.94	1.88	1.80	1.72	1.64
60	5.29	3.93	3.34	3.01	2.79	2.63	2.51	2.41	2.33	2.27	2.17	2.06	1.94	1.88	1.82	1.74	1.67	1.58	1.48
120	5.15	3.80	3.23	2.89	2.67	2.52	2.39	2.30	2.22	2.16	2.05	1.94	1.82	1.76	1.69	1.61	1.53	1.43	1.31
∞	5.02	3.69	3.12	2.79	2.57	2.41	2.29	2.19	2.11	2.05	1.94	1.83	1.71	1.64	1.57	1.48	1.39	1.27	1.00

$\alpha = 0.01$ degrees of freedom for the numerator (ν_1)

ν_2 \ ν_1	1	2	3	4	5	6	7	8	9	10	12	15	20	24	30	40	60	120	∞
1	4052	4999.5	5403	5625	5764	5859	5928	5982	6022	6056	6106	6157	6209	6235	6261	6287	6313	6339	6366
2	98.50	99.00	99.17	99.25	99.30	99.33	99.36	99.37	99.39	99.40	99.42	99.43	99.45	99.46	99.47	99.47	99.48	99.49	99.50
3	34.12	30.82	29.46	28.71	28.24	27.91	27.67	27.49	27.35	27.23	27.05	26.87	26.69	26.60	26.50	26.41	26.32	26.22	26.13
4	21.20	18.00	16.69	15.98	15.52	15.21	14.98	14.80	14.66	14.55	14.37	14.20	14.02	13.93	13.84	13.75	13.65	13.56	13.46
5	16.26	13.27	12.06	11.39	10.97	10.67	10.46	10.29	10.16	10.05	9.89	9.72	9.55	9.47	9.38	9.29	9.20	9.11	9.02
6	13.75	10.92	9.78	9.15	8.75	8.47	8.26	8.10	7.98	7.87	7.72	7.56	7.40	7.31	7.23	7.14	7.06	6.97	6.88
7	12.25	9.55	8.45	7.85	7.46	7.19	6.99	6.84	6.72	6.62	6.47	6.31	6.16	6.07	5.99	5.91	5.82	5.74	5.65
8	11.26	8.65	7.59	7.01	6.63	6.37	6.18	6.03	5.91	5.81	5.67	5.52	5.36	5.28	5.20	5.12	5.03	4.95	4.46
9	10.56	8.02	6.99	6.42	6.06	5.80	5.61	5.47	5.35	5.26	5.11	4.96	4.81	4.73	4.65	4.57	4.48	4.40	4.31
10	10.04	7.56	6.55	5.99	5.64	5.39	5.20	5.06	4.94	4.85	4.71	4.56	4.41	4.33	4.25	4.17	4.08	4.00	3.91
11	9.65	7.21	6.22	5.67	5.32	5.07	4.89	4.74	4.63	4.54	4.40	4.25	4.10	4.02	3.94	3.86	3.78	3.69	3.60
12	9.33	6.93	5.95	5.41	5.06	4.82	4.64	4.50	4.39	4.30	4.16	4.01	3.86	3.78	3.70	3.62	3.54	3.45	3.36
13	9.07	6.70	5.74	5.21	4.86	4.62	4.44	4.30	4.19	4.10	3.96	3.82	3.66	3.59	3.51	3.43	3.34	3.25	3.17
14	8.86	6.51	5.56	5.04	4.69	4.46	4.28	4.14	4.03	3.94	3.80	3.66	3.51	3.43	3.35	3.27	3.18	3.09	3.00
15	8.68	6.36	5.42	4.89	4.56	4.32	4.14	4.00	3.89	3.80	3.67	3.52	3.37	3.29	3.21	3.13	3.05	2.96	2.87
16	8.53	6.23	5.29	4.77	4.44	4.20	4.03	3.89	3.78	3.69	3.55	3.41	3.26	3.18	3.10	3.02	2.93	2.84	2.75
17	8.40	6.11	5.18	4.67	4.34	4.10	3.93	3.79	3.68	3.59	3.46	3.31	3.16	3.08	3.00	2.92	2.83	2.75	2.65
18	8.29	6.01	5.09	4.58	4.25	4.01	3.84	3.71	3.60	3.51	3.37	3.23	3.08	3.00	2.92	2.84	2.75	2.66	2.57
19	8.18	5.93	5.01	4.50	4.17	3.94	3.77	3.63	3.52	3.43	3.30	3.15	3.00	2.92	2.84	2.76	2.67	2.58	2.59
20	8.10	5.85	4.94	4.43	4.10	3.87	3.70	3.56	3.46	3.37	3.23	3.09	2.94	2.86	2.78	2.69	2.61	2.52	2.42
21	8.02	5.78	4.87	4.37	4.04	3.81	3.64	3.51	3.40	3.31	3.17	3.03	2.88	2.80	2.72	2.64	2.55	2.46	2.36
22	7.95	5.72	4.82	4.31	3.99	3.76	3.59	3.45	3.35	3.26	3.12	2.98	2.83	2.75	2.67	2.58	2.50	2.40	2.31
23	7.88	5.66	4.76	4.26	3.94	3.71	3.54	3.41	3.30	3.21	3.07	2.93	2.78	2.70	2.62	2.54	2.45	2.35	2.26
24	7.82	5.61	4.72	4.22	3.90	3.67	3.50	3.36	3.26	3.17	3.03	2.89	2.74	2.66	2.58	2.49	2.40	2.31	2.21
25	7.77	5.57	4.68	4.18	3.85	3.63	3.46	3.32	3.22	3.13	2.99	2.85	2.70	2.62	2.54	2.45	2.36	2.27	2.17
26	7.72	5.53	4.64	4.14	3.82	3.59	3.42	3.29	3.18	3.09	2.96	2.81	2.66	2.58	2.50	2.42	2.33	2.23	2.13
27	7.68	5.49	4.60	4.11	3.78	3.56	3.39	3.26	3.15	3.06	2.93	2.78	2.63	2.55	2.47	2.38	2.29	2.20	2.10
28	7.64	5.45	4.57	4.07	3.75	3.53	3.36	3.23	3.12	3.03	2.90	2.75	2.60	2.52	2.44	2.35	2.26	2.17	2.06
29	7.60	5.42	4.54	4.04	3.73	3.50	3.33	3.20	3.09	3.00	2.87	2.73	2.57	2.49	2.41	2.33	2.23	2.14	2.03
30	7.56	5.39	4.51	4.02	3.70	3.47	3.30	3.17	3.07	2.98	2.84	2.70	2.55	2.47	2.39	2.30	2.21	2.11	2.01
40	7.31	5.18	4.31	3.83	3.51	3.29	3.12	2.99	2.89	2.80	2.66	2.52	2.37	2.29	2.20	2.11	2.02	1.92	1.80
60	7.08	4.98	4.13	3.65	3.34	3.12	2.95	2.82	2.72	2.63	2.50	2.35	2.20	2.12	2.03	1.94	1.84	1.73	1.60
120	6.85	4.79	3.95	3.48	3.17	2.96	2.79	2.66	2.56	2.47	2.34	2.19	2.03	1.95	1.86	1.76	1.66	1.53	1.38
∞	6.63	4.61	3.78	3.32	3.02	2.80	2.64	2.51	2.41	2.32	2.18	2.04	1.88	1.79	1.70	1.59	1.47	1.32	1.00

Bibliography

ABS Group, Inc., *Root Cause Analysis Handbook: A Guide to Effective Incident Investigation*, ABS Group, Inc., Risk and Reliability Division, Rockville, MD, 1999.

Akao, Y., *Quality Function Deployment: Integrating Customer Requirements in Product Design*, Productivity Press, Portland, OR, 1989.

Altshuller, G., *Creativity as an Exact Science*, Gordon and Breach, New York, 1984.

Anderson, N., and Wilcoxon, R., "Framework for Prognostics of Electronic Systems," *Proceedings of International Military and Aerospace/Avionics COTS Conference*, Seattle, WA, August 3–5, 2004.

ASQC Glossary and Tables for Statistical Quality Control, American Society for Quality Control, Milwaukee, WI, 1983.

ASTM Publication STP-15D, Manual on the Presentation of Data and Control Chart Analysis, Philadelphia, PA, 1976.

Bamberger, J., "Essence of the Capability Maturity Model," *Computer*, June 1997, pp. 112–114.

Barlow, R., and Proschan, F., *Mathematical Theory of Reliability*, John Wiley & Sons, New York, 1965.

Barlow, R.E., and Proschan, F., *Statistical Theory of Reliability and Life Testing*, Holt, Rinehart & Winston, New York, 1975.

Barry, R., Murcko, A.C., and Brubaker, C.E., *The Six Sigma Book for Healthcare*, Health Administration Press, Chicago, IL, 2002.

Baxter, L.A., McConalogue, D.J., Scheuer, E.M., and Blischke, W.R., "On the tabulation of the renewal function," *Technometrics*, Vol. 24, 1982, pp. 151–156.

Berke, T.M., and Zaino, N.A., Jr., "Warranties: What Are They? What Do They Really Cost?" *Proceedings of the Annual Reliabiability and Maintainability Symposium*, 1991, pp. 326–331.

Bertels, T. (ed.), *Rath & Strong Six Sigma Leadership Handbook*, John Wiley and Sons, Hoboken, 2003.

Bhandarkar, S.M., et al., "Influence of Selected Design Variables on Thermomechanical Stress Distributions in Plated Through Hole Structures," *Transaction of the ASME—Journal of Electronic Packaging*, Vol. 114, 1992, pp. 8–13.

Bhattacharya, S.S., Bannerjee, S.K., Nguyen, B.-Y., and Tobin, P.J., "Temperature Dependence of the Anomalous Leakage Current in Polysilicon-on-Insulator MOSFETs," *IEEE Transactions on Electron Devices*, Vol. 41(2), 1994, pp. 221–227.

Reliability Engineering, First Edition. Kailash C. Kapur.
© 2014 John Wiley & Sons, Inc. Published 2014 by John Wiley & Sons, Inc.

Bibliography

Black, J.R., "Physics of Electromigration," *IEEE Proceedings of International Reliability Physics Symposium*, 1983, pp. 142–149.

Blischke, W.R., and Murthy, D.N.P., "Product Warranty Management—I. A Taxonomy for Warranty Policies," *European Journal of Operational Research*, Vol. 62, 1993, pp. 127–148.

Blischke, W.R., and Murthy, D.N.P., *Warranty Cost Analysis*, Marcel Dekker, Inc., New York, 1994.

Blischke, W.R., and Murthy, D.N.P. (eds.), *Product Warranty Handbook*, Marcel Dekker, Inc., New York, 1996.

Blischke, W.R., and Murthy, D.N.P., *Reliability: Modeling, Prediction and Optimization*, John Wiley & Sons, New York, 1998.

Blischke, W.R., and Scheuer, E.M., "Calculation of the cost of warranties as a function of estimated life distributions," *Naval Research Log. Quart*, Vol. 22, 1975, pp. 681–696.

Blischke, W.R., and Scheuer, E.M., "Applications of renewal theory in the analysis of the free-replacement warranty," *Naval Research Log. Quart*, Vol. 28, 1981, pp. 193–205.

Blischke, W.R., and Vij, S.-D., "Quality and warranty: Sensitivity of warranty cost models to distributional assumptions," in S. Ghosh, W.R. Sehucany, and W.B. Smith (eds.), *Statistics of Quality*, pp. 361–386, Marcel Dekker, Inc., New York, 1997.

Bodenhoefer, K., "Environmental Life Cycle Information Management and Acquisition—First Experiences and Results from Field Trials," *Proceedings of Electronics Goes Green 2004+*, Berlin, September 5–8, 2004, pp. 541–546.

Boehm, H., "Threads Cannot Be Implemented as a Library," November 12, 2004, http://www.hpl.hp.com/techreports/2004/HPL-2004-209.pdf (accessed on November 22, 2013).

Boeing, Electronic Component Management Program, Document D6-55583, 1996.

Boersma, J., Loke, G., Petkova, V.T., and Sander, P.C., "Quality of Information Flow in the Backend of a Product Development Process: A Case Study," *Quality and Reliability Engineering International*, Vol. 20(4), 2004, pp. 255–263.

Bollinger, T.B., and McGowan, C., "A Critical Look at Software Capability Evaluations," *IEEE Software*, Vol. 8(4), 1991, pp. 25–41.

Bowles, J.B., "Fundamentals of Failure Modes and Effects Analysis," Tutorial Notes Annual Reliability and Maintainability Symposium, 2003.

Bowles, J.B., and Bonnell, R.D., "Failure Modes, Effects and Criticality Analysis—What Is It and How to Use It," Tutorial Notes Annual Reliability and Maintainability Symposium, 1998.

Box, G., and Luceno, A., *Statistical Control by Monitoring and Feedback Adjustment*, John Wiley & Sons, New York, 1997.

Boyle v. United Technologies Corp., 487 U.S. 500 (1988), Hartford, CT.

Brennan, J.R., *Warranties: Planning Analysis and Implementation*, McGraw Hill, New York, 1994.

Brombacher, A.C., *Reliability by Design: CAE Techniques for Electronic Components and Systems*, John Wiley & Sons, New York, 1992.

Brombacher, A.C., "Maturity Index on Reliability: Covering Non-Technical Aspects of IEC61508 Reliability Certification," *Reliability Engineering and System Safety*, Vol. 66(2), 1999, pp. 109–120.

Bromstead, J.R., and Baumann, E.D., "Performance of Power Semiconductor Devices at High Temperatures," Transactions of the First International High Temperature Electronics Conference, Albuquerque, June, 1991, pp. 27–35.

Brown, D., Kalgren, P., Byington, C., and Orsagh, R., "Electronic Prognostics—A Case Study Using Global Positioning System (GPS)," IEEE Autotestcon, 2005.

Carey, S., "FAA Proposes $1.1 Million Fine Against a Southwest Air Maintenance Provider," *Wall Street Journal*, September 12, 2011.

Cassidy, K., Gross, K., and Malekpour, A., "Advanced Pattern Recognition for Detection of Complex Software Aging Phenomena in Online Transaction Processing Servers," *Proceedings*

of the International Performance and Dependability Symposium, Washington, DC, June 23–26, 2002.
Castelli, J., Nash, C., Ditlow, C., and Pecht, M., *Sudden Acceleration: The Myth of Driver Error*, CALCE EPSC Press, University of Maryland, College Park, MD, 2003.
Chan, H.A., and Englert, P.J., *Accelerated Stress Testing Handbook, Guide for Achieving Quality Products*, IEEE Press, New York, 2001.
Chiesa, V., Coughlan, P., and Voss, C., "Development of a Technical Innovation Audit," *Journal of Product Innovation Management*, Vol. 13(2), 1996, pp. 105–136.
Childs, J.A., and Mosleh, A., "A Modified FMEA Tool for Use in Identifying and Addressing Common Cause Failure Risks in Industry," Reliability and Maintainability Symposium, 1999, Proceedings Annual, January 18–19, 1999, pp. 19–24.
Chukova, S., Arnold, R., and Wang, D.Q., "Warranty Analysis: An Approach to Modeling Imperfect Repairs," *International Journal of Production Economics*, Vol. 89, 2004, pp. 57–68.
Condra, L.W., *Reliability Improvements with Design of Experiments*, Marcel Dekker, Inc., New York, 2001.
Crosby, P.B., *Quality Is Still Free: Making Quality Certain in Uncertain Times*, McGraw Hill, New York, 1996.
Cushing, M.J., Mortin, D., Stadterman, T.J., and Malhotra, A., "Comparison of Electronics-Reliability Assessment Approaches," *IEEE Transactions on Reliability*, Vol. 42(4), 1993, pp. 542–546.
Cutter, D., and Thompson, O., "Condition-Based Maintenance Plus Select Program Survey," Report LG301T6, January 2005, http://www.acq.osd.mil/log/mpp/cbm+/_LG301T6_FINAL.PDF (accessed January 24, 2014).
Daniel Ionescu, "Apple Responds to iPhone 4 Antenna Problem," *PCWorld*, June 25, 2010.
Dasgupta, A., and Pecht, M., "Material Failure Mechanisms and Damage Models," *IEEE Transactions on Reliability*, Vol. 40(5), 1991, pp. 531–536.
Dasgupta, A., Oyan, C., Barker, D., and Pecht, M., "Solder Creep-Fatigue Analysis by an Energy-Partitioning Approach," *ASME Transactions, Journal of Electronic Packaging*, Vol. 114(2), 1992, pp. 152–160.
Dasgupta, A., et al., "Does the Cooling of Electronics Increase Reliability?," Proceedings of Thermal Stresses 95 Conference, paper number: 231, 1995.
Dean, J., Shirouzu, N., and Areddy, J.T., "China's Wen Cautions on Rail Push," *Wall Street Journal*, July 29, 2011.
Deming, W.E., *Quality, Productivity and Competitive Position*, Massachusetts Institute of Technology, Cambridge, MA, 1982.
Denson, W., "A Tutorial: PRISM," *RAC Journal*, Vol. 7(3), 1999, pp. 1–6.
DoD 5000.2 Policy Document, Defense Acquisition Guidebook, Chapter 5.3—Performance Based Logistics, December 2004.
Dreike, P.L., Fleetwood, D.M., King, D.B., Sprauer, D.C., and Zipperian, T.E., "An Overview of High-Temperature Electronic Device Technologies and Potential Applications," *IEEE Transactions on Components, Packaging, and Manufacturing Technology—Part A*, Vol. 17(4), 1994, pp. 594–608.
Dummer, G.W.A., Tooley, M.H., and Winton, R.C., *An Elementary Guide to Reliability*, 5th ed., Butterworth Heinemann, Oxford, 1997.
Duncan, A.J., *Quality Control and Industrial Statistics*, 5th ed., Irwin, Homewood, IL, 1986.
Edwards, G.T., and Watson, I.A., "A Study of Common Mode Failures," SRD-R-146, United Kingdom Atomic Energy Authority, Safety and Reliability Directorate, July 1979.
Eskin, D.J., et al., "Reliability Derating Procedures," RADC-TR-84-254, Rome Laboratory, 1984.
Evans, H.R., "Superstition, Witchcraft, Reliability Prediction," *IEEE Transactions on Reliability*, Vol. 37, 1988, p. 457.

Fault Tree Handbook with Aerospace Applications, NASA Office of Safety and Mission Assurance, August 2002.

Federal Consumer Product Warranties Law, January 4, 1975, Title I of Magnusson–Moss Warranty Act—Federal Trade Commission, Public Law 93-637. U.S. Code, Title 15, Chapter 50, §2304, Federal minimum standards for warranties. http://www4.law.cornell.edu/uscode/15/2304.html (accessed on November 22, 2013).

Fink, J., Private Correspondence, January 1997.

FIDES Group, "FIDES Guide issue A, Reliability Methodology for Electronic Systems," 2004.

Foster, M., "Toyota recalls 550,000 cars for steering issue," *Yahoo! News*, November 9, 2011.

Foty, D., *MOSFET Modeling with SPICE: Principles and Practices*, Prentice Hall PTR, Upper Saddle River, NJ, 1997.

Foucher, B., Boullie, J., Meslet, B., and Das, D., "A Review of Reliability Predictions Methods for Electronic Devices," *Microelectronics Reliability*, Vol. 42(8), 2002, pp. 1155–1162.

Franceschini, F., and Galetto, M., "A New Approach for Evaluation of Risk Priories of Failure Modes in FMEA," *International Journal of Production Research*, Vol. 39(13), 2001, pp. 2991–3002.

Fraser, P., and Gregory, M., "A Maturity Grid Approach for the Assessment of Product Development Collaborations," *Proceedings of 9th International Product Development Management Conference*, Sophia Antipolis, France, May 27–28, 2002.

Fraser, P., Moultrie, J., and Holdway, R., "Exploratory Studies of a Proposed Design Maturity Model," Proceedings of 8th International Product Development Management Conference, University of Twentie, Holland, June 11–12, 2001.

Fushinobu, K., Majumdar, A., and Hijikata, K., "Heat Generation and Transport in Submicron Semiconductor Devices," *Transactions of the ASME: Journal of Heat Transfer*, Vol. 117(2), 1995, pp. 25–31.

Ganesan, S., Eveloy, V., Das, D., and Pecht, M., "Identification and Utilization of Failure Mechanisms to Enhance FMEA and FMECA," *Proceedings of the IEEE Workshop on Accelerated Stress Testing & Reliability (ASTR)*, Austin, TX, October 2–5, 2005a.

Ganesan, S., Eveloy, V., Das, D., and Pecht, M., "Identification and Utilization of Failure Mechanisms to Enhance FMEA and FMECA," *Proceedings of the IEEE Workshop on Accelerated Stress Testing and Reliability (ASTR)*, Austin, Texas, October 3–5, 2005b.

Gertman, D., and Blackman, H.S., *Human Reliability and Safety Analysis Data Handbook*, John Wiley and Sons, New York, 1994.

Goodman, D., Vermeire, B., Ralston-Good, J., and Graves, R., "A Board-Level Prognostic Monitor for MOSFET TDDB," IEEE Aerospace Conference, Big Sky, 2006.

Goodson, K.E., Filk, M.I., Su, L.T., and Antoniadis, D.A., "Prediction and Measurement of Temperature Fields in Silicon-on-insulator Electronic Circuits," *Transactions of the ASME: Journal of Heat Transfer*, Vol. 117(8), 1995, pp. 574–581.

Gorniak, M., "Rome Labs," Qualified Manufacturer's List Update," *Reliability Society Newsletter*, Vol. 40, 1994, p. 8.

Gu, J., and Pecht, M., "New Methods to Predict Reliability of Electronics," International Conference on Reliability, Maintainability, and Safety, Beijing, China, 2007, pp. 440–451.

Gu, J., Barker, D., and Pecht, M., "Prognostics Implementation of Electronics under Vibration Loading," *Microelectronics Reliability Journal*, Vol. 47(12), 2007, pp. 1849–1856.

Gu, J., Barker, D., and Pecht, M., "Uncertainty Assessment of Prognostics Implementation of Electronics Under Vibration Loading," 2007 AAAI Fall Symposium on Artificial Intelligence for Prognostics, Arlington, VA, November 9–11, 2007.

Guenzi, G., "Reliability Evaluation of Common Cause Failures and Other Interdependencies in Large Reconfigurable Networks," doctoral dissertation, University of Maryland, 2010.

Guidelines for Failure Mode and Effects Analysis, *Guidelines for Failure Mode and Effects Analysis for Automotive, Aerospace, and General Manufacturing Industries*, Dyadem Press, Ontario, Canada, 2003.

Guin, L., "Cumulative Warranties, Conceptualiza-tion and Analysis," doctoral dissertation, University of Southern California, Los Angeles, 1984.

Hakim, E., "Reliability Prediction: Is Arrhenius Erroneous?," *Solid State Technology*, Vol. 33, 1990, p. 57.

Harry, M., and Schroeder, R., *Six Sigma: The Breakthrough Management Strategy Revolutionizing the World's Top Corporations*, Currency, New York, 2000.

Haugen, E.B., *Probabilistic Approach to Design*, John Wiley & Sons, New York, 1968.

Hicks, C.R., and Turner, K.V., *Fundamental Concepts in the Design of Experiments*, 5th ed., Oxford University Press, New York, 1999.

Hoerl, R.W., "Six Sigma and the Future of the Quality Profession," *Quality Progress*, Vol. 31(6), 1998, pp. 35–42.

Hotelling, H., *Multivariate Quality Control, Techniques of Statistical Analysis*, C. Eisenhart, M.W. Hastay, and W.A. Wallis (eds.), McGraw Hill, New York, 1947.

Howard, R.T., "Electrochemical Model for Corrosion of Conductors on Ceramic Substrates," *IEEE Transactions on CHMT*, Vol. 4(4), 1981, pp. 520–525.

Hu, J., Barker, D., Dasgupta, A., and Arora, A., "Role of Failure-mechanism Identification in Accelerated Testing," *Journal of the IES*, Vol. 36(4), 1993, pp. 39–45.

Huang, C.J., Grotjohn, T.A., Sun, C.J., Reinhard, D.K., and Yu, C.-C.W., "Temperature Dependence of Hot-Electron Degradation in Bipolar Transistors," *IEEE Transactions on Electron Devices*, Vol. 40(9), 1993, pp. 1669–1674.

Hughes, G., Murray, J., Kreutz-Delgado, K., and Elkan, C., "Improved Disk-drive Failure Warnings," *IEEE Transactions on Reliability*, Vol. 51(3), 2002, pp. 350–357.

IEC Standard 60134, Ratings System for Electronic Tubes and Valves and Analogous Semiconductor Devices, Geneva, Switzerland, 1961 (last review date 1994).

IEC/PAS 62240, Use of Semiconductor Devices Outside Manufacturers' Specified Temperature Ranges, Edition 1, 2001-04, (Also being developed as GEIA 4900), 2001.

IEEE 100, *The Authoritative Dictionary of IEEE Standards Terms*, 7th ed., Standards Information Network IEEE Press, Piscataway, NJ, 2000.

IEEE Standard 1332–1998, IEEE Standards Board, "IEEE Standard Reliability Program for the Development and Production of Electronic Systems and Equipment," June 30, 1998.

IEEE Standard 1413.1-2002, IEEE Guide for Selecting and Using Reliability Predictions Based on IEEE 1413, IEEE, February 2003.

IEEE Standard 1413–1998, "IEEE Standard Methodology for Reliability Prediction and Assessment for Electronic Systems and Equipment," IEEE, December 1998.

IEEE Standards Project Editors, Reliability Program for the Development and Production of Electronic Products, IEEE Std. 1332, IEEE, New York, NY, 1998.

IEEE Std. 1332–1998, IEEE Standard Reliability Program for the Development and Production of Electronic Systems and Equipment, IEEE Reliability Society, 1998.

Intel, Military Product Data Book, 1990.

Intel, Application Note AP-480—Pentium® Processor Thermal Design Guidelines Rev 2.0, November 1995.

IPC-SM-785, "Guidelines for Accelerated Reliability Testing of Surface Mount Solder Attachments," November 1992.

Ishikawa, K., *What Is Total Quality Control? The Japanese Way*, Prentice-Hall, Inc., Englewood Cliffs, NJ, 1985.

Jackson, M., Mathur, A., Pecht, M., and Kendall, R., "Part Manufacturer Assessment Process," *Quality and Reliability Engineering International*, Vol. 15, 1999a, pp. 457–468.

Jackson, M., Sandborn, P., Pecht, M., Hemens-Davis, C., and Audette, P., "A Risk Informed Methodology for Parts Selection and Management," *Quality and Reliability Engineering International*, Vol. 15, 1999b, pp. 261–271.

Jayant, M., "Intel Recalls Fastest Pentium," *Electronic News*, September 4, 2000.

Bibliography

Jet Propulsion Laboratory, JPL Derating Guidelines, Electronic Parts Reliability, Section 507, JPL-D-8545, REV A, 1996.

JEDEC Publication JEP 122-B, "Failure Mechanisms and Models for Semiconductor Devices," August 2003.

JEDEC Publication JEP 148, "Reliability Qualification of Semiconductor Devices Based on Physics-of-Failure Risk and Opportunity Assessment," April 2004.

Johnson, D., "Review of Fault Management Techniques Used in Safety Critical Avionic Systems," *Progress in Aerospace Science*, Vol. 32(5), 1996, pp. 415–431.

Johnson, M., and Nilsson, L., "The Importance of Reliability and Customization from Goods to Services," *Quality Management Journal*, Vol. 10(1), 2003.

Johnson, P., and Rabe, D., "Detecting Intermittent Test Failures with the Aid of Environmental Stress Screening," AUTOTESTCON '95. Systems Readiness: Test Technology for the 21st Century. Conference Record, pp. 127–134, 1995.

Juran, J.M., and Gryna, F.M., *Quality Planning and Analysis*, McGraw-Hill, New York, 1980.

Kane, V.E., "Process Capability Indices," *Journal of Quality Technology*, Vol. 18(1), 1986, pp. 41–52.

Kanniche, M., and Mamat-Ibrahim, M., "Wavelet-based Fuzzy Algorithm for Condition Monitoring of Voltage Source Inverters," *Electronic Letters*, Vol. 40(4), 2004, pp. 267–268.

Kapur, K.C., "Quality Evaluation Systems for Reliability," *Reliability Review*, Vol. 6(2), 1986, pp. 19–20.

Kapur, K.C., and Feng, Q., "Integrated Optimization Models and Strategies for the Improvement of the Six Sigma Process," *International Journal of Six Sigma and Competitive Advantage*, Vol. 1(2), 2005, pp. 210–228.

Kapur, K.C., and Feng, Q., "Statistical Methods for Product and Process Improvement," in H. Pham (ed.), *The Springer Handbook of Engineering Statistics*, pp. 193–212, Springer, London, 2006.

Kapur, K.C., and Lamberson, L.R., *Reliability in Engineering Design*, John Wiley and Sons, New York, 1977.

Kar, T.P., and Nachlas, J.A., "Coordinated Warranty and Burn-In Strategies," *IEEE Transactions on Reliability*, Vol. 46, 1997, pp. 512–518.

Kara-Zaitri, C., Keller, A.Z., and Fleming, P.V., "A Smart Failure Mode and Effect Analysis Package," *Annual Reliability and Maintainability Symposium Proceedings*, 1992, pp. 414–421.

Kasley, K.L., Oleszek, G.M., and Anderson, R.L., "Investigation of the Kink and Hysteresis from 13 K to 30 K in n-Channel CMOS Transistors with a Floating Well," *Solid State Electronics*, Vol. 36(7), 1993, pp. 945–948.

Kececioglu, B.D., *Reliability Engineering Handbook*, Vol. 1 and 2, Prentice Hall, Englewood Cliffs, NJ, 1991.

Kececioglu, D., and Cormier, D., "Designing a Specified Reliability Directly into a Component," *Proceedings of Third Aerospace Reliability and Maintainability Conference*, 1968.

Kimseng, K., Hoit, M., and Pecht, M., "Physics-of-Failure Assessment of a Cruise Control Module," *Microelectronics Reliability*, Vol. 39(10), 1999, pp. 1423–1444.

Kirkland, L., Pombo, T., Nelson, K., and Berghout, F., "Avionics Health Management: Searching for the Prognostics Grail," *Proceedings of the IEEE Aerospace Conference*, Vol. 5, 2004, pp. 3448–3454.

Kirwan, B., *A Guide to Practical Human Reliability Assessment*, Taylor and Francis, Bristol, PA, 1994.

Klausner, M., Grimm, W., and Hendrickson, C., "Reuse of Electric Motors in Consumer Products," *Journal of Ecology*, Vol. 2(2), 1998a, pp. 89–102.

Klausner, M., Grimm, M., Hendrickson, C., and Horvath, A., "Sensor-Based Data Recording of Use Conditions for Product Take-Back," IEEE International Symposium on Electronics and the Environment, New York, 1998b, pp. 138–143.

Korsmeyer, D., "Discovery and Systems Health," National Aeronautics and Space Administration, http://ti.arc.nasa.gov/tech/techarea.php?ta=4 (accessed on November 21, 2013).

Kotz, S., and Lovelace, C.R., *Process Capability Indices in Theory and Practice*, Arnold, London, 1998.

Lall, P., Pecht, M., and Hakim, E., "Characterization of Functional Relationship between Temperature and Microelectronic Reliability," *Microelectronics and Reliability*, Vol. 35(3), 1995, pp. 377–402.

Lall, P., Pecht, M., and Hakim, E.B., *Influence of Temperature on Microelectronics and System Reliability: A Physics of Failure Approach*, CRC Press, Boca Raton, FL, 1997.

Lasance, C.J.M., "Accurate Temperature Prediction in Consumer Electronics: A Must but Still a Myth," in S. Kakaç, H. Yüncü, and K. Hijikata (eds.), *Cooling of Electronics Systems*, pp. 825–858, Kluwer Academic Publishers, Doedrecht, 1993.

Leemis, L.M., *Reliability: Probabilistic Models and Statistical Methods*, Prentice-Hall, Upper Saddle River, NJ, 1995.

Leonard, C., "Mechanical Engineering Issues and Electronic Equipment Reliability: Incurred Costs without Compensating Benefits," *Transactions of the ASME: Journal of Electronic Packaging*, Vol. 113(1), 1991a, pp. 1–7.

Leonard, C., "MIL-HDBK-217: It's Time to Rethink It," *Electronic Design*, 1991b, pp. 79–82.

Lewis, E.E., *Introduction to Reliability Engineering*, John Wiley and Sons, New York, 1996.

Lewis, N., "Reliability Through the Ages," Presented at Canadian Reliability and Maintainability Symposium, Ottawa, Canada, October 16–17, 2003.

Lieberman, F., "How to Select Parts and What Users Need to Know a Manufacturer's Perspective," CALCE Parts Selection and Management Workshop, June 16, 1998.

Liker, J., *The Toyota Way: 14 Management Principles from the World's Greatest Manufacturer*, McGraw Hill, New York, 2003.

Lowerre, J.M., "On warranties," *Journal of Industrial Engineering*, Vol. 19, 1968, pp. 359–360.

Lycoudes, N., (Motorola) Advanced Technology, Acquisition, Qualification and Reliability Workshop, Motorola Semiconductors, Newport Beach, CA, August 15–17, 1995.

Lycoudes, N., et al., "Considerations for Burn-In Tailoring of Integrated Circuits," *Proceedings of the Institute of Environmental Sciences*, 1990, pp. 27–41.

Macbeth, D., and Fergusson, N., *Partnership Sourcing: An Integrated Supply Chain Management Approach*, Pittman Publishing, London, 1994.

Maher, E., "No-Fault Finder," May 2000, http://www.aviationtoday.com/reports/avmaintenace/previous/0500/05nff.htm (accessed on October 3, 2002).

Maniwa, R.T., and Jain, M., "Focus Report: High Speed Benchmark," Integrated System Design, March 1996. Website: http://www.isdmag.com.

Martz, H.F., and Waller, R.A., *Bayesian Reliability Analysis*, John Wiley & Sons, New York, 1982.

Martz, H.F., and Waller, R.A., "Bayesian Reliability Analysis of Complex Series/Parallel Systems of Binomial Subsystems and Components," *Technometrics*, Vol. 32, 1990, pp. 407–416.

Mathew, S., Das, D., Osterman, M., Pecht, M., and Ferebee, R., "Prognostic Assessment of Aluminum Support Structure on a Printed Circuit Board," *ASME Journal of Electronic Packaging*, Vol. 128(4), 2006, pp. 339–345.

Mauri, G., "Integrating Safety Analysis Techniques, Supporting Identification of Common Cause Failures," PhD thesis, The University of York, Department of Computer Science, September 2000.

McCluskey, P.F., *High Temperature Electronics*, CRC Press, Boca Raton, Florida, 1996.

McCluskey, P.F., et al., "Uprating of Commercial Parts for Use in Harsh Environments," CALCE EPRC Report, November 1996.

McGrath, M.E. (ed.), *Setting the PACE in Product Development: A Guide to Product and Cycle Time Excellence*, Butterworth-Heinemann, Burlington, MA, 1996.

Bibliography

Meister, D., "Human Factors in Reliability," in G.W. Ireson, C.F. Coombs, and R.Y. Moss (eds.), *Handbook of Reliability Engineering and Management*, 2nd ed., pp. 9.1–9.30, McGraw Hill, New York, 1996.

Menke, W.W., "Determination of warranty reserves," *Management Science*, Vol. 15, 1969, pp. 542–549.

Micron Semiconductor, TN-00-07—IBIS Behavioral Models, 1998.

Middendorf, A., Griese, H., Reichl, H., and Grimm, W., "Using Life-Cycle Information for Reliability Assessment of Electronic Assemblies," IEEE International Integrated Reliability Workshop, Final Report, Piscataway, NJ, 2002, pp. 176–179.

Mishra, K., and Gross, K., "Dynamic Stimulation Tool for Improved Performance Modeling and Resource Provisioning of Enterprise Servers," Proceedings of the 14th IEEE International Symposium on Software Reliability Engineering (ISSRE'03), Denver, CO, November 2003.

Mishra, S., and Pecht, M., "In-Situ Sensors for Product Reliability Monitoring," *Proceedings of the SPIE*, Vol. 4755, 2002, pp. 10–19.

Mishra, S., Pecht, M., Smith, T., McNee, I., and Harris, R., "Remaining Life Prediction of Electronic Products Using Life Consumption Monitoring Approach," *Proceedings of the European Microelectronics Packaging and Interconnection Symposium*, Cracow, Poland, 2002, pp. 136–142.

Mishra, S., Ganesan, S., Pecht, M., and Xie, J., "Life Consumption Monitoring for Electronics Prognostics," *Proceedings of the IEEE Aerospace Conference*, Vol. 5, 2004, pp. 3455–3467.

Mitra, A., and Patankar, J.G., "Market Share and Warranty Costs for Renewable Warranty Programs," *International Journal of Production Economics*, Vol. 50, 1997, pp. 155–168.

MIL-STD-883, "Test Methods and Procedures for Microelectronics," Department of Defense, 1968.

MIL_STD_1629 (SHIPS), "Procedures for Performing a Failure Mode and Effects Analysis for Shipboard Equipment," U.S. Department of the Navy, Naval Ship Engineering Center, Hyattsville, MD, November 1974.

Mobley, R.K., *Root Cause Failure Analysis*, Butterworth-Heinemann, Boston, MA, 1999.

Montgomery, D.C., *Design and Analysis of Experiments*, 5th ed., John Wiley and Sons, New York, 2001.

Montgomery, D.C., *Introduction to Statistical Quality Control*, 5th ed., John Wiley & Sons, New York, 2005.

Montgomery, D.C., and Runger, G.C., *Applied Statistics and Probability for Engineers*, 4th ed., John Wiley & Sons, New York, 2007.

Morris, S.F., "Use and Application of MIL-HDBK-217," *Solid State Technology*, Vol. 33, 1990, pp. 65–69.

Murthy, D.N.P., and Blischke, W.R., "Product warranty management—II: An integrated framework for study," *European Journal of Operational Research*, Vol. 62, 1991a, pp. 261–280.

Murthy, D.N.P., and Blischke, W.R., "Product warranty management—III: A review of mathematical mod-," *European Journal of Operational Research*, Vol. 62, 1991b, pp. 1–34.

Myers, R.H., and Montgomery, D.C., *Response Surface Methodology: Process and Product Optimization Using Designed Experiments*, John Wiley and Sons, New York, 2002.

Naem, A.A., Deen, J., and Chee, L.Y., "Temperature Effects on the Resistivity of Polycrystalline Silicon Titanium Salicide," *Journal of Applied Physics*, Vol. 76(2), 1994, pp. 1071–1076.

Nakayama, W., "Information Processing and Heat Transfer Engineering: Some Generic Views on Future Research Needs," in S. Kakaç, H. Yüncü, and K. Hijikata (eds.), *Cooling of Electronics Systems*, pp. 911–944, Kluwer Academic Publishers, Doedrecht, 1993.

National Society of Professional Engineers, "NSPE Code of Ethics for Engineers," 1964, http://www.nspe.org/ethics/codeofethics/index.html (accessed on May 15, 2013).

Naval Air Systems Command, Department of the Navy, "Application and Derating Requirements for Electronic Components," AS-4613, 1976.

Newman, D.G., and Nesbitt, L.D., "USAF Experience with RIW," *Proceedings of the Annual Reliability and Maintainability Symposium*, 1978, pp. 55–61.

Ng, K.K., "A Survey of Semiconductor Devices," *IEEE Transactions on Electron Devices*, Vol. 43(10), 1996, pp. 1760–1766.

Nguyen, D.G., and Murthy, D.N.P., "Optimal Burn-In Time to Minimize Cost for Products Sold under Warranty," *IIE Transactions*, Vol. 14, 1982, pp. 167–174.

Nguyen, D.G., and Murthy, D.N.P., "A General Model for Estimating Warranty Costs for Repairable Products," *IEEE Transactions on Reliability*, Vol. 16, 1984, pp. 379–386.

O'Conner, P.D.T., "Commentary: Reliability—Past, Present, and Future," *IEEE Transactions on Reliability*, Vol. 49(4), 2000, pp. 335–341.

Ohmite, "Resistor Selection," http://www.ohmite.com/techdata/res_select.pdf, August 2002 (accessed January 24, 2014).

Ohno, T., *Toyota Production System: Beyond Large Scale Production*, Productivity Press, London, 1988.

OPS ALACARTE Company, "Reliability Services in the Design Phase: Sneak Circuit Analysis (Hardware or Software)," 2006, http://www.opsalacarte.com/Pages/reliability/reliability_des_sneakcircut.htm (accessed on November 22, 2006).

Pande, P.S., Newman, R.P., and Cavanagh, R.R., *The Six Sigma Way*, McGraw-Hill, New York, 2000.

Pasztor, A., and Landers, P., "Toshiba Agrees to Settlement on Laptops—Pact for $2.1 Billion Stems from Inquiries into Flaws That May Be Common," *Wall Street Journal*, November 1, 1999, p. A3.

Paulk, M.C., Weber, C.V., Garcia, S.M., Chrisis, M.B., and Bush, M., "Key Practices of the Capability Maturity ModelSM, Version 1.1," Technical Report CMU/SEI-93-TR-025, ESC-TR-93-178, Software Engineering Institute, Carnegie Mellon University, February 1993.

Pecht, M., *Integrated Circuit, Hybrid, and Multichip Module Package Design Guideline*, John Wiley & Sons, New York, 1994.

Pecht, M., "Issues Affecting Early Affordable Access to Leading Electronics Technologies by the US Military and Government," *Circuit World*, Vol. 22(2), 1996a, pp. 7–15.

Pecht, M., "Why the Traditional Reliability Prediction Models Do Not Work—Is There an Alternative?," *Electronics Cooling*, Vol. 2(1), 1996b, pp. 10–12.

Pecht, M., and Dasgupta, A., "Physics-of-Failure: An Approach to Reliable Product Development," *Journal of the Institute of Environmental Sciences*, Vol. 38, 1995, pp. 30–34.

Pecht, M., and Nash, F., "Predicting the Reliability of Electronic Equipment," *Proceedings of the IEEE*, Vol. 82(7), 1994, pp. 992–1004.

Pecht, M., and Ramakrishnan, A., "Development and Activities of the IEEE Reliability Standards Group," *Journal of the Reliability Engineering Association of Japan*, Vol. 22(8), 2000, pp. 699–706.

Pecht, M., and Ramappan, V., "Are Components Still a Major Problem? A Review of Electronic Systems and Device Field Failure Returns," *IEEE Transactions on Components, Hybrids, and Manufacturing Technology*, Vol. 15(6), 1992a, pp. 1060–1064.

Pecht, M., and Ramappan, V., "Are Components Still the Major Problem: A Review of Electronic System and Device Field Failure Returns," *IEEE Transactions CHMT*, Vol. 15(6), 1992b, pp. 1160–1164.

Pecht, M., Dasgupta, A., and Evans, J., *Quality Conformance and Qualification of Microelectronic Packages and Interconnects*, John Wiley & Sons, New York, 1994.

Pecht, M., Nguyen, L., and Hakim, E., *Plastic Encapsulated Microelectronics*, John Wiley & Sons, New York, 1995.

Pecht, M., Radojcic, R., and Rao, G., *Guidebook for Managing Silicon Chip Reliability*, CRC Press, Boca Raton, FL, 1999.

Pecht, M., Dube, M., Natishan, M., and Knowles, I., "An Evaluation of Built-in Test," *IEEE Transactions on Aerospace and Electronic Systems*, Vol. 37(1), 2001, pp. 266–272.

Bibliography

Pecuh, I., Margala, M., and Stopjakova, V., "1.5 Volts Iddq/Iddt Current Monitor," *Proceedings of the IEEE Canadian Conference on Electrical and Computer Engineering*, Vol. 1, 1999, pp. 472–476.

Pendsé, N., and Pecht, M., "Parameter Re-characterization Case Study: Electrical Performance Comparison of the Military and Commercial Versions of All Octal Buffer," in *Future Circuits International*, Vol. 6, pp. 63–67, Technology Publishing, London, 2000.

Pfaffenberger, R., and Patterson, J., *Statistical Methods*, Irwin Publishers, Toronto, ON, 1987.

Phadke, M.S., *Quality Engineering Using Robust Design*, Prentice-Hall, Edgewood Cliffs, NJ, 1989.

Philips, Family Specifications: HCMOS Family Characteristics, March 1988.

Plastic Package Availability Program, Final Report, sponsored by Defense Logistics Agency, Contract Number DLA900-92-C-1647, *Sharp Proceedings*, Indianapolis, IN, November 15, 1995.

Pyzdek, T., and Keller, P.A., *The Six Sigma Handbook: A Complete Guide for Greenbelts, Blackbelts, and Managers at All Levels*, McGraw Hill, New York, 2009.

Ramakrishnan, A., and Pecht, M., "A Life Consumption Monitoring Methodology for Electronic Systems," *IEEE Transactions on Components and Packaging Technologies*, Vol. 26(3), 2003, pp. 625–634.

Ramakrishnan, A., Syrus, T., and Pecht, M., "Electronic Hardware Reliability," in *Avionics Handbook*, pp. 22.1–22.21, CRC Press, Boca Raton, FL, 2000.

Rao, S.S., *Reliability-Based Design*, pp. 505–543, McGraw-Hill, Inc., New York, 1992.

Rausand, M., and Hoyland, A., *System Reliability Theory: Models, Statistical Methods and Applications*, 2nd ed., Wiley-Interscience, Hoboken, NJ, 2003.

Ridgetop Semiconductor-Sentinel Silicon™ Library, "Hot Carrier (HC) Prognostic Cell," August 2004.

Rofail, S.S., and Elmasry, M.I., "Temperature-Dependent Characteristics of BiCMOS Digital Circuits," *IEEE Transactions on Electron Devices*, Vol. 41(1), 1993, pp. 169–177.

Roland, N., "Toyota agrees to pay $32.4 million U.S. fine over timing of recalls," *Automotive News*, December 21, 2010a.

Roland, N., "Toyota to pay $32.4 million U.S. fine over timing of recalls," *Autoweek*, December 21, 2010b.

Rose, C., Beiter, A., and Ishii, K., "Determining End-of-Life Strategies as a Part of Product Definition," 1999 IEEE International Symposium on Electronics and the Environment, Piscataway, NJ, 1999, pp. 219–224.

Rouet, V., and Foucher, B., "Development and Use of a Miniaturized Health Monitoring Device," *Proceedings of the IEEE International Reliability Physics Symposium*, 2004, pp. 645–646.

Rudra, A.B., et al., "Electrochemical Migration in Multichip Modules," *Circuit World*, Vol. 22(1), 1995, pp. 67–70.

Sandborn, P., and Murphy, C., "A Model for Optimizing the Assembly and Disassembly of Electronic Systems," *IEEE Transactions on Electronics Packaging Manufacturing*, Vol. 22(2), 1999, pp. 105–117.

Sander, P.C., and Brombacher, A.C., "MIR: The Use of Reliability Information Flows as a Maturity Index for Quality Management," *Quality and Reliability Engineering International*, Vol. 15(6), 1999, pp. 439–447.

Sander, P.C., and Brombacher, A.C., "Analysis of Quality Information Flows in the Product Creation Process of High-Volume Consumer Products," *International Journal of Production Economics*, Vol. 67(1), 2000, pp. 37–52.

SAE Standard SAE J1739 "Potential Failure Mode and Effects Analysis in Design (Design FMEA) and Potential Failure Mode and Effects Analysis in Manufacturing and Assembly Processes (Process FMEA) and Effects Analysis for Machinery (Machinery FMEA)," August 2002.

Scalise, J., Honeywell Air Transport Systems Division, private communication, Summer 1996.

Scheidt, L., and Zong, S., "An Approach to Achieve Reusability of Electronic Modules," 1994 IEEE International Symposium on Electronics and the Environment, New York, 1994 pp. 331–336.

Searls, D., Dishongh, T., and Dujari, P., "A Strategy for Enabling Data Driven Product Decisions Through a Comprehensive Understanding of the Usage Environment," *Proceedings of IPACK'01*, Kauai, HI, July 8–13, 2001.

Self-Monitoring Analysis and Reporting Technology (SMART), "Self-Monitoring Analysis and Reporting Technology (SMART)," *PC Guide*, 2001, http://www.pcguide.com/ref/hdd/perf/qual/featuresSMART-c.html (accessed on November 21, 2013).

Shetty, V., Das, D., Pecht, M., Hiemstra, D., and Martin, S., "Remaining Life Assessment of Shuttle Remote Manipulator System End Effector," *Proceedings of the 22nd Space Simulation Conference*, Ellicott City, MD, October 21–23, 2002.

Sheu, S.H., and Chien, Y.H., "Optimal Burn-In Time to Minimize the Cost for General Repairable Products Sold under Warranty," *European Journal of Operational Research*, Vol. 163, 2005, pp. 445–461.

Shewhart, W.A., *Economic Control of Quality of a Manufactured Product*, D. Van Nostrand Company, New York, 1931.

Shingo, S., *A Study of the Toyota Production System*, Productivity Press, London, 1989.

Shishko, R., *NASA System Engineering Handbook*, SP-6105, June 1995.

Simon, M., Graham, B., Moore, P., JunSheng, P., and Changwen, X., "Life Cycle Data Acquisition Unit—Design, Implementation, Economics and Environmental Benefits," IEEE International Symposium on Electronics and the Environment, Piscataway, NJ, 2000, pp. 284–289.

Skormin, V., Gorodetski, V., and Popyack, L., "Data Mining Technology for Failure Prognostic of Avionics," *IEEE Transactions on Aerospace and Electronic Systems*, Vol. 38(2), 2002, pp. 388–403.

Slay, B., "Best Commercial Practices in Military Semiconductors," *Sharp Proceedings of the 4th Annual Commercial and Plastic Components in Military Applications Workshop*, November 16, 1995.

Smith, P., and Campbell, D., "Practical Implementation of BICS for Safety-Critical Applications," *Proceedings of the IEEE International Workshop on Current and Defect Based Testing-DBT*, April 2000, pp. 51–56.

Society of Automotive Engineers, Recommended Environmental Practices for Electronic Equipment Design, SAE J1211, Rev. November 1978.

Solomon, R., Sandborn, P., and Pecht, M., "Electronic Part Life Cycle Concepts and Obsolescence Forecasting," *IEEE Transactions on Components and Packaging Technologies*, Vol. 23(3), 2000, pp. 707–717.

Sorensen, B., "Digital Averaging—The Smoking Gun behind 'No-Fault-Found'," Air Safety Week, February 24, 2003, http://www.aviationtoday.com/regions/weur/2120.html#.UuncRrS1T2I (accessed on January 29, 2014).

Sorensen, B.A., Kelly, G., Sajecki, A., and Sorensen, P.W. "An Analyzer for Detecting Aging Faults In Electronic Devices," AUTOTESTCON '94. IEEE Systems Readiness Technology Conference. "Cost Effective Support into the Next Century," Conference Proceedings, pp. 417–421, September 20–22, 1994.

Steinberg, D.S., *Vibration Analysis for Electronic Equipment*, 2nd ed., John Wiley & Sons, New York, 1988.

Stoller, G., "Engineer Fears Repeat of 1988 Aloha Jet Accident," Honolulu Advertiser, January 18, 2001.

Stoumbos, Z.G., Reynolds, M.R., Ryan, T.P., and Woodall, W.H., "The State of Statistical Process Control as We Proceed into the 21st Century," *Journal of the American Statistical Association*, Vol. 95, 2000, pp. 992–998.

Stracener, J.T., "Mathematical Basis of Reliability," in T.A. Cruse (ed.), *Reliability-Based Mechanical Design*, pp. 91–122, Marcel Dekker, New York, 1997.

Bibliography

Strutt, J.E., "Reliability Capability Maturity Briefing Document," Report No. R-03/2/1, Reliability Engineering and Risk Management Centre, Cranfield University, UK, 2001.

Suh, N., *The Principles of Design*, Oxford University Press, New York, 1990.

Suh, N., *Axiomatic Design: Advantages and Applications*, Oxford University Press, New York, 2001.

Szakonyi, R., "Measuring R&D Effectiveness—I," *Research Technology Management*, Vol. 37(2), 1994a, pp. 27–32.

Szakonyi, R., "Measuring R&D Effectiveness—II," *Research Technology Management*, Vol. 37(3), 1994b, pp. 44–55.

Taguchi, G., *Introduction to Quality Engineering: Designing Quality into Products and Processes*, Asian Productivity Organization, Unipub/Quality Resources, White Plains, NY, 1986.

Taguchi, G., *System of Experimental Design*, Vol. I and II, Quality Resources, New York and American Supplier Institute, Dearborn, MI, 1987.

Talmor, M., and Arueti, S., "Reliability Prediction: The Turn-Over Point," *Proceedings of the Annual Reliability and Maintainability Symposium*, 1997, pp. 254–262.

Tang, S.M., "New Burn-in Methodology Based on IC Attributes, Family IC Burn-in Data, and Failure Mechanism Analysis," *Proceedings of the Annual Reliability and Maintainability Symposium*, 1996, pp. 185–190.

Telcordia Technologies, "Reliability Prediction Procedure for Electronic Equipment Issue 1," Special Report SR-332, Telcordia Customer Service, Piscataway, NJ, May 2001.

Thomas, D.A., Ayers, K., and Pecht, M., "The 'Trouble Not Identified' Phenomenon in Automotive Electronics," *Microelectronics Reliability*, Vol. 42(4–5 and 10), 2002, pp. 641–651.

Tibken, S., "Intel to Resume Shipments of Flawed Chip," *Wall Street Journal*, February 7, 2011.

Tiku, S., Azarian, M., and Pecht, M., "Using a Reliability Capability Maturity Model to Benchmark Electronics Companies," *Int'l Journal of Quality & Reliability Management*, Vol. 24(5), 2007, pp. 547–563.

Tuchband, B., and Pecht, M., "The Use of Prognostics in Military Electronic Systems," Proceedings of the 32nd GOMACTech Conference, Lake Buena Vista, FL, March 19–22, 2007, pp. 157–160.

Udy, J., "Recall Central: Defective Honda Airbag Component; Toyota Sienna Load Placard," *Automobile Magazine*, December 2, 2011.

United States Department of Defense, "Military Standard: Climatic Information to Determine Design and Test Requirements for Military Systems and Equipment—MIL 210C," Washington, DC, 1987.

United States Department of Defense, "Military Standard: Environmental Test Methods and Engineering Guidelines—MIL 810E," Washington, DC, 1989.

United States Department of Defense, MIL-STD-965: Parts Control Program, Columbus, OH, September 26, 1996.

University of Utah, Mechanical Engineering Department, "Kansas City Hyatt Regency Walkway Collapse Packet," 1981, http://www.mech.utah.edu/ergo/pages/educational/safety_modules/kc/index.html (accessed on May 15, 2013).

U.S. Department of Defense, MIL-HDBK 217, Military Handbook for Reliability Prediction of Electronic Equipment, Version A, 1965.

Vaidyanathan, K., and Gross, K., "MSET Performance Optimization for Detection of Software Aging," *Proceedings of the 14th IEEE International Symposium on Software Reliability Engineering*, Denver, CO, November 2003.

Vichare, N., and Pecht, M., "Prognostics and Health Management of Electronics," *IEEE Transactions on Components and Packaging Technologies*, Vol. 29(1), 2006, pp. 222–229.

Vichare, N., Rodgers, P., Eveloy, V., and Pecht, M., "In Situ Temperature Measurement of a Notebook Computer: A Case Study in Health and Usage Monitoring of Electronics," *IEEE Transactions on Device and Materials Reliability*, Vol. 4(4), 2004, pp. 658–663.

Vichare, N., Rodgers, P., and Pecht, M., "Methods for Binning and Density Estimation of Load Parameters for Prognostics and Health Management," *International Journal of Performability Engineering*, Vol. 2(2), 2006, pp. 149–161.

Vichare, N., Rodger, P., Eveloy, V., and Pecht, M., "Environment and Usage Monitoring of Electronic Products for Health Assessment and Product Design," *International Journal of Quality Technology and Quantitative Management*, Vol. 4(2), 2007, pp. 235–250.

Voas, J., Ghosh, A., Charron, F., and Kassab, L., "Reducing Uncertainty about Common-Mode Failures," *Proceedings of the Eighth International Symposium on Software Reliability Engineering*, November 2–5, 1997, pp. 308–319.

Washington Post, "Ready When Chip Lines Are Down," December 18, 1999, p. E01.

Weibull, W., "A Statistical Theory of the Strength of Material," *Proceedings Royal Swedish Institute for Engineering Research*, Vol. 151(1), 1939.

Weibull, W., "A Statistical Distribution Function of Wide Applicability," *Journal of Applied Mechanics*, Vol. 18, 1951, pp. 293–297.

Western Electric, *Statistical Quality Control Handbook*, Western Electric Corporation, Indianapolis, IN, 1956.

Whisnant, K., Gross, K., and Lingurovska, N., "Proactive Fault Monitoring in Enterprise Servers," 2005 IEEE International Multi-Conference in Computer Science and Computer Engineering, Las Vegas, NV, June 2005.

Williams, K., Robertson, N., Haritonov, C.R., and Strutt, J., "Reliability Capability Evaluation and Improvement Strategies for Subsea Equipment Suppliers," *Journal of the Society for Underwater Technology*, Vol. 25(4), 2003, pp. 165–174.

Wilson, P., Dell, L.D., and Anderson, G.F., *Root Cause Analysis*, ASQC Quality Press, Milwaukee, WI, 1993.

Wong, K., "The Bathtub and Flat Earth Society," *IEEE Transactions on Reliability*, Vol. 38, 1989, p. 403.

Wong, K.L., "What Is Wrong with the Existing Reliability Prediction Methods?," *Quality and Reliability Engineering International*, Vol. 6, 1990, pp. 251–258.

Wright, R., and Kirkland, L., "Nano-Scaled Electrical Sensor Devices for Integrated Circuit Diagnostics," *IEEE Aerospace Conference*, Vol. 6, 2003, pp. 2549–2555.

Wright, R., Zgol, M., Keeton, S., and Kirkland, L., "Nanotechnology-Based Molecular Test Equipment (MTE)," *IEEE Aerospace and Electronic Systems Magazine*, Vol. 16(6), 2001, pp. 15–19.

Wright, R., Zgol, M., Adebimpe, D., and Kirkland, L., "Functional Circuit Board Testing Using Nanoscale Sensors," IEEE Systems Readiness Technology Conference, September 2003, pp. 266–272.

Xue, B., and Walker, D., "Built-In Current Sensor for IDDQ Test," *Proceedings of the IEEE International Workshop on Current and Defect Based Testing-DBT*, 25 April 2004, pp. 3–9.

Yang, K., and andEl-Haik, B., *Design for Six Sigma: A Roadmap for Product Development*, McGraw-Hill, New York, 2003.

Yeh, L.T., "Review of Heat Transfer Technologies in Electronic Equipment," *Transactions of the ASME, Journal of Electronic Packaging*, Vol. 117(4), 1995, pp. 333–339.

Yun, W.Y., Lee, Y.W., and Ferreira, L., "Optimal Burn-In Time under Cumulative Free Replacement Warranty," *Reliability Engineering and System Safety*, Vol. 78, 2002, pp. 93–100.

Zaino, N.A., Jr., and Berke, T.M., "Some Renewal Theory Results with Application to Fleet Warranties," *Naval Research Logistics Quarterly*, Vol. 41, 1994, pp. 465–482.

Index

Accelerated life tests (ALT) 133, 134, 184, 248, 250, 251, 254–266
Accelerated stress testing (AST) 250, 255, 261, 265, 270, 331, 336
Active redundancy 376, 381, 392, 402
Assignable cause 297–300, 308, 313, 325
Average outgoing quality 182, 317–321

Bathtub curve 26, 56, 419
Bayesian network 422, 430
Bayesian reliability analysis 445, 446
Bernoulli trials 46, 50, 51
Binomial coefficients 50
Binomial distribution 27, 46, 47, 50, 51, 249, 283
Binomial expansion 46, 47
Binomial variable 312
Block diagram, *see* Reliability block diagram
Burn-in process 26, 248, 249, 285, 331–336, 439

c chart 302, 304, 312, 314–316
Canary 131, 418–420
Cascade failure 358, 359
Cause-and-effect diagram 344, 346, 352–354, 356
Censored data 267, 268, 287, 289
Chance (or common) cause(s) 179, 181, 182, 295, 297, 299, 300, 316, 319, 323
Chi-square distribution 461–465, 467
Coefficient of kurtosis 37, 38, 53
Coefficient of skewness 37, 53
Coefficient of variation 36–38, 213, 214, 220
Coherent structures 396
Common cause failure 355, 356, 358–361
Common mode failure(s) 355, 356, 359, 361, 362, 364, 366, 369, 372, 393
Conditional probability 29, 357, 395
Conditional reliability 19, 31–33, 39, 55, 62, 82
Confidence interval 65, 249, 276–283, 291
    on the mean, variance known 279
    on the mean, variance unknown 280
    on parameter for exponential distribution 287–291
    for reliability for success-failure testing 283
Continuous random variable 51, 73

Control chart 96, 98, 180, 279, 295–297, 299–325
Control limits 297, 300, 301, 304, 306–308, 311–315
Correlation 239, 254, 306, 323, 325, 424, 430
Critical-to-quality (CTQ) 92–93, 98, 105, 106
Cumulative distribution function 24, 45, 78, 210, 455
Customer satisfaction 2, 6, 10, 95, 105–107, 175, 188, 409
Cut sets 120, 121, 127, 128

Deductive method 116, 117, 119, 347
Deductive vs. inductive methods 116, 117, 119, 347
Degrees of freedom 280, 281, 284, 291, 457, 461, 467–472
Derating 128–130, 172, 223–245
Design review 106, 131, 132, 341, 348, 373
Design for six sigma (DFSS) 10, 89–108
Discrete random variable 45, 46
DMAICT or DMAIC process 92, 97–99, 104, 105, 107, 108
Dynamic system reliability models
    parallel model 390, 399
    series model 306, 325
    shared-load parallel model 357, 390, 391
    standby redundant system 381, 385, 402

Expected value 27, 35, 36, 100, 179, 217, 401, 413
    of a continuous random variable 35, 36
    (mean) of a discrete random variable 46
    of a function of random variables 100
Exponential distribution 32, 51, 55, 57, 61–64, 66, 76, 287–291, 377, 386–390, 442, 444, 446, 448

F-distribution 285, 467
Factor of safety 209, 211–214, 219
Failure models 79, 128–131, 173, 174, 183, 195–199, 202, 204, 206, 239, 252, 256, 258, 260, 261, 347, 413, 443
Failure precursors 415, 418, 420–430
Failure rate 26, 27, 29, 32, 34, 39, 54, 57, 60–62, 67, 71, 72, 74, 76, 77, 203, 204, 233, 335, 347, 376–379, 383–389, 412, 413, 435, 439, 443, 446 *see also* hazard rate

---

Reliability Engineering, First Edition. Kailash C. Kapur and Michael Pecht.
© 2014 John Wiley & Sons, Inc. Published 2014 by John Wiley & Sons, Inc.

Index

Failure mode and effect analysis (FMEA) 95, 97, 118, 193–195, 205, 206, 248, 341, 364, 366, 369, 373
Failure mode, effect and criticality analysis (FMECA) 91, 93, 98, 117–119, 193–195
Failure mode, mechanism and effect analysis (FMMEA) 193–206, 219, 342, 343, 345, 417, 425
Failure modes 117, 118, 123, 193–206
Failure truncated test 65, 290
False alarm on a control chart 325, 415, 422
Fault tree analysis (FTA) 91, 117, 119–128, 195, 341, 342, 343, 347, 360
Frequency histogram 21
Function of random variable 215–218
Fuses 130, 131, 261, 418–420, 431

Gamma distribution 75–77, 288
Gamma function 56, 76, 451, 453, 454
Gaussian distribution, see Normal distribution
Geometric distribution 50, 51

Hazard rate 19, 22, 26–34, 36, 39, 45, 54–57, 62–64, 69, 72, 74, 79, 376, 378, 383–385, 390, 392, 436; see also Failure rate
Health monitoring 137, 195, 409–431
Histogram 21, 23–25, 30, 83, 98, 427
Human factors 115–117, 131, 132, 160
Hypothesis testing 95

Imperfect switching 387–390
Independence 325, 442
Inductive methods 117, 119
Infant mortality 26, 56, 57, 82, 136, 249, 334
Instantaneous failure rate 27, see also Hazard rate
Interpretation of the confidence interval 277, 278

Key reliability practices 114, 170–175
Kurtosis 37, 38

Lack of memory property, exponential distribution 51
Legal liability 15
Life cycle conditions 4, 5, 8, 19, 111, 113, 131, 141, 149–166, 173, 177, 183, 184, 190, 198, 262, 347, 417
Life cycle cost 10, 106, 143, 414, 441, 446
Life cycle environment 156, 198, 250, 254, 412
Life cycle profile 111, 149, 150, 167, 198, 206, 353, 373, 413, 424
Lognormal distribution 45, 73–75, 79, 84, 210, 404, 446

Maintainability 2, 5, 8, 12, 37, 45, 73, 106, 112, 115, 116, 131, 132, 138, 143, 145, 160, 375, 401
Mean
   of continuous distribution 62
   of a continuous random variable 35, 51, 62
   of a discrete random variable 46
   of random variable 35, see also Expected value
Mean time between failures (MTBF) 36, 62–67, 77, 288–291, 378, 385, 386, 388–390, 392, 404
Mean time to failure (MTTF) 36, 39, 58, 61, 62, 67, 74, 75, 379, 384, 391, 440, 443
Median 34, 37, 56, 59, 60, 62, 68, 69, 74, 75, 78, 79, 277, 302–304
Minimum cut set (MCS) 120, 121, 127, 128, 396, 397, 399, 400

Minimum (minimal) path set 396, 397, 399, 400
Minimum variance unbiased estimator 65, 284, 287
Mode, of a random variable 56, 62, 68, 74
Moments, of a random variable 36
Moving range chart 303, 308, 309

np-chart 302, 304, 312–314
No-fault found (NFF) 351–373, 436
Normal distribution 38, 67–71, 73, 75, 89, 90, 180, 212–214, 279–282, 284, 318, 321, 322, 325, 455–457, 461

Occurrence rating 199, 200, 204, 205

p-chart for proportion 300, 302–304, 312
Parallel system 381–385, 391
Parameter estimation 74, 242
Part assessment process 114, 115, 177–185, 190
Patterns, on control charts 324, 326
Percentiles 19, 33–35, 39, 42, 61, 68
Physics of failure (PoF) 111, 128, 129, 138, 239–264, 268, 270, 275, 276, 412, 413, 415, 417, 418, 420, 428, 430
Point estimator 65, 66, 90, 94, 290
Poisson distribution 45, 50, 314
Probabilistic design 98, 207–220
Probability density function (pdf) 19, 23–25, 36, 37, 39, 51–58, 60, 61, 63, 67, 68, 73–75, 136, 208, 209, 215, 260, 299, 383–385, 392, 455, 457, 461, 467
Probability distribution 34, 46, 55, 68, 74, 79, 101, 121, 226, 249, 277, 412, 440
Probability mass function 45, 57
Probability plot 77–83
Process capability index 90, 94, 179, 180–182, 322, 323
Process mean 89, 90, 95, 179, 180, 299, 306, 317, 324
Product failure(s) 12, 14, 15, 23, 137, 138, 193, 195, 254, 339–341, 352, 355, 373, 409
Product qualification 132, 250–276, 353
Product screening 12, 248, 249, 295, 331–336
Prognostics 131, 206, 260, 261, 409–431
Prognostics and health management (PHM) 260, 261, 413–424, 429, 430

Qualification 111, 113, 115, 132–138, 145, 150, 173, 178, 183, 184, 186, 189, 190, 194, 195, 206, 227, 245, 248, 250–276, 331, 336, 353, 414, 420
Quality
   and customer satisfaction 6
   definition 1–2
Quality control 2, 11, 26, 132, 182, 288, 295, 299, 306, 312, 316, 317, 345, 348, 440, 444
Quality function deployment 2, 91, 97, 98

R chart 300, 303, 305–307
Redundancy 76, 121, 131, 182, 355, 359–361, 381, 392, 393, 401, 402
Reliability
   definition 2–5
   of system 9, 32, 76, 84, 169, 247, 347, 375–402
   and warranty 437–439
Reliability assessment 21, 45, 114, 115, 178, 182–184, 260, 415, 416, 425

Index

Reliability block diagram (RBD) 117, 125, 219, 375, 376, 382, 383, 389, 391, 397, 398, 401
   k-out-of-n system 391
   parallel system 382
   series system 376
   series-parallel system 383
Reliability capability 16, 114, 169–175
Reliability estimation 21, 64, 81, 247–291
Reliability function 19, 24, 25, 31, 32, 34, 36, 39, 55, 57, 60, 83, 84, 380, 383, 386, 387, 390–39
Reliability improvement 105, 171, 174, 175, 438
Reliability management 137, 172, 175, 448
Reliability testing 55, 107, 115, 169, 171–174, 288, 331
Requalification 415, 420
Risk management 137, 188–190
Root cause(s) 13, 91, 95, 96, 98, 108, 112, 137, 252, 339–351, 359, 362, 364, 373, 436
Root cause analysis 95, 97, 137, 172–174, 184, 185, 194, 195, 201, 206, 339–350, 353, 373, 424, 435
Rules, for out-of-control 301, 305, 308

S chart 303, 306, 307, 311
Safety factor 128, 207, 214
Sample mean 276, 279–282, 300
Sample range (R chart) 300, 306, 308
Sample size 279, 280
Sample standard deviation 276, 306
Sample variance 280, 461
Series system 376–382, 401, 402
Series-parallel systems 383
Severity rating 200, 205
Six sigma process 10, 89–108
Skewness 37, 53
Sources of variability 95, 297
SPC, *see* Statistical process control
Special (assignable) cause 96, 179, 296–300, 305, 308, 313, 325
Stable process 179, 318
Standard deviation 36–38, 52, 53, 56, 59, 62, 67–70, 73, 89, 90, 94, 134, 180, 181, 210–213, 217, 276, 277, 280–282, 299, 300, 302, 304, 306, 311, 318, 322, 324
   of a continuous random variable 36–38, 52
Standard normal random variable 68, 69, 455
Standby redundant systems 385
Standby systems 387

Statistical process control (SPC) 11, 91, 97, 104, 241, 295, 299, 325, 336
Statistical tolerances 216
Stress–strength interference models 209–212
Supply chain 16, 111, 112, 114, 137, 142, 143, 169, 171, 173, 175, 177, 185, 188, 190, 252, 428, 430, 436
System effectiveness 6, 115, 116
System reliability 9, 32, 76, 84, 169, 247, 347, 375–402

t-distribution 280, 281, 457–459
Time-truncated test 65
Total time on test 65–67, 287, 289, 291

U-chart, for defects per unit 302–304, 312, 314, 315
Unbiased estimator 65, 284, 287, 306
Unreliability function 19, 39, 76, 79
Uprating 179, 239–245

Variance
   of a continuous random variable 51
   of a discrete random variable 46
Variance transmission equation 95, 98, 100, 101
Virtual qualification 184, 194, 195, 206, 245, 250, 260, 261, 262, 274
Virtual testing 250, 260, 262, 270

Warranty, and reliability 437–439
Warranty analysis 433–449
Warranty cost 435, 437, 439–449
Wearout 26, 56, 57, 69, 82, 83, 130, 133, 134, 136, 157, 197, 199, 202–204, 258, 266, 270, 356, 365, 418, 435
Wearout failures 133, 157, 197, 199
Weibull distribution 45, 55–59, 61, 63, 79–84, 210, 249, 446, 448
   mean 56
   probability paper 81
   reliability function 57
   scale parameter 56
   shape parameter 56
   standard deviation 56

$\bar{X}$ control chart 305–308
$\bar{X}$ and R control chart 302–308
$\bar{X}$ and S control chart 304, 305, 311

Zone rules for control charts 300, 301

WILEY SERIES IN SYSTEMS ENGINEERING AND MANAGEMENT

Andrew P. Sage, Editor

ANDREW P. SAGE and JAMES D. PALMER
Software Systems Engineering

WILLIAM B. ROUSE
Design for Success: A Human-Centered Approach to Designing Successful Products and Systems

LEONARD ADELMAN
Evaluating Decision Support and Expert System Technology

ANDREW P. SAGE
Decision Support Systems Engineering

YEFIM FASSER and DONALD BRETTNER
Process Improvement in the Electronics Industry, Second Edition

WILLIAM B. ROUSE
Strategies for Innovation

ANDREW P. SAGE
Systems Engineering

HORST TEMPELMEIER and HEINRICH KUHN
Flexible Manufacturing Systems: Decision Support for Design and Operation

WILLIAM B. ROUSE
Catalysts for Change: Concepts and Principles for Enabling Innovation

LIPING FANG, KEITH W. HIPEL, and D. MARC KILGOUR
Interactive Decision Making: The Graph Model for Conflict Resolution

DAVID A. SCHUM
Evidential Foundations of Probabilistic Reasoning

JENS RASMUSSEN, ANNELISE MARK PEJTERSEN, and LEONARD P. GOODSTEIN
Cognitive Systems Engineering

ANDREW P. SAGE
Systems Management for Information Technology and Software Engineering

ALPHONSE CHAPANIS
Human Factors in Systems Engineering

YACOV Y. HAIMES
Risk Modeling, Assessment, and Management, Third Edition

DENNIS M. BUEDE
The Engineering Design of Systems: Models and Methods, Second Edition

ANDREW P. SAGE and JAMES E. ARMSTRONG, Jr.
Introduction to Systems Engineering

WILLIAM B. ROUSE
Essential Challenges of Strategic Management

YEFIM FASSER and DONALD BRETTNER
Management for Quality in High-Technology Enterprises

THOMAS B. SHERIDAN
Humans and Automation: System Design and Research Issues

ALEXANDER KOSSIAKOFF and WILLIAM N. SWEET
Systems Engineering Principles and Practice

HAROLD R. BOOHER
Handbook of Human Systems Integration

JEFFREY T. POLLOCK and RALPH HODGSON
Adaptive Information: Improving Business Through Semantic Interoperability, Grid Computing, and Enterprise Integration

ALAN L. PORTER and SCOTT W. CUNNINGHAM
Tech Mining: Exploiting New Technologies for Competitive Advantage

REX BROWN
Rational Choice and Judgment: Decision Analysis for the Decider

WILLIAM B. ROUSE and KENNETH R. BOFF (editors)
Organizational Simulation

HOWARD EISNER
Managing Complex Systems: Thinking Outside the Box

STEVE BELL
Lean Enterprise Systems: Using IT for Continuous Improvement

J. JERRY KAUFMAN and ROY WOODHEAD
Stimulating Innovation in Products and Services: With Function Analysis and Mapping

WILLIAM B. ROUSE
Enterprise Tranformation: Understanding and Enabling Fundamental Change

JOHN E. GIBSON, WILLIAM T. SCHERER, and WILLAM F. GIBSON
How to Do Systems Analysis

WILLIAM F. CHRISTOPHER
Holistic Management: Managing What Matters for Company Success

WILLIAM B. ROUSE
People and Organizations: Explorations of Human-Centered Design

MO JAMSHIDI
System of Systems Engineering: Innovations for the Twenty-First Century

ANDREW P. SAGE and WILLIAM B. ROUSE
Handbook of Systems Engineering and Management, Second Edition

JOHN R. CLYMER
Simulation-Based Engineering of Complex Systems, Second Edition

KRAG BROTBY
Information Security Governance: A Practical Development and Implementation Approach

JULIAN TALBOT and MILES JAKEMAN
Security Risk Management Body of Knowledge

SCOTT JACKSON
Architecting Resilient Systems: Accident Avoidance and Survival and Recovery from Disruptions

JAMES A. GEORGE and JAMES A. RODGER
Smart Data: Enterprise Performance Optimization Strategy

YORAM KOREN
The Global Manufacturing Revolution: Product-Process-Business Integration and Reconfigurable Systems

AVNER ENGEL
Verification, Validation, and Testing of Engineered Systems

WILLIAM B. ROUSE (editor)
The Economics of Human Systems Integration: Valuation of Investments in People's Training and Education, Safety and Health, and Work Productivity

ALEXANDER KOSSIAKOFF, WILLIAM N. SWEET, SAM SEYMOUR, and STEVEN M. BIEMER
Systems Engineering Principles and Practice, Second Edition

GREGORY S. PARNELL, PATRICK J. DRISCOLL, and DALE L. HENDERSON (editors)
Decision Making in Systems Engineering and Management, Second Edition

ANDREW P. SAGE and WILLIAM B. ROUSE
Economic Systems Analysis and Assessment: Intensive Systems, Organizations, and Enterprises

BOHDAN W. OPPENHEIM
Lean for Systems Engineering with Lean Enablers for Systems Engineering

LEV M. KLYATIS
Accelerated Reliability and Durability Testing Technology

BJOERN BARTELS, ULRICH ERMEL, MICHAEL PECHT, and PETER SANDBORN
Strategies to the Prediction, Mitigation, and Management of Product Obsolescence

LEVANT YILMAS and TUNCER ÖREN
Agent-Directed Simulation and Systems Engineering

ELSAYED A. ELSAYED
Reliability Engineering, Second Edition

BEHNAM MALAKOOTI
Operations and Production Systems with Multiple Objectives

MENG-LI SHIU, JUI-CHIN JIANG, and MAO-HSIUNG TU
Quality Strategy for Systems Engineering and Management

ANDREAS OPELT, BORIS GLOGER, WOLFGANG PFARL, and RALF MITTERMAYR
Agile Contracts: Creating and Managing Successful Projects with Scrum

KINJI MORI (editor)
Concept-Oriented Research and Development in Information Technology

KAILASH C. KAPUR and MICHAEL PECHT
Reliability Engineering